Quantum Well Intersubband Transition Physics and Devices

NATO ASI Series

Advanced Science Institutes Series

A Series presenting the results of activities sponsored by the NATO Science Committee, which aims at the dissemination of advanced scientific and technological knowledge, with a view to strengthening links between scientific communities.

The Series is published by an international board of publishers in conjunction with the NATO Scientific Affairs Division

A Life Sciences **B Physics**	Plenum Publishing Corporation London and New York
C Mathematical ** and Physical Sciences** **D Behavioural and Social Sciences** **E Applied Sciences**	Kluwer Academic Publishers Dordrecht, Boston and London
F Computer and Systems Sciences **G Ecological Sciences** **H Cell Biology** **I Global Environmental Change**	Springer-Verlag Berlin, Heidelberg, New York, London, Paris and Tokyo

NATO-PCO-DATA BASE

The electronic index to the NATO ASI Series provides full bibliographical references (with keywords and/or abstracts) to more than 30000 contributions from international scientists published in all sections of the NATO ASI Series.
Access to the NATO-PCO-DATA BASE is possible in two ways:

– via online FILE 128 (NATO-PCO-DATA BASE) hosted by ESRIN,
Via Galileo Galilei, I-00044 Frascati, Italy.

– via CD-ROM "NATO-PCO-DATA BASE" with user-friendly retrieval software in English, French and German (© WTV GmbH and DATAWARE Technologies Inc. 1989).

The CD-ROM can be ordered through any member of the Board of Publishers or through NATO-PCO, Overijse, Belgium.

Series E: Applied Sciences - Vol. 270

Quantum Well Intersubband Transition Physics and Devices

edited by

H. C. Liu
Institute for Microstructural Sciences,
National Research Council,
Ottawa, Ontario, Canada

B. F. Levine
AT&T Bell Laboratories,
Murray Hill, New Jersey, U.S.A.

and

J. Y. Andersson
Industrial Microelectronic Center,
Kista, Sweden

Kluwer Academic Publishers

Dordrecht / Boston / London

Published in cooperation with NATO Scientific Affairs Division

Proceedings of the NATO Advanced Research Workshop on
Quantum Well Intersubband Transition Physics and Devices
Whistler, Canada
September 7–10, 1993

A C.I.P. Catalogue record for this book is available from the Library of Congress.

ISBN 0-7923-2877-9

Published by Kluwer Academic Publishers,
P.O. Box 17, 3300 AA Dordrecht, The Netherlands.

Kluwer Academic Publishers incorporates the publishing programmes of
D. Reidel, Martinus Nijhoff, Dr W. Junk and MTP Press.

Sold and distributed in the U.S.A. and Canada
by Kluwer Academic Publishers,
101 Philip Drive, Norwell, MA 02061, U.S.A.

In all other countries, sold and distributed
by Kluwer Academic Publishers Group,
P.O. Box 322, 3300 AH Dordrecht, The Netherlands.

Printed on acid-free paper

CONTENTS

Preface xi

Detector Applications

Random Scattering Optical Couplers for Quantum Well Infrared Photodetectors 1
B. F. Levine, G. Sarusi, S. J. Pearton, K. M. S. Bandara, and R. E. Leibenguth

Performance of Grating Coupled AlGaAs/GaAs Quantum Well Infrared Detectors and
Detector Arrays 13
J. Y. Andersson, L. Lundqvist, J. Borglind, and D. Haga

Novel Grating Coupled and Normal Incidence III-V Quantum Well Infrared
Photodetectors with Background Limited Performance at 77 K 29
S. S. Li and Y. H. Wang

Background Limited 128x128 GaAs/AlGaAs Multiple Quantum Well Infrared Focal
Plane Arrays 43
L. J. Kozlowski

Imaging Performance of LWIR Miniband Transport Multiple Quantum Well Infrared
Focal Plane Arrays 55
W. A. Beck, J. W. Little, A. C. Goldberg, and T. S. Faska

Modeled Performance of Multiple Quantum Well Infrared Detectors in IR Sensor
Systems 69
R. L. Whitney, F. W. Adams, and K. F. Cuff

GaAs/AlGaAs QWIPs vs. HgCdTe Photodetectors for LWIR Applications 87
A. Rogalski

Detector Physics

The Physics of Emission-Recombination in Multiquantum Well Structures 97
E. Rosencher, F. Luc, L. Thibaudeau, B. Vinter, and Ph. Bois

Physics of Single Quantum Well Infrared Photodetectors 111
K. M. S. V. Bandara, B. F. Levine, G. Sarusi, R. E. Leibenguth, M. T. Asom, and
J. M. Kuo

A Three-color Voltage Tunable Quantum Well Intersubband Photodetector for Long
Wavelength Infrared 123
H. C. Liu, J. Li, Z. R. Wasilewski, M. Buchanan, P. H. Wilson, M. Lamm, and
J. G. Simmons

Multi λ Controlled Operation of Quantum Well IR Detectors Using Electric Field
Switching and Rearrangement 135
A. Shakouri, I. Gravé, Y. Xu, and A. Yariv

Infrared Hot-Electron Transistor Design Optimization 151
K. K. Choi, M. Z. Tidrow, M. Taysing-Lara, and W. H. Chang

16 μm Infrared Hot Electron Transistor 167
S. D. Gunapala, J. K. Liu, J. S. Park, and T. L. Lin

Long Wavelength λc=18μm Infrared Hot Electron Transistor 177
C. Y. Lee, M. Z. Tidrow, K. K. Choi, W. Chang, L. F. Eastman

A Novel Transport Mechanism for Photovoltaic Quantum Well Intersubband Infrared
Detectors 187
H. Schneider, S. Ehret, E. C. Larkins, J. D. Ralston, and P. Koidl

Intersubband Stark-Ladder Transitions in Miniband-Transport Quantum-Well Infrared
Detectors 197
J. W. Little and F. J. Towner

CO2-Laser Heterodyne Detection with GaAs/AlGaAs MQW Structures 207
E. R. Brown, K. A. McIntosh, K. B. Nichols, F. W. Smith, and M. J. Manfra

Intersubband Transition Experimental

Intersubband Absorption in n-Type Si and Ge Quantum Wells 221
K. L. Wang, C. Lee, and S. K. Chun

Intersubband Transitions in p-Type SiGe/Si Quantum Wells for Normal Incidence
Infrared Detection 237
R. P. G. Karunasiri

Large Energy Intersubband Transitions in High Indium Content InGaAs/AlGaAs
Quantum Wells 251
H. C. Chui, E. L. Martinet, M. M. Fejer, and J. S. Harris, Jr.

Applications of High Indium Content InGaAs/AlGaAs Quantum Wells in the 2-7 μm
Regime 261
E. L. Martinet, B. J. Vartanian, G. L. Woods, H. C. Chui, J. S. Harris, Jr.,
M. M. Fejer, B. A. Richman, and C. A. Rella

Spectroscopy of Narrow Minibands in the Continuum of Multi Quantum Wells 275
D. Gershoni, R. Duer, J. Oiknine-Schlesinger, E. Ehrenfreund, D. Ritter, R. A. Hamm,
J. M. Vandenberg, and S-N. G. Chu

Intersubband Absorption in Strongly Coupled Superlattices: Miniband Dispersion,
Critical Points, and Oscillator Strengths 291
M. Helm, W. Hilber, T. Fromherz, F. M. Peeters, J. Alavi, and R. N. Pathak

Electronic Quarter-Wave Stacks and Bragg Reflectors: Physics of Localized
Continuum States in Quantum Semiconductor Structures 301
C. Sirtori, F. Capasso, J. Faist, D. Sivco, and A. Y. Cho

Modulation of the Optical Absorption by Electric-Field-Induced Quantum
Interference in Coupled Quantum Wells 313
J. Faist, F. Capasso, A. L. Hutchinson, L. Pfeiffer, K. W. West, D. L. Sivco, and
A. Y. Cho

Phase Retardation and Induced Birefringence Related to Intersubband Transitions in
Multiple Quantum Well Structures 321
A. Sa'ar and D. Kaufman

The Interaction of Photoexcited e-h Pairs with a Two Dimensional Electron Gas
Studied by Intersubband Spectroscopy 331
Y. Garni, E. Cohen, E. Ehrenfreund, D. Gershoni, A. Ron, E. Linder, L. N. Pfeiffer,
K-K. Law, J. L. Merz, and A. C. Gossard

Photoinduced Intersubband Transitions in GaAs/AlGaAs Asymmetric Coupled
Quantum Wells 345
F. H. Julien, P. Vagos, P. Boucaud, L. Wu, and R. Planel

Far Infrared Spectroscopy of Intersubband Transitions in Multiple Quantum Well
Structures 361
W. J. Li and B. D. McCombe

Observation of Intersubband Transitions in Asymmetric δ-Doped GaAs, InSb, and
InAs Structures 371
C. C. Phillips, H. L. Vaghjiani, E. A. Johnson, P. J. P. Tang, R. A. Stradling,
J. J. Harris, M. J. Kane

Intersubband Transition Theory

Effects of Coupling on Intersubband Transitions 379
J. M. Xu

On Some Peculiarities of Intersubband Absorption in Semiconductor Quantum Wells ..389
Z. Ikonić and V. Milanović

The Relative Strengths of Interband and Intersubband Optical Transitions: Breakdown
of the Atomic Dipole Approximation for Interband Transitions 399
M. G. Burt

Optical Transitions and Energy Level Ordering for Quantum Confined Impurities 403
S. R. Parihar and S. A. Lyon

Subband Structures of Superlattices under Strong In-Plane Magnetic Fields 411
W. Tan, J. C. Inkson, and G. P. Srivastava

Intersubband Relaxation

Temperature Dependent Intersubband Dynamics in n-Modulation Doped Quantum
Well Structures 421
A. Seimeier, U. Plödereder, J. Baier, and G. Weimann

Ultrafast Dynamics of Electronic Capture and Intersubband Relaxation in GaAs
Quantum Well 433
D. Morris, B. Deveaud, A. Regreny, and P. Auvray

Structure-Dependent Electron-Phonon Interactions 443
B. K. Ridley

Nonlinear Phenomena

Second Harmonic Generation in p-Type Quantum Wells 457
Z. Xu, P. M. Fauchet, G. W. Wicks, M. J. Shaw, M. Jaros, B. Richman, and C. Rella

Resonant Harmonic Generation Near 100μm in an Asymmetric Double Quantum
Well 467
J. N. Heyman, K. Craig, M.. S. Sherwin, K. Campman, P. F. Hopkins, S. Fafard, and
A. C. Gossard

Second Harmonic Generation in GaAs-AlAs and Si-SiGe Quantum Well
Structures 477
E. Corbin, M. J. Shaw, K. B. Wong, and M. Jaros

Non-Resonant Two-Photon Absorption in Quantum Well Infrared Detectors 493
E. Dupont, P. B. Corkum, P. W. Dooley, H. C. Liu, P. H. Wilson, M. Lamm,
M. Buchanan, and Z. R. Wasilewski

Optical Saturation of Intersubband Transitions 501
L. C. West and C. W. Roberts

Novel Phenomena

Use of Classically Free Quasibound States for Infrared Emission 511
L. C. West, C. W. Roberts, J. Dunkel, M. T. Asom, G. N. Henderson, T. K. Gaylord,
E. Anemogiannis, and E. N. Glytsis

Evidence for LWIR Emission Using Intersubband Transitions in GaAs/AlGaAs MQW
Structures 525
A. G. U. Perera

Fast Data Coding Using Modulation of Interband Optical Properties by Intersubband
Absorption in Quantum Wells 533
V. B. Gorfinkel and S. Luryi

Theory of Terahertz Generation due to Quantum Beats in Quantum Wells 547
S. L. Chuang and M. S. C. Luo

Far-Infrared Study of an Antenna-Coupled Quantum Point Contact 553
Q. Hu, R. A. Wyss, C. C. Eugster, J. A. del Alamo, and S. Feng

Control of Electron Population by Intersubband Optical Excitation in a Novel
Asymmetric Double Quantum Well Structure 565
H. Sugawara, H. Akiyama, Y. Kadoya, A. Lorke, S. Tsujino, T. Matsusue, and
H. Sakaki

Index   575

PREFACE

This book contains papers presented at the NATO Advanced Research Workshop on "Quantum Well Intersubband Transition Physics and Devices" held on 7-10 September 1993 in Whistler, Canada.

Intersubband transitions in quantum wells have attracted tremendous attention in the past few years, mainly due to the promise of various applications in the mid- and far-infrared regions (2-20μm). A literature search result is given in the figure below, which shows the steady growth of interest in the topic.

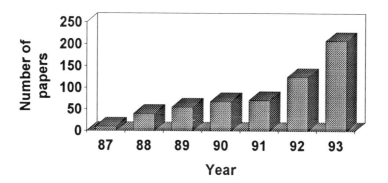

Papers presented in the workshop cover many aspects of the intersubband transitions. We have grouped them in seven categories according to their main emphases. The grouping, in several cases, is somewhat arbitrary. Many papers are on the basic linear intersubband transition processes, detector physics, and detector applications. This reflects the current status of understanding and of detector applications. Indeed, highly uniform large focal plane arrays have been demonstrated in several laboratories. With the improvement of performance, detectors and arrays should find many practical applications. Other areas of applications are still in their early stages, which include infrared modulation, harmonic generation, and emission. Undoubtedly, these areas will expand in the coming years.

We believe that all the participants would agree that the workshop has been a success: papers are of high quality, discussions were lively, and conference site was an excellent choice.

We would like to thank the generous financial support provided by the NATO Scientific Affairs Division, the Institute of Microstructural Sciences of the National Research Council of Canada, and the Defense Research Establishment Valcartier of the Canadian Department of National Defense. We also appreciate the help of J. R. Thompson and P. H. Wilson for their assistance in organizing the successful workshop.

The Editors

RANDOM SCATTERING OPTICAL COUPLERS FOR QUANTUM WELL INFRARED PHOTODETECTORS

B. F. LEVINE, G. SARUSI, S. J. PEARTON, K. M. S. BANDARA AND
R. E. LEIBENGUTH
AT&T Bell Laboratories
Murray Hill, New Jersey 07974

ABSTRACT. Detailed measurements on random scattering optical couplers for quantum well infrared photoconductors are presented. By optimizing the scattering parameters, we have achieved an order of magnitude improvement in optical absorption, responsivity and detectivity compared with the 45° angle of incidence geometry, or a one-dimentional grating coupler.

Quantum well infrared photodetectors (QWIPs), have achieved excellent imaging performance using large area 128×128, and 256×256 highly uniform pixel arrays.[1-6] Because of the quantum mechanical selection rule requiring a component of the optical electric field normal to the surface, these QWIP arrays have used 1-dimensional (1-D) or 2-dimensional (2-D) gratings to enhance the optical coupling.[1-10] Optimized 1-D gratings have achieved a normal incidence integrated black body responsivity equal to that of a standard 45° polished input face, while 2-D gratings have demonstrated twice this 45° optical coupling.[8,9] We discuss here our work on improved optical coupling using random scattering optical reflectors, which achieves an order of magnitude improvement in responsivity and detectivity compared with a one-dimensional grating or 45° angle of incidence geometry. The wavelength of the QWIPs used in this demonstration had a wavelength in the very long wave infrared (VWIR) spectral region (λ_p = 16 − 17 μm), which is desired for earth observing atmospheric satellite applications.

Although gratings have achieved relatively high quantum efficiency, the grating period strongly affects the spectral peak position λ_p, the spectral line width $\Delta\lambda$ and the coupling efficiency.[8-10] To demonstrate these effects we show a QWIP responsivity spectrum measured using the 45° angle of incidence geometry in Fig. 1. It is peaked at λ_p = 16.4 μm with a linewidth $\Delta\lambda$ = 2.9 μm. The same sample measured using a two-dimentional grating coupler with a period of Λ = 5.3 μm has a much narrower spectral width of only $\Delta\lambda$ = 1.7 μm. Furthermore as this grating period is varied, large spectral changes occur, as can be seen in Figures 2 and 3 where both λ_p and $\Delta\lambda$ are strongly dependent on Λ. In addition, multiple grating coupled modes can be observed.

1

H. C. Liu et al. (eds.), Quantum Well Intersubband Transition Physics and Devices, 1–12.

For example, Fig. 2 shows that with $\Lambda = 6.2$ μm two peaks can be seen at $\lambda_p \sim 14$ μm and $\lambda_p \sim 18$ μm. As the period is further increased Fig. 3 shows that the shorter wavelength spectral mode becomes dominant and shifts strongly with Λ. In addition to these normalized spectral shifts, the absolute responsivity shown in Fig. 4,

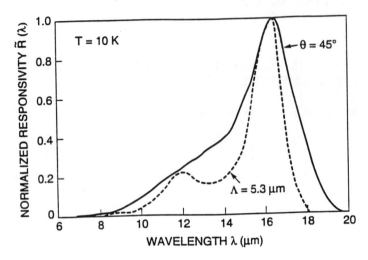

Fig. 1. Comparison of normalized responsivity for a $\lambda = 16$ μm QWIP for $\theta = 45°$ and for a 2D grating of period $\Lambda = 5.3$ μm.

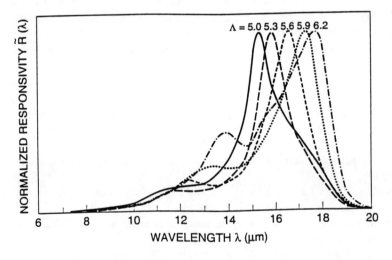

Fig. 2. Measured normalized responsivity spectra at $T = 20$K as a function of grating period Λ for $\Lambda = 5.0 - 6.2$ μm, at $V_b = -5$V.

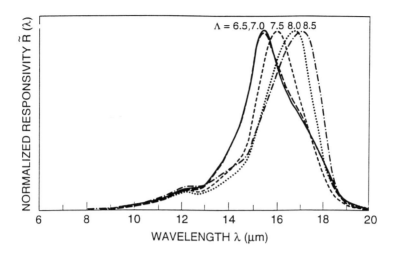

Fig. 3. Measured normalized responsivity spectra at T = 20K as a function of grating period Λ for Λ = 6.5–8.5 μm, at V_b = −5V.

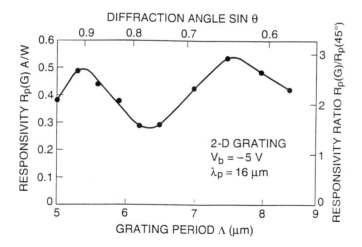

Fig. 4. Absolute responsivity (left hand side) of the 2-dimensional grating QWIPs as a function of grating period Λ, for a peak wavelength of λ_p = 16 μm and a bias of V_b = −5V. Ratio of 2D grating responsivity to θ = 45° responsivity is shown on right hand scale.

also varies substantially with Λ, again showing the two mode behavior. As we will show later the random scattering reflector does not show these spectral effects.

Another important advantage of random reflectors is that although gratings can efficiently diffract the incident radiation into a large angle, (as shown in Fig. 5a), a grating can only achieve *one* absorption pass without an optical cavity and *two* passes through the QWIP with an optical cavity[8,9] before the infrared beam is diffracted out of the substrate. Many more passes of the infrared radiation, and a significantly higher absorption can be achieved with a randomly roughened reflecting surface (shown in Fig. 5b). Yablonovitch and Cody[11] show that a large intensity enhancement of $2n^2$ can be achieved, (where n is the index of refraction), and that an even larger absorption enhancement of $4n^2$ can be obtained, (as a result of angle averaging effects which increase the effective path length by an additional factor of 2). That is, most of the

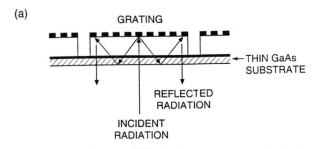

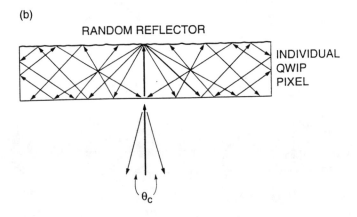

Fig. 5. Schematic illustration of a two pass grating coupler (a) and a random scattering reflector, showing the multipass trapping of the incident radiation (b).

radiation can be trapped inside the medium except for the small escape fraction which is within the critical reflection cone angle θ_c ($\theta_c = 18°$ in GaAs) given by $\sin\theta_c = 1/n$. Thus, much larger effective path lengths and higher absorptions coefficients, α, are to be expected. Since for QWIPs, $\alpha \propto \sin^2\theta$, angle averaging slightly reduces this enhancement by a factor of $\int_0^\pi \sin^3\theta d\theta / \int_0^\pi \sin\theta d\theta = 2/3$, resulting in a maximum absorption enhancement of $(8/3)n^2 = 30$ for GaAs. This is more than an order of magnitude larger than for optimized two dimensional gratings.

In order to design an optimal "random" scattering reflector several considerations must be addressed. Firstly to maximize the destructive interference of the reflected radiation, (and hence minimize the low angle back reflected power which can escape within the critical angle θ_c), the phase of the reflected optical electric field should be randomized. Secondly the "random" pattern should avoid clustering of the scattering centers which will reduce the effective number of scatterers. Finally the size and shape of the "random" scattering centers should be optimized with respect to the wavelength of the incident radiation, in order to maximize the scattering amplitude.

With these considerations in mind we designed a photolithography mask to fabricate the scattering surface, using reactive ion etching.[12] The advantage of photolithography over a "true" random process is the ability to accurately control the feature size and to insure that each pixel has the *identical* "random" pattern. This is crucial for the high uniformity required for imaging. A very simple procedure was chosen for making this mask. The "random" pattern was generated by dividing up the light scattering area into square unit cells of side U; (one representative cell is shown in Fig. 6). This unit cell contained three etched levels; a small inner square (region 1) of etch depth $\lambda_p'/2$ (where λ_p' is the peak wavelength in the medium) and an area $(1/4)U^2$; a middle square (region 2) of etch depth $\lambda_p'/4$ and area $(1/2)U^2$; and an outer unetched level (region 3) of area $(1/4)U^2$. The areas and depths of these three levels were chosen from the following considerations. After the incident radiation is reflected from the Au covered outer area (region 3) it will be in phase (i.e. λ_p' path difference) with that reflected from the small inner square (region 1). Both of these reflections will thus be 180° out of phase (i.e. $\lambda_p'/2$ path difference) with that from the middle level (region 2). The reflection areas were therefore chosen so that the in phase and out of phase scattering would be equal, in order to maximize the destructive interference.

The spatial position of the middle square was randomly chosen to be in one of the four corners of the unit cell, and the position of the small square was randomly chosen to be in one of the four corners of the middle square. Thus 16 equally likely random combinations are possible and Fig. 6(b) shows one of them. This unit cell randomization was used to avoid spatially overlapping the various reflection levels, which would have reduced the scattering. This is a well known[13] consideration for maximizing the scattering of solid particle suspensions by "encapsulating" them so as to avoid the scattering centers from "sticking together" and thereby reducing the net number of scatterers. An additional benefit of the unit cell structure is that by varying U the scattering angle distribution could be modified from the usual Lambertian $\cos\theta$ law to increase the large angle scattering, thus increasing the intersubband absorption

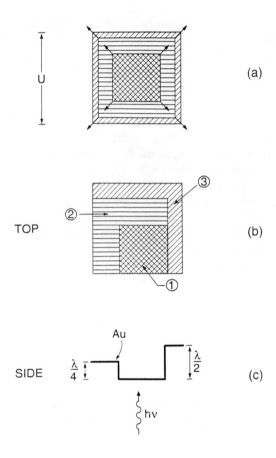

Fig. 6. Unit cell of the "random" scattering reflector; (a) three nested squares which are randomly assigned to the 16 possible corner positions; (b) one of the 16 equally likely possible combinations; (c) side view showing the different etching depths of the 3 levels, and the direction of the normally incident radiation.

$\alpha \propto \sin^2 \theta$, while decreasing the critical angle escape fraction. The unit cell size was varied from $U = 4.6$ µm to 9.1 µm and the normal incidence responsivity was measured for a QWIP having double stop etch top layers (two GaAs layers doped $N_D = 3 \times 10^{17}$ cm^{-3} of thickness $\lambda'_p/4 = 1.25$ µm, separated by 300 Å of $x = 0.11$ similarly doped etch stop layers, and followed by the active QWIP absorbing layers consisting of 50 periods of $N_D = 3 \times 10^{17}$ cm^{-3}, $L_w = 72$ Å quantum wells of GaAs, with barriers of $L_b = 600$ Å with $x = 0.11$). To form the different etch depths within the unit cells, a combination of selective etching of GaAs over AlGaAs (using $CCl_2 F_2$

discharges) and non-selective etching (using BCl_3 discharges) was employed. Low-pressure (1 mTorr), low-bias (−50V dc) conditions were used in both cases to produce highly anisotropic, damage-free etching.[12]

The top mesa contact (500 μm × 500 μm) consisted of a small alloyed area (25 μm × 25 μm) surrounded by a large nonalloyed highly reflecting Au contact. The nonalloyed Au contact was found, in agreement with Andersson et al.,[9] to increase the responsivity by ~20-30%, due to its higher reflectivity. The normal incidence "random" peak responsivity, $R_p(R)$, measurements are shown in Fig. 7 for several GaAs substrate thicknesses t. For the standard GaAs substrate thickness of t = 650 μm, note the very

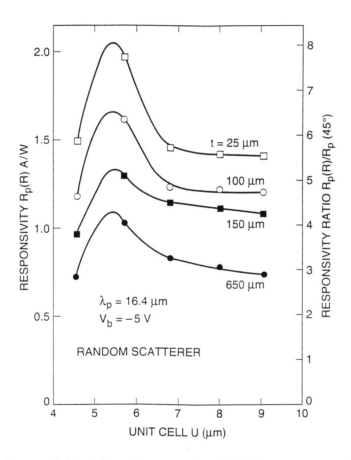

Fig. 7. Peak responsivity of the random scattering QWIP $R_p(R)$ as a function of random unit cell size, measured at V_b = −5V. Left hand scale absolute responsivity in A/W; right hand scale ratio relative to a standard 45° incidence geometry. The different curves are for several substrate thicknesses t.

large values of the responsivity with the maximum $R_p(R) = 1.1 A/W$ being nearly 4 times larger than that of the $\theta = 45°$ QWIP. This is approximately *twice* as large as the best grating responsivity of $R_p(G) = 0.5 A/W$ we have obtained on a similar QWIP. It is worth noting that the maximum "random" scattering responsivity $R_p(R)$ is achieved at a value of $U = 5.7 \, \mu m$ for which $U \simeq \lambda_p'$. Thus, this generation of "random" scattering using photolithographic patterns allows a high degree of control over the details of the scattering process. The maximum in the shape of the absolute responsivity as a function of the unit cell size U shown in Fig. 7, is a result of two factors. The increase in the scattering efficiency as U decreases (i.e. the increase in the number of scattering centers/cm^2), and the decrease in the scattering at small U. This latter effect is a result of $U < \lambda_p'$, for $U = 4.6$, resulting in weaker scattering, as well as the photolithography becoming more difficult (since the smallest feature sizes can now be $< 1 \, \mu m$) and thus the scattering surface actually becoming smoother.

Another important feature of the random scattering reflector is that (as shown in Fig. 8) the spectral shape of the responsivity curve is unaffected by the optical coupling process, and is the same for different unit cell size (e.g. $U = 5.7$ or $9.1 \, \mu m$). In particular the spectral width for the "random" responsivity is the same as that of the intrinsic spectrum as measured by the $45°$ angle of incidence geometry. This is strikingly different from the case of grating coupling (shown in Figs. 1-3) where there is a substantial dependence of the spectral shape and peak wavelength on grating period.[9]

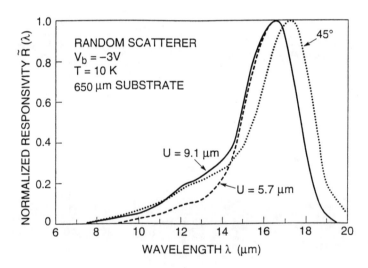

Fig. 8. Normalized responsivity spectra for two different unit cell sizes $U = 5.7 \, \mu m$ and $9.1 \, \mu m$ as well as the $45°$ incidence geometry for comparison. Note the similarity in spectral width in the three measurements.

For example, Fig. 1 shows that there can be a substantial spectral narrowing of a factor of 2. Thus although the peak grating responsivity R_p is dramatically increased over 45° coupling, the increase in the integrated blackbody responsivity is only half as much. For the random coupling discussed here the integrated responsivity and the peak responsivity both increase by similar amounts.

To further improve the optical absorption, the path length can be increased by the use of an optical cavity. Andersson *et al.*[8,9] used an AlAs layer grown under the active QWIP to create a total internal reflection and thus double the path length. Another simpler approach to create such a cavity without using AlAs is to thin down the QWIP substrate,[1,6] (as is done for example in InSb cameras since the substrate is highly absorbing). One of the advantages of such a thinned substrate is that the critical angle

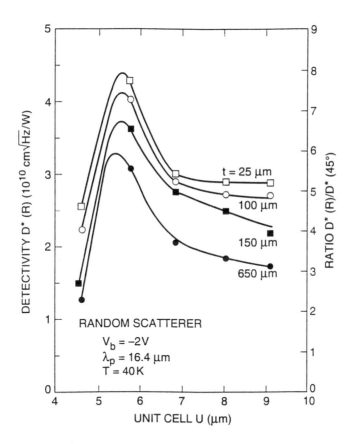

Fig. 9. Detectivity of the random scattering QWIP as a function of random unit cell size. Left hand scale is detectivity; right hand scale ratio relative to that of the standard 45° angle of incidence geometry. The different curves are for several substrate thicknesses t.

$\theta_c = 18°$ for the GaAs-air interface is much smaller than the value $\theta = 60°$ corresponding to the GaAs-AlAs interface. Thus it reflects significantly more of the scattered radiation back into the QWIP absorbing layers, resulting in a larger responsivity and additionally less cross talk between the imaging pixels. The results of our thinning experiments are shown in Fig. 7. The factor of two increase as the sample is thinned from $t = 650$ μm to 25 μm is clearly observed. The thinnest sample has a maximum improvement of a factor of 8 (right hand scale on Fig. 7) over that of a 45° angle of incidence. Further optimization of the "random" pattern can lead to an additional responsivity improvement towards the calculated maximum value of $(8/3)n^2 = 30$.

This order of magnitude improvement in optical coupling and blackbody responsivity directly leads to a similar increase in detectivity D^* shown in Fig. 9 and noise equivalent temperature difference NEΔT for imaging arrays since the dark current is unaffected by the "random" grating or substrate thinning. Our previous 128×128 pixel imaging demonstration[2] at $\lambda \simeq 10$ μm achieved an excellent value of NEΔT = 10 mK at T = 65K. This array used a one-dimensional linear grating for the optical coupling (equivalent in coupling efficiency to a 45° angle of incidence). Therefore, the use of optimized "random" scatters and thinned substrates would permit the same high NEΔT performance at T ≈ 80K, and thus allow the use of the same cooler technology as HgCdTe, InSb or PtSi arrays.

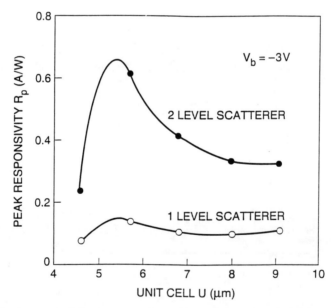

Fig. 10. Peak responsivity measured at $V_b = -3V$ for the 2-level random mask illustrated in Fig. 6 and a 1-level random mask.

As a further test of the optimization of the "random" scattering reflector and to see whether two etched levels are indeed necessary (as indicated in Fig. 6), we fabricated a "random" reflector having only a *single* etched level h = 1.25 μm deep. The result is shown on the lower curve in Fig. 10 for t = 650 μm. Note that the responsivity for this 1-level mask is over 4 times worse than that for the previously discussed 2-level mask. This clearly shows that a 1-level mask does not sufficiently randomize the scattered radiation and thus 2 levels are necessary.

In summary we have demonstrated a novel normal incidence optical coupling technique based on scattering from a random reflector monolithically etched into the top of a QWIP. We have achieved nearly an order of magnitude improvement in optical absorption, responsivity and detectivity as compared to a one-dimensional grating or 45° angle of incidence coupling.

ACKNOWLEDGMENT

The work done at AT&T Bell Laboratories was partially supported by NASA/Jet Propulsion Laboratory.

REFERENCES

1. B. F. Levine, J. Appl. Phys. *74*, R1 (1993).

2. B. F. Levine, C. G. Bethea, K. G. Glogovsky, J. W. Stayt and R. E. Leibenguth, Semicond. Sci. Technol. *6*, C114 (1991).

3. L. J. Kozlowski, G. M. Williams, G. J. Sullivan, C. W. Farley, R. J. Andersson, J. Chen, D. T. Cheung, W. E. Tennant and R. E. DeWames, IEEE Trans. Electron. Devices *38*, 1124 (1991).

4. T. S. Faska, J. W. Little, W. A. Beck, K. J. Ritter, A. C. Goldberg and R. LeBlanc, "Innovative Long Wavelength Infrared Detector Workshop", April 7-9, 1992 Pasadena, CA.

5. V. Swaminathan, J. W. Stayt Jr., J. L. Zilko, K. D. C. Trapp, L. E. Smith, S. Nakahara, L. C. Luther, G. Livescu, B. F. Levine, R. E. Leibenguth, K. G. Glogovsky, W. A. Gault, M. W. Focht, C. Buiocchi, and M. T. Asom, Proc. IRIS Specialty Group on Infrared Detectors, Moffet Field CA, Aug (1992).

6. W. A. Beck, J. W. Little, A. C. Goldberg and T. S. Faska, Proc. NATO Advanced Workshop an "Quantum Well Intersubband Transition Physics and Devices", Whistler, Canada Sept. 7-10, 1993 (Plenum, New York 1994) Edited by H. C. Liu, B. F. Levine and J. Y. Andersson p. 55.

7. G. Hasnain, B. F. Levine, C. G. Bethea, R. A. Logan, J. Walker and R. J. Malik, Appl. Phys. Lett. *54*, 2515 (1989).

8. J. Y. Andersson, L. Lundqvist and Z. F. Paska, Appl. Phys. Lett. *58*, 2264 (1991).

9. J. Y. Andersson and L. Lundqvist, Proc. NATO Advanced Workshop on "Intersubband Transitions in Quantum Wells" Cargèse France Sept. 9-14, 1991 (Plenum, New York 1992) Ed. E. Rosencher, B. Vinter and B. Levine p 1.

10. L. S. Yu, S. S. Li, Y. H. Wang and Y. C. Kao, J. Appl. Phys. *72*, 2105 (1992).

11. E. Yablonovitch and G. D. Cody, IEEE Trans. Electron Devices *29*, 300 (1982).

12. S. J. Pearton, F. Ren, T. R. Fullowan, J. R. Lothian, A. Katz, R. F. Kopf and C. R. Abernathy, Plasma Sources Sci. Technol. *1*, 18 (1992).

13. B. H. Kaye, "A Random Walk Through Fractal Dimensions", VCH Publishers N.Y. (1989).

PERFORMANCE OF GRATING COUPLED AlGaAs/GaAs QUANTUM WELL INFRARED DETECTORS AND DETECTOR ARRAYS

J. Y. ANDERSSON, L. LUNDQVIST, J. BORGLIND and D. HAGA
Industrial Microelectronic Center (IMC)
P.O. Box 1084
S-16421 Kista
Sweden

1. Introduction

Long wavelength (7.5-10 μm) quantum well infrared photodetectors (QWIPs) based on AlGaAs/GaAs quantum wells (QWs) have been shown to exhibit high peak detectivities $D*$ $= 1 \cdot 10^{10} - 9 \cdot 10^{10}$ cm Hz$^{1/2}$ W^{-1} at 80 K and close to background limited operation[1-3]. Due to the well established GaAs material and processing technology QWIPs are viable candidates for large, low cost LWIR (8-12 μm) focal plane arrays (FPAs)[4-5].

QWIPs operate on account of intersubband transitions in doped QWs which implies photoexcitation of charge carriers from a bound ground state to quasi-bound or extended excited states where the charge carriers are freely mobile perpendicularly to the QW planes, thus enabling photoconductive action. QWIPs with n-doped QWs offer the largest values of $D*$ mainly as a result of low effective mass. However, for this type of doping the quantum mechanical selection rules forbid absorption of radiation with incidence normal to the QW-layer plane. Therefore much work has been done to find optical geometries that overcome this fact and enhance the quantum efficiency η of the detectors[6-10].

The *absorptance* or *absorption quantum efficiency* is defined as the ratio of absorbed optical power and total incident optical power and is here denoted η.

Earlier we have shown that the absorptance for normally incident radiation can be enhanced and become polarisation independent by the use of a crossed (doubly periodic) grating with a waveguide (CGW) and reach values close to unity for large detector mesa sizes (500 μm diameter)[9]. The crossed grating is etched into the top of the detector mesa and the waveguide is defined by the grating, the active QW layer and a cladding layer, from top to bottom. Fig. 1 shows cross-sections of the grating coupled detector, with a cladding layer (a), and without (b). In this work we present an experimental investigation of grating coupled QWIPs, and the dependence of absorptance on various factors like mesa size, grating type, contact type and cladding layer is dealt with. A description of how to invoke finite detector mesa sizes in the theoretical modeling[11] of grating coupled QWIPs is also included. The question of optimizing for detectivity $D*$ instead of absorptance will also be dealt with.

Finally we investigate the performance of detector arrays in terms of noise equivalent temperature difference (*NETD*) for a few selected detector operating temperatures. The

H. C. Liu et al. (eds.), Quantum Well Intersubband Transition Physics and Devices, 13–27.
© 1994 *Kluwer Academic Publishers. Printed in the Netherlands.*

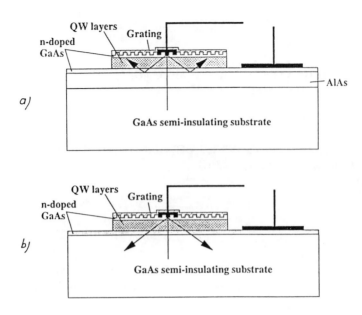

Fig. 1. *Cross-section of a detector with an etched grating a) without waveguide, and b) with a waveguide.*

experimental results of part 2 below will be used as input for the calculations of array performance.

2. Absorptance (absorption quantum efficiency) of grating coupled detectors of finite mesa area

2.1. Theory

The modal expansion method (MEM) applied to grating coupling is investigated in Ref. 11, for detectors with as well as without waveguide. Gratings of different symmetry (square or hexagonal) and different cavity shape (box-shaped or cylindrical) are dealt with. Only *phase gratings* are treated, where the phase of the scattered electromagnetic waves, and not their amplitude, is periodically modulated. Since the absorption and reflection losses are minor, phase gratings offer the largest coupling efficiencies.

Gratings are prepared by etching into the top of the detector mesa. Fig. 2 shows a detail of the case of an ideal etched grating of square symmetry and box-shaped cavities. The theory shows that the type of grating symmetry as well as the cavity shape is immaterial, the only important grating parameters being the ratio of etched area to total area, grating depth and grating period. Besides phase gratings, other promising methods of grating coupling are the use of deposited metalized grating patterns without any etching needed[10]. However, applying contacts to the detectors may be a delicate matter for the smaller mesa dimensions.

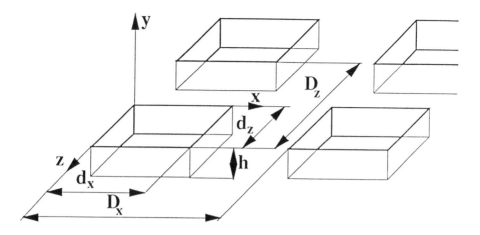

Fig. 2. *Detail of the geometry of the crossed (doubly periodic) grating. In the experiments* $D_x = D_z = D$, *and* $d_x = d_z = d$ *(grating of square symmetry).*

Ref. 11 assumes a grating of infinite extension (IEG) and a grating metal of infinite conductivity. It is however possible to approximately include finite extension of the grating (FEG) by a modification of the theory. It can be shown that for mesa sizes > ca 10 periods the quantum efficiency for a detector of square shape with the same grating constant in two orthogonal directions, can be approximated according to:

$$\eta_{fin}(\lambda, D_m) = \frac{1}{N\pi} \int_{-\infty}^{\infty} \eta_\infty(\lambda, D) \cdot \frac{\sin^2(N Q)}{Q^2} dQ, \qquad (1)$$

where η_{fin} is the absorption quantum efficiency of the detector in the FEG case, η_∞ the quantum efficiency for the case of IEG simulated according to Ref 11, N the number of grating periods, λ the wavelength, D_m the true grating constant of the detector, D a grating constant implicitly used as an integration variable (corresponding to IEG), and $Q(D, D_m) = \pi(D - D_m)/D_m$. In this approximate model, the more direct influence of mesa edges, e. g. transmission and reflection of radiation through the edges, as well as diffuse scattering down into the GaAs bulk induced by the presence of the mesa edge, are not taken care of.

2.2. Experimental

Two separate detector structures (sample A and B) are grown onto GaAs semi-insulating substrates using a low pressure metalorganic vapor phase epitaxy (MOVPE) reactor featuring wafer rotation. Both structures contain a 50 period (GaAs/Al$_x$Ga$_{1-x}$As) multi-QW stack sandwiched between a 1.0 μm bottom and a 1.3 μm top GaAs contact layer ($n = 7 \cdot 10^{17}$ cm^{-3}) and one of them (sample A) an additional 3 μm Al$_{0.78}$Ga$_{0.22}$As cladding layer located between the substrate and bottom contact. The structure parameters are characterized by capacitance voltage etch profiling, x-ray diffraction and fourier transform infrared spectroscopy (FTIR)[12]. The values for sample (A, B) are: GaAs QW doping concentration $n = (3.7 \cdot 10^{17}, 3.7 \cdot 10^{17})$ cm^{-3}, GaAs QW width a = (5.2, 4.9) nm, AlGaAs barrier width a_{barr} = (34.8, 37.7) nm and AlGaAs barrier Al-content x =(0.290, 0.286). The slightly different QW structure parameters lead to the type A structure possessing quasi-bound upper states, whereas the B type excited states are extended.

The detectors are fabricated as follows: gratings (linear and crossed with square symmetry) are etched by reactive ion etching (RIE) into the GaAs top contact layer. The processed grating dimensions viewed from the GaAs side are: grating depth h=0.75 μm, grating constant D=2.8 μm, crossed grating box shape cavity width d=1.6 μm and linear grating cavity width d=1.4 μm. For the definition of the grating parameters see Fig. 2 and Ref. 11. Square shaped mesas of various sizes (from 25 to 500 μm side length) are defined by etching (RIE) down to the bottom contact layer, and finally AuGe/Ni ohmic contacts are deposited onto the top and bottom contact layer. The rows of cavities of the grating were either aligned with the mesa edges or in a direction of 45^0 to these, and will be referred to as 0^0 or 45^0 alignment below. The reason for changing the alignment angle from 0^0 is an attempt to decrease possible radiation leakage through the mesa edges. Since the top contact also serves as a grating metal it is important that this contact possesses high reflectivity as well as good ohmic behaviour. Therefore two different types of top contacts are used, one (labelled N-type) where the whole mesa area is covered with AuGe/Ni and another (labelled S-type) with AuGe/Ni covering only the central 10 % of the mesa area. After alloying, all the mesas are finally covered with a sputtered high reflective Au-layer (below designated "mirror"). In order to apply bond wires to the detectors, gold metalization connecting detector mesas with bond pads is utilized. Silicon nitride is used for insulation to the GaAs substrate. Finally, for characterization of the detectors, a grating monochromator, a 1000 K glowbar source and a pyroelectric reference detector are used. A 1000 K blackbody source is exploited for determining blackbody responsivities and detectivities as well as for calibration purposes.

2.3. Results and discussion

Response measurements of different types of detectors are carried out and the absorption quantum efficiency (absorptance) calculated from $\eta = R_I\, hc/(\lambda q\, g)$, where R_I is the current responsivity, h the Planck constant, c the vacuum speed of light, q the electron charge and g the photoconductive gain. g is obtained from: $g = i_N^2/(4q\,I_d)$, where i_N is the generation-recombination (gr) noise current spectral density and I_d the DC detector current. In order to obtain the gr noise spectral density contribution, the Johnson and excess noise should be subtracted from the total noise. Strictly speaking, the measured value of quantum efficiency is not the absorptance but equal to $\eta \cdot p$, where p is the escape probability of electrons from the upper state of the well to the continuum. p approaches unity when the bias voltage increases, mainly due to facilitated tunneling. Consequently, in order to obtain the absorptance, the measurement should be conducted at a large enough bias voltage to allow $p = 1$. All the results in this work are compensated for the 29 % reflection loss at the air GaAs interface experienced by normal incidence radiation, thus simulating the presence of an ideal antireflective coating on the rear side of the GaAs detector wafer. Detector parameters for edge detectors will refer to unpolarized radiation. Finally, it should be mentioned that the detectors have been optimized with respect to D^* or signal current/dark current ratio and not quantum efficiency, primarily by decreasing the QW doping concentration. Consequently this results in lower quantum efficiency than previously presented in Ref. 9.

The respective edge detectors (EDGE) serve as reference detectors within each of the groups A and B. This is necessary since the QW structures are not identical, and possess slightly different absorption properties. Table 1 collects results on the measurement of detector performance including quantum efficiencies and detectivities D^* for different detector types and geometries. Both the mesa dimension as well as the lateral dimension of the gold covered grating are indicated in it, the latter defining the optical area used in the calculations. In addition to the peak quantum efficiency η, the integrated quantum efficiency η_{int} is also included. η_{int} is here defined according to

$$\eta_{int} = \int_{\lambda_1}^{\lambda_2} \eta \, d\lambda$$, with wavelengths $(\lambda_1, \lambda_2) = (7, 12)$ μm, and is a useful measure when discussing

LWIR (8 - 12 μm window) thermal imaging applications for QWIPs. All the presented data is for 45^0 alignment of the grating cavity rows with the mesa edges, however, the measured deviations between 45^0 and 0^0 alignment are minute and can be neglected in practice. Evidently, compared to the EDGE detector with η_{int} (A, B) = (9.5, 11) %, the quantum efficiency enhancement is substantial for detectors with crossed gratings (CG) and even more so for detectors with a crossed grating and waveguide (CGW). Considering large mesa areas of 500 μm, the enhancement ratio $r_{CG} = \eta_{int} (CG)/\eta_{int} (EDGE) = 2.1$, whereas the corresponding $r_{CGW} = 4.5$. The ratio r_{CGW}/r_{CG} = 2.1, which clearly demonstrates the effect of the waveguide (cladding layer). The detectors with lamellar gratings with and without waveguide, respectively, (LGW and LG), are evidently inferior compared to the corresponding crossed grating detectors.

It is noteworthy that the quantum efficiency η and η_{int} decreases with decreasing mesa area, and more so than is predicted by Eqn. 1. This is also evident from Fig. 3, which depicts the quantum efficiency η_{int} vs. mesa dimension. The aforementioned fact that the measured data for 45^0 and 0^0 grating alignment are close to each other, points to that direct leakage of radiation through the mesa edges may be neglected. Only diffuse scattering from the QW region to the GaAs substrate due to the disturbance by the mesa edge remains a possible explanation. It is, however, evident that even for smaller mesas the CGW detector type is superior as compared to the other types. Therefore, the described method of coupling radiation into QWIPs may advantageously be exploited in the fabrication of large focal plane arrays, where pixel dimensions < 50 μm are needed.

When comparing grating detectors with N and S type contacts, detectors with S type contacts generally give a factor of two higher response compared to detectors with N type contacts, emphasizing the importance of a high reflectivity grating metal. The change from AlAs (used in our previous work[3,8,9,11]) to $Al_{0.78}Ga_{0.22}As$ cladding layer material is justified by the better chemical stability of the latter[13] . According to simulations, the change in cladding layer composition

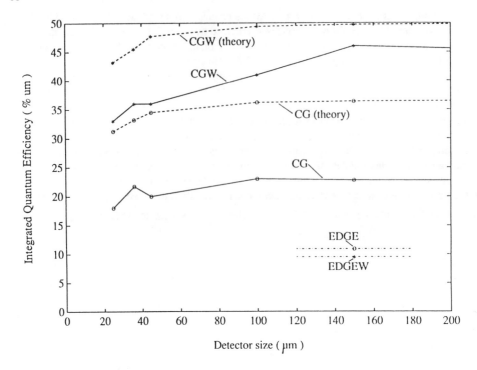

Fig. 3. *Integrated quantum efficiency* η_{int} *vs. detector mesa size for crossed grating detectors with a waveguide* (CGW), *as well as without waveguide* (CG), *according to experiment (full curve), and theory (dashed). The corresponding results of* 45^0 *polished edge detectors are included for the case of: with a waveguide* (EDGEW), *and without* (EDGE).

causes a decrease in η only in the short wavelength region of the response spectrum, leaving the remaining part of the spectrum unaffected.

Theoretical and experimental results of η vs. λ are presented in Fig. 4a and b. The simulation parameters were extracted by fitting to EDGE detector absorption and response curves. A comparison with the experimental curves shows that the main difference is the presence of a predominant evanescent response peak in the former which is lacking in the latter. We attribute this to broadening and absorption factors such as irregularities of the gratings and the finite conductivity of the metal which adds to the broadening due to mesa size included in the theory. It should be noted that broadening factors are taken into account for the smaller mesa sizes, but η_{int} is still too large compared to experimental data. Probably, the main reason for the minor contribution of the evanescent fields is considerable absorption losses in the metal as a result of a large electromagnetic field intensity close to the metal.

Finally, as concerns peak D^*, very large values of $1.4 \cdot 10^{11}$ cm Hz$^{1/2}$ W^{-1} are obtained for the 500 μm mesa side length CGW detector at $\lambda = 8.7$ μm and 77 K operation temperature, and the detector shielded from background radiation during noise measurement. To our knowledge this is the largest D^* reported to date for QWIPs at the wavelength and operation temperature specified.

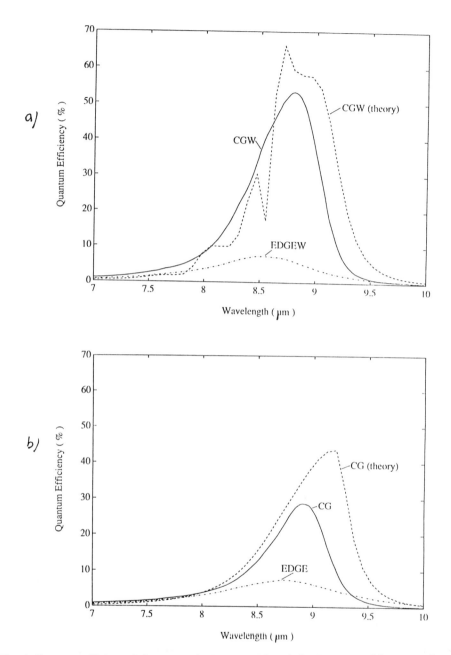

Fig. 4. *Quantum efficiency (absorptance) η vs. wavelength for detectors with a crossed grating, according to experiment (full curves), and theory (dashed). Results of 45^0 polished edge detectors are included for reference purposes. 4a) detectors with a waveguide, and 4b) without waveguide.*

To conclude, the absorptance of detectors with crossed gratings, with or without waveguides has been shown to be substantial as compared to EDGE detectors. Even for mesa sizes as small as 45 μm, suitable for large area detector arrays, the enhancement factors are 3.8 and 1.8 for the CGW and CG detectors, respectively. Finally, the importance of a cladding layer (dielectric mirror) as well as of a high conductivity grating metal is demonstrated, each contributing a factor of two to η_{int}.

3. Calculation of the performance of detector arrays

In order to determine the signal to noise performance of detectors and detector arrays, detector noise is essential. For *single detectors* the detectivity D^* is commonly used, and is a signal to noise ratio normalized with respect to detector area and noise bandwidth. Considering the optimum design of a single detector of specified wavelength response, the design parameters to consider are mainly: optical coupling efficiency and QW doping concentration. It has been shown by several authors that as long as the absorptance is proportional to doping concentration, which applies for the low doping region, the doping sheet concentration per QW that optimizes D^* is:

$n_C = 8\pi m^* k_B T/h^2$, which implies $n_C = 3.7 \cdot 10^{11}$ cm^{-2} . Here m^* is the electron effective mass, k_B the Boltzmann constant and T the detector operating temperature. On the other hand, the doping concentration that maximizes the ratio of photon current to dark current is half this value. It is evident that the sheet concentration of the detectors in part 2 above, concides with an optimum photon current/dark current ratio. However, since the optimum is fairly broad, D^* is also nearly optimum (about 85 % of the maximum possible value).

Instead of using detector current it is common praxis for focal plane arrays to use the total number of collected charge carriers N or the number of noise electrons n, during an integration period τ_{int}, instead of detector current and noise current. This is especially convenient when electrons are physically collected in an integration capacitor, as for the case of direct injection readout circuits. The relation between N (or n) and either direct current or noise current I is:

N (or $n) = I \cdot \tau_{int}/q$. The integration time τ_{int} is assumed to be 17 ms below, corresponding to 60 Hz frame rate.

Noise in a single detector element, consisting of temporal noise only, can be written:

$$n_{temp}^2 = 2gN_{tot} + n_J^2 + n_{1/f}^2 \qquad\qquad (2)$$

where N_{tot} is the total number of charge carriers (electrons) which equals the number of optically generated electrons N_{op} and the number of thermally generated electrons N_{th}. n_J is the number of Johnson noise electrons, and $n_{1/f}$ is the excess noise consisting of mainly 1/f noise. The latter noise component is usually negligible in QWIPs, whereas Johnson noise is important only at small values of detector bias.

For *detector arrays* the *noise equivalent temperature difference NETD* is a suitable measure of performance especially for thermal imaging applications. A calculation of *NETD* of detector arrays is performed below, taking measured detector data from the single detectors (mesa size length 36 μm) from part 2 above as input for the calculation.

For detector arrays, fixed pattern noise n_{fix} enters as a new type of noise component. The fixed pattern noise arises as a result of non-uniformity of detector dark current and responsivity of the detector array. For the case of QWIP, n_{fix} is the remaining number of noise electrons after two point compensation of the detector current (e. g. offset and gain correction), and can be written[14] :

$$n_{fix}^2 = \beta_{gn}^2 \cdot \left(N_{op} \sigma_{op} \right)^2 + \beta_{of}^2 \cdot \left[\left(N_{th} \sigma_{th} \right)^2 + \left(N_{op} \sigma_{op} \right)^2 \right] \qquad (3)$$

where σ_{op} and σ_{th} are the ratios of standard deviation to mean of photon current and dark current, respectively. The factors β_{of} and β_{gn} represent offset and gain corrections determined by sensor calibration.

Still another noise component of importance is noise arising due to readout circuits of a focal plane array. For a well designed circuit this noise should be smaller than the detector noise. In the calculations below it is assumed that this type of noise can be neglected. The optics $f\# = 2$, and 70 % optical transmission are assumed in the calculations throughout.

Fig. 5 shows the number of charge carriers collected during 17 ms for a) 70 K and b) 77 K. Each figure displays the optical, thermal and total number of electrons, respectively. The data at 70 K has been extrapolated from measured currents at 77 K and 85 K, using a semi-empirical expression for dark current and responsivity presented in Ref. 2. Evidently, at 70 K both optically and thermally generated electrons contribute approximately equally. At 77 K the thermally generated electrons dominate.

Fig. 6 shows the number of noise electrons for a detector array consisting of CGW detectors at a) 70 K and b) 77 K operating temperature. The figures show photon noise, dark current noise, Johnson noise, fixed pattern noise, and total noise for the detector array, respectively[14]. The uniformity and calibration parameters of Eq. 3 have been chosen according to Ref. 14: $\sigma_{th} = 0.1$, $\sigma_{op} = 0.01$, $\beta_{of} = 10^{-3}$ and $\beta_{gn} = 0.1$, which are reasonable values for detector arrays. It is found that at both 70 K and 77 K the fixed pattern noise largely dominate the total noise behavior at all values of bias voltage except for low values (< ca 0.2-0.3 V). In the latter case the Johnson noise contributes significantly. It should be noted that neither the photon noise nor the g-r noise of the detector contributes.

The noise equivalent temperature difference (NETD) of the detector array is obtained from:

$$NETD = T_C \cdot \left(n_{tot} / N_{opt} \right)$$

where n_{tot} is the total number of noise electrons during T_{int}, and T_C a temperature equal to

$$T_C = \left[\frac{\partial \ln M}{\partial T_B} \right]^{-1},$$

where M is the exitance of the surrounding photon background assuming an ambient temperature of T_B. T_C is about 65 K in the 8 - 12 μm wavelength region.

22

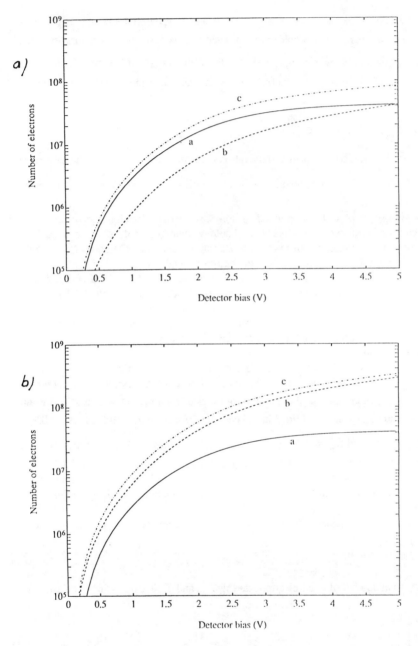

Fig. 5. *Number of collected charge carriers for a detector with a crossed grating and waveguide of mesa size 36 μ m: a) optically generated, b) dark current, c) total number of carriers. Fig. 5a: 70 K operating temperature, and 5b: 77 K. Integration time 17 ms, optics f# = 2, optical transmission = 70 %.*

NETD vs. detector bias is shown in Fig. 7a and b at the operating temperatures 70 K and 77 K, respectively. Each figure shows *NETD* for the case of fixed pattern noise not included, as well as included. In addition, the case of finite charge capacity of the integration capacitor of *e. g.* a direct injection readout is taken into account for some of the curves presented. In the latter case, for large detector currents, the integration time has been decreased, in order for the total number of charge carriers not to exceed $1.6 \cdot 10^7$ (which is a reasonable value of the number of maximum charge carriers of an integration capacitor of an FPA). It is found that at 70 K operating temperature, the obtained value of *NETD* is 55 mK with fixed pattern noise included, and 10 mK without (i. e. only temporal noise). At 77 K one obtains 60 mK and 20 mK, respectively. It should be noted that at 70 K it is possible to operate the detector up to about 1.7 V bias voltage, but at 77 K only up to 1.2 V, in order not to exceed the maximum charge capacity of the integration capacitor.

4. Conclusions

To conclude, the efficiency of grating coupling of radiation into quantum well infrared detectors is studied theoretically and experimentally, with special emphasis on the influence of detector mesa size. It is found that compared to a 45^0 polished edge detector, the integrated quantum efficiency for crossed grating detectors of mesa sizes as small as 45 μm is about four times larger with a waveguide included, and two times larger without waveguide. Furthermore, the importance of a high conductivity grating metal, and the use of crossed gratings instead of lamellar gratings is demonstrated. To show the superior coupling efficiency, the detectivity is measured for a detector with a crossed grating and a waveguide, and of mesa size 500 μm, resulting in $D* =$

$1.4 \cdot 10^{11}$ cm Hz$^{1/2}$ W^{-1} at 77 K operating temperature with the detector shielded from background.

Finally, the performance of a detector array in terms of noise equivalent temperature difference *NETD* is investigated. Here fixed pattern noise is taken into account. It is found that at 77 K operating temperature, *f#* = 2 optics, and an optical transmission of 70 %, the obtained value of *NETD* is ca 60 mK with fixed pattern noise included, and 20 mK otherwise (i. e. only temporal noise). In order not to exceed the maximum charge capacity of the integration capacitor, the bias voltage of the detector should not exceed about 1 V.

24

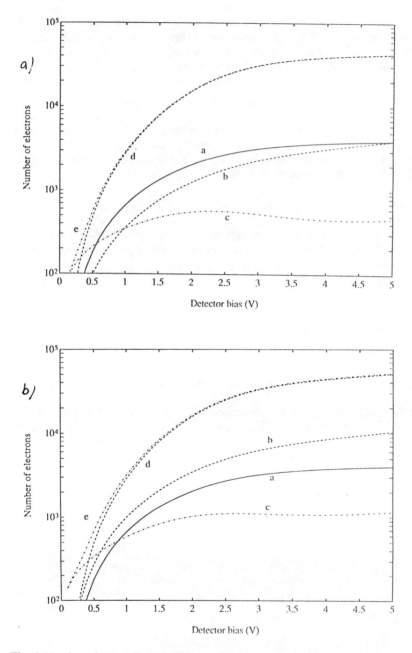

Fig. 6. *Number of noise electrons for a detector with a crossed grating and waveguide of mesa size 36 μ m: a) photon, b) thermally generated (gr noise), c) Johnson, d) fixed pattern, and e) total noise. Fig. 6a: 70 K operating temperature, and 6b: 77 K. Integration time 17 ms, optics f# = 2, and optical transmission = 70 %.*

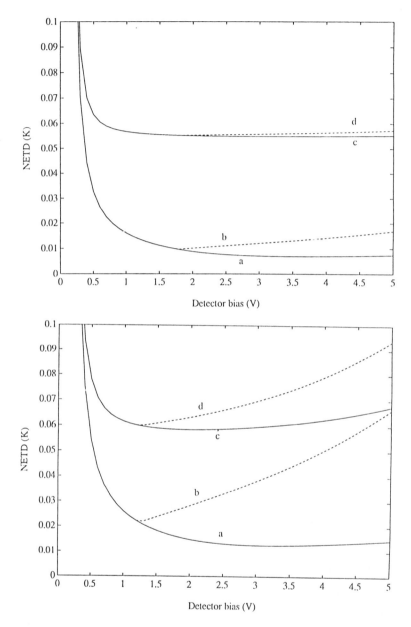

Fig. 7. *Noise equivalent temperature difference NETD vs. detector bias voltage for a) only temporal noise, b) only temporal noise considering the limited charge capacity of the integration capacitor, c) total noise, and d) total noise considering the limited charge capacity of the integration capacitor. Fig. 7a: 70 K operating temperature, and 7b: 77 K. Maximum integration time 17 ms, optics f# = 2, and optical transmission = 70 %. Assumed maximum charge capacity:* $1.6 \cdot 10^7$ *electrons.*

Sample No	Detector type	Detector size [Mirror size] (μm)	Bias voltage, U_d (V)	Dark current density, j_d (A/cm²)	Photoconductive gain, g	Peak current responsivity, R_i (A/W)	Peak detectivity, D^* (cm√Hz W⁻¹)	Peak quantum efficiency, η (%)	Integrated quantum efficiency, η_{int} (% μm)
A	EDGEW	150 [-]	4	2.0e-4	0.19	0.10	1.7e10	7.4	9.5
"	CGW	500 [494]	4	2.1e-4	0.19	0.71	1.4e11	53	43
"	"	150 [144]	4	2.0e-4	0.18	0.63	1.3e11	49	46
"	"	100 [94]	4	2.1e-4	0.18	0.55	1.0e11	42	41
"	"	45 [40.5]	4	1.9e-4	0.18	0.42	8.0e10	32	36
"	"	36 [31.5]	4	2.1e-4	0.18	0.39	6.9e10	31	36
"	"	25 [20.5]	4	2.2e-4	0.17	0.31	5.2e10	26	33
"	LGW	150 [144]	4	2.1e-4	0.18	0.32	6.2e10	25	20
B	EDGE	150 [-]	4	8.6e-4	0.28	0.14	1.0e10	7.4	11
"	CG	500 [494]	2.5	4.0e-4	0.38	0.76	7.6e10	29	23
"	"	150 [144]	2.5	4.0e-4	0.36	0.62	6.0e10	23	23
"	"	100 [94]	2.5	3.9e-4	0.37	0.55	5.4e10	22	23
"	"	45 [40.5]	2.5	4.0e-4	0.34	0.43	4.0e10	21	20
"	"	36 [31.5]	2.5	4.0e-4	0.36	0.42	3.8e10	16	22
"	"	25 [20.5]	2.5	4.2e-4	0.38	0.36	2.9e10	13	18
"	LG	150 [144]	2.5	4.0e-4	0.34	0.17	1.8e10	7.1	8.6

Table 1. Detector data for the different detector configurations utilized: CG = crossed grating, CGW = crossed grating and waveguide, LG = lamellar grating, LGW = lamellar grating and waveguide, EDGE = 45⁰ polished edge detector, EDGEW = edge detector with a cladding layer (waveguide). A and B refer to different QW structures. The measurements have been conducted at 77 K throughout.

References

1. B. F. Levine, C. G. Bethea, G. Hasnian, V. O. Shen, E. Pelve, R. R. Abbot, and S. J. Hsieh, Appl. Phys. Lett. 56, 851 (1990).
2. B. F. Levine, A. Zussman, S. D. Gunapala, M. T. Asom, J. M. Kuo, and W. S. Hobson, J. Appl. Phys. 72, 4429 (1992) and references therein.
3. J. Y. Andersson and L. Lundqvist, "Intersubband Transitions in Quantum Wells": Proceedings of the NATO Workshop, Cargèse, Corsica, France. Eds. E. Rosencher, B.F. Levine, B. Vinter.
4. L. J. Kozlowski, G. M. Williams, G. J. Sullivan, C. W. Farley, R. J. Andersson, J. K. Chen, D. T. Cheung, W. E. Tennant, and R. E. DeWames, IEEE Trans Electron. Devices 38, 1124 (1991).
5. B. F. Levine, C. G. Bethea, K. G. Glogovsky, J. W. Stayt, and R. E. Leibenguth, "Narrow Band Gap Semiconductors": Proceedings of the NATO Workshop, Oslo, Norway, June 25-27 (1991).
6. K. W. Gossen, S. A. Lyon, and K. Alavi, Appl. Phys. Lett. 53, 1027 (1988)
7. G. Hasnain, B. F. Levine, C. G. Bethea, R. A. Logan, J. Walker, and R. J. Malik, Appl. Phys. Lett. 54, 2515 (1989).
8. J. Y. Andersson, L. Lundqvist, and Z. F. Paska, Appl. Phys. Lett. 58, 2264 (1991).
9. J. Y. Andersson, L. Lundqvist, Appl. Phys. Lett. 59, 857 (1991).
10. L. S. Yu, S. S. Li, and Y. H. Wang, J. Appl. Phys. 72, 2105 (1992).
11. J. Y. Andersson, and L. Lundqvist, J. Appl. Phys. 71, 3600 (1992).
12. Z. F. Paska, J. Y. Andersson, L. Lundqvist, and C-O. A. Olsson, J. Cryst. Growth 107, 845 (1991).
13. J. M. Dallesasse, N. El-Zein, N. Holonyak, Jr., and K. C. Hsieh, J. Appl. Phys. 68, 2235 (1990).
14. R. L. Whitney, K. F. Cuff, and F. W. Adams, Semiconductor Quantum Wells and Superlattices for Long-Wavelength Infrared Detectors, Ed. M. O. Manasreh, Artech House, Boston, London 1993, p. 55.

NOVEL GRATING COUPLED AND NORMAL INCIDENCE III-V QUANTUM WELL INFRARED PHOTODETECTORS WITH BACKGROUND LIMITED PERFORMANCE AT 77 K

SHENG S. LI AND Y. H. WANG
Dept. of Electrical Engineering
University of Florida
Gainesville, Florida 32611

ABSTRACT. We report four novel III-V quantum well infrared photodetectors (QWIPs) fabricated on the MBE grown n-type GaAs/AlGaAs, AlAs/AlGaAs, and p-type strained layer (PSL) InGaAs/InAlAs material systems. The detection schemes for these QWIPs are based on bound-to-continuum (BTC) and bound-to-miniband (BTM) intersubband transitions with wavelengths covering from 2-6 μm and 8 -14 μm. QWIP-A is a dual mode (PV & PC) operation GaAs/AlGaAs BTC QWIP with heavily-doped enlarged (11 nm) GaAs quantum wells and thick AlGaAs barrier layers (87.5 nm) grown on the semi-insulating (SI) GaAs substrate for 7.7 and 12 μm two-color detection. QWIP-B is a GaAs/AlGaAs BTM QWIP using an enlarged GaAs quantum well (8.8 nm) and a short-period GaAs/AlGaAs superlattice barrier layer grown on (100) SI GaAs. QWIP-C is a normal incidence type-II AlAs/AlGaAs QWIP grown on the (110) SI GaAs for 2 -14 μm multicolor detection. QWIP-D is a normal incidence p-type strained layer (PSL) InGaAs/InAlAs QWIP grown on the (100) SI InP. For QWIP-A and B, a planar square mesh metal grating coupler is used to achieve high coupling efficiency under normal incidence illumination. Detectivities ranging from 10^9 to 10^{12} cm-Hz$^{1/2}$/W have been obtained for these QWIPs at 77 K operation. The PSL-InGaAs/InAlAs QWIP shows the best overall performance, featuring the lowest dark current density ever reported for the QWIPs and a background limited performance (BLIP) at $\lambda_p = 8.1 \mu$m and T = 77 K.

1. Introduction

III-V quantum well infrared photodetectors (QWIPs) based on intersubband transition schemes for detection in the 3-5 μm mid-wavelength infrared (MWIR) and 8 -14 μm long-wavelength infrared (LWIR) atmospheric spectral windows have been extensively investigated in recent years[1-18]. A great deal of work has been reported on the lattice-matched GaAs/AlGaAs multiple quantum well and superlattice systems using bound-to-bound (BTB)[1,2], bound-to-miniband (BTM)[3-5], bound-to-quasi-continuum

29

H. C. Liu et al. (eds.), Quantum Well Intersubband Transition Physics and Devices, 29–42.
© 1994 *Kluwer Academic Publishers. Printed in the Netherlands.*

(BTQC)[6] and bound-to-continuum (BTC)[7,8] band intersubband transition mechanisms. Although a majority of the study on intersubband absorption QWIPs has been based on the photoconductive (PC) mode operation[9-11], studies of the photovoltaic (PV) mode detection have also been reported in the literature. [12-14] For n-type QWIPs, a dielectric or metal grating coupled structure is needed in order to achieve intersubband absorption under normal incidence illumination. On the other hand, for type-II and p-type QWIPs, no grating coupler is required to induce vertical intersubband absorption under normal incidence illumination. In this paper we present the physical principles and performance characteristics of the grating coupled n-type GaAs/AlGaAs BTC and BTM QWIPs, a normal incidence type-II AlAs/AlGaAs QWIP, and an ultra-low dark current normal incidence p-type strained layer InGaAs/InAlAs QWIP operating at 77 K.

2. Physical Principles and QWIP Structures

In this Section we describe the basic principles and structures of four novel III-V QWIPs fabricated on the lattice-matched n-type GaAs/AlGaAs, AlAs/AlGaAs, and the lattice-mismatched p-type strained layer (PSL) InGaAs/InAlAs material systems. The optical absorption and detection for these QWIPs are based on the bound-to-miniband (BTM) and bound-to-continuum (BTC) state intersubband transition mechanisms. Figure 1 shows the energy band diagrams for the four different types of QWIPs (A, B, C, and D). The spectral responses of these QWIPs extend from 2-6 μm (MWIR) to 8-14 μm (LWIR) bands.

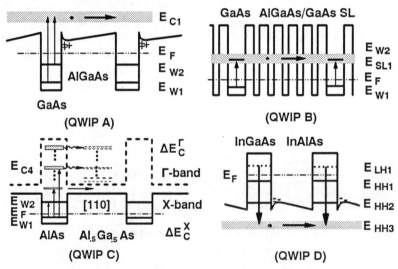

Figure 1 Energy band diagrams for four III-V QWIPs: (a) GaAs/AlGaAs DM-QWIP, (b) GaAs/AlGaAs BTM-QWIP, (c) type-II AlAs/AlGaAs QWIP, and (d) InGaAs/InAlAs PSL-QWIP.

QWIP-A and B were grown on the semi-insulating (SI) (100) GaAs substrate by the molecular beam epitaxy (MBE) technique, while QWIP-C and D were grown on SI (110) GaAs and SI (100) InP, respectively. The QWIP-A structure consists of 40 periods of thick (875 Å) undoped $Al_{.25}Ga_{.75}As$ barrier layers and heavily doped ($5 \times 10^{18} cm^{-3}$) enlarged (110 Å) GaAs quantum wells. Due to the heavy doping both the ground state and first excited state inside the well are filled by electrons at 77 K. As a result, electrons in each of the populated localized states in the well will undergo intersubband transition to the first continuum band states upon excited by the IR radiation. Due to the larger absorption strength from the excited state to the continuum states as compared to that from the ground state to the continuum states, a larger enhanced absorption of IR radiation is expected from the excited state to the continuum states transition. The device parameters of QWIP-A are chosen so that there are two bound states inside the enlarged well (i.e., E_{W1} and E_{W2}), and the first continuum band (E_{C1}) is just slightly above the top of the barrier layer. To eliminate the undesirable tunneling current through the barrier layers, a thick (875 Å) undoped $Al_{0.25}Ga_{0.75}As$ barrier layer was used in this QWIP structure to suppress the tunneling current from the heavily populated ground state E_{EW0} and the first excited state E_{W2} in the quantum well. As shown in Fig.1(a), the first transition scheme for QWIP-A is from the localized ground-state E_{W1} in the GaAs quantum well to the first continuum band states E_{CN} above the AlGaAs barrier layer, while the second transition is taking place from the first excited state E_{W2} to the first continuum states E_{C1} above the barrier. These intersubband transitions will lead to two color detection (i.e., λ_p at 7.7 and 12 μm) for QWIP-A. Due to the thick barrier layer and enlarged quantum well, very low dark current and high responsivity are expected for this QWIP.

QWIP-B is a bound-to-miniband (BTM) transition GaAs/AlGaAs QWIP, which consists of 40 periods of doped GaAs quantum wells and undoped GaAs/AlGaAs superlattice barrier layers. Each barrier layer is composed of 6-period of undoped GaAs/$Al_{0.25}Ga_{0.75}As$ (2.9/5.8 nm) superlattice layers. A wide and highly-degenerated global miniband is formed by the superlattice barrier layer inside the quantum well, which is aligned with the first excited state of the quantum well. The transition scheme for this BTM QWIP is from the ground state to the global miniband states inside the quantum well. Current conduction is via thermionic-assisted resonant tunneling through the global miniband. Low dark current and low noise are expected for this BTM QWIP. Both QWIP-A and B exhibit the photovoltaic (PV) and photoconductive (PC) dual mode detection. The QWIP mesa structures were created by chemical etching through the quantum well active layers and stopped at the 1 μm thick heavily doped GaAs buffer layer for ohmic contacts. The active area of the QWIPs is $200 \times 200 \mu m^2$. To increase the light coupling efficiency in the quantum well, a planar square mesh metal grating coupler was developed for both QWIP-A and B for normal incidence illumination[15,16]. The planar metal grating coupler consists of regularly spaced square mesh metal grating of 0.2 μm thick. By using optimum grating period and grating aperture dimensions we can achieve maximum coupling quantum efficiency in the wavelengths of interest for both QWIP-A and B.

QWIP-C is a normal incidence type-II AlAs/AlGaAs QWIP. The QWIP structure is composed of 20 periods of AlAs/Al$_5$Ga$_5$As quantum wells, which are grown on the [110] SI GaAs by the MBE technique. Since the conduction band minima of AlAs are located at X-points (i.e. along (110) direction) of the Brillouin zone (BZ), the indirect bandgap AlAs X-conduction band becomes the quantum well, while the indirect bandgap Al$_5$Ga$_5$As forms the barrier layer. As a result, a type-II energy band alignment is created [17-20] in this QWIP. The key energy states involved in the intersubband transitions include the two bound states, $E_{W1,2}$, in the AlAs X-band well, the four continuum states, E_{C2-5}, in the X-band, which can find their resonant pair levels in the Γ-band, and the continuum state E_{C1} located just below the Γ-band minima of the Al$_5$Ga$_5$As layers, as shown in Fig. 1(c). Due to the anisotropic band structures and the tilted growth direction with respect to the principal axes of ellipsoidal valleys, one can realize a multi-band and multi-color normal incidence IR detection in this QWIP without using the grating coupler.

QWIP-D is a normal incidence p-type strained layer (PSL) In$_{0.3}$Ga$_{0.7}$As/In$_{0.52}$Al$_{0.48}$As QWIP. The optical absorption and detection of this QWIP is based on the ground light-hole state (E_{LH1}) to heavy-hole band (E_{HH3}) valence intersubband transition [21,22]. The PSL-QWIP structure was grown on a (100) SI InP by the MBE technique. It consists of 20 periods of 4-nm Be-doped In$_{0.3}$Ga$_{0.7}$As quantum well with a dopant density of 1×10^{18} cm^{-3} separated by a 45-nm In$_{0.52}$Al$_{0.48}$As undoped barrier layer. A 0.3 μm thick cap-layer and a 1μm thick buffer layer of Be-doped In$_{0.53}$Ga$_{0.47}$As with a dopant density of 2×10^{18} cm^{-3} were grown as the top and bottom ohmic contact layers. The contact and barrier layers are lattice-matched to the SI InP substrate, and the quantum well layer is in biaxial tensile strain created by the lattice-mismatch ($\sim 1.5\%$) between the well and the barrier layer. The active area for this PSL-QWIP is 200×200 μm^2, which was formed by mesa etched process. Au/Zn alloy was used to form p-type ohmic contacts on this QWIP. The Au/Zn ohmic contact rings were thermally evaporated onto the QWIP mesas with a film thickness of 0.12 μm, followed by thermal annealing at 480 °C for 5 minutes to obtain stable and low contact resistance.

The tensile strain created by the lattice-mismatch (1.5 % in the present case) between the well and the barrier layer in the PSL-QWIP can greatly affect the energy band structure, and causes the splitting between the heavy-hole and light-hole bands in the valence band zone-center[21] (i.e., in-plane wavevector $k_{\parallel} = 0$), which is degenerated for the unstrained case. When tensile strain is applied between the quantum well and the barrier layer[21,22] along the quantum well growth direction, the tensile strain can push the light-hole state upward and pull the heavy-hole state downward. As a result, the heavy-hole and light-hole states can be inverted at a certain strain and quantum well thickness. This in turn will cause the intersubband transition from the populated light-hole ground state to the upper energy band states. Since the light hole has a small effective mass (i.e., comparable to the electron effective mass) and a large in-plane density of states, the optical absorption and responsivity in the p-type strained layer QWIPs can be greatly enhanced by using this new approach.

3. Results and Discussion

To analyze the intersubband transition schemes, we performed theoretical calculations of the energy levels E_{Wn}, E_{SLn} and E_{Cn} (n = 1,2,...) in the quantum well(W), superlattice (SL), and the continuum (C) states as well as the transmission probability T*T for these QWIPs using a multi-layer transfer matrix method (TMM)[23]. Design of the square mesh metal grating coupler for n-type QWIPs (A & B) was carried out by using modal expansion technique.[16] Maximum coupling quantum efficiency of around 50 % has been achieved in a square mesh grating coupled BTM QWIP.[24] The device parameters and the calculated energy levels for these QWIPs are summarized in Table I.

Table I. Device parameters and calculated energy levels for the four III-V QWIPs

QWIP (Type)	L_W (Å)	L_B (Å)	N_D (cm^{-3})	QW Periods	Energy Levels[†] (meV)
A (I)	GaAs 110	Al$_{.25}$Ga$_{.75}$As 875	5×10^{18}	40	E_{W1}=24 E_{W2}=90 E_{C1}=190
B (I)	GaAs 88	GaAs/AlGaAs 29/58	2×10^{18}	40	E_{W1}=27 E_{W2}=163
C (II)	AlAs 30	Al$_{.5}$Ga$_{.5}$As 500	2×10^{18}	20	E_{W1}=20 E_{C4}=365 E_{C6}=600
D (P)	In$_{.3}$Ga$_{.7}$As 40	In$_{.52}$Ga$_{.48}$As 450	1×10^{18}	20	E_{LH1}=58 E_{HH3}=215

† In reference to the energy band edge of the quantum well.

3.1 OPTICAL ABSORPTION CONSTANT

The room temperature absorbance curves for QWIP-A and B were measured by using a Perkin-Elmer Fourier transform infrared (FTIR) spectrometer. Since the intersub-band resonance is expected to vanish under normal incidence for n-type QWIPs, both

QWIP-A and B were oriented at Brewster's angle during the measurements to maximize the intersubband absorption. The direct measured quantity is the absorbance
A = -log$_{10}$(transmission), which can be converted to the absorption coefficient α for
45° incident angle. The main lobes of the absorption spectra for QWIP-A and B are
shown in Fig.2. For QWIP-A the main absorption peak is centered at $\lambda_p = 12.3\mu m$
with $\alpha = 4,000 cm^{-1}$. For QWIP-B the absorption peak occurs at $\lambda_p = 9.6\mu m$ with
$\alpha = 4,350 cm^{-1}$. For QWIP-C (type-II), the absorption spectra are more complicated than QWIP-A and B. The absorbance measurements for QWIP-C were carried
out by using a BOMEM interferometer. In order to eliminate substrate absorption,
we performed absorbance measurements on QWIP-C samples with and without the
quantum well layers. The absorbance data were taken under normal incidence illumination at 77 K and room temperature. The absorption coefficients deduced from the
absorbance data are also shown in Fig. 2. Six absorption peaks at $\lambda = 2.3, 2.7, 3.5,$
4.8 6.8, and 14 μm were detected in QWIP-C. The absorption coefficients measured
at 77 K were found to be about a factor of 1.2 higher than that of the room temperature values for this QWIP. From our theoretical analysis, the 14 μm peak ($\alpha \sim$
2000 cm^{-1}) is attributed to the transition between the ground state E_{W1} and the first
excited state E_{W2} in the quantum well, while the 6.8 μm peak ($\alpha \sim 1600$ cm^{-1}) is due
to transition between E_{W1} and E_{C1} above the barrier layer. The absorption peaks at
$\lambda_p = 2.3, 2.7, 3.5$ and 4.8 μm are attributed to the transition from E_{W1} to the higher
$\Gamma - X$ continuum resonant states, E$_{Cn}$ (n = 3,4,5,6), as shown in Fig. 1(C).

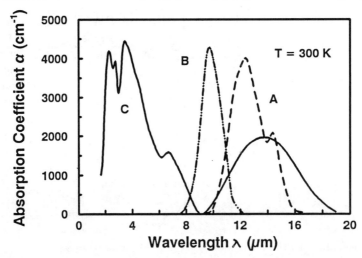

Figure 2 Optical absorption coefficient versus wavelength for the
four QWIPs shown in Fig. 1

3.2 DARK CURRENT

Measurements of the dark currents for all the QWIP samples reported here were made between 30 K and 90 K using a HP-4145 Semiconductor Parameter Analyzer. Figure 3 shows the current density versus bias voltage curves measured at 77 K for these QWIPs.

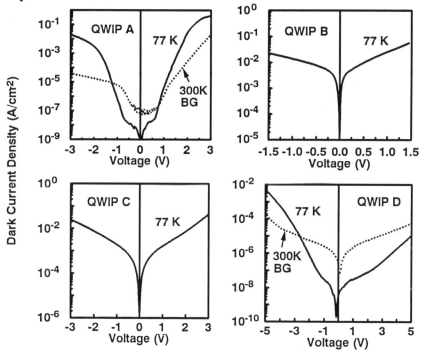

Figure 3 Dark current density versus bias voltage for the four QWIPs shown in Fig. 1.

In addition, the 300 K background photocurrent densities for QWIP- A and D were measured by using a cylindrical cold shielding window with a FOV of 90 $^\circ$, and the results are also shown Fig.1, which shows the background limited performance (BLIP) for these two QWIPs at $V_b = 1$ and 2.5 V, respectively, and T = 77 K. For QWIP-B and C, the device dark current is higher than the 300 K background photocurrent, and hence is not in BLIP condition at 77 K. The asymmetrical I-V characteristics (with I_d being larger for positive bias than negative bias) were observed in all four QWIPs reported here. In QWIP-B, the asymmetrical behavior is mainly ascribed to two effects: (i) superlattice-formed miniband asymmetrical conduction[12] which was observed in optical processes for thin multilayer semiconductor structures[24], and (ii) dopant impurity migration from the doped quantum well to the undoped barrier region, which further modifies the asymmetrical conduction miniband. In other QWIPs (A, C, and D), the asymmetrical I-V characteristics are mainly due to the band bending effect resulting from dopant impurity migration and heavy doping effect. It is interesting to note that the PSL-InGaAs/InAlAs QWIP (D) has the lowest dark current density ever reported for the QWIPs operating at 77 K. Our results reveal that QWIP-D is under BLIP conditions for $|V_b| \leq 2$ V, while QWIP-A meets BLIP condition for $|V_b| \leq 1$ V, at 77 K operation.

3.3. SPECTRAL RESPONSIVITY AND DETECTIVITY

The responsivity was measured as a function of temperature, bias voltage V_b, polarization direction, and wavelength using a globar IR source and an automatic PC controlled single-grating monochrometer spectral measurement system under normal incidence illumination. The normalized responsivity versus wavelength for the four QWIPs (A - D) reported here are shown in Fig. 4. The absolute peak responsivities R_A (or R_V) for these QWIPs were measured and calibrated using a room temperature pyroelectric detector. Table.2 summarizes the measured and calculated peak wavelengths, cutoff wavelengths, responsivities, and detectivities for these four QWIPs.

Table II. Summary of the peak and cutoff wavelengths,
responsivities, detectivities at 77 K for the four QWIPs.

QWIP	λ_p (μm)	λ_c (μm)	R (A/W)	D^* cm-Hz$^{1/2}$/W
	12 (PC)	13.2	0.48	2.0×10^{10}
A	7.7 (PV)	8.5	11,000 (V/W)	1.5×10^9
	8.9 (PC)	9.3	0.23	1.2×10^{10}
B	8.9 (PV)	9.2	0.15	7.5×10^9
	2.2 (PC)	2.45	110	1.1×10^{12}
	2.2 (PV)	2.45	64,000 (V/W)	1.4×10^{10}
C	3.5 (PC)	4.3	18.3	3.0×10^{11}
	3.5 (PV)	4.3	58,000 (V/W)	1.2×10^{10}
	12.5 (PC)	14.8	0.024	1.1×10^9
D	8.1 (PC)	8.8	0.018	5.9×10^{10}

QWIP-A is a dual-mode (PV & PC) operation GaAs/AlGaAs QWIP. As shown in Fig. 4 (curve-A), QWIP-A has two response peaks: one at $\lambda_p = 7.7$ μm under PV mode operation and the other at $\lambda_p = 12$ μm under PC mode detection ($V_b > 1$ V). In the PV mode detection, the peak voltage responsivity, R_V, is equal to 11,000 V/W at $\lambda_p = 7.7$ μm with a spectral bandwidth of $\Delta\lambda/\lambda_p = 18$ %. This response peak is attributed to the transition from the ground state E_{W1} to the first continuum states E_{C1} above the barrier. An internal photovoltage is induced from the spatial charge separation when the asymmetrical energy band bending occurs as a result of dopant migration and heavy doping effects. When a bias voltage is applied to the detector, the PV response at $\lambda_p = 7.7$ μm disappears, and the PC mode conduction becomes dominant for $V_b \geq 1V$. In the PC detection mode, the peak response wavelength occurs at $\lambda_p = 12$ μm. A maximum responsivity, $R_A = 0.48$ A/W and detectivity D* $= 2 \times 10^{10}$ cm$\sqrt{Hz}$/W was obtained for this QWIP at $V_b = 2$ V, $\lambda_p = 12$ μm and T = 77 K. The cutoff wavelength was found to be $\lambda_c = 13.2$ μm with a spectral bandwidth $\Delta\lambda/\lambda = 18.3$ %. Thus, QWIP-A can be used as a two-color long wavelength infrared detector for the 6-9 and 10-14 μm detection.

QWIP-B is a bound-to-miniband (BTM) transition GaAs/AlGaAs QWIP. Peak response wavelength occurs at $\lambda_p = 8.9$ μm with cutoff wavelength of $\lambda_c = 9.3$ μm for both the PV and PC detection modes. The photoresponse of this QWIP is attributed to the transition from the ground state E_{W1} to the global miniband states E_{SL1}. In this QWIP, the resonant state E_{W2} lies near the bottom of E_{SL1}, which gives rise to a narrow bandwidth absorption peak with $\Delta\lambda/\lambda_p = 8.5$ %. A peak detectivity D* equal to 1.6×10^{10} cm$\sqrt{Hz}$/W was obtained at $\lambda_p = 8.9\mu$m and T = 77 K. The normalized spectral response curve (curve-B) for this QWIP is also shown in Fig. 4.

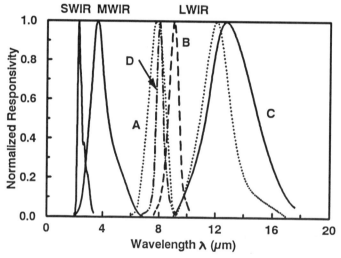

Figure 4 Normalized responsivity versus wavelength for the four
QWIPs shown in Fig. 1.

QWIP-C is a normal incidence type-II AlAs/AlGaAs QWIP, which features multicolor and multiband detection for 2 μm $\leq \lambda \leq 14$ μm. As shown in Fig. 4 (curve-C), three main photoresponse peaks covering the wavelengths from 2 to 16 μm were observed. The long wavelength infrared (LWIR) peak response occurs at $\lambda_p = 12.5$ μm with a responsivity of $R_A = 24$ mA/W and a detectivity of D* $= 1.1 \times 10^9$ cm (Hz)$^{1/2}$/W at $V_b - 2$ V and T = 77 K. Two other normalized PV response peaks occured at $\lambda_{p4} = 3.5$ μm and $\lambda_{p6} = 2.2$ μm are also illustrated in Fig. 4 (curve-C). These two response curves cover the mid-wavelength infrared (MWIR) wavelengths from 2.2 to 6.5 μm with peak wavelength $\lambda_{p4} = 3.5$ μm and from 2.0 to 3.25 μm with peak wavelength $\lambda_{p6} = 2.2$ μm. Two additional peaks at $\lambda_{p5} \sim 2.7$ μm and $\lambda_{p3} \sim 4.8$ μm were also observed in these two bands. The positions of these four peak wavelengths λ_{p3-6} are in excellent agreement with the values deduced from FTIR optical absorption measurements, which correspond to the transitions from the ground state E_{W1} to the continuum states E_{Cn}, where n = 3, 4, 5, 6 in the $X - \Gamma$ resonant levels. The main response peaks occured at $\lambda_{p4} = 3.5$ μm and $\lambda_{p6} = 2.2$ μm with responsivities of $R_A = 29$ mA/W and 32 mA/W, respectively, at $V_b = 0$ and T = 77 K. The responsivity for λ_{p4} increases rapidly with bias voltage for $V_b > 0.5$ V, and reaches a constant value of 18.3 A/W at $V_b \geq 3$ V. On the other hand, the responsivity for λ_{p6} remains nearly constant for $V_b \leq 2$ V, and then increases exponentially to $R_A = 110$ A/W at $V_b \sim 6$ V. Huge photoconductivity gains of 630 and 3,200 were obtained at λ_{p4} and λ_{p6} for $V_b = 3$ V and 6 V, respectively. In a type-II indirect AlAs/AlGaAs QWIP, free carriers are confined in the AlAs quantum well formed in the X-conduction band minimum, which has a larger electron effective mass than that in the Γ-band valley. When a normal incidence radiation impinges on this QWIP, electrons in the ground-state of the X-well are excited to either the first excited state E_{W2} or one of the continuum states E_{C2} to E_{C6}. If the continuum state in the X-band valley is resonantly aligned with a state in the Γ-band valley, the photon-generated electrons in the X-band will undergo resonant transport to the resonant state in the Γ-band provided that the barrier layer (in the present case, AlAs layer) is thin enough that it is transparent to the conduction electrons.[25,26] This resonant transport from the X-band to Γ-band is expected to be a coherent resonance which can greatly enhance the transmission if the electron lifetime τ_L^{Γ} in these continuum states is much shorter than the X-band to Γ-band scattering time constant τ_S. The τ_L^{Γ} can be estimated from the uncertainty principle, $\tau_L^{\Gamma} = \frac{\hbar}{\Delta E_{FWHM}} \sim 10$ fs (where ΔE_{FWHM} is the spectral full width at half maximum), while $\tau_S \sim 1$ ps[27], hence $\tau_L^{\Gamma} \ll \tau_S$. The peak transmission at resonance is expected to be increased by the ratio of $\tau_S/\tau_L^{\Gamma} \sim 100$. In addition, due to the effective mass difference between the X-band and the Γ-band, electron velocity and mobility in the Γ-valley will be much higher than the value in the X-band valley. Since the photocurrent is proportional to the electron velocity and mobility, a large increase in the photocurrent is expected when photon-generated electron resonant transition from the X-band to Γ-band takes place under certain bias conditions. The optical gain is given by $G = \tau_L/\tau_T$, where τ_T is the transit time ($\tau_T = \frac{w_l}{\mu F}$, w_l is the superlattice thickness, μ is the electron mobility, and F is the electric field). In the coherent resonance and under certain bias condition, G will be significantly increased as well. The large responses observed at λ_{p4} and λ_{p6} are due to coherent alignment of these resonant levels, while the relatively lower responses for the λ_{p3} and λ_{p5} are ascribed to a slightly misalignment of the resonant levels, which results from the X-Γ coupling strength difference[28]. However, no photoconductivity gain is expected at λ_{p1} and λ_{p2} peak wavelengths due to the absence of resonant transition from the X-band to Γ-band in the electronic conduction. The PV mode operation at peak wavelengths of λ_{p3-6} in QWIP-D is resulted from the macroscopic polarization field (i.e. Hartree potential) caused by the energy band bending effect and spatial separation of electrons and holes[29,30]. However, the PV mode detection was not observed at wavelengths λ_{p1-2}.

QWIP-D is a normal incidence p-type strained layer InGaAs/InAlAs QWIP. The normalized spectral response curve (D) is also shown in Fig. 4. The peak response

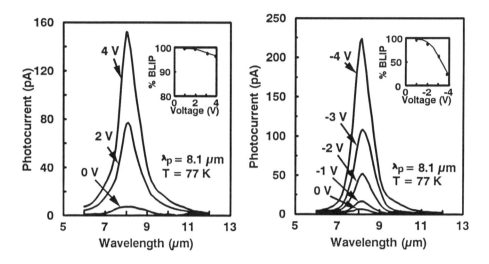

Figure 5 Photocurrent versus wavelength for the PSL-QWIP under different positive and negative biases

wavelength was found to be at $\lambda_p = 8.1$ μm, which is due to the intersubband transition between the confined ground light-hole state (E_{LH1}) and the continuum heavy-hole band states (E_{HH3}), as illustrated in Fig. 4. The cutoff wavelength for this QWIP was found to be $\lambda_c = 8.8$ μm with a spectral bandwidth of $\Delta\lambda/\lambda_p = 12\%$. Since the two heavy-hole subbands are confined inside the wells with very low tunneling probability due to the thicker barrier layer, no photoresponse from these two heavy-hole states was detected. The photocurrents versus wavelength for different positive and negative biases measured at 77 K are shown in Fig. 5(a) and (b), respectively. Responsivities of 34 mA/W at $V_b = 4$ V and 51 mA/W at $V_b = $ - 4 V were obtained for this PSL-QWIP. The maximum BLIP detectivity, D^*_{BLIP}, was found to be 5.9×10^{10} cm Hz$^{1/2}$/W (with a responsivity, $R_A = 18$ mA/W) at $\lambda_p = 8.1$ μm, $V_b = 2$ V, FOV = 90° and T = 77 K. The quantum efficiency was estimated to be $\sim$ 50 % from the responsivity measurement, assuming a photoconductive gain, g = 0.015. It should be noted that the dark current density in this QWIP was found to be one to two orders of magnitude smaller than that of 300 K background photocurrent density at $|V_b| \leq 2V$ and 77 K. Thus, a nearly 100 % BLIP condition was achieved for $|V_b| \leq 2V$ and T = 77 K for this PSL-QWIP, as shown in the inset of Fig. 5(a) and (b).

4. Conclusions

In conclusion, we have demonstrated four novel QWIPs in this paper: the grating-coupled n-type GaAs/AlGaAs BTC and BTM QWIPs, the normal incidence type-II AlAs/AlGaAs and p-type strained layer InGaAs/InAlAs QWIPs for the MWIR and LWIR detection. QWIP-A, B, and C all showed the PV and PC dual-mode detection features. Using square mesh metal grating coupler in QWIP-A and B, high coupling quantum efficiency ($\sim$ 50 %) has been obtained. Excellent detectivity and low dark current have been achieved for QWIP-A at $\lambda = 12$ μm and T = 77 K. QWIP-A can be operated under BLIP condition for $V_b \leq 1V$ and T = 77 K. By optimizing the device and grating structures, further improvement in the performance of QWIP-B can be realized. The normal incidence type-II AlAs/AlGaAs QWIP grown on (110) GaAs showed multicolor and multiband detection features with wavelengths extending from 2 to 14 μm. By further reduction of dark current in this QWIP, the detector should find practical applications for the multicolor IR image sensor systems. The p-type strained layer InGaAs/InAlAs QWIP has achieved the best overall performance, and is under BLIP conditions at 77 K with high detectivity and the lowest dark current density ever reported for the QWIPs.

Acknowledgement

This work was supported by the Advanced Research Project Agency (ARPA) under the Navy grant No.N00014-91- J-1976. The QWIP structures used in this study were grown by Dr. Pin Ho of Electronics Laboratory, Martin Marietta, Syracuse, NY.

References

1. J. S. Smith, L. C. Chiu, S. Margalit, A. Yariv, and A. Y. Cho, J. Vac. Sci. Technol., **B1**, 376 (1983).

2. B. F. Levine, K. K. Choi, C. G. Bethea, J. Walker, and R. J. Malik, Appl. Phys. Lett. 50, 1092 (1987).

3. L. S. Yu, S. S. Li, Appl. Phys. Lett. **59**(11), 1332 (1991).

4. L. S. Yu, Y. H. Wang, S. S. Li, and P. Ho, Appl. Phys. Lett. **60**(8), 992 (1992).

5. Y. H. Wang, S. S. Li, and P. Ho, Appl. Phys. Lett. **62**(6), 621 (1993).

6. D. D. Coon and R. P. Karunasiri, Appl. Phys. Lett. **33**, 495 (1984).

7. B. F. Levine, G. Hasnain, C. G. Bethea, and N. Chand, Appl. Phys. Lett. **54**, 2704 (1989).

8. B. F. Levine, C. G. Bethea, G. Hasnain, V. O. Shen, E. Pelve, R. R. Abbott, and S. J. Hsieh, Appl. Phys. Lett. **56**, 851 (1990).

9. L. S. Yu, S. S. Li, and P. Ho, Appl. Phys. Lett. **59**(21), 2712 (1991).

10. E. Martinet, F. Luc, E. Rosencher, P. Bois, and S. Delaitre, Appl. Phys. Lett. **60**, 895 (1992).

11. I. Grave, A. Shakouri, N. Kuze, and A. Yariv, Appl. Phys. Lett. **60**, 2362 (1992).

12. A. Kastalsky, T. Duffield, S. J. Allen, and J. Harbison, Appl. Phys. Lett.**52** 1320 (1988).

13. B. F. Levine, S. D. Gunapala, and R. F. Kopf, Appl. Phys. Lett. **58**, 1551 (1991).

14. G. Hasnain, B. F. Levine, C. G. Bethea, R. A. Logan, J. Walker, and R. J. Malik, Appl. Phys. Lett. **54**, 2515 (1989).

15. L. S. Yu, S. S. Li, Y. H. Wang, and Y. C. Kao, J. Appl. Phys. **72**(6), 2105 (1992).

16. Y. C. Wang and S. S. Li, J. Appl. Phys. August 15 (1993).

17. E. R. Brown and S. J. Eglash, Phys. Rev. **41**, 7559 (1990).

18. J. Katz, Y. Zhang and W. I. Wang, Appl. Phys. Lett. **61**, 1697 (1992).

19. L. S. Yu, S. S. Li, and P. Ho, Elec. Lett.,**28**, 1468 (1992).

20. Y. H. Wang, S. S. Li, P. Ho, and O. M. Manasreh, J. Appl. Phys.,,**74**(2), 1382 (1993).

21. H Xie, J. Katz, and W. I. Wang, Appl. Phys. Lett. **59**(27), 3601 (1991).

22. R. T. Kuroda and E. Garmire, Infrared Phys. **34**(2), 153 (1993).

23. K. Ghatak, K. Thyagarajan, and M. R. Shenoy, IEEE J. Quantum Elec., **24**, 1524 (1988).

24. Y. C. Wang and S. S. Li, J. Appl. Phys., to be published (1993).

25. M. Dutta, H. Shen, D. D. Smith, K. K. Choi, and P. G. Newman, Surface Science, **267**, 474 (1992).

26. D. Z. Y. Ting and T. C. McGill, J. Vac. Sci. Technol. **B 10**, 1980 (1992).

27. H. C. Liu, Appl. Phys. Lett. **51**, 1019 (1987).

28. B. F. Levine, S. D. Gunapala, and R. F. Kopf, Appl. Phys. Lett. **58**, 1551 (1991).

29. J. Feldmann, E. Gobel, and K. Ploog, Appl. Phys. Lett. **57**, 1520 (1990).

30. W. S. Fu, G. R. Olbright, J. F. Klem, and J. S. Harris Jr., Appl. Phys. Lett. **61**, 1661 (1992).

BACKGROUND LIMITED 128×128 GaAs/AlGaAs MULTIPLE QUANTUM WELL INFRARED FOCAL PLANE ARRAY

L.J. Kozlowski
Rockwell International Science Center
1049 Camino Dos Rios
Thousand Oaks, CA 91360

ABSTRACT. The performance of 9 μm GaAs/AlGaAs multiple quantum well focal plane arrays (FPA) is reported including the first achievement of background-limited infrared photodetector (BLIP) sensitivity at low photon backgrounds (<10^{12} photons/cm^2-s) with the GaAs-based quantum well infrared photodetector (QWIP) technology. Though high and nonuniform detector dark current often precludes this objective using other LWIR detector materials,[1] near-theoretical 128x128 FPA peak detectivity (D*) with >99.4% pixel operability has been achieved at 35-40K operating temperature at background as low as 1.2x10^{10} photons/cm^2-s. Mean D* of 4.51x10^{13} cm-Hz$^{1/2}$/W was measured at 35K, which corresponds to 75% of BLIP for the measured effective quantum efficiency of ~1.0%. Typical noise equivalent temperature difference of ≈8 mK and detectivity >4x10^{10} cm-Hz$^{1/2}$/W were measured at conventional imaging backgrounds at up to 65K operating temperature. Also reported is progress in improving the effective quantum efficiency to about 10% with low optical crosstalk using two-dimensional gratings, resonant cavity structure and mechanical thinning. This work was performed under DARPA Contract N00014-91-C-0163 for Mr. Raymond Balcerak.

1. Introduction

Several infrared camera applications require large staring FPAs offering moderate sensitivity at reasonable cost. Camera manufacturers are thus exploiting PtSi Schottky barrier detector (SBD) array technology for MWIR (3 to 5 μm) applications.[2] The long wavelength spectral region (LWIR; 7.5 to 12 μm) currently has no comparable array technology. IrSi SBD[3] and SiGe heterojunction internal photoemission (HIP) detectors[4] are being developed, but both currently have very high dark current translating to a need for cooling well below 80K. Another candidate cost-effective LWIR detector technology is the quantum well infrared photodetector (QWIP), particularly in the GaAs/AlGaAs material system.

The quantum well infrared photodetector is an extrinsic photoconductor. Infrared detection is via intersubband or bound-to-extended state transitions within the multiple quantum well superlattice structure. Due to the polarization selection rules for transitions between the first and second quantum wells, the photon electric field must have a component parallel to the superlattice direction. Light absorption is thus anisotropic with

H. C. Liu et al. (eds.), Quantum Well Intersubband Transition Physics and Devices, 43–54.
© 1994 *Rockwell International Corporation. Printed in the Netherlands.*

zero absorption at normal incidence. The QWIP detector's spectral response is also narrowband, with peak response near the absorption energy. The wavelength of peak response can be adjusted via quantum well parameters and can be made bias dependent. Because of the extensive scientific and commercial exploration of the GaAs/AlGaAs material system, the application of this technology to infrared detector array fabrication is already well advanced despite its relative novelty. The materials are well understood, readily available, and highly advanced.

Possible disadvantages of the QWIP include its relatively low quantum efficiency and narrow-band spectral characteristics. Increasing the quantum efficiency mandates development of optimized optical gratings and resonant cavities. Another key limitation is the short lifetime translating to a need for lower operating temperatures than intrinsic detector materials. The operating temperature constraint is aggravated by the low quantum efficiency. However, the reported performance of GaAs/AlGaAs quantum well detectors generally approaches theoretical limits closer than intrinsic LWIR detectors such as $Hg_{1-x}Cd_xTe$. When coupled with readout-imposed charge-handling limitations, the devices are competitive for niche applications. Possible advantages in uniformity, controllability, and yield are also significant factors for producing large arrays.

We fabricated hybrid 128x128 FPAs by mating multiple quantum well detector arrays supplied by AT&T Bell Laboratories to either of two CMOS readout multiplexers via indium interconnects. The detector arrays were fabricated by AT&T Bell Laboratories at the Solid State Technology Center, Breinigsville, PA.[5] Detector performance improvement work was partially funded by S3-Maxwell (Dr. J.D. Boisvert, San Diego, CA) for the Phillips Laboratory under Contract F29601-88-C-0025. Subsequent FPA-specific detector development was performed for DARPA (Mr. R. Balcerak) under Contract N00014-91-C-0163. The two CMOS readouts have direct injection[6] or gate modulation[7] input circuits. The former was used for high background tests including conventional imaging. The latter was used for low background characterization since it has self-adjusting current gain for attaining low read noise at low background as long as the detector current is dominated by the photocurrent, i.e., the detectors have low dark current. FPA data implies QWIP detector dark current of ≤ 0.3 fA at 2V bias, 60 μm pitch and 40K operating temperature for material having ~8.7 μm peak wavelength.

2. Quantum Well Infrared Photodetector Arrays

The multiple quantum well arrays consist (Figure 1) of 50 periods of 40Å GaAs quantum wells doped with Si to give a nominal electron concentration, n, of ~1×10^{18} cm^{-3} and 500Å of undoped $Al_{0.25}Ga_{0.75}As$ barriers. The quantum wells are sandwiched between n-type GaAs layers at the bottom and top. The layers are grown on a semi-insulating (100) GaAs substrate by molecular beam epitaxy (MBE) in a Varian Gen II reactor. Initial measurements were performed on arrays having either one-dimensional or two-dimensional lamellar gratings for increased optical coupling efficiency, but we also report results achieved with mechanically-thinned arrays not having optical gratings. The reflective gratings were etched on top of the detectors to disperse the normally incident radiation through the backside of the substrate to give a component of light with electric vector perpendicular to the quantum wells for maximizing optical coupling efficiency. Both mechanical thinning and addition of an AlAs waveguide layer were tested as alternative schemes for increasing optical coupling (c.f., Figure 1).

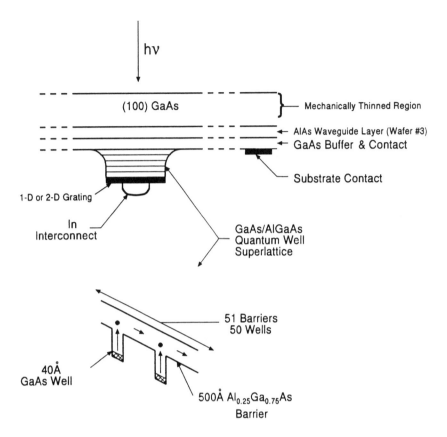

Figure 1. GaAs/AlGaAs Detector Cross-section

In one of the two wafers (to be referred as #1), one-dimensional (1-D) lamellar gratings with a grating constant of 3.5 μm were used. In the other wafer (referred to as #2), two-dimensional (2-D) square gratings with a grating constant of 3.0 μm were used. In both types of gratings, the grating height was 0.7 μm. To further improve optical coupling and reduce crosstalk, additional coupling-improvement techniques were undertaken including adding an AlAs layer on a third wafer (#3) to create a resonant optical cavity and mechanically thinning the various hybrid FPAs. The results of these efforts are summarized in Table 1 for thinned and unthinned FPAs. The effective quantum efficiency is defined as the absorption quantum efficiency, η, which is typically 20%, multiplied by a parameter C_p, which is the polarization dependent coupling efficiency and quantifies the grating effectiveness. Though the highest effective quantum efficiency was achieved with unthinned devices having 2-D grating and optical cavity, the best overall performance was achieved on thinned devices having 2-D grating and optical cavity due to much lower optical crosstalk. The relative crosstalk is described in qualitative units since the actual values can vary greatly due to experimental setup and apparatus. The thinned devices, however, have low crosstalk (<3%) that is relatively independent of measurement conditions.

Table 1. QWIP FPA Quantum Efficiency

Detector Configuration	QWIP Wafer	Relative Crosstalk	Typical ηC_p (%)
1-D Grating	#1	Very High	1.0
2-D Grating	#2	High	2.3
2-D Grating & Optical Cavity	#3	High	15.0
Thinned with 1-D Grating	#1	Low	0.4
Thinned with 2-D Grating	#2	Low	1.2
Thinned with 2-D Grating & Optical Cavity	#3	Low	8.0

Radiometric testing of individual test pixels of 200 µm x 200 µm size were made at 30-77K. Optical response measured at 60K using a calibrated blackbody source at 900K under flood illumination indicated a peak responsivity for the arrays with 2-D gratings that is a factor of ~2 higher than that for the arrays with 1-D gratings. This is consistent with the 2-D gratings being essentially polarization independent as opposed to the 1-D gratings.[8] The measured responsivity for focused spot illumination corresponded to effective quantum efficiencies (internal quantum efficiency × optical coupling efficiency, $\eta \cdot C_p$)[9] of 1.0% and 2.3% at -2V bias (the top of the detector biased negative) for wafer #1 and #2, respectively. Spectral response showed peak wavelengths of 8.6 µm and 8.8 µm, respectively, for #1 and #2. The 3 dB cut-off wavelength was approximately 9.5 µm for both wafers.

The QWIP can be treated as a standard photoconductor. Figure 2 is a plot of the noise power spectral density, which was earlier measured on a 7.7 µm (λ_{pk}) 128x128 QWIP FPA without optical gratings for basic material characterization, versus applied bias at 1.4×10^{16} photons/cm^2-sec background and 78K, 70K and 60K operating temperatures. Clearly visible are the Johnson noise regimes at low bias and the generation-recombination

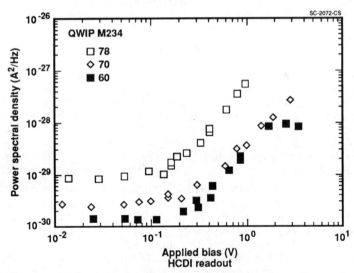

Figure 2. Noise Power Spectral Density vs. Applied Bias at 1.4×10^{16} photons/cm^2-sec background and 78K, 70K and 60K operating temperatures.

regimes at high bias at the various temperatures. A QWIP readout thus must have sufficient charge-handling capacity for biasing the devices in the generation-recombination regime to achieve the highest possible sensitivity.

Figure 3 shows the current-voltage (I-V) traces from wafers #1 (a) and #2 (b) measured on discrete detectors at zero degree field-of-view with the detectors looking at 20–30K background. Note that the dark current at low bias voltages (-2 to +2V) is limited by the measuring instrument.

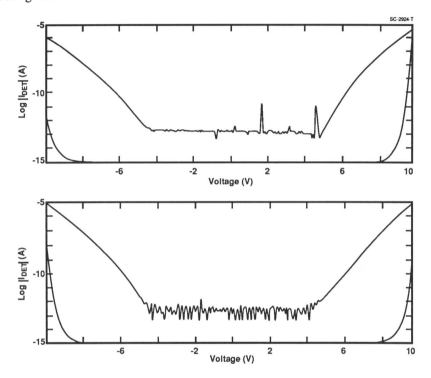

Figure 3. Current-Voltage traces from wafers #1 (a) and #2 (b) measured at zero degree field-of-view with the detector looking at 20-30K background.

3. CMOS 128x128 Readout

The 128x128 readout used for QWIP FPA evaluation at low backgrounds has gate modulation input with p-MOSFET load (Figure 4). Detector current, including photocurrent and dark current, flows through the load device while a proportional current flows into the integration capacitor via a p-MOSFET input FET (20 μm width and 10 μm length). The gain and dc level are adjusted by varying the source voltage (GMODS) of the input FET (6 μm width and 49 μm length). Cell access, cell reset, and video multiplexing are performed by enabling and disabling switch-FETs via static CMOS shift registers. The switched-FET readout has a single output with minimum slew rate of 3.9 V/μsec and maximum output excursion of ≤ 2 V. Maximum power dissipation is 8.4 mW.

48

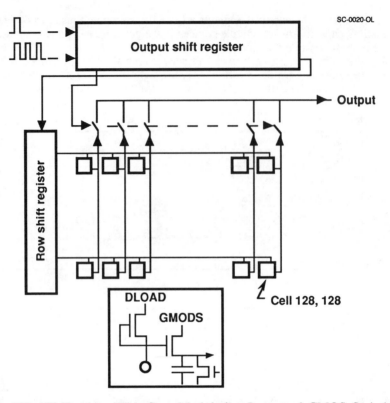

Figure 4. 128x128 Readout With Gate Modulation Input and CMOS Switched-FET Readout Architecture.

The 128x128 readout was fabricated using a custom 2 μm CMOS process at Orbit Semiconductor. Good fabrication yield was achieved, but p-MOSFET characteristics were peculiar due to the presence of a parasitic edge transistor. The main impact of the short-channel edge transistor on FPA performance was increased, but manageable, 1/f noise at operating temperatures <100K.

Gate Modulation Current Gain. The current gain of the gate modulation input circuit is approximately:

$$A_1 = \frac{g_m}{g_{m,LOAD}}\, \eta_{inj}; \quad \eta_{inj} = \frac{g_{m,LOAD}R_{det}}{1+g_{m,LOAD}R_{det}}\left[\frac{1}{1+\dfrac{j\omega C_{det}R_{det}}{1+g_m R_{det}}}\right]$$

where η_{inj} is the injection efficiency of detector current into the load transistor, g_m is the input FET transconductance, $g_{m,LOAD}$ is the load FET transconductance, and C_{det} and R_{det} are the detector resistance and capacitance, respectively.

The signal integrated in the integration capacitor is a dc-suppressed, gain-proportioned facsimile of the total detector current. Any changes in load current modulate the integrated current via the gate of the input FET. For highest current gain and best circuit noise figure, the detector dark current must be small relative to the photocurrent for minimization of load FET thermal noise and maximization of current gain to reduce the impact of input FET thermal noise. The circuit not only is capable of near-zero bias operation via the low threshold nonuniformity achieved, but prefers it since excess detector dark current generates additional detector shot noise and load FET thermal noise.

The operating point is established by concurrently adjusting the detector substrate voltage (DSUB) and the source voltage (DLOAD) of the load MOSFET. The current gain and the integration capacitor percentage fill is then set by either adjusting GMODS, or by equivalently shifting the DLOAD and DSUB biases. Using a laboratory mechanization of this adaptive bias control, minimum total dynamic range of 200 dB was measured.

The use of a MOSFET as an active load device provides dynamic range management via automatic gain control in combination with adaptive background pedestal suppression.[10,11] The current gain self-adjusts by orders of magnitude depending upon the total detector current. Input-referred read noise of tens of electrons[12] can thus be achieved with high impedance QWIP detectors (at low temperature and background) since the read noise is approximately:

$$N_{read} = \frac{\left(I_{load}^2 + I_{input,ir}^2 + I_{mux,ir}^2\right)^{1/2} t_{int}}{q}$$

where I_{load}^2, which is the composite thermal and 1/f noise of the load device, is

$$I_{load}^2 = \int^{\Delta f} \frac{1}{\eta_{inj}^2}\left(\frac{8}{3}kTg_{m,loadFET} + \frac{K_{load}^2 g_{m,loadFET}^2}{f^{\alpha}}\right) df$$

and K_{load} is the noise spectral density of the load FET at 1 Hz in units $V/\sqrt{Hz}$.

The input FET is often subthreshold; the input-referred input circuit noise, $I_{input,ir}^2$, is consequently

$$I_{input,ir}^2 = \int^{\Delta f} \left[\frac{1}{A_I^2}\left(2qI_{input} + \frac{K_{inputFET}^2 g_{m,inputFET}^2}{f^{\alpha}}\right)\right] df$$

The dominant input-referred readout noise, $I_{mux,ir}^2$, is

$$I_{mux,ir}^2 = \frac{1}{A_I^2}kTC_{input}\Delta f$$

where C_{input} is the total input capacitance including the capacitance of the bus line servicing the entire column.

For the limiting low background case where the current gain is sufficiently high for the load FET noise to dominate the total readout noise and where the detector impedance is satisfactorily high, the minimum read noise is approximately

$$N_{min} \approx \sqrt{\frac{4}{3} \frac{I_{det} t_{int}}{q \, n_{loadFET}}}$$

where n_{load} FET is the subthreshold ideality of the load MOSFET, I_{det} is the total detector current, and t_{int} is the integration time. The minimum read noise is thus proportional to the photo-generated shot noise (N_{shot}):

$$N_{min} \approx N_{shot} \sqrt{\frac{4}{3} \frac{1}{n_{loadFET}}}$$

The circuit's current gain thus compensates at low background to maintain a relatively constant percentage of BLIP if the detectors have satisfactory quality.

Summarized in Table 2 are the characteristics of the readout. It offers large total dynamic range at the expense of gain uniformity and strict requirements on MOSFET threshold uniformity and dc bias noise. The minimum instantaneous dynamic range is typically >65 dB.

Table 2. 128x128 Multiplexer Characteristics

Parameter	Value	Units
Nominal Supply Voltage	6	V
Maximum Charge Capacity	61	10^6 e-
Total Dynamic Range	>1	10^5
Minimum Instantaneous DR	>1	10^3
Total Dynamic Range (Adaptive)	>1	10^7
Responsivity Nonuniformity	1.6:1	Max:Min Ratio
Maximum Data Rate	≥ 4	MHz
Input Offset Nonuniformity	<4	mV p-p
Transfer Ratio	29.6	nV/e-

4. Experimental Results

Hybrid FPAs were made from the two wafers by mating them via indium interconnects to the CMOS readout multiplexer with gate modulation input circuit rather than the direct injection circuit previously used in similar QWIP arrays.[13,14] The gate modulation circuit has self-adjusting current gain for maintenance of noise figure at low background as long as the detector current is dominated by the photocurrent, i.e., the detectors have low dark current. This is the case for these QWIP arrays at temperatures below about 65K depending upon the background flux.

Figure 5(a) is a histogram of the peak D* measured from one of the QWIP FPAs from wafer #1 at 1.29×10^{10} photons/cm^2-s background and 40K temperature. The D* is detector-limited in this case to 61% of BLIP. At 35K [Figure 4(b)] mean D* of 4.51×10^{13} cm-Hz$^{1/2}$/W is obtained, which corresponds to 75% of BLIP for the measured $\eta \cdot C_p$

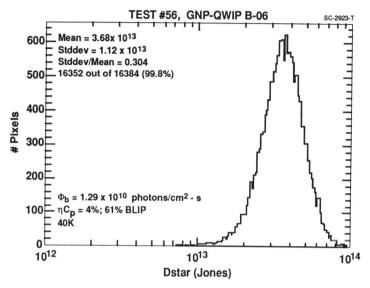

Thermally Stabilized (1 hr. after cooldown to 40K)
Flood $\eta C_p = 4\%$

(a)

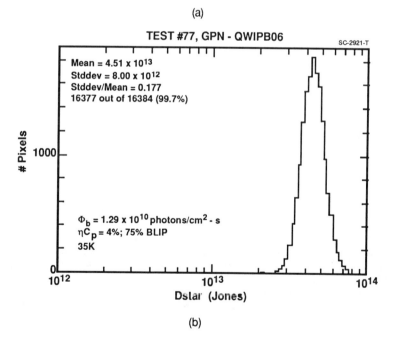

(b)

Figure 5. Histograms of the peak D* measured on QWIP FPA from wafer #1 at 1.29×10^{10} photons/cm^2-s background, (a) 40K temperature and (b) 35K temperature.

~1.0% under flood illumination. The measured D*'s and a comparison of the measured (≈34,000) gate modulation current gain to the theoretical value imply very low QWIP detector dark current of ≤ 0.3 fA at 2V bias and 40K operating temperature. Note that the inter-array D* uniformity improved by ~70% at the lower operating temperature due to lower noise. The uncorrected response uniformity for this array is ~4%.

Figure 6 is a histogram of the peak D* measured at 1.68×10^{11} photons/cm^2-s background and 40K temperature with a hybrid which uses a detector array from wafer #2. The mean of 1.44×10^{13} cm-Hz$^{1/2}$/W is about 55% of BLIP assuming $\eta \cdot C_p$ ~2.3%, measured under condition of flood illumination. Further tests at lower background revealed that the FPA performance was detector-limited at roughly this level. The uncorrected response uniformity for this array is ~3%.

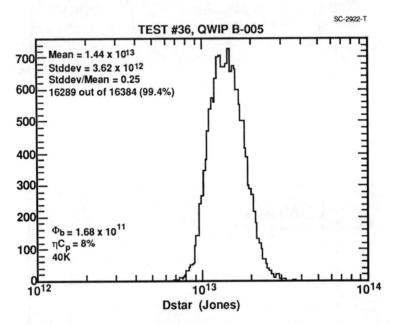

Figure 6. Histogram of the peak D* measured at 1.68×10^{11} photons/cm^2-s background and 40K temperature with a hybrid which uses a detector array from wafer #2.

Figure 7 summarizes FPA measurements taken at higher backgrounds on several QWIP FPAs in the configurations earlier described. The higher backgrounds enabled easier attainment of BLIP sensitivity; the various D*'s are commensurate with the appropriate BLIP limits for the external quantum efficiencies that were achieved. The mechanically thinned devices with AlAs waveguide layer offer the best combination of D* and low optical crosstalk; these QWIP FPAs generated detectivities consistent with quantum efficiency of about 10%.

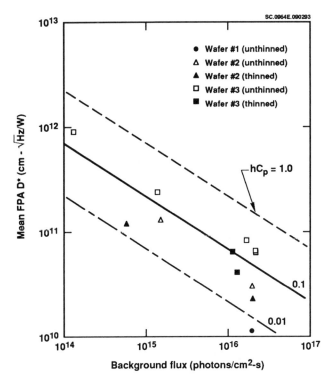

Figure 7. Mean D* vs. Background Flux at 40K for several thinned and unthinned hybrids using detector arrays from wafers #1, #2 and #3.

5. Summary and Conclusion

In summary, we report the low background and low operating temperature performance of 128x128 GaAs/AlGaAs QWIP FPAs. Peak detectivity that is near the theoretical background limit with high pixel operability of 99.7% was achieved with a GaAs/AlGaAs quantum well infrared photodetector (QWIP) focal plane array (FPA) at low background. Mean D* of 3.2×10^{13} cm-Hz$^{1/2}$/W was measured at a background of 1.2×10^{10} photons/cm^2-s and 40K operating temperature. Both the sensitivity and uniformity of the 8.8 μm peak wavelength device were improved by cooling to 35K. The mean D* at 35K was 4.51×10^{13} cm-Hz$^{1/2}$/W.

The continuing improvements in quantum efficiency along with the uniform and low dark current of the GaAs/AlGaAs detectors confirm a niche for this detector technology at photon backgrounds <10^{12} photons/cm^2-s and ~40K operating temperature. Incorporating the improvements achieved in the optical coupling using resonant optical cavities[15,16,17] should further enhance the performance of this LWIR detector material. Attainment of BLIP sensitivity at high backgrounds, however, presently falls short of the nominal goal of 80K; operating temperatures ≤ 65K are required for BLIP achieving BLIP sensitivity at conventional imaging backgrounds.

References

[1] R.E. DeWames, J.M. Arias, L.J. Kozlowski, and G.M. Williams, "An assessment of HgCdTe and GaAs/AlGaAs technologies for LWIR infrared imagers," SPIE Vol. 1735, (1992).

[2] T. S. Villani, W. F. Kosonocky, F. V. Shallcross, J. V. Groppe, G. M. Meray, J. J. O'Neill, III, and B. J. Esposito, "Construction and Performance of a 320 x 244-Element IR-CCD Imager with PtSi SBDs," SPIE Vol. 1107-01, pp. 9-21 (1989).

[3] B.Y. Tsaur, M.M. Weeks, R. Trubiana, P.W. Pellegrini and T.R. Yew, "IrSi Schottky-barrier infrared detectors with 10 μm cutoff wavelength," IEEE Electron Device Lett., vol. 9, p. 650, 1988.

[4] T.L. Lin, A. Ksendzov, T.N, Krabach, J. Maserjian, M.L. Huberman, and R. Terhune, "Novel $Si_{1-x}Ge_x$/Si heterojunction internal photemission long-wavelength infrared detectors," Appl. Phys. Lett., 57, (1990).

[5] L.J. Kozlowski, G.M. Williams, R.E. DeWames, J.W. Stayt, Jr., V. Swaminathan, K.G. Glogovsky, R.E. Leibenguth, L.E. Smith, and W.A. Gault, "128 x 128 GaAs/AlGaAs QWIP Infrared Focal Plane Array with Background Limited Sensitivity at 40K," submitted for publication.

[6] L.J. Kozlowski, G.M. Williams, G.J. Sullivan, C.W. Farley, R.J. Anderson, J.K. Chen, D.T. Cheung, W.E. Tennant, and R.E. DeWames, "LWIR 128x128 GaAs/AlGaAs Multiple Quantum Well Hybrid Focal Plane Array," IEEE Trans. on Electron Devices, Vol. 38, No. 5, May 1991.

[7] L.J. Kozlowski, S. A. Cabelli, D.E. Cooper and K. Vural, "Low Background Infrared Hybrid Focal Plane Array Characterization," SPIE Vol. 1946, (1993).

[8] J.Y. Andersson and L. Lundqvist, J. Appl. Phys., 71, 3600 (1992).

[9] R.E. DeWames, J.M. Arias, L.J. Kozlowski, and G.M. Williams, "An assessment of HgCdTe and GaAs/AlGaAs technologies for LWIR infrared imagers," SPIE Vol. 1735, (1992)

[10] L.J. Kozlowski, S.L. Johnston, W.V. McLevige, A.H.B. Vanderwyck, D.E. Cooper, S.A. Cabelli, E.R. Blazejewski, K. Vural and W.E. Tennant, SPIE Vol. 1685, April 1992.

[11] L.J. Kozlowski, W.V. McLevige, A.H. Vanderwyck, D.E. Cooper, S.L. Johnson, K. Vural and W.E. Tennant, Proceedings of the IRIS Specialty Group on Infrared Detectors, NIST, Boulder, CO, 13-16 August 1991.

[12] L.J. Kozlowski, S. A. Cabelli, D.E. Cooper and K. Vural, "Low Background Infrared Hybrid Focal Plane Array Characterization," SPIE Vol. 1946, (1993).

[13] L.J. Kozlowski, G.M. Williams, G.J. Sullivan, C.W. Farley, R.J. Anderson, J. Chen, D.T. Cheung, W.E. Tennant and R.E. DeWames, IEEE Trans. Electron Dev. 38, 1124 (1991).

[14] R.E. DeWames, J.M. Arias, L.J. Kozlowski, and G.M. Williams, SPIE Vol. 1735, (1992).

[15] J.Y. Anderrson, L. Lundqvist and Z.F. Paska, Appl. Phys. Lett. 58, 2264 (1991).

[16] J.Y. Anderrson and L. Lundqvist, Appl. Phys. Lett. 59, 857 (1991).

[17] V. Swaminathan, J.W. Stayt Jr., J.L. Zilko, K.D. C. Trapp, L.E. Smith, S. Nakahara, L.C. Luther, G. Livescu, B.F. Levine, R.E. Leibenguth, K.G. Glogovsky, W.A. Gault, M.W. Focht, C. Buiocchi and M.T. Asom, IRIA-IRIS Proceedings, 1992 Meeting of the IRIS Specialty Group on Infrared Detectors, Aug. 1992, Moffet Field, CA.

IMAGING PERFORMANCE OF LWIR MINIBAND TRANSPORT MULTIPLE QUANTUM WELL INFRARED FOCAL PLANE ARRAYS

W. A. Beck, J.W. Little, A.C. Goldberg, and T.S. Faska
Martin Marietta Laboratories,
1450 South Rolling Road
Baltimore, MD 21227

ABSTRACT. The performance of 128x128 and 256x256 miniband-transport multiple quantum well focal plane arrays (FPAs) is presented. The detectors typically have a peak wavelength of 9.1 µm with spectral bandwidth of 1.2 µm. Laboratory measurements using the NVL 3D noise analysis technique indicate that the best MBT MQW FPAs have a temporal noise equivalent temperature difference (NETD) of 15 mK and spatial NETD of 30 mK (for a 10°C calibration interval) when operated at 60K in an f/2 camera system.

1. Introduction

Multiple quantum well (MQW) infrared detectors have received much interest because of their potential for large, uniform focal plane arrays (FPAs) with high sensitivity. Several MQW configurations have been reported based on transitions from bound-to-extended states,[1] bound-to-miniband states,[2] and between miniband states.[3] This paper presents recent progress in development of long wavelength infrared (LWIR) miniband transport (MBT) MQW detectors and FPAs.

As a class these detectors have several advantages. Availability of large wafers (up to 4 inches in diameter), reproducible and precise growth using molecular beam epitaxy (MBE) and a simple fabrication process have produced high yields in both detectors and FPAs. The large wafers and uniformity that can be achieved with MBE offer a unique capability for low cost production of large, high-sensitivity staring FPAs. In addition, measurements made on Martin Marietta devices have shown that MQW detectors have very high tolerances to both nuclear[4] and laser[5] radiation.

2. MBT Quantum Well Infrared Detector

The MBT detector (Fig. 1), which is functionally equivalent to the bound-to-miniband design,[2] uses doped QWs containing two bound states separated by short-period superlattice barrier layers.

H. C. Liu et al. (eds.), Quantum Well Intersubband Transition Physics and Devices, 55–68.

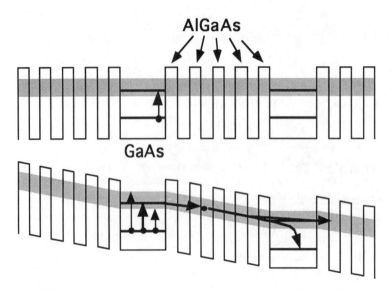

Fig. 1   Schematic band diagram of the MBT detector structure.

The MBT detector structure combines the desirable attributes of the early bound-to-bound (BB) transition and more recent bound-to-extended (BE) transition detectors. The use of two bound states in the QW (as in the BB structures) removes the requirement imposed by the BE design for a unique solution for the well width and barrier height for a given operating wavelength (i.e., it is possible to obtain the same operating wavelength with a continuous range of well widths and barrier heights). Since the excited state of the well is designed to be in resonance with the miniband, the absorption of a photon leaves the electron in a conducting channel (as in the BE designs) through which it can flow and generate photocurrent.

All the detectors reported here were grown by MBE. The active region of the detectors consisted of 40 Si-doped GaAs QWs separated by superlattice barriers comprising alternate layers of undoped $Al_xGa_{1-x}As$ and GaAs. A highly doped GaAs layer grown at each end of the active region provided ohmic contacts. Diffraction gratings were formed by reactive ion etching (RIE) into the top contact layer followed by metallization. The results reported here were from detectors with one-dimensional (etched stripes) gratings with 3- to 4- μm periods. The active detector areas were defined by RIE down to the lower doped contact thereby defining square mesas. Indium bumps were deposited onto the metalized gratings and the arrays were then hybridized either to a multiplexer (MUX) or silicon fanout. A scanning electron micrograph of a typical detector array is shown in Figure 2.

The imagery from early MQW FPAs displayed ghost images associated with total internal reflection of diffracted light within the detector substrate. Therefore, in some cases, we mechanically thin the detector substrate to a thickness of 20–30 μm after hybridization. As shown in Fig. 3, the thinning traps the diffracted light inside the pixel so that the responsivity for focused targets is increased and the ghost images are eliminated.

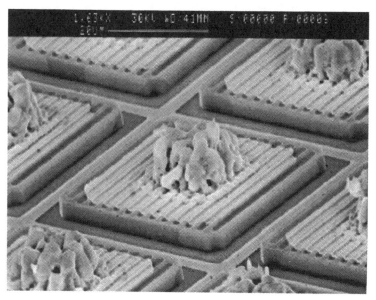

Fig. 2   Scanning electron micrograph of a typical MQW detector array.

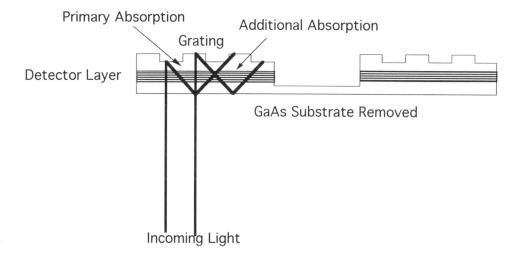

Fig. 3   Thinned MQW detector structure showing light trapped in pixel by total internal reflection.

58

Figure 4 shows the normalized spectral response of several 40 μm-square MBT detectors across a fanned out 128x128 array, displaying the quasi-Gaussian shape typical of all MQW detectors and the excellent spectral uniformity that is achieved.

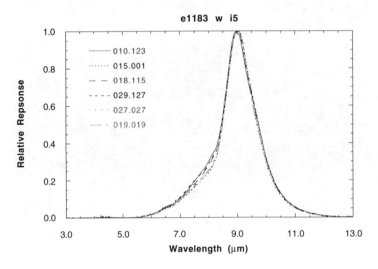

Fig. 4   Measured spectral response for a set of detectors across a 128x128 MBT detector array. The legend indicates row.column of the detector elements.

The measured peak responsivity, R_p, for a detector from the same array is shown in Fig. 5 . The values are for directly measured unpolarized light and are not corrected for the 30% reflection at the air/GaAs substrate interface. The peak D* for the detector was 2 x 10^{10} cm $Hz^{1/2}$/W at 77K. The photoconductive gain, g, was determined from the measured noise power w(f) at 80K using[6]

$$w(f) = 4e\bar{I}g\left(1 - \frac{1}{2N_w g}\right)$$

where e is the electronic charge, $\bar{I}$ is the average current and N_w is the number of (doped) wells. Note that the noise was reduced from the standard expression for generation-recombination noise in a homogeneous photoconductor ($w(f) = 4e\bar{I}g$) because carriers excite and decay only at the doped QWs. (The noise power reduction is about 5% in a 40-well device at normal operating bias in which g is typically 0.25–0.3 at the bias that yields maximum R_p.) Finally, Fig. 6 shows the peak quantum efficiency determined from the measured R_p and g by:

$$\eta_p = \frac{hcR_p}{eg\lambda_p}$$

where h is Planck's constant, c is the speed of light, and λ_p is the peak wavelength.

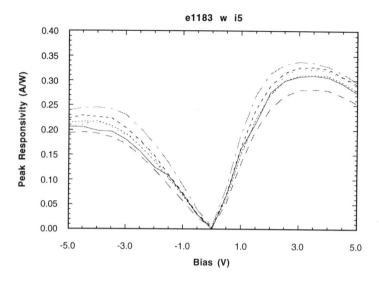

Fig. 5 Peak responsivity vs. bias for a detector from the same array as in Fig. 4. Detector temperature is 80K.

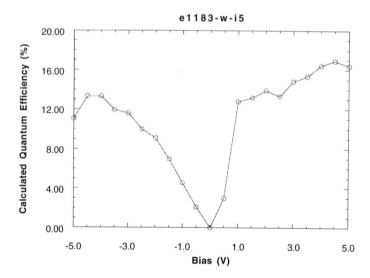

Fig. 6 Peak quantum efficiency deduced from measured responsivity and photoconductive gain.

There was initially some concern that the relatively high electric fields needed for optimum responsivity of MQW devices would also cause high 1/f noise. However, as shown in Fig. 7, the devices show no discernible 1/f noise down to 1 Hz. Some devices were measured to 0.1 Hz and also failed to show any 1/f noise.

a)

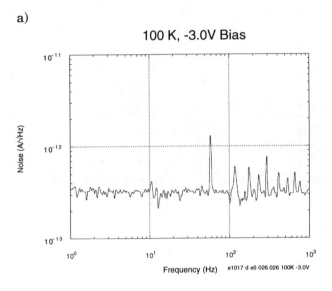

b)

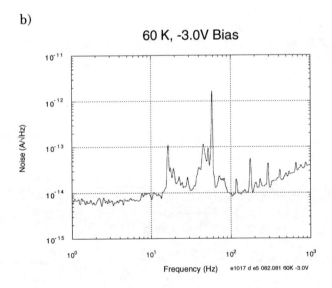

Fig. 7   Measurements of typical spectral noise from 1 Hz to 1 kHz at -3.0 volts bias on elements from fanned out 128x128 MBT detector array.

3. MBT Focal Plane Arrays

3.1. 128 X 128 FPAS

Imaging focal plane arrays (FPAs) were formed by indium bump bonding 128x128 MBT detector arrays to MUXs from Santa Barbara Focalplane, followed by mechanical thinning of the GaAs detector substrate to a thickness of 20–30 μm. The MUXs were originally designed for InSb photovoltaic detectors and use a 1.2 μm silicon CMOS process with 50×10^6 carrier charge well capacity. The detectors were operated at a bias of −2V (applied to the indium bump) for all measurements reported here.

The multiplexed FPAs were mounted in a camera with an f/2.0, 100 mm focal length lens that had an optical transmission of 60%, and the camera was aimed at an extended blackbody source. A buffered video board was used to collect 16 or 32 sequential digital frames of data at blackbody temperatures of 25, 27, 29, 30, 31, 33, and 35°C. In some cases, four sets of 32 sequential frames were collected and then spliced to form a 128-frame data set. The data in each frame were then converted to apparent temperature, T, by computing gain and offset coefficients from the mean response at 25°C and 35°C. The uniformity of the gain coefficients reflects the uncorrected response uniformity of the FPA. For example, the uncorrected response uniformity of one of the 128x128 FPAs reported in this paper is shown in Fig. 8. The standard deviation of the uncorrected response divided by the mean response was 3.1%, and the operability was 99.7%, defined as the percentage of pixels whose responsivity was within 20% of the median responsivity.

Figure 9 shows three-dimensional (3D) views of T as a function of the row (v), column (h), and frame index (t). At the calibration temperatures, the spatial fluctuations in T are entirely due to temporal noise. However, at intermediate temperatures, the fixed-pattern spatial noise can be seen.

Figure 10 shows an analysis of the temporal noise-equivalent temperature difference (NETD) in each pixel, determined as the standard deviation of the apparent temperature through a 128-frame data set. The narrow width of the histogram demonstrates that the high sensitivity extends uniformly over the entire array. Figure 11 shows the residual fixed-pattern spatial noise at 30 °C, determined as the mean of the temperatures through a 16-frame data set.

The NVL 3D noise analysis technique[7] was used to further separate the temporal and spatial noise components of the FPAs. The 3D noise analysis separates noise into seven statistically independent components that indicate the degree of correlation between the temporal and spatial noise. The components are referred to as N_t, N_v, N_h, N_{th}, N_{tv}, N_{vh}, and N_{tvh}, where the subscript indicates the variables on which the component is dependent.; the other components have been "averaged out." For example, N_v, N_h, and N_{vh} all represent fixed-pattern spatial noise from which the temporal noise has been separated. N_t represents spatially uniform temporal noise, i.e. fluctuations in T that extend across entire frames of data. N_{tvh} is conventional uncorrelated temporal noise that is not correlated with position. Figure 12 shows results of the 3D noise analysis for a 16-frame data set measured from one of the FPAs operated at 60K, with an integration time of 12 ms. The N_{tvh} component is 15 mK, consistent with the results shown in Fig. 10. The fixed-pattern

62

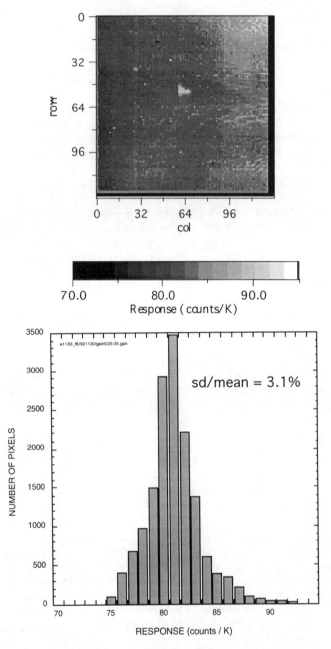

Fig. 8   Uncorrected response uniformity in a 128x128 MQW FPA.

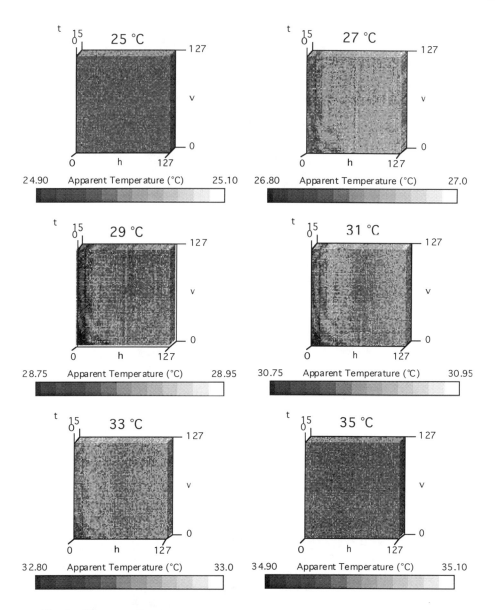

Fig. 9   3D arrays of apparent temperature vs row (v), column (h), and frame (t), for 16 sequential frames of data collected with the camera aimed at an extended blackbody set at the indicated temperatures. The gain and offset coefficients were calibrated at 25 °C and 35 °C. The temperature range for all images is 0.2 °C.

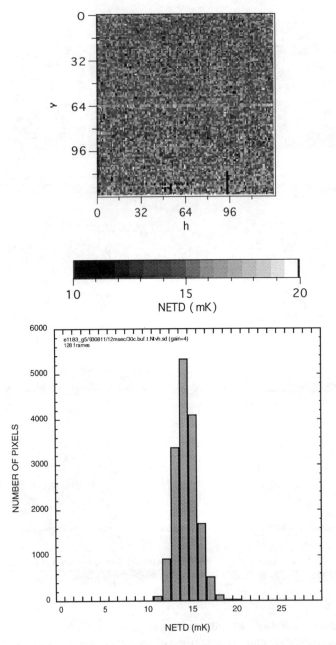

Fig. 10 Temporal NETD in a 128x128 MQW FPA operated at 60K with f/2.0, 60% transmission optics.

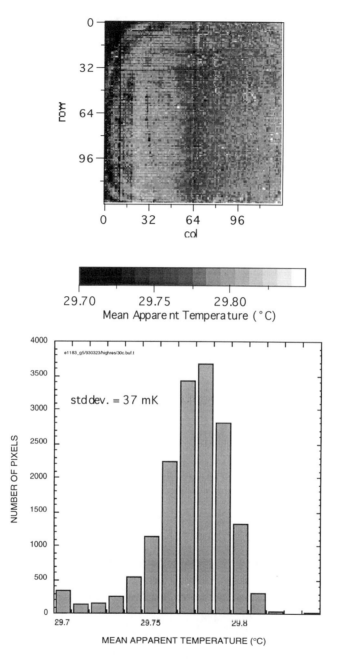

Fig. 11 Residual spatial noise in a 128x128 MQW FPA viewing an extended blackbody at 30 °C. The FPA is calibrated at 25 °C and 35 °C.

66

spatial components reach a maximum of around 30 mK between the calibration temperatures. This residual (after correction) spatial noise is believed to be due to non-linearities in the readout associated with operating the MUX at a detector bias that is much larger than originally anticipated in the MUX design. The readout nonlinearity also increases as the charge well is filled so that the fixed pattern noise components in Fig. 12 increase with scene temperature up to just below the upper calibration temperature. The residual spatial noise is almost certainly *not* due to spectral response non-uniformity, since the spectral variation seen in Fig. 2 would be associated with a fixed pattern equivalent temperature difference of less than 0.002 °C over the 10°C calibration interval used here.[8]

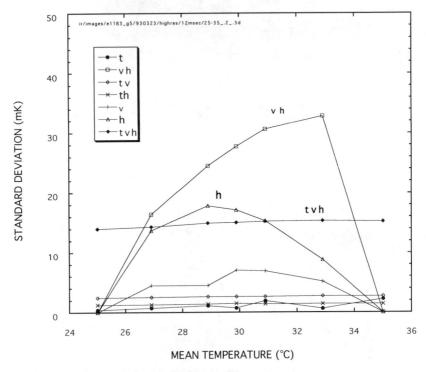

Fig. 12   3D noise analysis of a 128x128 MBT FPA with f/2.0, 60% transmission optics. The "tvh" component is uncorrelated temporal noise. The "vh" and "h" components represent fixed-pattern spatial noise.

The 3D noise measurement was also performed as a function of FPA temperature and integration time. The performance was approximately 50% background limited at an operating temperature of 68K, at which the maximum integration time was 5.1 ms, and the temporal NETD was 42 mK.

Imagery using the thinned FPAs is free of the ghost images that are often seen with unthinned MQW FPAs. For example, Fig. 13 shows a Bausch & Lomb Stereozoom microscope with a hot illuminator. The illuminator is strongly saturated but displays no ghost images.

Fig. 13 Digital image of a Bausch & Lomb Stereozoom microscope taken with a thinned MBT MQW FPA. The fully saturated illuminator does not generate ghost images.

3.2.   256 X 256 FPAS

Recently, imaging FPAs were formed by indium bump bonding 256x256 MBT detector arrays to MUXs designed by Martin Marietta. The MUXs use a CMOS process with a 100×10^6 electron charge well. The FPAs yield good imagery, and quantitative measurements of performance will be reported later.

4.   Conclusion

The results presented here demonstrate that MBT MQW FPAs are capable of excellent sensitivity when operated at 60K and that the performance remains mainly background limited up to 68K. The temporal noise equivalent temperature difference (NETD) of 15 mK is very competitive with other long wave infrared (LWIR) technologies, as is the fixed pattern NETD of 30 mK. Improved linearity of the readout should reduce the fixed pattern NETD to below the 15 mK temporal value.

The 128x128 MBT MQW FPAs are maturing rapidly and we have recently demonstrated our first 256x256 FPAs. Through planned improvements to both the detectors and MUX, we expect to increase the operating temperature of LWIR FPAs to 77K. In addition, MQW detectors can be structured to produce pixel registered multicolor FPAs [9] in the 3-to-19 µm wavelength region.

The valuable contributions of others to this work are gratefully acknowledged. In particular, the MBT detectors were built by Mike Taylor, Toby Olver, Lisa Lucas, and Kerri Wright. The detectors were formed on MBE structures grown by Fred Towner and Dave Gill. Single-pixel tests were performed by Chris Cooke and Kim Olver with software developed by Phil Conner. The 256x256 MUX was designed by

Kirk Reiff and Bob Martin. The 256x256 video was provided by Jon Albritton. Helpful discussions with Ken Ritter, Frank Crowne, Rich LeBlanc, Casey Contini, Frank Warren, Bob Landrum and Mark Stegall are also gratefully acknowledged. This work was supported by funding from the Advanced Research Projects Agency through Office of Naval Research contract N00014-91-C-0132, monitored by Max N. Yoder and Raymond Balcerak.

5. References

[1]B.F. Levine, A. Zussman, S.D. Gunapala, M.T. Asom, J.M. Kuo, and W.S. Hobson, J. Appl. Phys. **72**, 4429 (1992).

[2]L. S. Yu and S.S. Li, Appl. Phys. Lett. **59**, 1332 (1991).

[3]B. O, J.-W. Choe, M.H. Francombe, K.M.S.V. Bandara, D.D. Coon, Y.F. Lin, and W.J. Takei, Appl. Phys. Lett. **57**, 503 (1991).

[4]"Investigation of the Feasibility of Developing Hardened Superlattice Detectors," Final Report on contract F29601-88-C-0025 to United States Air Force, Air Force Weapons Laboratory, January 1993

[5]C.A. Hoffman, J.R. Meyer, F.J. Bartoli, J.R. Lindle and E.R. Youngdale, 1991 IRIS Proceedings on Infrared Countermeasures, Vol 1, 11 (1991).

[6]W.A. Beck, submitted to Appl. Phys. Lett..

[7]L.B. Scott and J.A. D'Agostino, SPIE Proceedings Vol. 1689.

[8]W.A. Beck, Proc. 1992 IRIS Specialty Group on Infrared Detectors, Vol 1, 167 (1992).

[9]T.S. Faska, J.W. Little, W.A. Beck, K.J. Ritter, and A.C. Goldberg, JPL Innovative Long Wavelength Infrared Detector Workshop, April 7-9, 1992

MODELED PERFORMANCE OF MULTIPLE QUANTUM WELL INFRARED DETECTORS IN IR SENSOR SYSTEMS

R. L. Whitney, F. W. Adams, and K. F. Cuff
Lockheed Palo Alto Research Laboratory
3251 Hanover Street
Palo Alto, CA 94304

ABSTRACT. We present a model for scaling performance of multiple-quantum-well infrared detector arrays with operating temperature and spectral wavelength band, and discuss the optimization of multiple-quantum-well infrared detector arrays for thermal imaging applications. We show that the specifics of the infrared sensor application can significantly affect the design and operation of these focal plane arrays. Background flux, desired operating temperature, sensor wavelength band, readout considerations, and calibration accuracy strongly affect design and optimization of the quantum well structure. Consequently, predictions of MQW focal plane performance and optimization of detector parameters must be done with the system requirements in mind, and not based purely on single detector performance.

1. Introduction

Photoconductive GaAs/AlGaAs multiple-quantum-well (MQW) LWIR detectors have been rapidly developed, but low quantum efficiency and high thermal generation rate have caused skepticism as to their eventual utility. For systems that operate in low-background conditions where low focal plane temperatures (30–40 K) are required, the signal-to-noise ratio of MQW detectors is quite good, despite these limitations. However, for thermal imaging systems where higher operating temperatures (70–80 K) are desired, dark currents are the dominant noise source, and therefore limit the both sensitivity and practical operation of the system. The dark current can be reduced by using lower doping in the wells or altering the device structure, but these approaches also lower the response of the detectors, so that it is not always clear whether there is a net gain in performance.

Successful optimization of MQW detectors for specific applications will require a performance model that easily scales dark current and quantum efficiency with operating temperature, cutoff wavelength, spectral waveband, doping density, bias and voltage. In addition this model must include the effects of background flux, readout considerations, the system optics, and calibration accuracy. In the following discussions, we will present such a model for GaAs/AlGaAs bound-to-continuum MQW focal plane arrays. We will choose detector parameters that provide optimized system performance for a nominal MQW focal plane array, and then show how changes in operating conditions, and improvement in detector properties will affect system performance.

69

H. C. Liu et al. (eds.), Quantum Well Intersubband Transition Physics and Devices, 69–85.
© 1994 *Kluwer Academic Publishers. Printed in the Netherlands.*

2.   System Figures of Merit

To evaluate the performance of MQW detectors for IR sensor applications, an appropriate system level figure of merit must be used. For comparison of individual detectors, detectivity (D*) is a suitable figure of merit. However, in staring sensors using two-dimensional detector arrays, uniformity of the array can be, and often is, more important than single-detector performance. And in certain cases it is readout noise rather than detector performance that limits the sensor as a whole. Therefore *sensor* figures of merit should include nonuniformity effects and readout noise as well as detector noise.

Our approach is to define the total noise in terms of spatial as well as temporal averaging. Specifically, we define the noise n corresponding to a variable N by

$$n^2 \equiv \left\langle \overline{\left(N - \langle \overline{N} \rangle\right)^2} \right\rangle$$
$$= \left\langle \overline{\left(N - \overline{N}\right)^2} \right\rangle + \left\langle \left(\overline{N} - \langle \overline{N} \rangle\right)^2 \right\rangle \tag{1}$$

where $\langle X \rangle$ denotes the spatial (array) average of X, and $\overline{X}$ denotes the average of X over time. The second form of Eq. (1) separates the spatiotemporal variance into two terms: the spatial average of the usual temporal noise, and the spatial variance of the time-averaged values (the temporally correlated or "pattern" noise). The fact that this is possible allows us to give a first-order treatment of nonuniformity effects by simply adding them to the detector noise in the usual sensor figures of merit. (To improve on this when the scale of image information is important, one can regard array nonuniformity as equivalent to noise in the focal-plane transfer function.)

For thermal imagers, the standard figure of merit is the noise equivalent temperature difference (NETD), which can be expressed as a scaled noise-to-signal ratio

$$\text{NETD} = \left[\frac{\partial \left(\ln N_{\text{op}} \right)}{\partial T_{\text{B}}} \right]^{-1} \frac{n_{\text{T}}}{N_{\text{op}}} \tag{2}$$

where T_{B} denotes scene (background) temperature (K), N_{op} is the number of optically generated carriers, and n_{T} is the total number of noise electrons. Sensors for other applications are sometimes characterized in terms of noise-equivalent radiance or noise-equivalent flux density (NEFD), which can likewise be written as a scaled noise-to-signal ratio

$$\text{NEFD} = \left[\frac{\partial \left(\ln N_{\text{op}} \right)}{\partial \Phi} \right]^{-1} \frac{n_{\text{T}}}{N_{\text{op}}} \tag{3}$$

where Φ is the scene spectral radiance (ph cm^{-2} sr^{-1} sec^{-1} μm^{-1}). When the sensor spectral bandwidth $\Delta\lambda$ is not too broad, so that the total scene radiance may be approximated by the product $\Phi \, \Delta\lambda$, these two figures of merit become approximately

$$\text{NETD} \approx \left[\frac{\Phi}{\Phi'}\right] \frac{n_T}{N_{op}} \tag{4}$$

and

$$\text{NEFD} \approx \Phi \frac{n_T}{N_{op}} \tag{5}$$

where $\Phi' = \partial\Phi/\partial T_B$ (ph cm^{-2} sr^{-1} sec^{-1} μm^{-1}K^{-1}).

In both these expressions, the total noise n_T is a sum of contributions from different physical sources. For photoconductive detector arrays, we use the decomposition of Eq. (1) to model n_T as

$$n_T^2 = n_{op}^2 + n_{th}^2 + n_{ro}^2 + \beta_{gn}^2 \left(N_{op}\sigma_{op}\right)^2 + \beta_{of}^2 \left[\left(N_{th}\sigma_{th}\right)^2 + \left(N_{op}\sigma_{op}\right)^2 + n_{ex}^2\right]. \tag{6}$$

The first three of these terms represent the usual single-detector noise currents, which are evaluated as variances over time. They are the array mean of the temporal variance of the number of optically generated carriers, $n_{op}^2 = \left\langle \overline{\left(N_{op} - \overline{N_{op}}\right)^2} \right\rangle$, the similarly defined variance of the number of thermally generated carriers, and the input-referred readout noise expressed as a number of noise charges n_{ro}.

The remaining four terms represent additional correlated or fixed-pattern noise that appear when array averages are taken—the array variances of mean values. They are the "optical correlated noise" $N_{op}\sigma_{op}$ due to nonuniform integrated response, the "thermal correlated noise" $N_{th}\sigma_{th}$ due to nonuniform dark current, and the pattern noise arising from low-frequency excess noise ($1/f$ noise) in the detectors, n_{ex}. The factors σ_{op} and σ_{th} are the ratios of the standard deviation to mean, over the array, of the optically and thermally generated charges, respectively. We have approximated the behavior of a two-point calibrated sensor by multiplying these measures of spatial noise by factors β_{gn} and β_{of} that represent gain and offset corrections determined by sensor calibration. Each is defined as the ratio of the residual, after-calibration correlated noise from one source to the raw, uncalibrated pattern noise from the same source; for no correction, $\beta=1$. Therefore, the remaining correlated noise terms after calibration are $\beta_{gn}N_{op}\sigma_{op}$, $\beta_{of}N_{op}\sigma_{op}$, $\beta_{of}N_{th}\sigma_{th}$, and $\beta_{of}n_{ex}$. These can be significant factors for staring systems. Even for small corrected nonuniformities (i.e., small $\beta\sigma$ products), the correlated noise contributions can be quite large if N_{op} and N_{th} are large.

3. Multiple Quantum Well Detector Model

To apply these figures of merit to sensors containing MQW focal planes, we need a simple analytical model of the dark current in MQW detectors that allows scaling of these parameters with operating temperature, wavelength, and doping density. We have developed such a model that gives results in agreement with the limited amount of detailed published data available. In addition, we have adapted quantum-efficiency and gain models from the literature. The resulting MQW detector model provides a tool for evaluating performance of MQW detectors in specific systems and for optimizing the structure of the device to achieve maximum system performance. It is important to note that this is a simplified model, and specific conclusions deduced from it may change if a more accurate model is substituted. However, it suffices to illustrate the process that must be followed to optimize MQW focal plane arrays, and it does provide important insight into the critical parameters affecting system operation. We shall not discuss details here, but will merely state the final expressions before using them to discuss thermal imaging applications.

3.1. DARK CURRENT

Our dark-current model is based on the work of Andrews and Miller,[1] who systematically compare the model of Levine et al.[2] with measurement, and on the work of Williams et al.,[3] who perform a similar service for the model of Choi et al.[4]

3.1.1. *Thermally Assisted Tunneling.* The thermally activated current is calculated by Levine et al.[7] from the Wentzel-Kramers-Brillouin (WKB) transmission coefficient by integrating the Fermi distribution over carrier energy. The behavior of the tunneling integrand is a complicated function of temperature and carrier energy. Numerical calculation of the integral is too involved for use in sensor design trades, and simple approximations to the integral are not valid for all the temperatures and electric fields at which detector behavior must be characterized.

By replacing the WKB transmission coefficient with an exactly integrable approximation, we obtained an approximate analytic expression valid over the entire interesting range of temperatures and electric fields.
The dark current expression that results is

$$J_{\mathrm{D}} \approx e v_{\mathrm{d}} \frac{kT}{L_{\mathrm{P}}} \frac{m_{\mathrm{W}}^*}{\pi \hbar^2} e^{-b} \left\{ 1 + \left[\frac{2\gamma}{1 - e^{-2\gamma}} e^{\gamma} - 1 \right] + 2\Gamma\left(\tfrac{5}{3}\right)\left(\tfrac{3}{4}\gamma\right)^{1/3} \left[1 + \left(\frac{3\gamma}{4a^2}\right)^{1/4} \right] e^{\gamma} \right\}. \tag{7}$$

with

$$b = \frac{\Delta E - E_{\mathrm{F}}}{kT}, \quad a^2 = \frac{2m_{\mathrm{B}}^* L_{\mathrm{B}}^2}{\hbar^2} kT, \quad \gamma = \frac{1}{3}\left(\frac{eV_1}{2akT}\right)^2 \tag{8}$$

where v_{d} is carrier drift velocity, L_{P} and L_{B} are the superlattice period and barrier width, m_{W}^* and m_{B}^* are the well and barrier effective masses, Γ is Euler's gamma function, ΔE is the barrier height, E_{F} is the Fermi energy, and V_1 is the potential drop across the barrier. This approximate expression fails at low temperatures and high fields, when thermally activated currents are small compared to temperature-independent tunneling currents. Following Andrews and Miller,[6] we account for image-charge effects by reducing the effective barrier height ΔE by an amount $\delta\phi$ that depends on the electric field F according to

$$\delta\phi = e\sqrt{\frac{eF}{4\pi\varepsilon}} \tag{9}$$

where ε is the static dielectric constant.

3.1.2. *Temperature-Independent Tunneling.* At low temperatures, the temperature-independent well-bottom tunneling becomes significant. For this situation, Williams *et al.*[8] have compared measured low-temperature dark currents to the model of Choi *et al.*[9]

$$I_{srt} = \frac{eA_D}{\hbar L_W^2} kT D_0 \ln\left(\frac{1+e^{u_F}}{1+e^{u_F-u_1}}\right) \tag{10}$$

where A_D is detector area, L_W is the well width, $u_F = E_F/kT$, $u_1 = V_1/kT$, and

$$D_0 \approx \exp\left\{-\frac{2\sqrt{m_B^*}L_B}{\hbar}(\Delta E - E_1)^{1/2}\left[1-\frac{1}{4}\left(\frac{eV_1}{\Delta E - E_1}\right)\right]\right\}. \tag{11}$$

We denote the energy of the confined state as E_1. The expression given in Eq. (11) is valid unless the applied field is so large that impact ionization has to be considered or so small that the reverse current must be added. The quantity D_0 is the WKB transmission through the barrier at the energy E_1 of the confined state.

3.2 COLLECTION EFFICIENCY

The collection efficiency depends first on absorption of a photon, then on escape of the excited carrier (without recombining in its original well), and finally on how far it travels without recombining in another well. The escape and recombination probabilities are not in general simply related,[5,6] and unlike the collection efficiency, the dark current depends only on the recombination probability. Since the photocurrent noise is also independent of the escape probability, we shall follow Levine *et al.*[12] in absorbing the escape probability into the quantum efficiency. The collection efficiency is then the product ηg of quantum efficiency and gain, where the gain depends only on the recombination probability and not on the escape probability.

3.2.1. *Quantum Efficiency.* The quantum efficiency may be written

$$\eta = (1-R)\left(1-e^{-\kappa\alpha L}\right)\chi p_e \approx (1-R)\kappa\alpha L\chi p_e, \tag{12}$$

where R is the reflection coefficient, α is the absorption coefficient in cm^{-1}, L is the length of the multiple-quantum well structure, κ is the number of passes the infrared radiation makes through the structure, p_e is the probability that carrier, once excited, is emitted from the well into the continuum, and χ is a polarization correction factor. The second form holds when $\kappa\alpha L \ll 1$, which is true in most cases of interest. For p-type quantum wells $\chi=1$, but because only one polarization is absorbed, $\chi=0.5$ in n-type GaAs/AlGaAs quantum wells. The transmission coefficient $(1-R)$ is close to 0.72 for a polished GaAs substrate.

For the absorption coefficient, we shall use the expression given by Choi *et al.*[7] for bound-to-bound transitions in n-type quantum wells. Written in terms of relevant detector parameters, it is

$$\alpha = \frac{\rho_s}{L_p} \frac{e^2}{2\pi\varepsilon_0 m_w^* c^2} f_0 \frac{\sin^2 \theta}{n} \frac{\lambda^2}{\Delta\lambda_{qw}} K_\alpha \tag{13}$$

where ρ_s is the two-dimensional sheet carrier density, θ is the polarization angle of the infrared radiation relative to the plane of the quantum wells, n is the refractive index, f_0 is the oscillator strength, λ is the peak response wavelength, and $\Delta\lambda_{qw}$ is the spectral bandwidth of the quantum well detector. The factor $K_\alpha \approx 4$ is inserted to scale this expression to agree approximately with measured absorption in bound-to-continuum-transition QWIPs. We shall even use this expression for p-type QWIPs by setting $\theta = 90°$. We do this because Eq. (13) gives in a simple form the major dependencies on sheet-charge density, superlattice period, refractive index, wavelength, effective mass, and spectral bandwidth, and because the literature still does not contain an equally simple expression for bound-to-continuum absorption.

For p-type GaAs/AlGaAs devices, we use $n = 3.27$, $m_w^* = 0.48\, m_0$, and $\kappa = 2$. We eliminate the dependence on polarization by setting $\theta = 90°$. For both cases we assume $f_0 \approx 0.5$. For n-type wells, $m_w^* = 0.067\, m_0$, $\theta = 45°$ and $\kappa = 1$. Using these values of θ and κ ignores the subtleties of coupling radiation into the n-type structure, which absorbs no normally incident photons (for which $\theta = 0$). The most efficient approach is to use a reflection grating on each detector.[8,9,10] The effective value of θ will then be determined by the type of grating employed. Also, multiple passes—that is, an effective $\kappa > 1$—may be obtained in n-type devices by using a waveguide.[11]

For both n- and p-type GaAs/AlGaAs quantum wells, the absorption coefficient is proportional to the carrier density. This fact alone would drive one to doping the wells to a high level as possible, to achieve the best absorptance. However, since detector dark current is exponentially dependent on carrier density at high density, the optimum doping will be determined by the operating conditions for specific applications.

3.2.2 *Photoconductive Gain*. In addition to quantum efficiency, the other factor in the collection efficiency is photoconductive gain. Liu[12] has shown that

$$g \approx N_w^{-1}\left(\frac{1-p_c}{p_c}\right) \tag{14}$$

where p_c is the probability of capture of a carrier in crossing a single well. An order-of-magnitude estimate for p_c, *neglecting* quantum-mechanical effects, is

$$p_c = \frac{L_w}{\tau_R v_d}, \tag{15}$$

where L_w is the well width, v_d is carrier drift velocity, and τ_R is the recombination time. Using $L_w = 50$ Å and plausible values for lifetime and drift velocity, $\tau_R = 10^{-12}$ s and $v_d = 10^7$ cm/s at an applied bias voltage (V_B) of 3 V, we obtain $p_c \approx 0.05$ from Eq. (15). This is generally consistent with experimental values.[11,18] At low electric fields, which turn out to be best for the applications we shall consider, we can approximate the drift velocity by the mobility-field product to obtain

$$g = \frac{\tau_R \mu V_B}{N_w^2 L_w L_p} , \tag{16}$$

Over the temperature range of interest for operating these detectors, we will assume the appropriate low field mobilities for AlGaAs are $\mu = 2000$ cm^2 V^{-1} s^{-1} for n-type and $\mu = 200$ cm^2 V^{-1} s^{-1} for p-type QWIPs. Equation (16) predicts that adjusting N_w can increase gain at fixed V_B or decrease the value of V_B required to obtain a given value of g. This relationship can be used to match detector gain and operating voltage to readout characteristics.

The important relation in Eq. (16) is $g \propto \tau_R F / N_w$ for low electric field F. We would not trust that the proportionality constant may in general be estimated as μ / L_w, though this estimate is consistent in order of magnitude with the few cases that we have checked, for all of which the well width is near $L_w = 50$ Å.

4. Applications

To illustrate the performance potential of GaAs/AlGaAs MQW detectors, and to demonstrate how the detectors can be optimized with respect to an infrared sensor system, we will study a typical thermal imaging system.

4.1. FIGURE OF MERIT

With our scaling equations for the dark current and collection efficiency in hand, we can expand our expression for NETD to provide more insight into the optimization of MQW focal planes. By substituting

$$N_{op} = \frac{\pi}{4} \eta g \eta_o \Delta\lambda \Phi (\theta_R D)^2 \tau_i \tag{17}$$

for N_{op}, we can rewrite Eqs. (5–6) as

$$\text{NETD} = \left[\frac{\Phi}{\Phi'} \right] \frac{4 n_T}{\pi \eta \eta_o \Phi (\theta_R D)^2 g \tau_i \Delta\lambda} \tag{18}$$

where

$\Delta\lambda \equiv$ sensor spectral bandwidth (μm),
$\tau_i \equiv$ integration time over which the sensor accumulates photons (s),
$\eta \equiv$ quantum efficiency,
$g \equiv$ photoconductive gain,
$D \equiv$ optics aperture diameter (cm),

η_o ≡ optics throughput, i.e., the fraction of scene photons incident on the aperture which are focused onto the correct pixel of the focal plane,

θ_R ≡ system IFOV (rad).

The detector collection efficiency, given by the product ηg, is the number of signal charges collected for each in-band photon incident on the detector. For photovoltaic detector arrays, the notation η is conventionally used for the collection efficiency, and the expression corresponding to Eq. (18) would accordingly lack the gain g. The internal field of view (IFOV) and aperture diameter are related to the pixel area, A_p (cm²), and the optics f-number, f_n, by

$$(\theta_R D)^2 = \frac{A_p}{f_n^2}. \qquad (19)$$

The lowest possible NETD for a given sensor is the background-limited (BLIP) value, which occurs when the number of noise carriers n_T is dominated by the shot noise of the optically generated carriers

$$n_T = n_{op} = \sqrt{2gN_{op}}. \qquad (20)$$

The shot-noise conversion factor is $2g$ for photoconductors.[13,14] The BLIP value of NETD is then

$$NETD_{BLIP} = \frac{\Phi}{\Phi'} \frac{2\sqrt{2}}{\theta_R D} \left(\pi \eta \eta_o \Delta\lambda \Phi \tau_i \right)^{-1/2}. \qquad (21)$$

Note that photoconductive gain is not present in Eq. (21) since it affects signal and noise carriers equally. The parameters in these equations are determined by the system design and the application, except for the quantum efficiency of the detector. In some system designs, the system wavelength band will be defined by the detector spectral bandwidth.

It is convenient to express NETD in terms of the BLIP value multiplied by a degradation factor that includes detector thermal noise, readout noise and correlated noise. To do this we first develop expressions for the thermal noise and the correlated noise. The number of thermally generated carriers can be determined from the detector dark current I_D, by

$$N_{th} = \frac{I_D \tau_i}{e}, \qquad (22)$$

and the resulting shot noise is

$$n_{th} = \sqrt{\frac{2 g I_D \tau_i}{e}}. \qquad (23)$$

This shot-noise approximation is a slight overestimate of the true generation-recombination noise of GaAs/AlGaAs QWIPs,[15,16,17,18] but is sufficient for our purpose.

The total uncorrelated noise can now be expressed as

$$n^2_{\text{uncorr}} = \frac{\pi\,\eta\,g^2\,\eta_o\,\Phi\,\Delta\lambda\,(\theta_R D)^2\,\tau_i}{2} + \frac{2g\,I_D\,\tau_i}{e} + n^2_{\text{ro}}. \tag{24}$$

Similarly for the correlated noise

$$n^2_{\text{corr}} = \beta^2_{\text{gn}}\left(\sigma_{\text{op}}\frac{\pi\,\eta\,g\,\eta_o\,\Phi\,\Delta\lambda\,(\theta_R D)^2\,\tau_i}{4}\right)^2$$
$$+ \beta^2_{\text{of}}\left(\left(\sigma_{\text{th}}\frac{I_D\,\tau_i}{e}\right)^2 + \left(\sigma_{\text{op}}\frac{\pi\,\eta\,g\,\eta_o\,\Phi\,\Delta\lambda\,(\theta_R D)^2\,\tau_i}{4}\right)^2\right). \tag{25}$$

We have neglected the term N_{ex} because for GaAs/AlGaAs QWIPs, this term is insignificant. By normalizing to the background shot noise we can express the NETD as

$$\text{NETD} = \frac{\Phi}{\Phi'}\frac{2\sqrt{2}}{\theta_R D}\left(\pi\eta\eta_o\Delta\lambda\Phi\tau_i\right)^{-1/2}\left\{1 + \frac{4\,I_D}{e\pi(\theta_R D)^2\eta g\eta_o\Phi\Delta\lambda} + \frac{2n^2_{\text{ro}}}{\pi(\theta_R D)^2\eta g^2\eta_o\Phi\Delta\lambda\,\tau_i}\right.$$
$$\left.+ \frac{\pi(\theta_R D)^2\eta\eta_o\Phi\Delta\lambda\tau_i}{8}(\beta_{\text{gn}}\,\sigma_{\text{op}})^2\left(1 + \left[\frac{\beta_{\text{of}}}{\beta_{\text{gn}}}\right]^2\left[1 + \left(\frac{\sigma_{\text{th}}}{\sigma_{\text{op}}}\right)^2\left(\frac{4\,I_D}{e\pi(\theta_R D)^2\eta g\eta_o\Phi\Delta\lambda}\right)^2\right]\right)\right\}^{1/2}. \tag{26}$$

Now we have an expression that is simply the background limited NETD (the first term) multiplied by a degradation factor that arises from thermal generation (I_D), readout noise (n_{ro}), and correlated noise effects. The maximum sensor performance is attained when the degradation factor is minimized. In designing a sensor system, it is desirable to optimize the detector parameters to balance readout noise against detector-array noise. For quantum well detectors, this involves a combination of detector design and bias-voltage adjustment, such that the photoconductive gain is large enough to make the combined thermal and optical noise contributions larger than the readout noise. To gain insight into this, we will rewrite Eq. (26) as

$$NETD \equiv NETD_{BLIP}\left[\left(1 + H_2 + \frac{1}{2}N_w H_1 H_3\right)(1+\Omega)\right]^{\frac{1}{2}} \tag{27}$$

where Ω is the square of the ratio of readout noise to total detector-array noise,

$$\Omega \equiv \frac{n_{ro}^2}{n_T^2 - n_{ro}^2} = \frac{n_{ro}^2}{2g^2 N_w H_1 \left(1 + H_2 + \frac{1}{2} N_w H_1 H_3\right)}$$ (28)

and the H_i are given by

$$H_1 \equiv \frac{N_{op}}{g N_w},$$ (29)

$$H_2 \equiv \frac{N_{th}}{N_{op}},$$ (30)

$$H_3 \equiv (\beta_{gn} \sigma_{op})^2 \left\{ 1 + \left(\frac{\beta_{of}}{\beta_{gn}}\right)^2 \left[1 + \left(\frac{\sigma_{th}}{\sigma_{op}}\right)^2 H_2^2 \right] \right\}.$$ (31)

The important property of Eq. (29–31) is that like NETD$_{BLIP}$, H_1 and H_2 are independent of photoconductive gain g. (In addition, they have been made independent of bias, V_B, and number of wells, N_w, which may be varied to adjust g.) Thus the dependence of NETD on g has been confined to Ω, the square of the ratio of readout noise to total detector-array noise. This makes it clear that photoconductive gain affects sensor performance only in relation to readout properties, and that the relative contribution of the readout to total noise, expressed as Ω, can be optimized by adjusting gain. Optimizing Ω means minimizing not Ω but NETD; heuristically, this is generally accomplished as soon as $\Omega < 1$, and no great benefit is derived from making $\Omega \ll 1$. Typical values for Ω are 0.1 to 0.4, depending on the bias voltage required to meet constraints imposed by the readout amplifier. These constraints arise from limitations of the readout's charge-storage capacity, from the requirement for minimizing readout noise with respect to total noise, and from the requirement for high injection efficiency in the case where a direct-injection readout amplifier is employed.

There are a number of factors involved in defining the optimum bias voltage values for quantum well detector arrays. It is desired to reduce the bias voltage to minimize the image charge barrier lowering effects (and the corresponding increase in dark current), and also to minimize tunneling effects. Maintaining a low bias (≈ 0.1-0.5 V) will also minimize readout storage capacity limitations. However, the quantum efficiency will be reduced at low bias voltages because some of the photogenerated carriers will not escape the quantum well.[10] At the same time, we must minimize the readout noise contribution to the total focal plane noise. Also to maintain high injection efficiency, we desire $V_B > 0.05$ V. The optimum bias will ultimately depend on the type of application.

4.2. THERMAL IMAGING SYSTEM

To illustrate the performance potential of MQW detectors, we will evaluate the performance of a terrestrial thermal imager that looks at background and scene temperatures around 300 K. In this example, we use typical parameters that apply to MQW detectors to provide semi-quantitative models for the principal purpose of outlining the multiple factors which must be included, and also to provide an example of the first order parametric analysis required to design MQW focal planes for various applications. All cases include thermionic emission effects, bias voltage and gain considerations, and tunneling contributions.

Table 1. System Parameters Used for Thermal Imaging System.

Parameter	Value	Units
Φ'	8.0×10^{14}	ph cm^{-2} (sr s μm K)$^{-1}$
Φ	4.3×10^{16}	ph cm^{-2} s^{-1} sr^{-1} μm^{-1}
$A_p = A_D$	2.5×10^{-5}	cm^2
D	10	cm
θ_R	3.0×10^{-4}	rad
η_o	.5	
τ_i	0.033	sec
$\Delta\lambda$	2.0	μm
λ	9.0	μm
n	3.27	
τ_R n-type	10^{-12}	sec
τ_R p-type	10^{-13}	sec
$1 - R$	0.72	

We will assume the parameters listed in Table 1 for this example. These are typical values for MQW properties and for sensor parameters. Included in this example is the comparison of n and p-type detectors, the parameters for which are listed in Table 2.

Table 2. MQW Detector Parameters Used for Thermal Imager Example.

Parameter	n-type	p-type
χ	0.5	1
$\sin^2\theta$	0.5	1
f_o	0.5	0.5
m_w^*	$0.067 \, m_o$	$0.48 \, m_o$
m_B^*	$0.088 \, m_o$	$0.21 \, m_o$
N_W	50	50
L_B	500 Å	500 Å
ρ_s	$2.7 \times 10^{11} \, cm^2$	$1.9 \times 10^{12} \, cm^2$
K_α	4	4

For the n-type focal plane we will initially assume the use of a lamellar reflection grating on the top of each detector so that the incident infrared radiation makes one pass through the device at an angle of 45° after reflecting off the grating and only one polarization is absorbed. For the p-type focal plane, we assume that the top of each detector is covered by a reflecting metal so that the infrared radiation makes 2 passes through the detector and both polarizations are absorbed.

We also assume the following values for the calibration parameters, which are consistent with $\beta_{gn}\sigma_{op} \geq 10^{-3}$.

Offset-Related	Gain-Related
$\sigma_{th} = 0.1$	$\sigma_{op} = 0.01$
$\sigma_{op} = 0.01$	$\beta_{gn} = 0.1$
$\beta_{of} = 10^{-4}$	

NETD is plotted versus focal plane temperature for n- and p-type QWIP focal plane arrays for the terrestrial thermal imager in Figs. 1-4. For all of these cases, thermal generation is the dominant uncorrelated noise source, and we have chosen the doping level so as to minimize the single detector NETD. Also, we have assumed $N_w = 50$ because of the following reasoning. When correlated noise dominates, NETD is independent of N_w. On the other hand, if uncorrelated noise dominates, NETD $\propto N_w^{-1/2}$ so that a large number of wells would be of some advantage in that case. We will assume in these sensor examples that $\beta_{gn}\sigma_{op}$ is limited to 10^{-3}, which, in our experience, is the best one can practically achieve due to limitations imposed by calibration source nonuniformity. In addition, we assume that $\beta_{of} = 10^{-4}$, which is a value that can be easily achieved with a 12-bit analog-to-digital convertor. It is important to note that when pattern noise due to residual response nonuniformity is the dominant noise term, NETD $= \frac{\Phi}{\Phi'}\beta_{gn}\sigma_{op}(1+\Omega)^{1/2}$, independent of all other factors. Included in each of the figures are the single detector limits for the system, which represent the ultimate performance that can be achieved if the spatial (correlated) noise can be suppressed to a level below the temporal (uncorrelated) noise.

We have chosen a bias voltage of 0.5 volts for most the cases we present here because of several factors. The most important is the fact that the integration capacitor in practical readout circuits will have a charge storage capacity limit $\approx 10^7$ carriers. To maintain practical data rates off the focal plane, the dark current must be limited to avoid saturation of the integration capacitor within an integration time. A practical limit of 1 ms is assumed in these cases, along with digital frame summation to achieve the desired system frame time. The low bias was chosen so that in combination with practical operating temperatures, the dark current level could meet the integration time restriction. Even though the quantum efficiency is reduced at low bias voltages due to lower escape probabilities, the sensor performance is not significantly degraded. This low voltage also limits power the dissipation of the FPA, and is more than sufficient to minimize the effects of readout noise.

The terrestrial thermal imager example in Fig. 1, shows NETD for n- and p-type detectors. We assume all of the following cases that the recombination and escape times (which we call the lifetime) are approximately the same for both n- and p-type detectors. At temperatures below 65 K the performance of the system with n-type detectors is limited by residual response nonuniformity (after calibration) to an NETD ≈ 60 mK. The single detector performance clearly exceeds this NETD, but this level cannot be reached without further improvement in calibration techniques. Above 65 K, the spatial noise due to offset nonuniformity becomes important. This effect is more pronounced in the p-type case. Because of the shorter recombination time assumed for the p-type detectors, the resulting degradation in NEFD that ocurrs through the H_2 term in Eq. 27 is larger for p-type focal planes than for n-type. This also translates into a requirement for more offset correction than the n-type system to achieve the same system performance level. With the level of offset correction we have assumed here, the p-type system does not reach the residual response nonuniformity limit until well below 60 K. Obviously, both types can

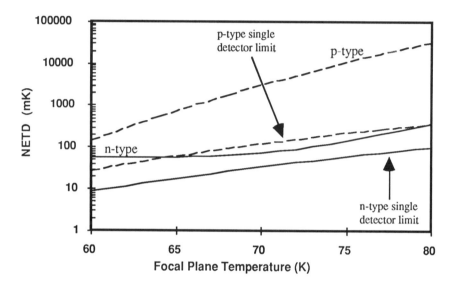

Figure 1. Example NETD versus GaAs/AlGaAs MQW Focal Plane Operating Temperature for Terrestrial Thermal Imager.

benefit considerably from improvement in β_{of}. For practical applications, most single stage mechanical coolers designed for 80 K operation can operate down to 60 K. The limit on operating temperature due to dark current saturation is ≈ 65 K. The drawback of having to operate at these temperatures is that the mean time to failure (MTTF) of the cooler is significantly reduced. For many IR sensor applications, it is desirable to operate the focal plane at temperatures ≥ 80 K, especially for those that require fast cooldown times.

In Fig. 2, we have assumed the detector dark current has somehow been reduced by a factor of 10. In this case, the single detector NETD is improved for both types, but the system is still limited by residual response nonuniformity to NETD≈ 60 mK. Clearly, the temperature at which this limit is reached increases as the dark current is lowered. From a system perspective, this provides significant increase in operating temperature because the integration capacitor does not reach dark current saturation until ≈77 K. This will increase the lifetime of a mechanical refrigerator, or allow the use of a liquid nitrogen system for cooling the focal plane.

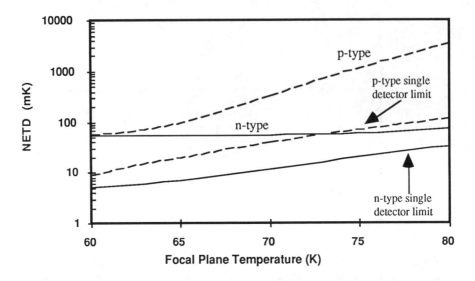

Figure 2. NETD vs Focal Plane Operating Temperature with a factor of ten reduction in dark current.

Since it may not be possible to achieve lower dark currents in these structures without using a more complex device structure, it is useful to investigate the effects of improved quantum efficiency. Figure 3 shows the performance of the system with the n-type focal plane now having a 2 dimensional grating on each detector so that both polarizations are absorbed. For the p-type focal plane we have assumed that the detectors have been constructed with a resonant optical cavity (ROC) so that the infrared radiation makes 4 passes through each detector. In this case, the single detector limits for both n- and p-type systems have improved, but the system performance remains limited by residual response nonuniformity to NETD≈60 mK. Here the temperature at which this limit is reached has increased, but we are still limited to operating temperatures of approximately 65 K due to charge storage limitations.

Now it is evident that some tradeoffs can be made between doping density and dark current. Using a lower doping will reduce both the dark current and quantum efficiency. The doping can be lowered to achieve a higher operating temperature without degrading the system performance when the noise is dominated by residual response non-uniformity. By employing a combination of gratings and waveguides, the number ofpasses that the IR radiation makes through the detector can be increased, thus providing futher room for reducing the doping concentration. Ultimately, the lower limit on doping will depend on how many passes through the detector can actually be achieved.

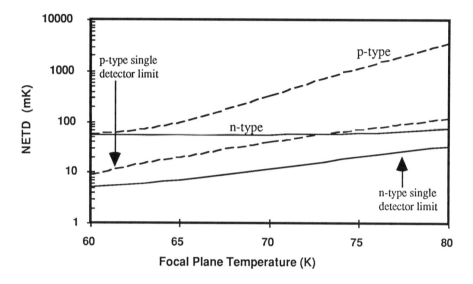

Figure 3. NETD vs. Focal Plane Operating Temperature with Improved Quantum Efficiency

Finally, it is prudent to look at the effects of improved response calibration. Improvements in calibration techniques will undoubtably occur in the near future as the use of starring arrays becomes more widespread. In Fig. 4, we show the effect of improving the response calibration to achieve a residual nonuniformity of 5×10^{-4}. In this case, using the same detector parameters as in Fig. 3, the system performance limit has been reduced to NETD ≈ 30 mK. This linear reduction occurs because the system NETD limit is proportional to the residual response nonuniformity in this example. Clearly, the system NETD is not strongly affected by improvements in single detector performance unless improved calibration techniques can be implemented.

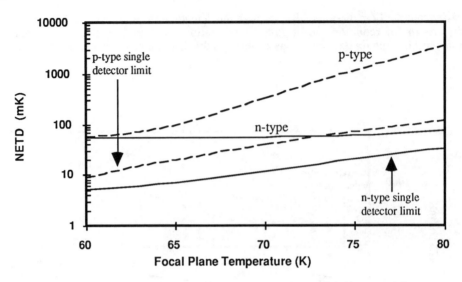

Figure 4. NETD vs. Focal Plane Operating Temperature with improved detector quantum efficiency and improved response calibration.

5. Conclusion

We have shown that the specifics of infrared sensor applications can significantly affect the design and operation of QWIP focal plane arrays. Background flux, desired operating temperature, sensor wavelength band, readout considerations, and especially calibration accuracy will strongly affect design and optimization of the quantum well structure. This dictates that predictions of MQW focal plane performance and optimization of detector parameters must be done with the system requirements in mind, and not based purely on single detector performance. In particular, for a terrestrial thermal imaging system, the performance of the system is limited by response calibration. Operating the focal plane at low bias, and improving the quantum efficiency by a modest amount will provide very adequate performance.

We have also discussed the basic framework for evaluating the performance of MQW focal plane arrays in staring infrared sensor systems. The fundamental concepts for the incorporation of spatial noise into the system noise model presented here are applicable to any type of focal plane array. We used these basic examples to analyze the performance GaAs/AlGaAs multiple-quantum-well photoconductive arrays in staring infrared sensors. We showed that the correlated noise of the GaAs/AlGaAs detector array dominates sensor performance over much of the temperature range of interest. This result is also likely to hold for other types of focal plane, including focal planes using HgCdTe detectors. In view of this, and the fact that large area MQW detectors may be much easier to fabricate at low cost, makes them potential alternatives for some specific applications.

Numerical values in our examples should be considered illustrative. The true values of important parameters may differ from our assumptions, and that could change some of our quantitative results. But the process of focal-plane design—understanding which of the parameters are important and how to optimize them—is independent of particular parameter values. The general expressions presented here can be adapted to other infrared-sensor applications, using our examples as a guide.

References

[1] S. R. Andrews and B. A. Miller, J. Appl. Phys. **70**, 993 (1991).

[2] B. F. Levine, C. G. Bethea, G. Hasnain, V. O. Shen, E. Pelvé, R. R. Abbott, and S. J. Hsieh, Appl. Phys. Lett. **56**, 851 (1990).

[3] G. M. Williams, R. E. DeWames, C. W. Farley, and R. J. Anderson, Appl. Phys. Lett. **60**, 1324 (1992).

[4] K. K. Choi, B. F. Levine, R. J. Malik, J. Walker, and C. G. Bethea, Phys. Rev. B **35**, 4172 (1987).

[5] B. F. Levine, A. Zussman, S. D. Gunapala, M. T. Asom, J. M. Kuo, and W. S. Hobson, J. Appl. Phys. **72**, 4429 (1992).

[6] H. C. Liu, Appl. Phys. Lett. **61**, 2703 (1992).

[7] K. K. Choi, L. Fotiadis, M. Taysing-Lara, W. Chang, and G. J. Iafrate, Appl. Phys. Lett. **59**, 3303 (1991).

[8] K. W. Goossen and S. A. Lyon, Appl. Phys. Lett. **47**, 1257 (1985).

[9] G. Hasnain, B. F. Levine, C. G. Bethea, R. A. Logan, and J. Walker, Appl. Phys. Lett. **54**, 2515 (1989).

[10] J. Y. Andersson and L. Lundqvist, Appl. Phys. Lett. **59**, 857 (1991).

[11] J. Y. Andersson, L. Lundqvist, and Z. F. Paska, Appl. Phys. Lett. **58**, 2264 (1991).

[12] H. C. Liu, Appl. Phys. Lett. **60**, 1507 (1992).

[13] R. J. Keyes, ed.: *Optical and Infrared Detectors*, 2nd edition (Springer-Verlag, New York, 1980).

[14] P. W. Kruse, in R. K. Willardson and A. C. Beer, eds.: *Semiconductors and Semimetals*, vol. 5: *Infrared Detectors* (Academic Press, New York, 1970), pp.15ff.

[15] B. K. Janousek, M. J. Daugherty, W. L. Bloss, M. L. Rosenbluth, and M. J. O'Loughlin, J. Appl. Phys. **67**, 7608 (1990).

[16] W. A. Beck, J. Appl. Phys. **69**, 4129 (1991).

[17] B. K. Janousek, M. J. Daugherty, W. L. Bloss, R. Lacoe, M. J. O'Loughlin, H. Kanter, F. J. de Luccia, and L. E. Perry, J. Appl. Phys. **69**, 4130 (1991).

[18] M. Rosenbluth, M. O'Loughlin, W. Bloss, F. De Luccia, H. Kanter, B. Janousek, E. Perry, and M. Daugherty, SPIE Proceedings 1283: Quantum-Well and Superlattice Physics III, 82 (1990).

GaAs/AlGaAs QWIPs VS HgCdTe PHOTODIODES FOR LWIR APPLICATIONS

A. ROGALSKI
Institute of Technical Physics, WAT
P.O. Box 49
01-489 Warsaw
Poland

ABSTRACT. Investigation of fundamental physical limitation of HgCdTe photodiodes indicate on better performance of this type detectors in comparison with GaAs/AlGaAs QWIPs operated in temperature range 40-77 K. Only at temperature 40 K, QWIPs with cutoff wavelength about 8 μm indicate higher detectivity. Advantage of QWIPs increases in wider spectral region in lower temperatures - below 40 K.

Usually, in temperature range below 50 K the performance of the n^+-p HgCdTe photodiodes is determined by trap-assisted tunneling. As a result, advantage of QWIPs increases in wider spectral region ($\lambda_c \approx$ 8-12 μm) and temperature (below 50 K). Above comparison with p^+-n HgCdTe photodiodes is more complicated for lack of precisely modeled current transport in these junctions.

GaAs/AlGaAs QWIPs at 40 K are background limited in low background conditions. This observation plus the maturity of GaAs/AlGaAs technology and its radiation hard characteristics promise that QWIPs technology can produce high quality focal plane arrays for space applications.

1. Introduction

At present $Hg_{1-x}Cd_xTe$ (HgCdTe) is the most important semiconductor alloy system for infrared detectors. However, in spite of achievements in HgCdTe material and device quality, difficulties still exit due to lattice, surface, and interface instabilities. This realization, together with continued progress in the growth of new ternary alloy systems and artificial semiconductor heterostructures, have intensified the search for alternative infrared materials.

In the class of infrared photon detectors we can distinguish three new groups of detectors which are developed recently [1]:

 (i) free carrier detectors; metal silicide-Schottky barriers,
 (ii) intrinsic detectors made from new (alternative to HgCdTe) ternary alloy systems such as InAsSb, HgZnTe and HgMnTe,
(iii) quantum well infrared detectors.

Between different types of quantum well infrared photodetectors (QWIPs), technology of GaAs/AlGaAs is the most mature. Rapid progress has recently been made in the performance of these detectors [2-8]. Detectivities have improved dramatically and are now high enough so that large 128×128 focal plane arrays (FPA's) with long wavelength infrared imaging performance comparable to state of art HgCdTe are fabricated [9,10]. With respect to HgCdTe detectors, GaAs/AlGaAs quantum well devices have a number of potential advantages including the use of

H. C. Liu et al. (eds.), Quantum Well Intersubband Transition Physics and Devices, 87–96.
© *1994 Kluwer Academic Publishers. Printed in the Netherlands.*

standard manufacturing techniques based on mature GaAs growth and processing technologies (monolithic integration of these detectors with GaAs FET's, charge coupled devices and high speed signal processing electronics is possible), highly uniform and well controlled MBE growth on over 3" GaAs wafers, high yield and thus low cost, more thermal stability and intrinsic radiation hardness.

Kinch and Yariv have presented an investigation of the fundamental physical limitations of individual GaAs/AlGaAs multiple quantum well infrared detectors as compared to ideal HgCdTe photoconductors with cut-off wavelengths λ_c=8.3 μm and 10 μm [11]. It appears, that for HgCdTe in temperature range 40-100 K the thermal generation rate is approximately five orders of magnitude smaller than the corresponding GaAs/AlGaAs superlattice. The dominant factor favoring HgCdTe in this comparison is the excess carrier lifetime, which for n-type HgCdTe is above 10^{-6} s at 80 K, compared to about 10^{-11} s for the GaAs/AlGaAs superlattice. In superlattice the confined carriers are free to move within the plane (the is no energy gap separating confined from unconfined states), so the carrier recombination rate is very high.

HgCdTe photoconductors represented by the common-module FPA and operated in a scanning mode make the dominant detector technology. At present however, second generation systems with electronic scanning on the focal plane are under intensive research and development efforts. The HgCdTe photodiodes are preferred over photoconductors due to their very low power dissipation and relatively high impedance matched directly into input stage of a silicon CCD in hybrid two-dimensional FPAs. Despite a great research and development effort, large photovoltaic HgCdTe FPAs remain expensive, primarily because of the low yield of operable arrays. Competitive GaAs/AlGaAs technology with advantages in uniformity, controllability, and yield is useful for producing large arrays. However to our knowledge, a comparative study of QWIPs and HgCdTe photodiodes has been not carried out in the literature.

In the present paper investigations of the performance of GaAs/AlGaAs QWIPs as compared to HgCdTe photodiodes operated at temperatures below 77K in long wavelength spectral region are presented.

2. Performance of QWIPs

It appears that detectivity of GaAs/AlGaAs QWIP depends only weakly on both doping concentration (in the range 10^{16} to 10^{18} cm^{-3}) and bias voltage (less than a few volts). The dark current can be reduced by three orders of magnitude, without significantly reducing the detectivity. As discussed in detail in Ref. 4, to a good approximation the detectivity can be described as

$$D^* = D_o^* \exp\left(\frac{E_c}{2kT}\right) = D_o^* \exp\left(\frac{hc}{2kT\lambda_c}\right) \tag{1}$$

where E_c is the cut-off energy $E_c = hc/\lambda_c$. We can notice the extremely rapid increase of D^* as the temperature is reduced. This is due to exponential decrease of the thermionic emission over the barriers and hence exponential reduction of dark current and noise.

According to Levine et al. [7,8], the best fit T=77 K detectivity for the n-type GaAs/AlGaAs QWIPs is

$$D^* = 1.1 \times 10^6 \exp\left(\frac{hc}{2kT\lambda_c}\right) \qquad \text{cmHz}^{1/2}\text{W}^{-1} \tag{2}$$

It should be noted that although Eqs (2) is fitted to data taken at 77 K, this equation is expected to be valid over a wide range of temperature.

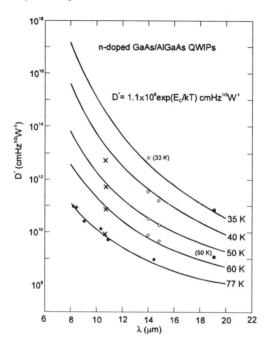

Figure 1. Detectivity vs cutoff wavelength for n-doped QWIPs at temperatures ≤ 77 K. The solid lines are calculated from Eqn. (2). The experimental data are taken from Refs: 4(×), 5(o), 6(■), 7(•), and 12(+).

Figure 1 shows the dependence of detectivity on the long wavelength cut-off for QWIPs at different temperatures. We can see very good agreement between curves calculated according to equation (2) and experimental data in wide wavelength ranges of cut-off wavelength $8 \leq \lambda_c \leq 19\,\mu m$ and temperature $35 \leq T \leq 77$ K. It should be stressed that agreement is satisfying considering the samples have different doping, different methods of crystal growth (MBE, MOCVD, and gas source MBE), different spectral widths, different excited states (continuum, bound, and quasicontinuum) and even in the case a different materials systems (InGaAs) [12].

3. Performance of HgCdTe photodiodes

Let us consider a typical high-quality n^+-p-p^+(p^+-n-n^+) photodiode structure which can be fabricated using modern epitaxial growth techniques (LPE, MBE or MOCVD). In such a type of photodiode the base p(n)-type with resultant carrier concentration about $5 \times 10^{15}(10^{15})$ cm^{-3} and thickness about 20 μm is sandwiched between high-doped regions. To obtain high quantum efficiency, a 1-2 μm thick n^+(p^+)-type cap layer is usually grown with carrier concentration to 10^{18} cm^{-3}. By thinning the base p(n)-type region of photodiode to a thickness smaller than the minority carrier diffusion length, the corresponding R_oA product increases, provided that the back junction is characterized by a low recombination rate. The backside p-p^+(n-n^+) junction is "blocking" in nature; a more intensely doped region causes a built-in electric field that repels minority carriers, thereby reducing recombination.

For simplicity, we take the one-dimensional model for the photodiode with ohmic electrical contacts. The influence of assuming doping profile on photodiode performance has been solved by forward-condition steady-state analysis [13]. More exact description of this type of calculations can be found in Ref. [14].

Hitherto, HgCdTe photodiodes have usually been based on an n^+-p structure. The most common technique for producing n^+-p junctions is by ion implantation into a p-type substrate.

In 1985, Rogalski and Larkowski indicated that, due to lower mobility of holes in the n-type region of p^+-n junctions with a thick n-type region, the diffusion-limited R_oA product of such junctions is larger than for n^+-p ones [15,16]. The above theoretical predictions have recently been confirmed by experimental results obtained for p^+-n HgCdTe junctions [17-22]. One important advantage of the p^+-n device is that the n-type HgCdTe carrier concentration is easy to control in the 10^{14}-10^{15} cm^{-3} range using extrinsic doping; while for n^+-p device, the control of p-type carrier concentration at these low levels is difficult. Because of the lower carrier concentrations achieved in the n-type HgCdTe base regions, they have longer minority-carrier lifetimes than in p-type base regions.

3.1. N^+-p-p^+ PHOTODIODES

Following the considerations carried out in Ref. 23, the optimum doping concentration in the base p-type layer of n^+-p-p^+ HgCdTe photodiodes is below 10^{16} cm^{-3}. On account of this the dependence of the $(R_oA)_D$ product (determined by diffusion current) on the long wavelength spectral cutoff for temperatures below 77 K was calculated assuming $N_a = 5 \times 10^{15}$ cm^{-3} using a procedure described in Ref. 24. However, in LWIR n^+-p-p^+ HgCdTe photodiodes at operating temperatures below 77 K, two other distinct mechanisms dominate the dark current: trap-assisted tunneling and band-to-band tunneling [25-31]. P-type base material is characterized by relatively high trap concentration, which dominates the excess carrier lifetime by the Shockley-Read-Hall recombination mechanism. Its influence depends on technological limits. Usually for $Hg_{1-x}Cd_xTe$ photodiodes with $x \approx 0.22$, in the zero-bias and low-bias region, diffusion current is the dominant current down to 60 K. For medium reverse bias, trap-assisted tunneling produces the dark current, and also dominates the dark current at zero bias below 50 K. For a high reverse bias, bulk band-to-band tunneling dominates [29].

The trap-assisted tunneling is modeled in a way described by Kinch et al.[32,33] and Nemirovsky et al. [30]. The model is based on two assumptions: that the dominant energy level in trap-assisted tunneling coincides with the Fermi level, and that the tunneling proceeds via thermally excited bulk Shockley-Read centers. Then the dynamic resistance-area product due to trap-assisted tunneling is equal

$$(R_oA)_{TAT}^{-1} = qN_t\left(c_p p_1\right)^2 \frac{6 \times 10^5}{E_g - E_t} \exp\left[-\frac{1.7 \times 10^7 E_g^{1/2}\left(E_g - E_t\right)^{3/2}\left(\varepsilon_o \varepsilon_s\right)^{1/2}}{\left(2qN_a V_b\right)^{1/2}} \right]$$

$$\times \left\{ c_1p_1 + \frac{6 \times 10^5 (2qN_aV_b)^{1/2}}{(\varepsilon_0\varepsilon_s)^{1/2}} \exp\left[-\frac{1.7 \times 10^7 E_g^{1/2}(E_g - E_t)^{3/2}(\varepsilon_0\varepsilon_s)^{1/2}}{(2qN_aV_b)^{1/2}} \right] \right\}^{-2} \quad (3)$$

$$\times \left[1 + \frac{1.7 \times 10^7 E_g^{1/2}(E_g - E_t)^{3/2}(\varepsilon_0\varepsilon_s)^{1/2}}{(2qN_aV_t)^{1/2}} \right]$$

The dominant trap energy is calculated with

$$E_t = E_F = \frac{E_g}{2} + \frac{kT}{q}\ln\left(\frac{m_h^*}{m_e^*}\right)^{3/4} - \frac{kT}{q}\ln\left(\frac{N_a}{n_i}\right) \quad (4)$$

where N_a is the effective doping in the depletion region, and (m_h^* / m_e^*) is the ratio of the effective masses of holes and electrons. In equation (3), V_b is the built-in potential, N_t and c_p are the trap density and hole capture coefficient, respectively. Since the Fermi energy is measured relative to the valence band and $E_t = E_F$

$$p_1 = N_v \exp\left(-\frac{E_F}{kT}\right) \quad (5)$$

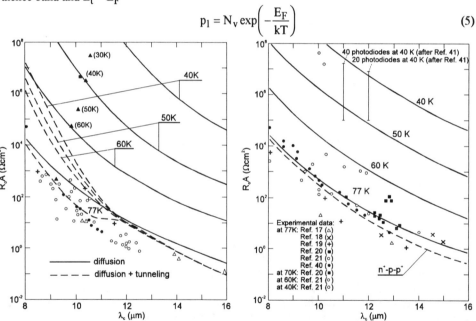

Figure 2. Dependence of the R_oA product on the long wavelength cutoff for LWIR n^+-p-n^+ HgCdTe photodiodes at temperatures $\leq$77 K. The theoretical lines are calculated assuming diffusion current mechanism (solid lines) and calculated from Eqn. (7) (dashed lines). The experimental values are taken from Refs:27($\blacktriangle$), 34(o), 35($\bullet$), 36(+), 37($\blacksquare$), and 38 ($\triangle$).

Figure 3. Dependence of the R_oA product on the long wavelength cutoff for LWIR p^+-n-n^+ HgCdTe photodiodes at temperatures $\leq$77 K. The theoretical lines are calculated assuming diffusion caurrent mechanism. The experimental values are taken from Refs (17)-(21), (40) and (41).

Direct band-to-band tunneling is modeled with the simple approach of a triangular barrier determined by the bandgap [32,33]. The R_oA product associated with direct band-to-band tunneling is given by

$$(R_oA)_{BTB}^{-1} = 10^{-2} N_a^{1/2} \exp\left(-\frac{4 \times 10^{10} E_g^2}{(N_a V_b)^{1/2}}\right)\left(\frac{3}{2} V_b^{1/2} + \frac{3 \times 10^{10} E_g^2}{2 N_a^{1/2}}\right) \tag{6}$$

The total dynamic resistance-area product is calculated and modeled with

$$\left(\frac{1}{R_oA}\right)_{TOT} = \left(\frac{1}{R_oA}\right)_D + \left(\frac{1}{R_oA}\right)_{TAT} + \left(\frac{1}{R_oA}\right)_{BTB} \tag{7}$$

The dependence of the R_oA product on the long wavelength cutoff for n^+-p-p^+ HgCdTe photodiodes at temperatures ≤ 77 K is shown in Fig. 2. In the calculation of trap-assisted tunneling contribution in R_oA product, $N_t = 10^{15}$ cm^{-3} and $c_p = 5 \times 10^{-10}$ cm^3/s have been assumed [30]. The experimental data are mainly taken for n^+-p bulk junctions fabricated by ion implantation. The experimental data at 77 K show greater spread, probably due to additional currents in the junctions, such as the surface leakage current. However, a satisfactory consistence between the theoretical curve of $(R_oA)_D$ and the upper experimental data has been achieved at 77 K.

Several studies have been reported in the literature on the low-temperature (below 77 K) n^+-p LWIR HgCdTe photodiodes and several orders of magnitude spread in R_oA products have been observed for a given cutoff wavelength and temperature [27]. The current mechanisms controlling the behavior at temperatures below 60 K are dominated by defects of unknown origins. Multistep recombination via these defects plus variations in local electrical parameters are probably the cause for not observing the expected strong band to band gap effect on devices as evidenced in the 77 K data for the best arrays of photodiodes, where the currents are diffusion limited. In Refs 25 and 39, the diodes were of relatively poor quality, as evidenced by the magnitude and temperature dependence of R_o. In the best quality photodiodes, the generation-recombination currents are not dominant current components of the dark current [27-29]. The experimental R_oA data in spectral range below 11 μm and temperatures < 77 K concerns the best quality photodiodes. In temperature range below 50 K, the R_oA product is limited by trap-assisted tunneling modeled above with $N_t = 10^{15}$ cm^{-3} and $c_p = 5 \times 10^{-10}$ cm^3/s.

3.2. P^+-n-n^+ PHOTODIODES

Figure 3 shows the dependence of the R_oA product on the long wavelength cutoff for p^+-n-n^+ LWIR HgCdTe photodiodes at different temperatures. Theoretical curves are calculated assuming diffusion dark current. Comparing Figs 2 and 3 we can see that at 77 K n^+-p-p^+ structure is inferior in comparison with p^+-n-n^+ one especially in more long wavelength spectral range. Wang's experimental results indicate that the performance of p-on-n and n-on-p heterostructure photodiodes exhibit similar R_oA products at 77 K for $\lambda_c < 10$ μm. Conversely, for $\lambda_c > 10$ μm, the R_oA products of n-on-p photodiodes are generally inferior; their quantum efficiency is lower and the impedance is higher [21]. Wang suggest that these problems are probably caused by the relative immaturity in fabricating n-on-p heterostructure, especially in the surface control on the p-side for diodes with a longer wavelength cutoff.

In Fig. 3 are also included the experimental data reported by many authors for p^+-n-n^+ structures. A satisfactory agreement between the theoretical curves and experimental data has been achieved for the temperature range 77-60 K. The experimental data according to Ref. 20 especially are in very good agreement with theoretical calculations. Several experimental points for temperature 77 K in the wavelength range $7<\lambda_c<10$ μm are situated above the theoretical curve [40]. Such experimental data measured in the Santa Barbara Research Center, concern photodiodes fabricated with HgCdTe/CdTe heterostructures. In this case a gradient of x-composition in proximity of CdTe substrate can improve blocking properties of n-n$^+$ backside junction, which results in increasing R_oA product.

With a lowering of the operation temperature of photodiodes, the discrepancy between theoretical curves and experimental data increase, what is probably due to additional currents in the junctions (such as tunneling current or surface leakage current) which are not considered. However, it should be noticed that the upper experimental data at 40 K coincides very well with theoretical curve. But this agreement is accidental, since the experimental values were obtained for double-layer heterojunction (DLHJ) structures. In these structures the built-in electric fields at heterointerfaces repel minority carriers and the R_oA product is enhanced. Thus theoretically predicted R_oA product for DLHJ structures is higher than shown in Fig. 3 for p$^+$-n-n$^+$ structures.

4. GaAs/AlGaAs QWIPs vs HgCdTe photodiodes

In this section a comparative study between GaAs/AlGaAs QWIPs and HgCdTe photodiodes in LWIR spectral region is presented. This comparison of the performance of both type of detectors should be treated as approximative, because the above assumed models of dark currents in n$^+$-p-p$^+$ and p$^+$-n-n$^+$ junctions and material parameters taken in calculations are also approximative.

The detectivity of photodiode determined by the Johnson-Nyquist noise of the zero bias resistance R_0 and the shot noise of the current generated by the background photon flux Φ_B is given by:

$$D^* = \frac{\eta\lambda_c q}{hc}\left[\frac{4kT}{R_oA} + 2q^2\eta\Phi_B\right]^{-1/2} \quad (8)$$

This formula is valid for ideal detector. Additional noise contributions, such as 1/f noise, are not included in equation (8). For thermal noise limited performance, $4kT/R_oA \gg 2q^2\eta\Phi_B$, and then

$$D^* = \frac{\eta\lambda q}{2hc}\left(\frac{R_oA}{kT}\right)^{1/2} \quad (9)$$

Comparing Figs 2 and 3 we can see that the values of $(R_oA)_D$ products for n$^+$-p-p$^+$ and p$^+$-n-n$^+$ HgCdTe junctions at given cutoff wavelength and operating temperature are comparable. Figure 4 compares detectivity of GaAs/AlGaAs QWIPs with theoretical ultimate performance of n$^+$-p-p$^+$ HgCdTe photodiodes limited by diffusion current. We can see that for $\lambda_c\approx8$ μm, the detectivity of both types of detectors is comparable at 40 K, and at lower temperatures QWIPs exhibit an advantage of performance. In the range of longer wavelength cutoff, the detectivity of HgCdTe photodiodes is higher. This advantage of HgCdTe photodiodes increases with increasing of long wavelength cutoff. Usually however, in temperature range below 50 K the limiting performance of the n$^+$-p-p$^+$ HgCdTe photodiodes is determined by trap-assisted tunneling. Including influence of trap-assisted tunneling (as modeled above in calculations of R_oA product), above comparison of detectivity is more advantageous for GaAs/AlGaAs QWIPs in wide spectral region (see dash-dotted lines in Fig. 4). It is very important, because in more long wavelength region compositional uniformity of HgCdTe FPAs is worse.

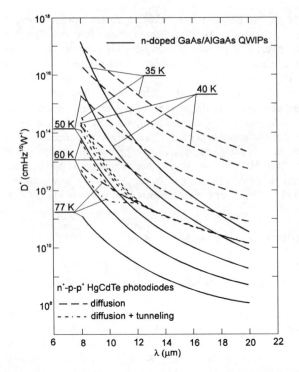

Figure 4. Dependence of detectivity on the long wavelength cutoff for QWIPs and n^+-p-p^+ HgCdTe photodiodes at temperatures $\leq$ 77 K. The solid lines are calculated from Eqn. (2) for n-doped QWIPs. The theoretical lines for photodiodes are calculated from Eqn. (9) assuming diffusion current mechanism (dashed lines) and diffusion and tunneling mechanisms (dash-dotted lines).

Above comparison with p^+-n-n^+ HgCdTe photodiodes is more complicated for lack of precisely modeled current transport in these junctions at lower temperature. At 40 K detectivity of p^+-n HgCdTe photodiodes is comparable with detectivity of GaAs/AlGaAs QWIPs.

In the above considerations we compared performance of QWIPs and HgCdTe photodiodes assuming thermal noise limited detectivity as a criterion of comparison. It should be noticed however, that for large FPAs the relevant figure of merit is the noise equivalent temperature difference (NEΔT), the temperature change of a scene required to produce a signal equal to the rms noise. An array having $D^*=10^{10}$ cmHz$^{1/2}$W^{-1} would achieve a very sensitive NEΔT = 0.01 K. This would produce an excellent image, comparable or superior to that of present arrays which are uniformity limited.

The important strengths of GaAs/AlGaAs QWIPs are highly advanced material systems and the capability to rapidly fabricate new designs by programming the MBE system. Recently, Levine et al. [10] have presented thermal imaging data of hybrid 128×128 GaAs/AlGaAs FPA consists of 50 ×50-μm square photoconductors having peak response at λ_p=9 μm. The 99 % yield of this array technology is a result of the excellent MBE grown uniformity (1 %) in thickness and the mature processing technology. After correction, measured non-uniformity of the array was better than 0.1%, and an NEΔT of 0.01 K was observed at 60 K.

At present, the performance of HgCdTe FPAs are uniformity limited. At 77 K, the operability of planar p-on-n heterostructures grown by MBE, determined by cumulative distribution functions for R_oA product, is $\geq$ 95 %, but 40 K the sport population limits the operability to 70-80 % [41]. This spread in values of R_oA product is caused by defects of unknown origins and limits applications of HgCdTe FPAs in low background conditions. The uniformity of QWIPs is superior to HgCdTe photodiode arrays operated at 40 K. This observation plus radiation hard characteristics promise that QWIPs technology can produce high quality FPAs for space applications.

For QWIPs detectors to be background limited, they will need to operate at temperatures below 77 K. It should be noticed that QWIP FPAs are far from optimum. Improvements which include the use of two-dimensional gratings, cavity design and optimizations of dopants, magnitude and profile could cumulatively result in a factor of five improvement in responsivity and raise the operation temperature [41,42]. Then QWIP FPAs with peak responsivity 8-10 μm, should be operated at temperature 77 K.

References

[1]. Rogalski, A. 'New Ternary Alloy Systems for Infrared Detectors', SPIE Optical Engineering Press, will be published.
[2]. Levine, B. F. et al. (1988) 'High detectivity D = 1.0×10^{10} cmHz$^{1/2}$/W GaAs/AlGaAs multiquantum well λ = 8.3 μm infrared detectors', Appl. Phys. Lett. 53, 296-298 (1988).
[3]. Levine, B. F. et al. (1990) 'High sensitivity low dark current 10 μm GaAs quantum well infrared photodetectors', Appl. Phys. Lett. 56, 851-853.
[4]. Gunapala, S. D. et al. (1991) 'Dependence of the performance of GaAs/AlGaAs quantum well infrared photodetectors on doping and bias', J. Appl. Phys. 69, 6517-6520.
[5]. Zussman, A. et al. (1991) 'Extended long-wavelength λ = 11-15 μm GaAs/Al$_x$Ga$_{1-x}$As quantum-well infrared photodetectors', J. Appl. Phys. 70, 5101-5107.
[6]. Levine, B. F. et al. (1992) '19 μm cutoff long-wavelength GaAs/AlGaAs quantum-well infrared photodetectors, J. Appl. Phys. 71, 5130-5135.
[7]. Levine, B. F. et al. (1992) 'Photoexcited escape probability, optical gain, and noise in quantum well infrared photodetectors', J. Appl. Phys. 72, 4429-4443.
[8]. Levine, B. F. et al (1993) 'Device physics of quantum well infrared photodetectors', Semicond. Sci. Technol. 8, S400-S405.
[9]. Kozlowski, L. J. et al. (1991) 'LWIR 128×128 GaAs/AlGaAs multiple quantum well hybrid focal plane array', IEEE Trans. Electron Devices 38, 1124-1130.
[10]. Levine, B. F. et al (1991) 'Long-wavelength 128×128 GaAs quantum well infrared photodetector arrays', Semicon. Sci. Technol. 6, C114-C119.
[11]. Kinch, M. A. and Yariv, A. (1989) 'Performance limitations of GaAs/AlGaAs infrared superlattices,' Appl. Phys. Lett. 55, 2093-2095.
[12]. Gunapala, S. D. et al (1991) 'InGaAs/InP long wavelength quantum well infrared photodetectors', Appl Phys. Lett. 58, 2024-2026.
[13]. Kurata, M. (1982) Numerical Analysis of Semiconductor Devices, Lexington Books.
[14]. Jóźwikowska, A. et al. (1991) 'Performance of mercury cadmium telluride photoconductive detectors', Infrared Phys. 31, 543-554.
[15]. Rogalski, A. and Larkowski, W. (1985) 'Comparison of photodiodes for the 3-5.5 μm and 8-14 μm spectral regions', Electron Technology 18 (3/4) 55-69.
[16]. Rogalski, A. and Piotrowski J. (1988) 'Intrinsic infrared detectors', Prog. Quant. Electron. 12, 87-289.

[17]. Arias, J. M. et al. (1989) 'Long and middle wavelength infrared photodiodes fabricated with $Hg_{1-x}Cd_xTe$ grown by molecular beam epitaxy', J. Appl. Phys. 65, 1747-1753.

[18]. Bubulac, L. O. et al. (1990) 'P-on-n arsenic-activated junctions in MOCVD LWIR HgCdTe/GaAs', Semicon. Sci. Technol. 5, S45-S48.

[19]. Arias, J. M. et al. (1991) 'Molecular-beam epitaxy growth and in situ arsenic doping of p-on-n HgCdTe heterojunctions', J. Appl. Phys. 69, 2143-2148.

[20]. Pultz, G. N. et al. (1991) 'Growth and characterization of p-on-n HgCdTe liquid-phase epitaxy heterojunction material for 11-18 μm applications', J. Vac. Sci. Technol. B9, 1724-1730.

[21]. Wang, C. C. (1991) 'Mercury cadmium telluride junctions grown by liquid phase epitaxy', J. Vac. Sci. Technol. B9, 1740-1745.

[22]. Arias, J. M. et al. (1993) 'Planar p-on-n HgCdTe heterostructure photovoltaic detectors', Appl. Phys. Lett. 62, 976-978 (1993).

[23]. Rogalski, A. (1988) 'Analysis of the R_oA product in n^+-p $Hg_{1-x}Cd_xTe$ photodiodes', Infrared Physics 28, 139-153.

[24]. Rogalski, A. et al. (1992) 'Performance of p^+-n HgCdTe photodiodes', Infrared Physics 33, 463-473.

[25]. Wong, J. Y. (1980) 'Effect of trap tunneling on the performance of long-wavelength $Hg_{1-x}Cd_xTe$ photodiodes', IEEE Trans. Electron Devices ED-27, 48-57.

[26]. Anderson, W. W. and Hoffman, H. J. (1982) 'Field ionization of deep levels in semiconductors with applications to $Hg_{1-x}Cd_xTe$ p-n junctions', J. Appl. Phys. 53, 9130-9145.

[27]. DeWames, R. E. et al. (1988) 'Current generation mechanisms in small band gap HgCdTe p-n junctions fabricated by ion implantation', J. Crystal Growth 86, 849-858.

[28]. DeWames, R. E. et al. (1988) 'Current generation mechanisms and spectral noise current in long-wavelength infrared photodiodes', J. Vac. Sci. Technol. A6, 2655-2663.

[29]. Nemirovsky, Y. et al. (1989) 'Tunneling and dark currents in HgCdTe photodiodes', J. Vac. Sci. Technol. A7, 528-535.

[30]. Nemirovsky, Y. et al. (1991) 'Trapping effects in HgCdTe', J. Vac. Sci. Technol. B9, 1829-1839.

[31]. Nemirovsky, Y and Unikovsky, A. (1992) 'Tunneling and 1/f noise currents in HgCdTe photodiodes', J. Vac. Sci. Technol. B10, 1602-1610.

[32]. Kinch, M. A. (1981) 'Metal-insulator-semiconductor infrared detectors', in R. K. Willardson and A. C. Beer (eds.), Semiconductors and Semimetals, Academic Press, New York, Vol. 18, pp. 313-378.

[33]. Blanks, D. K. et al. (1988) 'Band-to-band processes in HgCdTe: Comparison of experimental and theoretical studies', J. Vac. Sci. Technol. A6, 2790-2794.

[34]. Reine, M. B. et al. (1981) 'Photovoltaic infrared detectors', in R. K. Willardson and A. C. Beer (eds.), Semiconductors and Semimetals, Academic Press, New York, Vol. 18, pp. 201-313.

[35]. Destefanis, G. L. et al. (1986) 'Recent developments in infrared detector at LIR', in Third International Conference on Advanced Infrared Detectors and Systems, IEE, London, Conf. Publ. No. 263, pp. 44-49.

[36]. Lanir, M. and Riley, K. J. (1982) 'Performance of PV HgCdTe arrays for 1 - 14 μm applications', IEEE Trans. Electron Devices ED29, 274-279.

[37]. Bubulac, L. O. (1988) 'Defects, diffusion and activation in ion implanted HgCdTe', J. Cryst. Growth 86, 723-734.

[38]. Smith, L. M. et al. (1991) 'The growth of MCT on GaAs for high quality linear arrays of MWIR and LWIR photodiodes', J. Cryst. Growth 107, 605-609.

[39]. Placzek-Popko, E. and Pawlikowski, J. M. (1985) 'Tunneling spectroscopy of $Cd_xHg_{1-x}Te$ detectors', J. Cryst. Growth 72, 485-489.

[40]. Norton, P. R. (1991) 'Infrared image sensors', Optical Engineering 30, 1649-1663.

[41]. DeWames, R. E. et al. (1993) 'An assessment of HgCdTe and GaAs/GaAlAs technologies for LWIR infrared imagers', Proc. SPIE 1735, 121-132 (1993).

[42]. Adams, F. W. et al. (1991) 'A critical look at AlGaAs/GaAs multiple-quantum-well infrared detectors for thermal imaging applications', Proc. SPIE 1541, 24-37.

THE PHYSICS OF EMISSION-RECOMBINATION IN MULTIQUANTUM WELL STRUCTURES

E. ROSENCHER, F. LUC, L. THIBAUDEAU,
B. VINTER and Ph. BOIS
Laboratoire Central de Recherches
THOMSON-CSF
Domaine de Corbeville
F-91404 Orsay, FRANCE

ABSTRACT.- The emission-recombination mechanisms at quantum wells (QWs) have been studied by Impedance Spectroscopy (IS), Deep Level Optical Spectroscopy (DLOS) and Photoconductive gain measurements for different QW thicknesses, barrier widths and applied electric fields. These studies show that those QWs behave as giant two-dimensional traps for which capture rates and emission probabilities are determined. Those measurements explain very satisfactorily the QW detector performances. Particularly, the concept of QW capture velocity is shown to be more relevant than the capture time one. The measured values of capture velocities compare well with the theoretical results on optical phonon mediated transitions from barrier to well states.

1-INTRODUCTION

The performances of Quantum Well Infrared Photodetectors (QWIP) are determined by the rate at which electrons are ionized out of the Quantum Well (QW) either by photoionization or by thermal emission; and by the rate at which the excited electrons are recaptured into the QWs[1]. Indeed, in QWIPs, the density of photoexcited carriers n_{3d} is given by the balance equation between recombination R and generation G_{op} rates at each QW. G_{op} is given by $p_e \, \alpha_{1qw} \, \Phi$ where α_{1qw} is the absorption coefficient of each quantum well, p_e is the emission probability and Φ is the photon flux. R is given by n_{2d}/τ_c, where n_{2d} is the two-dimensional density of photoexcited carriers and τ_c is the electron lifetime in the AlGaAs barrier before recombination. Since the photocurrent J_{op} is given by $J_{op} = n_{3d} \, q \, v$, where v is the carrier velocity and n_{3d} is the three-dimensional density of carriers, one needs a relationship between n_{3d} and n_{2d}. *This one is usually chosen as* $n_{3d} = n_{2d} / L$, where L is the period of the QWs, i.e the barrier width plus that of the QW [2,3]. This leads to the following expression of the QWIP responsivity $R_{op} = J_{op}/q \, \Phi$:

$$R_{op} = p_e \, \alpha_{1qw} \, g \qquad (1)$$

with the photoconductive gain:

$$g = \frac{\tau_c}{L/v} \qquad (2)$$

97

H. C. Liu et al. (eds.), Quantum Well Intersubband Transition Physics and Devices, 97–110.
© 1994 *Kluwer Academic Publishers. Printed in the Netherlands.*

98

In Sec. II, we show how the emission probability p_e may be extracted from Deep Level Optical Spectroscopy measurements. In Sec. III, the capture time τ_c is determined by Impedance Spectroscopy and the results compared with theoretical results. In Sec. IV, photoconductive gain *vs* barrier width measurements validate the physical understanding brought by those spectroscopic experiments.

2-DEEP LEVEL OPTICAL SPECTROSCOPY OF QWs

In order to measure the rate at which electrons will be ionized out of the QW without the complications due to photoconductive gain and dark current, we have made measurements of the transient photocurrent from a multiquantum well structure in a Schottky diode. This technique is very similar to the Deep Level Optical Spectroscopy method developed to study the optical cross section of deep levels in semiconductors[4].

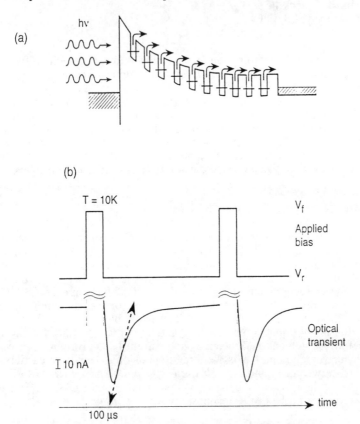

Fig.1: (a) Schematic energy band diagram of the sample. Note that the Schottky barrier blocks the refilling current. (b) Polarization cycle applied to the device and experimental transient current due to the photoionization of the QWs. The photoionization cross section is measured as the initial slope of the transients.

The sample (see Fig.1), grown by molecular beam epitaxy, consists in ten periods of a 6 nm GaAs 5×10^{11} cm^{-2} δ–doped quantum well sandwiched between 35 nm $Ga_{0.25}Al_{0.75}As$ barriers on the top of a 500 nm 10^{18} cm^{-3} Si-doped GaAs contact layer.The sample has been designed so that there are two bound levels in the well (E_1 and E_2) with $E_{21} = E_2 - E_1 = 138$ meV and that the second energy level E_2 is a few meV below the $Ga_{0.25}Al_{0.75}As$ barrier E_B ($E_{B2} = E_B - E_2 \approx 8$ meV). The depth of the potential well as well as the barrier thickness have been chosen sufficiently high to suppress tunnel emission[5] as well as thermal emission for temperatures up to 60 K. Mesa structures 200 μm x 100 μm were processed using standard lithography techniques. An evaporated AuGeNi layer provides an ohmic contact at the bottom of the structure while Cr was evaporated to provide a semi-transparent Schottky diode on the top. *The Schottky diode insures that no refilling current flows through the structure* for reverse biases down to –1 Volt in the considered temperature range (up to 60 K). Finally, only 10 QWs have been used since, as only a few 10^{12} electrons per cm^2 may be extracted from a depletion layer[6], a larger number of wells would just have increased the series resistance without providing any additional transient photosignal.

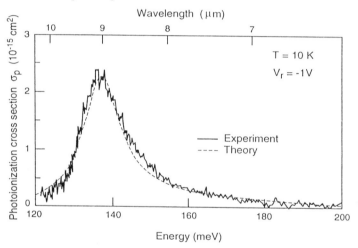

Fig.2: Photoionization cross section σ_P of the QWs as a function of photon energy. The experimental curve (full line) is fitted by the simple quantum mechanical model for intersubband absorption with a linewidth of 5 meV (dotted line). The agreement is excellent.

The experiment follows the usual voltage cycle of transient optical spectroscopy[4]. Under infrared illumination, the structure is first forward biased so that electrons from the GaAs contact are injected in the $Ga_{0.25}Al_{0.75}As$ barrier and captured in the wells. The diode is then reverse biased and the trapped electrons are emitted from the well to the lower contact. The density of electrons N_i as a function of time in each well is determined by :

$$\frac{\partial N_i}{\partial t} = - \sigma_P (i) \, \Phi \left(N_i - N_i^{eq} \right) \qquad (3)$$

where i stands for the well index, *eq* stands for equilibrium, Φ is the photon flux and $\sigma_P(i)$ is the photoionization cross section which may depend on the electric field at the ith QW.

Recent studies[7,8,9] have shown that the photoionization cross section can be phenomenologically written $\sigma_P = p_e \sigma_O$, where p_e and σ_O are the escape or emission probability and the optical capture cross section respectively. σ_O is the quantity measured by absorption spectroscopy which is well accounted for by quantum mechanics[10]: for this sample, $\sigma_0 = 2.10^{-15}$ cm^{-2} at resonance. p_e is the probability for a photoexcited electron to be scattered into a diffusive energy state (instead of being scattered back to the fundamental state of the well). The internal quantum efficiency is equal to the product of the escape probability and the sample absorption.

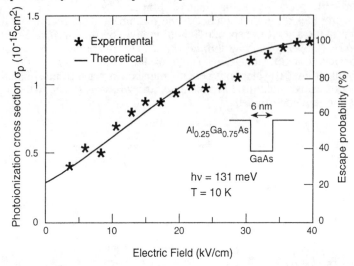

Fig.3: Experimental variation of the photoionization cross section and of the deduced escape probability from QWs as a function of applied electric field (points) for photon energy hν = 131 meV. This variation is fitted by a model based on the statistical gaussian distribution of the ionization energy (variance of 5 meV). This variance corresponds to a one monolayer fluctuation of the QW width on both sides.

Summing over the 10 wells and assuming a cross section $\sigma_P(i)$ independent of i, we find that integration of Eq.(3) leads to a current exponentially decreasing with time. Note that, since the current has a transient behavior, the current is the *sum of the contribution of all the wells*. This is markedly different from steady state photoconduction where the current is constant through the structure. Figure 1 shows a typical transient signal obtained at Brewster angle for a 1.4 Volt V_f filling pulse (120 μs duration) and a reverse bias V_r of −1 Volt with a polarized tunable CO$_2$ laser source at λ = 9.47 μm (130 meV) for an incident power of 100 mW/cm^2. As a matter of fact, since the escape probability is dependent on the local electric field at each quantum well [8], this signal is not purely exponential since the electric field changes as a function of time. Nevertheless, when the reverse bias V_r has just been applied (t=0), the electric field F is homogeneous in the structure: F=(V_r+V_0)/L, where V_0 is the Schottky offset ($V_0 \approx$ 750 meV) and L is the total structure length. $\sigma_P\Phi$ can therefore be measured directly as the initial slope of the transient current. For the case of Fig. 1, the experimental photoionization cross section is found to be σ_p= 1.3 × 10^{-15} cm^{-2} at 9.47 μm.

We then measure the variation of the photoionization cross section with the photon energy hν. The sample is illuminated by the light of a glowbar dispersed by a two-prism spectrometer. In order to get rid of the transient due to the background flux, we have used a boxcar technique. The experiment is performed at 10 K with a filling pulse of 1.8 V during 80 μs, a reverse bias of –1 V (*i.e.* an initial electric field of 40 kV/cm) and a chopping frequency of 200 Hz. Fig. 2 shows the resulting σ_P versus hν curve. It exhibits the usual resonance feature, with a maximum peak value of 2.2×10^{-15} cm^2 in total agreement with the quantum mechanical calculation [10] $\alpha(h\nu)$ *assuming an escape probability* p_e *close to 100%* (independent of hν).

We are now in position to study the electric field dependence of the escape probability. Indeed, since the QWs are symmetric, the variation of the optical cross section with the electric field F is negligible [10] so that the variation of σ_P directly reflects the influence of F on the probability p_e. The experimental results shown in Fig.3 for a laser beam at 9.47 μm (130 meV), indicate a smooth variation of p_e from 100% down to 30% when the electric field is decreased from 40 kV/cm down to 4 kV/cm.

The escape probability is usually related to the electron escape time τ_e and the recombination time τ_r through the relation[9] $p_e = (1+\tau_e / \tau_r)^{-1}$. When the electric field is enhanced, τ_r should not be significantly affected by orders of magnitude (see below) while τ_e should vary with the tunnel barrier transparency. However, the theoretical variation of the transparency with the electric field is very steep, from practically 0 to 1 for electric field values close to F_s given by $qF_sL_w/2 = E_B - E_2$ where L_w is the quantum well width. *The variation of p_e should then be described by a Heaviside function at $F = F_s$,* which is clearly not in agreement with the experimental results. We therefore believe that the smooth variation of p_e has a statistical origin, *i.e.* the energy E_2 fluctuates in each QW with a Gaussian distribution, so that the p_e vs F curve is the convolution of the Heaviside function with this Gaussian distribution. The best fit of the experimental results[11], taking into account this distribution, is obtained by a deviation of $\Delta E = 3.5$ meV around a mean value 4 meV below the barrier (see Fig.3). The former value can be reasonably accounted for by a *one monolayer fluctuation* in the QWs on both sides.

3- CAPTURE RATE INTO A QUANTUM WELL

3-1: IMPEDANCE SPECTROSCOPY OF QWs

The second set of experiments determine the capture of electrons from the continuum in the barrier layers into a QW. The method used is that of impedance spectroscopy on a sample containing only one QW in a semiconductor-insulator-semiconductor (SIS) structure.

The samples (see Fig.4) consist of a sandwich of a 94 nm $Al_{0.4}Ga_{0.6}As$, 50 nm $Al_{0.22}Ga_{0.78}As$, a GaAs (undoped) QW of width L_w, 50 nm $Al_{0.22}Ga_{0.78}As$ structure. It is clad between two 500 nm 10^{18} cm^{-3} Si doped GaAs contacts. In a first set of samples, the well parameters (length, barrier height) have been chosen so that for $L_w = 6$ nm the first energy level E_1 is bound (59 meV above the GaAs conduction band) and the second level E_2 is quasi resonant with the $Al_{0.22}Ga_{0.78}As$ conduction band (E_2-E_1=117 meV). Other QW widths (4 nm and 5 nm) has also been studied, though not in the quasi-resonant case. Moreover, on the one hand, the Al concentration of the $Al_{0.4}Ga_{0.6}As$ layer has been chosen sufficiently large so that this barrier is totally insulating at temperatures below 120 K, which has been experimentally verified. On the other hand, the Al concentration of the $Al_{0.22}Ga_{0.78}As$ intermediate barrier is low enough so that, in the temperature range of our experiments (80 – 120 K), the QW is not far from the thermodynamical equilibrium with the electron reservoir of the lower contact. Consequently, *the Fermi level in the well is set by the lower contact.* This condition is the basis of the equivalent circuit analysis of interface traps in MOS structures so that we can extend this analysis to our SIS system[12]. Mesa structures (265 μm × 265 μm) were fabricated using standard photolithography techniques and AuGeNi alloy contacts. The sample capacitance is independent of the applied bias in the +1,−1 V range at the temperature of 10 K and the value of 33 pF is in perfect agreement with the nominal total thickness of 200 nm. Admittance measurements are performed, using a lock-in technique, as a function of the frequency for different temperatures between 85 K and 115 K and for different applied biases.

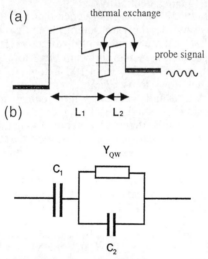

Figure 4: Schematic energy band diagram of the investigated sample (a) and equivalent electrical circuit (b).

From a *direct analogy* with MOS devices, the electrical equivalent circuit of the structure for a sinusoidal signal can be immediately deduced (see Fig. 4) by introducing the "quantum well admittance" Y_{QW} as[12]:

$$Y_{QW} = \frac{\partial I_{QW}}{\partial V_{QW}} = G_{QW} + j\omega\, C_{QW} \qquad (4)$$

where V_{QW} is the potential drop between the well and the lower contact, ω is the signal radial frequency and I_{QW} is the current flowing into the well: $I_{QW} = -qA\, \partial N_{QW}/\partial t$, N_{QW} is the electron density in the well and A is the sample area. The value of G_{QW} is deduced from the measured sample admittance and the known values of the geometrical capacitances C_1 and C_2. In our case the values of C_1 and C_2 are respectively 44 pF and 132 pF.

Considering the whole well as a single 2D trap, one can write the emission-recombination equation as:

$$A\, \frac{\partial N_{QW}}{\partial t} = c_n n - e_n A N_{QW} \qquad (5)$$

where n is the 3D distribution of carriers above the well. e_n (s^{-1}) is the thermal emission rate from the QW while c_n (cm^3 / s) is the classical capture coefficient in the QW. A classical treatment for small sinusoidal perturbations leads then to the expression of the equivalent impedance for such a mechanism[12]:

$$\frac{G_{QW}(\omega)}{\omega} = C_T \frac{\omega\tau}{1 + \omega^2\tau^2} \qquad (6)$$

with

$$C_T = \frac{q^2 A\ m^*}{\pi\ \hbar^2} \exp\left(\frac{E_F + q\ V_{QW} - E_1}{kT}\right) \qquad (7)$$

where $\tau = A N_{eq} / c_n n_{eq}$ is the emission time, N_{eq} (resp. n_{eq}) is the 2D-carrier density in the well (resp. 3D density above the well) in steady state. Since the well is undoped, N_{eq} is given by $N_{eq} = (m^* / \pi\hbar^2)\ kT \exp((E_F + qV_{QW} - E_1)/kT)$ where m^* is the effective mass of the electron in GaAs.

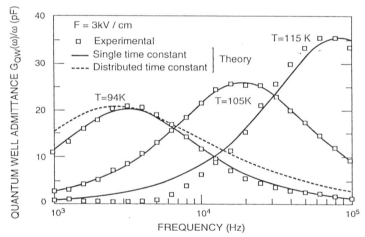

Figure 5: Experimental conductance $G(\omega)/\omega$ vs ω curves of a quantum well with the best theoretical fits obtained using a single and a distributed time constant models. The quantum well effectively behaves as a single time constant recombination centre.

Figure 5 shows the comparison between experimental curves $G_{QW}(\omega)/\omega$ vs ω, the theoretical curves given by Eq. 6 and a theoretical curve obtained by using a distributed time constant approach[12]. The excellent agreement between the single state theory and the experiment is the signature of an emission-recombination process occuring between *a single trap* and the Al$_{0.22}$Ga$_{0.78}$As conduction band.

In our case of a 2D trap, the capture coefficient c_n is proportional to the sample area so that the significant capture parameter is the recombination velocity v_{QW} defined as $v_{QW} = c_n / A$ [13]. From Eq. 5, this quantum well recombination velocity is related to the time constant τ through the relation $v_{QW} = N_{eq} / \tau\, n_{eq}$. If we make the assumption that the 3D equilibrium density of states is not affected by the presence of the quantum well, so that n_{eq} is given by the classical expression[6] $n_{eq} = 2\ (m^*/2\pi \hbar^2)^{3/2}\ (kT)^{3/2}$ $\exp\ ((E_F + qV_{QW} - E_2)\,/\,kT)$ we find:

$$\tau = \frac{1}{v_{QW}} \left[\frac{2\pi\, \hbar^2}{m^*\, kT}\right]^{\frac{1}{2}} \exp\left(\frac{E_2 - E_1}{kT}\right) \qquad (8)$$

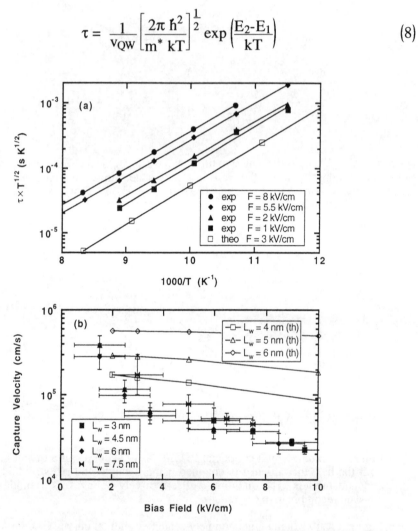

Fig. 6: (a) Arrhenius plot of ln $(\tau T^{1/2})$ vs 1000/T (τ is the resonance time constant) deduced from the frequency dependence of the admittance for a 6 nm QW. The corresponding theoretical results are also shown for one field. The intersection of the extrapolation lines with the abscissa yields the recombination velocity in the QW. This velocity is shown in (b) as a function of the applied electric field and for several different QWs. Rectangles indicate the experimental uncertainties.

In Fig. 6a the corresponding Arrhenius plot is shown for a 6 nm wide quantum well. For the activation energy $E_2 - E_1$ a value of 113 meV is found, independent of the field. This supports our assumption that the observed emission-recombination process occurs between the bound state of the well and the $Al_{0.22}Ga_{0.78}As$ conduction band. We also find $v_{QW} \approx 8 \times 10^4$ cm/s for an electric field of 3 kV/cm. Moreover, in Fig. 6b we show that the experiments indicate a strong decrease of the recombination velocity for an increasing applied electric field. This can be seen to be the case for the samples of different well width, too. In contrast to what has been reported from time-dependent photoluminescence measurements[14], we find no clear dependence of the capture velocity on well width.

The capture coefficient may be connected to the capture probability p only if a velocity v is introduced, $i.e.$ $c_n = A p v$. If we take a usual drift velocity of a QWIP $v \approx 1 - 5 \times 10^6$ cm / s with a minimum value of $v_{QW} = 3 \times 10^4$ cm / s, we find a capture probability $p = v_{QW} / v \approx 0.03 - 0.006$. This is in good agreement with the values obtained by photoconductive gain in QW infrared detectors[9].

3-2: THEORY OF QW IMPEDANCE

We have performed calculations of the net capture rate into a QW in a constant electric bias field. The hypotheses of the calculation are that the polar optical phonon scattering processes from levels of the continuum to the fundamental bound state subband of the QW control the capture time; the continuum states are assumed to be in thermal equilibrium with the Fermi energy of the contact, and the electrons in the QW are in thermal equilibrium in the ground state QW subband. In the electric field the wavefunctions can be written as linear combinations of Airy functions in the barriers and in the QW. Using Fermi's golden rule to calculate the scattering rates $S(\varepsilon_z, \mathbf{K} \rightarrow \varepsilon_0, \mathbf{K}_0)$ between states of the continuum (denoted $\varepsilon_z, \mathbf{K}$) and states in the QW (denoted $\varepsilon_0, \mathbf{K}_0$) we have:

$$
\frac{\partial N_{QW}}{\partial t} = \sum_{\varepsilon_z, \mathbf{K}, \mathbf{K}_0} S_{\substack{emi \\ abs}} \left(\varepsilon_z, \mathbf{K} \rightarrow \varepsilon_0, \mathbf{K}_0\right) f\left(\varepsilon_z, \mathbf{K}\right) \left(1 - f\left(\varepsilon_0, \mathbf{K}_0\right)\right)
$$
$$
- \sum_{\varepsilon, \mathbf{K}, \mathbf{K}_0} S_{\substack{abs \\ emi}} \left(\varepsilon_0, \mathbf{K}_0 \rightarrow \varepsilon_z, \mathbf{K}\right) f\left(\varepsilon_0, \mathbf{K}_0\right) \left(1 - f\left(\varepsilon_z, \mathbf{K}\right)\right) \tag{9}
$$

which by using detailed balance between phonon emission processes in one direction and phonon absorption processes in the opposite direction can be reduced to $\partial N_{QW} / \partial t = -(N_{QW} - N_{eq}) / \tau$, where N_{eq} is the number of electrons in the quantum well when no ac voltage is applied, and

$$
\frac{1}{\tau} = \frac{\pi \hbar^2}{m^* kT} \sum_{\varepsilon_z, \mathbf{K}, \mathbf{K}_0} S_{\substack{emi \\ abs}} \left(\varepsilon_z, \mathbf{K} \rightarrow \varepsilon_0, \mathbf{K}_0\right) \exp\left(\frac{\varepsilon_0 - \left(\varepsilon_z + \hbar^2 K^2/2 \, m^*\right)}{kT}\right) . \tag{10}
$$

Results of this calculation are also shown for a 6 nm wide QW in Fig. 6a. It can be seen that time constants in the 100 μs to 10 ms range are indeed reproduced by the theory and that the activation energy is the same as experimentally observed. However, the dependence on bias field is much smaller than observed. This can be seen in the theoretical recombination velocity (derived by making the same fitting procedure as for the experiment) which is also plotted in Fig. 6b for different QW widths. In addition, one can also see a stronger QW width dependence. It remains an interesting question which transport processes create competing time constants in the experiment in order to explain these less important discrepancies.

4- CAPTURE TIME vs BARRIER WIDTH MEASURED BY INFRARED PHOTOCONDUCTIVE GAIN

Eq. (1) and (2) predict a τ_c/L dependence of the photoconductive gain. On the other hand, in a photoemissive approach, Liu has shown that the photocurrent could be written as $J_{op} = p_e \, \alpha_{1qw} \, \Phi / p_c$, where p_c is the capture probability, i.e. the ratio between the captured and the incident electron flows[9]. In the latter approach, the barrier thickness plays no role so that , for the sake of consistency, the equality $g = 1/p_c = v \, (\tau_c / L)$ should mean that *the ratio τ_c/ L is constant vs barrier width.*We now show that this is indeed the case and that it is consistent with a quantum mechanical description of the capture process[15].

A series of 4 samples has been grown by molecular beam epitaxy under identical growth conditions. They consist of 25 periods of 6 nm GaAs QWs δ-doped with a Si concentration of 5×10^{11} cm^{-2}, sandwiched between $Al_{0.22}Ga_{0.78}As$ barriers of thicknesses 25, 50, 75 and 100 nm respectively, yielding absorption peaks in the 110 meV range. No thinner barrier structures have been used because they exhibit a high tunneling dark current [16]. Higher AlGaAs thicknesses have been tried but yielded degraded structures. These structures are clad between two 500-nm-thick 2×10^{18} cm^{-3} Si-doped contact layers. The samples are processed into mesa structures using conventional photolithographic techniques. The different quantities necessary to determine the ratio τ_c/L through Eq. (1) and (2) have been determined by the following experiments.

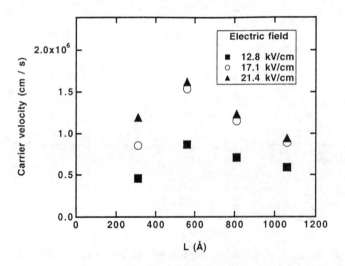

Figure 7: Carrier velocity in QWIPs as a function of the period of the QWs and for different applied electric fields. These values are obtained by fitting the dark current vs 1/T curves in the temperature range of 40 to 120 K.

The velocity v of the different samples for different applied electric field is determined by dark current measurements. The measurements of dark current J_{dark} vs applied electric field F and temperature T are done in a Helium blind cryostat .The carrier velocity v is obtained from the relation $J_{dark} = q\, v\, n_{th}$ where n_{th} is the 3-dimensional density of thermal carriers. Somewhat similarly to Kinch and Yariv[17], one considers a quantum well of thickness d and potential depth V placed in an infinite potential box of dimension L equal to the period. The density of thermal free carriers is given by the integrated contribution of the population on each *barrier* subband i , i.e:

$$n_{th} = \frac{m^*}{\pi \hbar^2} \frac{kT}{L} \, e^{-q\frac{V - E_F}{kT}} \, U\left(L, T\right) \tag{11}$$

with

$$U\left(L, T\right) = \sum_{i=1}^{\infty} e^{-i^2 \frac{\hbar^2 \pi^2}{2\, m^*\, L^2\, kT}} \tag{12}$$

E_F is the Fermi energy given by the condition that the total amount of carriers in the structure (the thermal and the trapped ones) is equal to the doping level. This formula is somewhat different from the one of Kinch and Yariv,[17] who considered that all thermal carriers are generated by the first subband in the barrier, i.e. U(L,T) = 1. However, their approximation is no more valid for large value of period L, which is our case, where $U \approx L\sqrt{\,}\,(m^*\, kT/\, 2\pi\, h^2\,)$ for which Eq.(11) tends to the 3D case. The thermal activation energies V - E_F are thus obtained by fitting the J_{dark} vs 1/T curves through Eqs. (11) and (12) in the interval 40 K to 120 K The carrier velocity is then extracted from J_{dark} = q v n_{th} and is shown in Figure 7 for different thicknesses and three different applied electric fields. The slight decrease of the carrier velocity, i.e. mobility, with increasing barrier thicknesses is ascribed to the degradation of the AlGaAs material with thicker barrier. Indeed, because of a progressive roughening of the AlGaAs growth, thicker AlGaAs barriers are of lower quality than the thinner ones. The low value of v for small barriers is probably due to the high density of QW interface scattering experienced by the electrons for such a high volume density of QWs .

The QWIP responsivities are then measured between 10 K and 80 K in a Fourier transform infrared spectrometer FTIR apparatus. The photocurrent is found to be little affected by the temperature. For a more accurate determination of the absolute value of the photoconductive gain, the values obtained in the FTIR apparatus are checked at a given wavelength to the ones obtained with a CO_2 laser. From Eq. 1, one can extract the photoconductive gain once p_e and α_{1qw} are known. *The applied electric fields are high enough so that p_e is close to 1* (see Sec. II) and α_{1qw} is obtained by absorption measurements in the FTIR. The photoionization energies are also extracted from FTIR measurements and are in good agreement with the activation energies taking the Fermi energy into account. Figure 8 shows the variation of the photoconductive gain, obtained from the responsivity measurements, with the barrier thickness for different electric fields. As stated in Eq. 2, the electron lifetime is thus obtained from the ratios between the points of Fig.7 and Fig.8. The resulting points are shown in Fig. 9 which exhibits the values of the electron recombination times τ_c as a function of barrier thickness L for different applied electric fields. It is clear from Fig.9 that *the proportionality between the QW recombination time and the barrier thickness is thus unambiguously demonstrated.* Moreover, at rather low electric fields, the recombination times range from 8 ps in the 25 nm barrier up to 80 ps in the 100 nm one. These values are far higher than those measured by picosecond photoluminescence spectroscopy. We find mainly two reasons for this. Firstly, photoluminescence measurements are obscured by the small hole capture time. Secondly, as observed in this experiment and in impedance spectroscopy ones (see Sec. III), the carrier recombination time decreases for decreasing electric field; and this photoconductive gain experiment cannot be performed at zero electric field where it could be compared to photoluminescence ones.

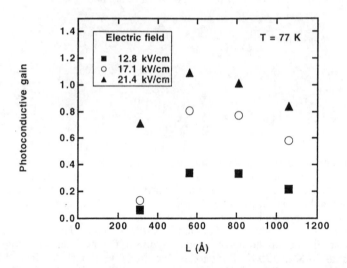

Figure 8: Photoconductive gain measured in QWIPs at 77 K as a function of the QW periods and for different applied electric fields.

The constancy of the τ_c/L ratio can be easily taken into account in a quantum approach. Indeed, the capture rate of an electron in a free electron state of energy E_k with normalized delocalized Bloch wave function $1/\sqrt{L}\, u_c(\mathbf{r})\, e^{i\mathbf{k}\,\mathbf{r}}$ to a bound state of energy E_1 of wavefunction $\chi_1(z)\, u_c(\mathbf{r})\, e^{i\mathbf{k}_{//}\mathbf{r}_{//}}$ ($k_{//}$ is the wavevector parallel to the interfaces) is:

$$\tau^{-1} = \frac{2\pi}{\hbar}\,\left|\left\langle \frac{1}{\sqrt{L}}\, u_c(\mathbf{r})\, e^{i\mathbf{k}\mathbf{r}}\,\middle|\, H_{per}\,\middle|\, u_c(\mathbf{r})\, \chi_1(z)\, e^{i\mathbf{k}_{//}\mathbf{r}_{//}}\right\rangle\right|^2 \qquad (13)$$

where H_{per} is the perturbation Hamiltonian describing electron-phonon, impurity scattering,[15,18] ... From Eq.(5), it is clear that, if the barrier width is significantly larger than the QW one, the delocalized quantum nature of the electron leads to a proportionality between τ and L [15]: In other words the greater the barrier width, the smaller the presence probability of the free electron above the barrier. The fundamental quantity describing the electron capture is the quantum well capture velocity $v_{qw} = L/\tau_c$ already introduced in Sec. II. It is the natural quantity which describes the recombination of a three-dimensional electron gas at a two-dimensional discontinuity, such as e.g. the Si/SiO$_2$ interface [19].

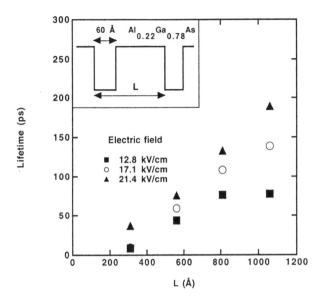

Figure 9: Recombination lifetime in 6 nm QWs of electrons in a $Al_{0.22}Ga_{0.78}As$ barrier as a function of QW period and for different applied electric fields.

5-CONCLUSION

The picture of the QWIP which emerges from these several independent experiments is very close to that of an extrinsic photoconductor in which the quantum wells simply act as giant traps from which electrons are ionized and retrapped. We have measured the important parameters of these processes in specifically designed experiments and have found that they can be consistently explained by microscopic theory, and that they explain very satisfactorily the responsivity and detectivity performance of QWIPs. The optical emission probability p_e, measured by DLOS, has a field dependence well taken into account by a one-monolayer statistical fluctuation of the QW width. The QW capture velocity v_{qw}, shown to be the relevant parameter, is measured by Impedance Spectroscopy and photoconductive gain. The values of v_{qw} are ranging from 5×10^4 up to 5×10^5 cm/s, depending on the applied electric fields and the QW width. The experimental zero field value of v_{qw} compare well with the theoretical results on optical phonon mediated transitions from barrier to well states. The discrepancies at high electric fields are probably due to hot electron effects on the capture mechanisms and are currently under study.

110

REFERENCES

1.   E. Rosencher, B. Vinter, and B. Levine, eds., "Intersubband Transitions in Quantum Wells," Plenum, London (1992).
2.   E. Rosencher, F. Luc, P. Bois, and S. Delaitre, Injection mechanism at contact in a quantum well intersubband infared detector, Appl. Phys. Lett. 61:468 (1992).
3.   H. C. Liu, A. G. Steele, M. Buchanan, and Z. R. Wasilewski, Dark current in quantum well infrared photodtectors, Appl. Phys. Lett. 2029 (1993).
4.   A. Chantre, G. Vincent, and D. Bois, Deep Level Spectroscopy in GaAs, Phys. Rev. 23:5335 (1981).
5.   E. Martinet, E. Rosencher, F. Chevoir, J. Nagle, and P. Bois, Determination of electron tunneling time from a GaAs/AlGaAs quantum well by capacitance transient spectroscopy, Phys. Rev.B 44:3157 (1991).
6.   S. M. Sze, "Physics of Semiconductor Devices," Wiley Interscience, New York (1981).
7.   B. F. Levine, A. Zussman, S. D. Gunapala, M. T. Asom, J. M. Kuo, and W. S. Hobson, Photoexcited Escape Probability, Optical Gain, and Noise in Quantum Well Infrared Photodetectors, J. Appl. Phys. 72:4429 (1992).
8.   E. Martinet, F. Luc, E. Rosencher, P. Bois, E. Costard, S. Delaître, and E. Böckenhoff, Electric Field Effects on Bound to Quasibound Intersubband Absorption and Photocurrent in GaAs/AlGaAs Quantum Wells, in: "Intersubband Transitions in Quantum Wells," E. Rosencher, B. Vinter, and B. Levine, eds., Plenum, London (1992) p.299.
9.   H. C. Liu, Photoconductive gain mechanism of quantum well intersubband infrared detectors., Appl. Phys. Lett. 60:1507 (1992).
10.   E. Rosencher, E. Martinet, F. Luc, P. Bois, and E. Böckenhoff, Dicrepancies between photocurrent and absorption spectroscopies in intersubband photoionization from GaAs/AlGaAs multiquantum wells, Appl. Phys. Lett. 59:3255 (1991).
11.   F. Luc, E. Rosencher, and P. Bois, Intersubband optical transients in quantum well structure, Appl. Phys. Lett. 2542 (1993).
12.   E. H. Nicollian and A. Gotzberger, The Si/SiO2 interface - Electrical properties as determined by the metal-insulator-silicon conductance technique, Bell Syst. Tech. J. 46:1055 (1967).
13.   W. Schockley and W. T. Read, Statistics of the recombination of holes and electrons, Phys. Rev. 87:835 (1952).
14.   P. W. M. Blom, C. Smit, J. E. M. Haverkort, and J. H. Wolter, Carrier capture into a semiconductor quantum well, Phys. Rev. 47:2072 (1993).
15.   J. A. Brum and G. Bastard, Resonant carrier capture by semiconductor quantum well, Phys. Rev. B 33:1420 (1986).
16.   B. F. Levine, Quantum Well Infrared Photodetectors (QWIPs), J. Appl. Phys. (to be published) (1993).
17.   M. A. Kinch and A. Yariv, Performance limitations of GaAs/AlGaAs infrared superlattices, Appl. Phys. Lett. 55:2093 (1989).
18.   V. D. Shadrin and F. L. Serzhenko, The theory of multiquantum well GaAs/AlGaAS infrared detectors, Infrared Phys. 33:345 (1992).
19.   A. S. Grove, "Physics and Technology of Semiconductor Devices," John Wiley and Sons, New York (1967).

PHYSICS OF SINGLE QUANTUM WELL INFRARED PHOTODETECTORS

K. M. S. V. Bandara*, B. F. Levine, G. Sarusi, R. E. Leibenguth,
M. T. Asom and J. M. Kuo
AT&T Bell Laboratories
600 Mountain Avenue
Murray Hill, New Jersey 07974

I. Introduction

Although there has been extensive research[1-8] on multi-quantum well infrared photodetectors (QWIPs), which typically contain many (~50) periods, there has been relatively limited experimental work[9-11] (using a fixed frequency CO_2 laser) done on QWIPs containing only a single quantum well. Recently, we have performed a complete series of experiments[12-14] on single quantum well structures with n-type doped well, p-type doped well and un-doped well. These doped-single well detectors are, in fact, particularly interesting since they have exceptionally high optical gain compared to multi well detectors. The optical response of the undoped single quantum well detector is strongly dependent on the bias voltage because the well is filled by tunneling through a thin emitter barrier. In addition, their simple band structures allow accurate calculations of the bias voltage dependence of the potential profiles of each of the two barriers, band bending effects in the contacts, as well as charge accumulation (or depletion) in the quantum well. Therefore by comparing theory with experiment, one can achieve a better overall understanding of the optical and transport physics in these quantum well infrared photodetectors.

II. Directly Doped Well n-type

We have grown a single QWIP via molecular beam epitaxy consisting of a $L_w = 40$ Å GaAs quantum well (doped $N_D = 1 \times 10^{18}$ cm^{-3} with Si) surrounded by two $L_e = L_c = 500$ Å undoped $Al_{0.27}Ga_{0.73}$ As barriers, and sandwiched between a 0.5 μm thick top and 1 μm thick bottom doped ($N_D = 1 \times 10^{18}$ cm^{-3}) GaAs contact layer. Sample was processed into 200 μm diameter mesas (area $A = 3.1 \times 10^{-4}$ cm^2) using wet chemical etching. Under an applied bias voltage of V_b across the entire structure (in Fig. 1) a current $I_e(V_e)$ will be injected, via field emission, from the emitter contact by the voltage drop V_e across the emitter barrier. In addition, a current $I_w(V_c)$ will also be generated from the quantum well composed of both a field emission component $I_w^{fe}(V_c)$ and an optically excited component $I_w^{opt}(V_c)$ by the voltage drop V_c across the collector barrier. Following Rosencher et al.,[11] these currents and voltages can be related using

* Present Address: Department of Physics, University of Peradeniya, Peradeniya, Sri Lanka.

H. C. Liu et al. (eds.), Quantum Well Intersubband Transition Physics and Devices, 111–122.

112

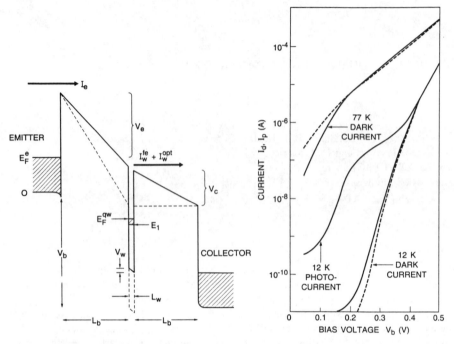

Fig. 1. Schematic conduction band diagram of a single quantum well infrared photodetector. The solid line is for T = 12 K in the dark; the dashed line is under illumination.

Fig. 2. Dark current I_d, and photocurrent I_p vs. bias voltage V_b at T = 12 K and 77 K. The solid curves are experiment; the dashed lines are theory.

$$I_e(V_e) = I_w^{fe}(V_c) + I_w^{opt}(V_c) \tag{1}$$

$$V_b = V_e + V_c + V_w \tag{2}$$

$$e(N_D - n) L_w A = C_e V_e - C_c V_c = 2C_w V_w \tag{3}$$

n is the 3-dimensional free carrier density in the well, V_w is the voltage drop across the well, C is the capacitance of the emitter (C_e) and collector (C_c) given by $C_e = C_c = \varepsilon A/L_b$, and C_w is the capacitance of the quantum well $C_w = \varepsilon A/L_w$. The expressions for the 3-dimensional emitter current I_e, and 2-dimensional quantum well field emission current I_w^{fe}, are given by[1,11-13]

$$I_e(V_e) = \left[\frac{e m^* kTA}{2\pi^2 \hbar^3} \right] \int_0^\infty T_e(E, V_e) \ln \left[1 + e^{(E_F^e - E)/kT} \right] dE \tag{4}$$

and

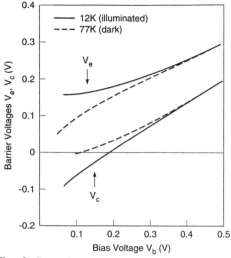

Fig. 3. Potential drops across the emitter V_e and collector V_c barriers, as a function of total applied bias V_b, at T = 12K (illuminated) solid lines, and T = 77K (dark), dashed lines.

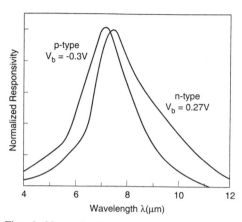

Fig. 4. Normalized responsivity spectrum for n-type and p-type single QWIP.

$$I_w^{fe}(V_c) = \left[\frac{em^* A}{\pi \hbar^2 \ell} \right] v(V_c) \int_{E_1}^{\infty} T_c(E, V_c) \left[1 + e^{(E-E_1 - E_F^{qw})/kT} \right]^{-1} dE \qquad (5)$$

where $\ell = (L_b + L_w)$ and E_F^e and E_F^{qw} are the Fermi energies in the emitter (relative to the bottom of the conduction band) and quantum well (relative to the ground state energy E_1) respectively. The energy dependent transmission coefficients of the emitter and collector barriers are $T_e(E, V_e)$ and $T_c(E, V_c)$ which were calculated numerically by matching boundary conditions (across the entire structure for T_e, and across the collector barrier for T_c using matrix techniques). The collector velocity is given by[1]

$$v(V_c) = \mu F_c / \sqrt{1 + (\mu F_c / v_s)^2} \qquad (6)$$

At zero bias these equations require the equality of the Fermi levels $E_F^e = E_1 + E_F^{qw}$ and thus determine the $V_b = 0$ band alignment, which has $V_e > 0$ and $V_c < 0$. As V_b is increased, the potential drop on both barriers increases. Using Eqs. (1)-(6) the potentials V_e and V_c, as well as the currents can be calculated. These theoretical dark currents at T = 12K and 77K (dashed curves) are compared with experiment (solid curves) in Fig. 2, and the (dark) potential drops V_e and V_c at T = 77K are shown in Fig. 3 (dashed). The excellent agreement in the dark currents (over several orders of magnitude) are obtained by fitting μ and v_s in Eq. (6). The resulting values of mobility and saturation velocity $\mu = 3000$ cm^2/Vs, $v_s = 1 \times 10^7$ cm/s at

$T = 12K$ and $\mu = 2000 \text{ cm}^2/\text{Vs}$, $v_s = 5 \times 10^6$ cm/s at $T = 77K$ are similar to those found previously[1] in 50 period *bound-to-continuum* QWIPs. Also shown in Fig. 2 is the large increase in current with the $T=300K$ background illumination through the dewar window (with the QWIP at $T = 12K$). By using this experimental photocurrent I_w^{opt} in Eq. (1), the $T = 12K$ illuminated values for V_e and V_c shown in Fig. 3 have been determined.

The responsivity spectrum has been measured using a $45°$ facet polished on the substrate and is shown in Fig. 4 for $T = 12K$ and a bias of $V_b = 0.35$ V. The peak at $\lambda_p = 7.4$ μm is in agreement with that expected for a $L_w = 40 \text{Å}$ quantum well with $Al_{0.27}Ga_{0.73}As$ barriers. The wide spectral width $\Delta\lambda/\lambda = 36\%$ and long cutoff wavelength $\lambda_c = 9.4$ μm are also consistent with the *bound-to-continuum* intersubband transition in this structure.[15] This spectral shape as well as the peak responsivity $R_p^o = 0.325$ A/W (corrected for the $\phi = 45°$ angle of incidence) are similar to previous[15] 50 period QWIPs. The responsivity can in turn be expressed as[15]

$$R(V_b) = (e/h\nu)\eta_a(V_b)p_e(V_c)g(V_b) \qquad (7)$$

where ν is the optical frequency, η_a is the optical absorption quantum efficiency, p_e is the escape probability of the photoexcited electrons out of the well, and g is the optical gain. Here, $R \propto \eta g$ and the increase of the quantum efficiency η with number of quantum wells N ($\eta \propto N$), together with the decrease of the gain g ($g \propto N^{-1}$) will explain values of the responsivity similar to 50 period qwips.

The spectral shape of this responsivity spectrum was essentially unchanged for different biases, but the peak value R_p^o (measured at $\lambda = \lambda_p$) has the unusual behavior shown in Fig. 5. Note the large voltage offset of $V_b \approx 0.1$ V before the photoresponse increases, the maximum in R_p as a function of bias, and the larger response at lower temperature (with maximum values of $R_p^o = 0.325$ A/W at $T = 12K$ and 0.15 A/W at $T = 77K$). All of these features are significantly different than those for the usual 50 period *bound-to-continuum* QWIPs, for which R_p increases nearly linearly with V_b from $V_b = 0$, and then saturates with V_b becoming essentially constant at high bias. Furthermore, the 50 period QWIP responsivity is only weakly dependent on temperature in this range of $T = 12K - 77K$.

In order to understand this unusual behavior we also measured the current noise i_n and experimentally determined the optical gain[15] at $T = 77K$ from the dark current and at $T = 12K$ from the photocurrent produced from a blackbody source focussed through the dewar window; (the dark current noise was too small to measure directly at $T = 12K$). Note the striking maxima in Fig. 6, for the gain vs. bias curves at both temperatures. The initial increase of g is due to the usual increase in velocity with field, i.e. $g(V_b) = v(V_b)\tau_L/\ell \approx \mu\tau_L V_b/\ell^2$, where τ_L is the hot electron lifetime. However, at high bias where the gain for the usual 50 period QWIP saturates,[15] g decreases strongly for this single quantum well detector. This is a result of high field tunneling of electrons from the emitter contact directly into the quantum well and thus "short circuiting" the hot electron transport process. This tunneling process is more important in these single QWIPs than in the multiquantum well detectors, since most of the bias voltage drop occurs across the emitter tunnel barrier (i.e. $V_e > V_c$ as shown in Fig. 3).

In order to further understand this device, we have plotted in Fig. 7 the 3-dimensional free electron density, n, in the quantum well calculated from Eqs. (1)-(6), at $T = 12K$ (for both illuminated and dark conditions) and at $T = 77K$ (in the dark). Note that in the dark the free carrier density decreases with increasing bias. This can be understand from the following considerations. From Eqs. (4) and (5) we can see since $E_F^{qw} + E_1 > E_F^e$, that for the *same* voltage drop across the barriers (i.e. $V_e = V_c = V$), $I_w^{fe}(V) \geq I_e(V)$. Thus, as discussed by

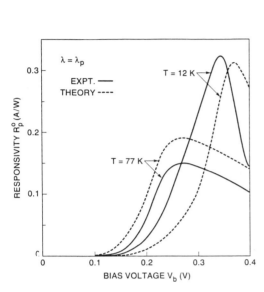

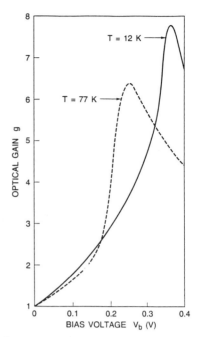

Fig. 5. Experimental and theoretical (dashed line) responsivity, at the peak wavelength λ_p, as a function of bias V_b for $T = 12\,K$ and $77\,K$ for n-type doped well.

Fig. 6. Experimental optical gain g as a function of bias V_b for $T = 12\,K$ and $77\,K$.

Rosencher *et al.*[11] the emitter contact will *not* be able to supply a sufficient current to replenish that generated from the quantum well. Therefore to satisfy the dark current requirement $I_w^{fe}(V_c) = I_e(V_e)$, necessitates $V_e > V_c$, as is in fact clear from Fig. 3. In order to achieve this voltage inequality, Eq. (4) shows that $(N_D - n)$ must increase and hence n decreases. A similar *photoinduced* drop in n also occurs at $T = 12\,K$. That is, in order to extract the large required field emission current out of the emitter to balance the large photocurrent generated by the quantum well, V_e must increase, and hence according to Eq. (3), n dramatically decreases.[11] (The strong potential shift in the barrier under illumination is indicated in Fig. 1). This photoinduced carrier depletion in the well also explains the *rise* in the 12K photocurrent free carrier density curve in Fig. 7. As V_b increases the photocurrent becomes a smaller and smaller fraction of the total current and thus the photoinduced depletion becomes less and less, eventually becoming negligible for $V_b > 0.4\,V$ where the dark and photocurrent curves become essentially identical.

We can now combine all of these effects together (e.g. the potential drops V_e, V_c, the 3-dimensional carrier density depletion the well n, and the optical gain) and calculate the responsivity as a function of V_b from Eq. (7). In order to do this we need $\eta_a(V_b)$ and $p_e(V_c)$. The escape probability p_e has been shown to be related to the ratio of the time, τ_e, for the photoexcited electron to escape from the well, and the recombination time, τ_r, for recapture into the well. This relation is[15]

$$p_e(V_c) = [1 + \tau_e(V_c)/\tau_r(V_c)]^{-1} \tag{8}$$

where

$$\tau_e(V_c)/\tau_r(V_c) = (\tau_e/\tau_r)_0 e^{-V_c/V_p} . \tag{9}$$

For *bound-to-continuum* QWIPs, the zero bias ratio has been shown to be[12] typically $(\tau_e/\tau_r)_0 \simeq 2$, with an exponential barrier lowering potential of typically $V_p \sim 30$ meV. Finally, we now relate the absorption quantum efficiency $\eta_a = \alpha\ell$ (which is proportional to the 2-dimensional free carrier density ρ in the well) using

$$\eta_a = 2\% (\rho/10^{12}) \tag{10}$$

where the 3-dimensional electron density n is related to ρ by $\rho = n \, L_w$. Equation (10) for η_a in n-doped single well QWIPs and was obtained by scaling the measured 50 period multiquantum well absorption by the number of wells and also including the factor of 2 enhancement in the absorption due to the single quantum well being at a maximum in the optical intensity (i.e. a quarter wavelength $\lambda'/4$ from the metallic reflecting contact).

Combining the above results together with Fig. 6, we have calculated the bias dependence of the responsivity as shown in Fig. 5 (dashed line). Note that all of the features of Fig. 5 are now explained. The bias offset of $V_b = 0.1$ V is a result of $V_c < 0$ for low V_b and hence from Eqs. (8) and (9) $p_e \sim 0$ for $V_b \leq 0.1$ V. The increase in the maximum responsivity at $T = 12$K over that for 77K is due to several factors: the increase in the gain g, the larger electron density n in the well (at $V_b > 0.4$ V and $T = 12$K) resulting in a larger η_a, and the increase in the escape probability p_e at higher bias. Finally, the broadening of the $T = 77$K responsivity vs. bias curve can be understood as due to the increase of n with V_b (Fig. 7). The reason that 50 quantum well QWIPs do not show these interesting effects is that except for a few quantum wells near the contacts, their periodic and symmetrical structure ensures that the input and output barriers have identical potential drops (i.e. $V_e = V_c$ in the present notation) and hence from Eq. (3) that $n = N_D$.

III. Directly doped well p-type

The structure, schematically shown in Fig. 8(a) was grown on a semi-insulating GaAs substrate via gas source molecular beam epitaxy, and consisted of a single p-doped ($p = 2 \times 10^{18}$ cm^{-3} with Be) $L_w = 40$ Å quantum well surrounded by two $L_b = 300$ Å undoped $Al_{0.3}Ga_{0.7}As$ barriers. The GaAs contact layers (top 1/2 μm and bottom 1μm) were also p-doped to the same carrier density.

The responsivity was measured at *normal incidence*[8] since the strong mixing between the light and heavy holes at $k \neq 0$ allows this advantageous geometry. The resulting spectrum is shown in Fig. 4 for a bias of $V_b = -0.4$V (i.e. mesa top negative), and was found to be independent of temperature from $T = 10$K to 80K to within experimental error (< 10%). The peak position $\lambda_p = 7.2$ μm as well as the absolute magnitude of the responsivity are, as expected, quite similar to those of a previously discussed[8] 50 period p-QWIP having the same values of $L_w = 40$ Å, $L_b = 300$ Å and $Al_{0.3}Ga_{0.7}As$ barrier composition. The bias dependence of the responsivity is shown in Fig. 9 as the solid curve together with that of the previously measured single n-QWIP (shown as the dashed curve) for comparison. Note that the p-QWIP

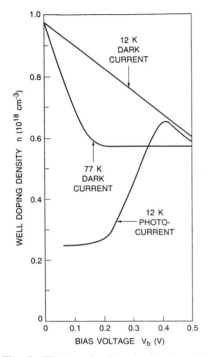

Fig. 7. Electron density n in the quantum well as a function of bias V_b for $T = 12\,K$ (both illuminated and dark conditions)

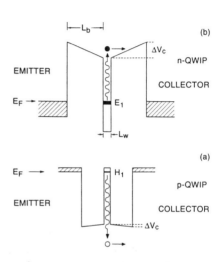

Fig. 8 Schematic band diagram of a (a) single p-doped QWIP (b) a single n-doped QWIP. The ground state hole and electron levels are indicated as H_1 and E_1.

responsivity increases approximately linearly with bias at low voltage demonstrating the *bound-to-continuum*[15] nature of the intersubband transition, whereas the n-QWIP requires a large bias offset (i.e. the responsivity is essentially zero for $V_b < 0.1\,V$). In addition, for the p-QWIP there is no saturation of the responsivity with increasing bias even for very large voltages of $V_b = -0.5\,V$ (i.e. a voltage drop per barrier of $-0.25\,V$ corresponding to 12.5V for a 50 period QWIP). This is also strikingly different from that of the n-QWIP.

The responsivity bias offset for p-single QWIPs is less than n-single QWIPs (see Fig. 5 and 9) because as indicated in Fig. 8, the n-QWIP has a much stronger band bending of the barriers than the p-QWIP. This is a result of the larger hole effective mass m_h^* relative to the electron mass m_e^* and the consequent small ground state hole energy $H_1 = 25$ meV [in Fig. 8 (a)] with respect to the ground state electron energy $E_1 = 88$ meV [in Fig. 8(b)]. That is, the alignment of the Fermi level in the doped quantum well with the Fermi levels in the emitter and collector contact layers, requires a large drop in the E_1 electron level (and thus a substantial barrier band bending) whereas for H_1 this adjustment is much less. The barrier band bending for the single n-QWIP is $\Delta V_c = 50$ meV whereas it is only $\Delta V_c = 15$ meV for the p-QWIP discussed above. Furthermore (due to the $k \neq 0$ valence band hole intersubband selection rules) the responsivity peak at $\lambda_p = 7.2\,\mu m = 172$ meV is sufficiently high in the *continuum* ($\Delta E = 22$ meV) *above* the $V_b = 150$ meV barrier height that $\Delta E > \Delta V_c$. In strong contrast, the n-QWIP band bending

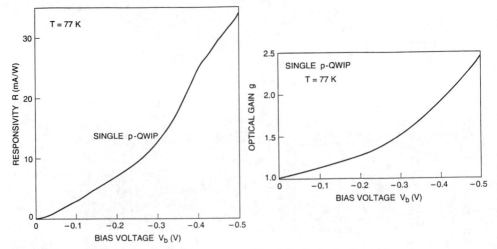

Fig. 9. Responsivity vs bias voltage measured at peak wavelength ($\lambda_p = 7.2$ μm) and T = 77 K for p-type single QWIP.

Fig. 10. Optical gain vs bias voltage measured at T = 77 K for p-type single QWIP.

of $\Delta V_c = 50$ meV is so large that the photoexcited electron ($\Delta E = 13$ meV) remains *bound* (i.e. $\Delta E < \Delta V_c$) and thus cannot escape without a large applied bias voltage to eliminate ΔV_c. That is, for the p-QWIP p_e is large even at low bias and thus remains essentially constant as the bias is increased, where for the n-QWIP p_e is very small at low V_b and thus requires a substantial bias to overcome the band bending ΔV_c and allow the photoexcited electrons to escape.

The other important difference shown in Fig. 9 is the continuing increase of the p-QWIP responsivity with bias, while the n-QWIP reaches a maximum after which it strongly decreases with increasing bias. This n-QWIP maximum was found to be due to a maximum in the gain, as shown in the dotted curve in Fig. 7. Thus, we measured the current noise of this single p-QWIP at T = 77 K and obtained the gain shown in Fig. 10. Note that the gain increases monotonically with increasing bias, thereby explaining the similar increase of the responsivity with bias in Fig. 10. That is, the responsivity and gain of this single well p-QWIP both behave like the 50 period p-QWIP[8] (i.e. *not* like the single well n-QWIP). This difference is again a result of the large effective hole mass m_h^*. For both n- and p-QWIPs the initial increase in g is due to the decrease in the photoexcited carrier transit time across the quantum well. However, at high bias the n-QWIP gain decreases due to direct tunneling of the low m_e^* electrons from the emitter contact which effectively "short circuits" the hot carrier transport process. For the high mass holes this direct tunneling process is inhibited and thus the gain continues to increase as shown in Fig. 10.

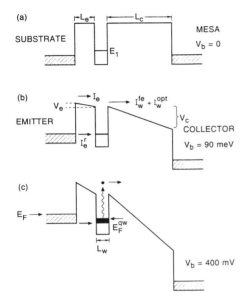

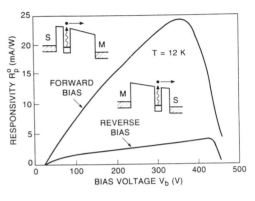

Fig. 11. Schematic conduction band diagram of the undoped single well QWIP.
(a) $V_b = 0$, (b) $V_b = 90$ meV corresponding to resonance between the Fermi energy and the bound state energy E_1, (c) $V_b = 400$ mV corresponding to resonance between the bottom of the conduction band and E_1.

Fig. 12. Measured responsivity R_p^o vs. bias voltage V_b at T = 12K for both forward ($V_b > 0$) and reverse ($V_b < 0$) bias. The inserts show the conduction band structure under bias where "S" is the substrate side and "M" is the mesa top.

IV. Undoped Well

The structure shown schematically in Fig. 11a, consists of a 1 μm GaAs contact layer (doped $n = 1.3 \times 10^{18}$ cm^{-3}) grown on a semi-insulating substrate followed by a $L_e = 155$ Å $Al_{0.27}Ga_{0.73}As$ emitter barrier, a $L_w = 42$ Å undoped GaAs quantum well, a $L_c = 500$ Å $Al_{0.27}Ga_{0.73}As$ collector barrier and a top 0.5 μm $n = 1.3 \times 10^{18}$ cm^{-3} GaAs contact layer.

The bias dependence of the responsivity for forward and reverse bias is shown in Fig. 12. As expected,[9] the photoresponse for forward bias (where the electrons from the emitter contact can readily tunnel through the thin L_c barrier and populate the quantum well) is nearly an order of magnitude larger than for reverse bias (where the injected electrons must tunnel through the thick L_c barrier and thus the electron density in the well is low). Two features of this responsivity vs. bias curve are noteworthy, namely the finite bias voltage required to get a nonzero responsivity (i.e. the responsivity does not start to increase from $V_b = 0$ as expected for a *bound-to-continuum* QWIP) and the sharp drop at high bias.[9] In order to understand this behavior we have done a self consistent theoretical calculation[13] of the voltage drops across the individual barriers, the electron density in the well populated by emitter resonant tunneling through the thin barrier, and the resulting dark current. The equations relating these quantities are similar to Eqs. (1)-(3) except that both resonant and non-resonant tunneling contributions to

120

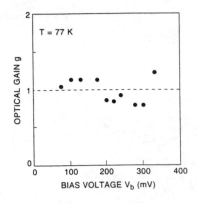

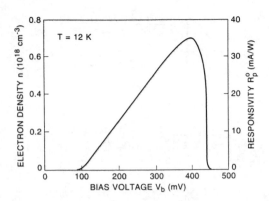

Fig. 13. Measured optical gain g at
T = 77K as a function of bias voltage V_b.

Fig. 14. Theoretical responsivity and
electron density n in the quantum well at
T = 12 K as a function of bias voltage V_b.

emitter current must be included.[13] As indicated in Fig. 11c, the origin of this striking drop in n is that at this bias the quantum well level E_1 drops below the conduction band edge of the emitter and thus the emitter resonant tunneling contribution is suppressed. For higher biases the nonresonant tunneling contribution becomes dominant.

For the n-doped single QWIP gain is so large (maximum g = 7.4) since g = τ_L/τ_T where τ_L is the photoconductive carrier lifetime, and τ_T is the transit time across the structure (i.e. $\tau_T = \ell/v$). Thus, for this single well QWIP ℓ and τ_T are more than an order of magnitude smaller, and thus g more than an order of magnitude larger, than for the usual 50 period n-doped multiple well QWIPs. However, this scaling[19,20] g $\propto \ell^{-1}$ in well doped QWIPs results from the lifetime τ_L being independent of the transit time τ_T. This is not true for the undoped single QWIP for which the quantum well is filled by tunneling through the emitter barrier. That is, after the photoexcited electron escapes from the well it is immediately refilled from the emitter contact and thus $\tau_L = \tau_T$, and thus g = 1. That is, instead of making many passes through the single quantum well before being recaptured (as for the doped QWIPs), the photoelectron is recaptured by the well in only a single pass. To check this expectation that g = 1, we measured the current noise i_n on a spectrum analyzer and determined the gain from[21] $i_n = \sqrt{2eI_d g\Delta f}$ where Δf is the noise bandwidth. Note that a *factor of 2* in the noise expression rather than the usual *factor of 4*. This is because for the usual *photoconductor* $\tau_L \neq \tau_T$ and the hot carrier "recirculates" through the quantum well. Thus, in addition to the noise due to the *photogeneration* processes there is also the noise due to the *recombination* (i.e. recapture) process. However, when the well is immediately refilled by emitter tunneling, there is no uncertainty in the recapture process and no excess noise is produced. This is analogous to p-n junction photodetectors, where because of the non-ohmic contacts the photocarriers are not reinjected. The results of our noise measurements taken using photocurrent at T = 77K are shown in Fig. 13. Note that as expected g = 1 to within experimental error.

The responsivity can now be calculated from Eqs. (7)-(10) using the calculated carrier density n together with $p_e = 1$ and g = 1. Since the field across the collector is so large (e.g. 6×10^4 V/cm at V_c = 300 mv) that to an excellent approximation we can take $p_e \approx 1$ (at

$V_c \simeq 200$ mv in Eq. 8 $p_e > 0.99$). The result is shown on the right hand scale in Fig. 14 since $R(V_b) \propto n(V_b)$. This theoretical responsivity can now be compared with the experimental responsivity shown in Fig. 12. Although the measured bias dependent responsivity is somewhat broader than that calculated, the over all shape and absolute magnitude are in excellent agreement. Note that the responsivity does not increase immediately at low bias but requires a significant bias offset. This is because the quantum well is essentially empty at low bias since $E_1 > E_F^e$, thus requiring a substantial V_b to lower this ground level and allow filling of the well, [Fig. 11(b)]. The other feature of the data that is reproduced in the theory is the drop in responsivity for $V_b > 400$ mV, as a result of E_1 dropping below the conduction band edge at high bias and thus cutting off the resonant tunneling [Fig. 11(c)].

V. Conclusions

In conclusion we have presented a detailed series of measurements on the optical, electrical and transport properties of single quantum well infrared photodetectors involving n-type, p-type and undoped well. The different bias voltage behavior of n-type single QWIP can be understood by calculating the voltage drops as the individual barriers and the photoinduced carrier depletion in the well as a function of applied bias voltage. The bias behavior of the responsivity, optical gain, and quantum well escape probability of p-type single QWIPs were found to be different from that of n-type single QWIPs, as a result of the large effective hole mass relative to the electron mass. The optical and transport properties of undoped single-well QWIP has several noval features such as the filling of the well via resonant tunneling through the emitter barrier, a unity optical gain and a dramatic drop in the carrier density inside the well at high bias when the well ground state drops below the emitter conduction band edge.

Acknowledgment

The work done at AT&T Bell Laboratories was partially supported by NASA/jet Propulsion Laboratory.

REFERENCES

1. B. F. Levine, C. G. Bethea, G. Hasnain, V. O. Shen, E. Pelve, R. R. Abbott and S. J. Hseih, Appl. Phys. Lett. *56*, 851 (1990).

2. B. F. Levine, Proc. NATO Advanced Research Workshop on ''Intersubband Transitions in Quantum Wells'', Cargèse, France Sept. 9-14 1991, (Plenum, New York 1992) edited by E. Rosencher, B. Vinter, B. F. Levine, p. 43.

3. J. Y. Andersson and L. Lundqvist, Appl. Phys. Lett. *59*, 857 (1991).

4. B. K. Janousek, M. J. Daugherty, M. L. Bloss, M. L. Rosenbluth, M. J. O'Loughlin, H. Kanter, F. J. De Luccia and L. E. Perry, J. Appl. Phys. *67*, 7608 (1990).

5. S. R. Andrews and B. A. Miller, J. Appl. Phys. *70*, 993 (1991).

6. B. F. Levine, C. G. Bethea, K. G. Glogovsky, J. W. Stayt and R. E. Leibenguth, Semicond. Sci. Technol. *6*, C114 (1991).

7. L. J. Kozlowski, G. M. Williams, G. J. Sullivan, C. W. Farley, R. J. Anderson, J. K. Chen, D. T. Cheung, W. E. Tennant and R. E. DeWames, IEEE Trans Electron. Devices *38*, 1124 (1991).

8. B. F. Levine, S. D. Gunapala, J. M. Kuo, S. S. Pei and S. Hui, Appl. Phys. Lett. *59*, 1864 (1991).

9. H. C. Liu, G. C. Aers, M. Buchanan, Z. R. Wasilewski and D. Landheer, J. Appl. Phys. *70*, 935 (1991).

10. H. C. Liu, M. Buchanan, G. C. Aevs and Z. R. Wasilewski, Semicond. Science Technol. *6*, C124 (1991).

11. E. Rosencher, F. Luc, Ph Bois and S. Delaitre, Appl. Phys. Lett. *61*, 468 (1992).

12. K. M. S. V. Bandara, B. F. Levine, R. E. Leibenguth and M. T. Asom, J. Appl. Phys. *74*, 1826 (1993).

13. K. M. S. V. Bandara, B. F. Levine and M. T. Asom, J. Appl. Phys. *74*, 346 (1993).

14. K. M. S. V. Bandara, B. F. Levine and J. M. Kuo, Phys. Rev. B*48*, 7999 (1993).

15. B. F. Levine, A. Zussman, S. D. Gunapala, M. T. Asom, J. M. Kuo and W. S. Hobson, J. Appl. Phys. *72*, 4429 (1992).

16. L. Pfeiffer, E. F. Schubert and K. W. West, Appl. Phys. Lett. *58*, 2258 (1991).

17. H. C. Liu and G. C. Aers, J. Appl. Phys. *65*, 4908, (1989).

18. M. C. Payne, J. Phys. C*19*, 1145 (1986).

19. G. Hasnain, B. F. Levine, S. Gunapala and N. Chand, Appl. Phys. Lett. *57*, 608 (1990).

20. A. G. Steele, H. C. Liu, M. Buchanan and Z. R. Wasilewski, J. Appl. Phys. *72* 1062 (1992).

21. R. H. Kingston, "Detection of Optical and Infrared Radiation", Springer Verlag (Berlin) 1978.

A three-color voltage tunable quantum well intersubband photodetector for long wavelength infrared

H. C. Liu,   Jianmeng Li,   Z. R. Wasilewski, and M. Buchanan
Institute for Microstructural Sciences
National Research Council Canada
Ottawa
Ontario K1A 0R6
Canada

and

P. H. Wilson,   M. Lamm, and J. G. Simmons
Centre for Electrophotonic Materials and Devices
McMaster University
Hamilton
Ontario L8S 4M1
Canada

ABSTRACT. A three-color voltage tunable quantum well long wavelength infrared photodetector based on GaAs - AlGaAs is realized. A detailed study of the detector characteristics is carried out, including dark current, responsivity, and detectivity. An equivalent circuit model is used in understanding the observed behavior. The basic principle of operation can be used to produce detectors with many response colors.

1. Introduction

Quantum well infrared photodetectors (QWIPs)(Levine 1993) have recently attracted a great deal of attention. Although the absolute performance of single element detector is somewhat inferior than the standard HgCdTe technology for the long wavelength infrared (IR), advantages of the QWIP approach based on GaAs and Si multiple quantum wells (MQWs) are obvious, including mature materials and processing technologies, and extremely uniform and low defect density samples suitable for large array fabrication. Moreover we believe that with the advances in our understanding of the physics involved, the QWIP's performance related, e.g, to the dark current(Liu et al. 1993d) and operating temperature(Liu 1992) will be improved. Another distinctive feature of the QWIP approach is the capability of producing voltage

H. C. Liu et al. (eds.), Quantum Well Intersubband Transition Physics and Devices, 123–133.
© *1994 National Research Council of Canada. Printed in the Netherlands.*

124

tunable multicolor detectors.(Gravé et al. 1992; Tsai et al. 1993; Köck et al. 1992; Choi et al. 1989; Martinet et al. 1992; Liu et al. 1993a) In this paper, we present a detailed study of our multicolor voltage tunable QWIP using a three-color version. The idea and a preliminary experimental demonstration has been given in Ref. (Liu et al. 1993a). Our multicolor QWIP is achieved by simply stacking the usual one-color QWIPs having different detection wavelengths with heavily doped conducting layer between them which does not need to be thick as it is not used for making an electrical contact.

Our approach is similar to that employed by Gravé et al.(Gravé et al. 1992) and by Tsai et al.(Tsai et al. 1993) but with an important difference. Their devices did not have conducting layers separating the different one-color QWIPs, and consequently, the division of the applied voltage among the one-color QWIPs is similar to the case of high-low field domain formations as discussed in Ref. (Choi et al. 1987; Grahn et al. 1991; Helgesen et al. 1991; Liu et al. 1993c). A quantitative model for the high-low field formation problem does not yet exist;(Liu et al. 1993c) whereas for our device the voltage division is trivially predicted, as our device is simply three one-color QWIPs in series. The voltage tunability in our device is achieved by making use of the dark current characteristics(Liu et al. 1993d) of QWIPs with different peak detection wavelengths. The shorter the detection wavelength the higher the device dc resistance; and the dependence of the dc resistance on the detection photon energy is exponential.(Liu et al. 1993d) When a voltage is applied to QWIPs in series, the voltage division between different QWIPs will be determined by their dc resistances, i.e., the one with the shortest detection wavelength will have the largest fraction of the bias dropped across it. In a multicolor QWIP, the different colors turn on sequentially with bias, and in a particular voltage range, one of one-color QWIPs dominates the spectral response because of device differential resistance characteristics (as seen later).

2. Sample

The sample structure for our three-color QWIP grown by molecular beam epitaxy is shown in Fig. 1. The layer thicknesses and the alloy fractions were reconfirmed by X-ray measurement of an extra AlAs - GaAs superlattice grown just before all device structures. The simple 45°-facet geometry for optical measurement is also shown in Fig. 1. Two terminal mesa devices were processed using standard wet chemical etching and ohmic contacts were formed with alloyed NiGeAu. We used a detector with an active area of $120 \times 120 \ \mu m^2$. We use the labels MQW1, MQW2, and MQW3 to indicate the three different MQW structures in order of increasing detection wavelength.

125

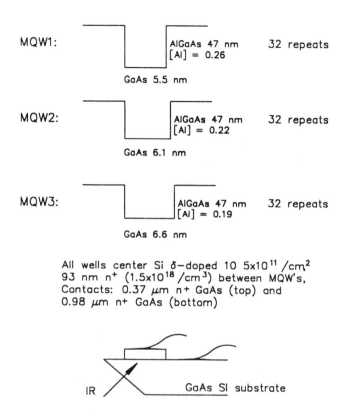

Fig. 1. Sample parameters of the three-color detector.

3. Detector characteristics

Detector dark current-voltage (I-V) characteristics at 77 and 82 K are shown in Fig. 2. The voltage polarity is referenced to the bottom contact. Temperature calibration is important, as the difference of 5 K results in a factor of more than two in dark current over the entire range of voltages. The 77 K data were taken with the device immersed in liquid nitrogen, while 82 K was the actual temperature of the device mounted in the optical dewar.

To show the voltage tunability, spectral response curves at selected voltages are shown with an arbitrary scale in Fig. 3. The spectra were taken using a Bomem MB series Fourier transform IR spectrometer which uses a glowbar source and a KBr beam splitter. The detector temperature was about 82 K.

We have also measured the detector responsivity vs. both voltage and

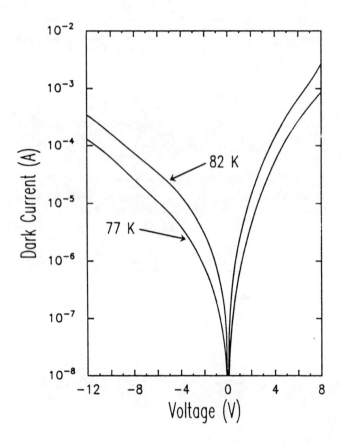

Fig. 2. Measured detector dark current-voltage characteristics at 77 and 82 K.

photon energy. For this measurement, we used a calibrated 1000 K black-body source, a variable narrow band infrared filter, an optical chopper, and a lock-in amplifier. We have corrected for reflections in the optical path between the blackbody source and the detector mesa (i.e., reflections at the dewar window and GaAs-vacuum interface), and corrected for the effective area by $1/\sqrt{2}$ due to the 45° tilt angle with respect to the incident IR beam. The optical dewar window was ZnSe anti-reflection coated for the 9–11 μm band, and it consequently had a substantial reflection at shorter wavelengths, in the 5–7 μm region. This modifies slightly the raw spectra shown in Fig. 3. The unpolarized responsivity vs. voltage at 8.5 μm and a contour plot of responsivity vs. both voltage and wavelength are shown in Fig. 4. Because the responsivity in some voltage regions involves an interplay of differential resistances of more than one of the QWIP and is therefore somewhat

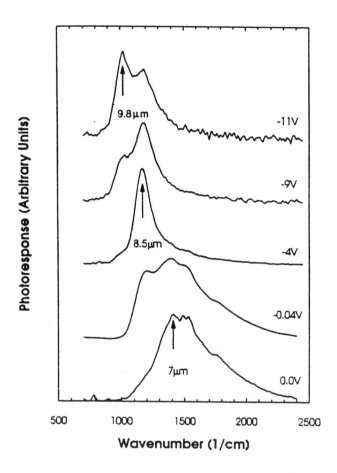

Fig. 3. Raw spectral response curves at 82 K. The bias polarity is defined by choosing the bottom mesa contact as the ground.

complicated, we give the precise definition used here to obtain the results in Fig. 4: the responsivity is defined by the voltage change detected on the lock-in amplifier divided by the resistance resulting from the parallel of the series load resistor and the device differential resistance, for unity incident monochromatic IR power.

There are two ways to evaluate the detectivity $D^\star$: (a) using measured absorption quantum efficiency together with measured responsivity and dark current, and (b) using measured noise together with the measured responsivity. Since measurements using the two methods yield identical results at larger voltages,(Levine et al. 1992) we present the data using scheme (a). At low bias voltages, when the excited electron escape probability is substantially less than unity, this scheme overestimates the $D^\star$. (Liu 1992; Levine et al. 1992)

128

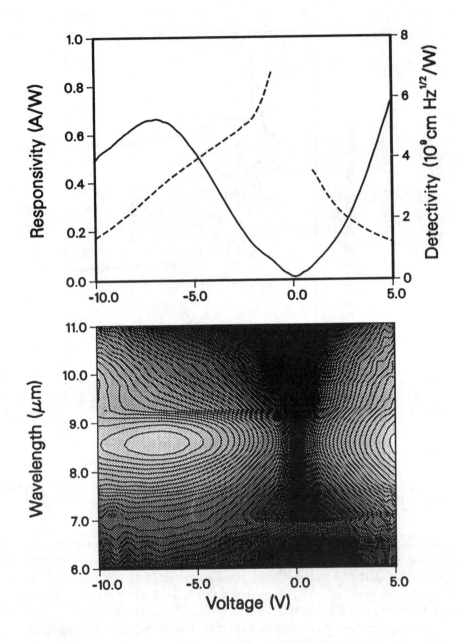

Fig. 4. Top: responsivity (solid) and detectivity (dashed) vs. voltage at 8.5 μm; bottom: contour plot of responsivity at all voltages and wavelengths, both at 82 K.

The absorption was measured at room temperature using P-polarized light and Brewster angle (72° for GaAs) incidence, and then was converted (Steele et al. 1992) for our detector geometry (82 K, 45° incidence, double-path, and unpolarized). This experiment was performed on a large (6 × 7 mm^2) sample, backside polished and unprocessed. The photoconductive gain g was then calculated from $\mathcal{R} = eg\eta/h\nu$, where $\mathcal{R}$ is the responsivity, η is the absorption efficiency, and ν is the photon frequency. Finally, the detectivity $D^\star$ is given by(Liu 1992) $D^\star = \mathcal{R}\sqrt{A}/\sqrt{4egI_D}$, where A is the device area and I_D is the measured dark current. Note that the $D^\star$ here is dark current limited. The result at 8.5 μm is shown in the top part of Fig. 4.

4. Discussions

Three clearly resolved spectral peaks at about 7, 8.5, and 9.8 μm are displayed in the spectra given in Fig. 3. These response curves show several interesting features. First, the shortest wavelength detector is turned on at zero bias, due to a built-in field caused by dopant segregation (which we now can correct for.(Liu et al. 1993b) Second (and most interesting) is the relative strength of different spectral peaks as a function of voltage: at low voltages (zero to −1 V) the 7-μm response dominates; in the range from −4 to −8 V, the 8.5 μm response dominates (and is the only spectral peak visible on the linear scale); and at high voltages (from −9 V on) the response at 9.8 μm starts to dominate.

To understand this behavior, we use the equivalent circuit shown in Fig. 5, where each one-color QWIP ($i = 1$, 2, and 3) is modeled by a photocurrent source $i_{p,i}$ and a parallel differential resistance r_i. The measured photoresponse is the voltage developed across the load resistor due to the IR illumination:

$$v_{meas} = R_s \frac{i_{p1}r_1 + i_{p2}r_2 + i_{p3}r_3}{R_s + r_1 + r_2 + r_3}, \tag{1}$$

where R_s is the resistance of the series load resistor. From the above expression, the measured response is not only related to the intrinsic photocurrents (i_{p1}, i_{p2}, and i_{p3}) but is also weighted by the differential resistances (r_1, r_2, and r_3). If in a given voltage range one of the differential resistances dominates, the photoresponse will be dominated by this one-color QWIP. Specifically, if r_1 is much larger than both r_2 and r_3 at low voltages, r_2 is much larger than both r_3 and r_1 at intermediate voltages, and r_3 is much larger than both r_1 and r_3 at high voltages, one obtains the behavior shown in Fig. 3.

To test if this is true experimentally, we have measured two individual one-color QWIPs with similar device parameters to those of MQW2 and MQW3. The results for I vs. V and dV/dI vs. I at 77 K are shown in Fig. 6.

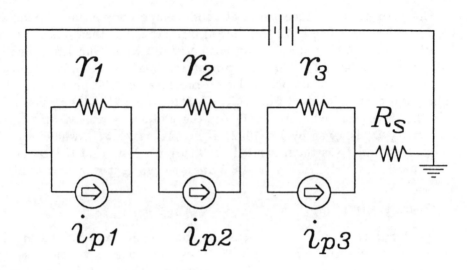

Fig. 5. Equivalent circuit model. The three one-color detectors are modeled by three photocurrent sources in parallel with three resistances which are the differential resistances of the corresponding one-color detector. The series load resistor is used for measurement.

The plot of dV/dI vs. I gives the differential resistance, an important parameter since our multicolor QWIP is made by serially connecting one-color QWIPs, and the dark current through each QWIP is constant at a given bias. Figure 6(b) (dV/dI vs. I) clearly shows that the differential resistances from the two one-color QWIPs cross each other: at low currents one is larger, and at high currents the other. To further substantiate our interpretation we have calculated I vs. V and dV/dI vs. I using the dark current model given in Ref. (Liu et al. 1993d). The calculated results are consistent with the measured results.

This feature that one of the differential resistances dominates in a particular voltage region is not only advantageous for the tuning of the response

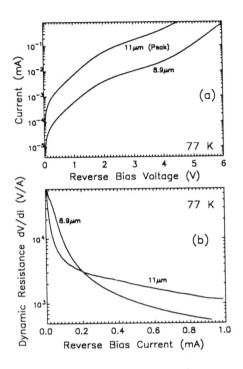

Fig. 6. (a) Current vs. voltage (I-V) and (b) dV/dI vs. current characteristics of two individual one-color detectors at 77 K.

spectra by the voltage but also for the noise performance. In actual use as a detector, the noise contribution will follow the same equivalent circuit shown in Fig. 5, but with the photocurrent sources replaced by noise sources. Over an appropriate voltage range, therefore, the noise contribution from *one* of the three QWIPs dominates. Consequently, our device will *not* have a degraded performance compared with the usual one-color QWIP for a given color.

An accurate value for the actual one color QWIP responsivity is difficult to obtain, at all voltages and for a given wavelength. In our measurement of responsivity, we use a monochromatic light. For small IR signal, the responsivity at wavelength λ is

$$\mathcal{R}_\lambda = \frac{v_{meas}/R_\parallel}{P_\lambda}, \tag{2}$$

where v_{meas} (as before) is the voltage change measured by the lock-in amplifier, $R_\parallel$ is the value of R_s and r in parallel, $r = r_1 + r_2 + r_3$, which is the (total) device differential resistance, and P_λ is the incident monochromatic IR power onto the detector. Let us assume that λ is set to the peak detection wavelength of MQW2 at $\lambda 2$. Only i_{p2} is large and therefore using Eq. (1), Eq. (2) becomes

$$\mathcal{R}_{\lambda 2} = \frac{i_{p2}}{P_{\lambda 2}} \frac{r_2}{r_1 + r_2 + r_3}, \tag{3}$$

Therefore, we cannot obtain a pure measure of the responsivity of MQW2, as instead of $\mathcal{R}_{\lambda 2} = i_{p2}/P_{\lambda 2}$, there is a factor involving all the r: only in the voltage range where the factor $r_2/(r_1 + r_2 + r_3) \approx 1$ we can obtain a true measure of $\mathcal{R}_{\lambda 2}$. This means that only in the range where $r_2 \gg r_1$ and $r_2 \gg r_3$ does the measured responsivity shown in Fig. 4 reflect a pure response from MQW2. This range for the 8.5 μm response from MQW2 is approximately -4 to -8 V. The same restriction applies to $D^\star$, i.e., the result shown in the top part of Fig. 4 for 8.5 μm is only an accurate, measure of $D^\star$ for MQW2 in the range of -4 to -8 V.

5. Conclusions

We have presented a detailed study on our three-color QWIP. We believe that we have a quantitative understanding of the operation of this multi-color voltage tunable QWIP and therefore can design improved detectors for specific wavelengths and applications. Further studies are underway to fabricate a low resolution (≈ 1 μm) compact voltage tuned spectrometer (with no moving parts), and to design experiments utilizing this multicolor QWIP for target recognition and passive ranging.

Acknowledgements

The authors thank P. Chow-Chong, P. Marshall, J. Stapledon, and J. R. Thompson for sample preparation, and Dr. W. Tam for discussions. This work was supported in part by DND DREV.

133

References

Choi, K. K., Levine, B. F., Malik, R. J., Walker, J., and Bethea, C. G., (1987). Phys. Rev. B **35**, 4172.

Choi, K. K., Levine, B. F., Bethea, C. G., Walker, J., and Malik, R. J., (1989). Phys. Rev. B **39**, 8029.

Grahn, H. T., Haug, R. J., Müller, W., and Ploog, K., (1991). Phys. Rev. Lett. **67**, 1618.

Gravé, I., Shakouri, A., Kruze, N., and Yariv, A., (1992). Appl. Phys. Lett. **60**, 2362.

Helgesen, P., Finstad, T. G., and Johannessen, K., (1991). Appl. Phys. Lett. **69**, 2689.

Köck, A., Gornik, E., Abstreiter, G., Böhm, G., Walther, M., and Weimann, G., (1992). Appl. Phys. Lett. **60**, 2011.

Levine, B. F., (1993). J. Appl. Phys. **74**, R1.

Levine, B. F., Zussman, A., Gunapala, S. D., Asom, M. T., Kuo, J. M., and Hobson, W. S., (1992). J. Appl. Phys. **72**, 4429.

Liu, H. C., (1992). Appl. Phys. Lett. **61**, 2703.

Liu, H. C., Li, J., Thompson, J. R., Wasilewski, Z. R., Buchanan, M., and Simmons, J. G., (1993a). IEEE Elect. Dev. Lett., (in press).

Liu, H. C., Wasilewski, Z. R., Buchanan, M., and Chu, H., (1993b). Appl. Phys. Lett. **63**, 761.

Liu, H. C., Li, J., Buchanan, M., Wasilewski, Z. R., and Simmons, J. G., (1993c). Phys. Rev. B **48**, 1951.

Liu, H. C., Steele, A. G., Buchanan, M., and Wasilewski, Z. R., (1993d). J. Appl. Phys. **73**, 2029.

Martinet, E., Rosencher, E., Luc, F., Bois, Ph., Costard, E., and Delaitre, S., (1992). Appl. Phys. Lett. **61**, 246.

Steele, A. G., Liu, H. C., Buchanan, M., and Wasilewski, Z. R., (1992). J. Appl. Phys. **73**, 1062.

Tsai, K. L., Chang, K. H., Lee, C. P., Huang, K. F., Tsang, J. S., and Chen, H. R., (1993). Appl. Phys. Lett. **62**, 3504.

Multi λ Controlled Operation of Quantum Well IR Detectors Using Electric Field Switching and Rearrangement

A. SHAKOURI, I. GRAVÉ, Y. XU, A. YARIV
Department of Applied Physics, 128-95
California Institute of Technology
Pasadena, California 91125

ABSTRACT.   We describe a new mode of operating quantum well infrared photodetectors which is based on electric field domain formation and readjustment. The domain readjustment and electric field spatial redistribution results from an interplay between the fixed applied voltage and the quantum tunneling physics. The paper will describe the relevant theoretical background and present supporting data. Also described will be the application of domain switching to attain multi λ response of such detectors.

1.   Introduction

Quantum well infrared photodetectors (QWIP) have recently been the subject of a considerable research effort [1,2,3,4,5,6,7]. These detectors have usually a narrow spectral range of detection ranging in width around $10-20\%$ of the peak wavelength [3]. In some applications this could be considered an advantage; in many other situations one would like to achieve as broad a range of detection as possible. A second issue which can be seen as limiting in many applications is the lack of flexibility, as for the peak wavelength and the whole spectral range, once the device has been designed and grown [8,9].

We discuss here the operation of a new type of bound-to-continuum GaAs/AlGaAs multiquantum well infrared detector, consisting of different stacks of quantum wells arranged in series All the wells in a given stack are identical, but each stack is designed for absorption and detection at a different wavelength, featuring distinct well widths and barrier heights. This detector can operate as a multicolor voltage controlled IR detector or as a broad band detector depending on the bias [10,11]. In the following sections we will discuss design and characterization of the detector, then we will explain different modes of operation of this device by formation of electric field domains in the multiquantum well region.

135

H. C. Liu et al. (eds.), Quantum Well Intersubband Transition Physics and Devices, 135–150.
© *1994 Kluwer Academic Publishers. Printed in the Netherlands.*

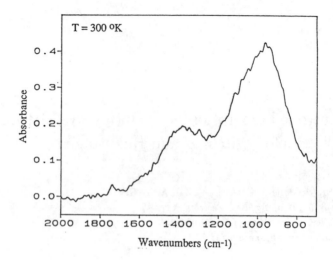

Figure 1: Absorption spectrum at room temperature. The measurement was performed with a Fourier transform spectrometer using a 45° multipass geometry; the spectrum is normalized to reflect the contribution of the intersubband absorption alone. An absorption coefficient α_{45}=600 cm^{-1} for the peak at 1364 cm^{-1} was derived.

2.   Design and characterization of the multi-stack infrared detector

The structure was grown by molecular beam epitaxy on a semi-insulating GaAs substrate. The multiquantum well region, clad by two n-doped contact layers, consisted of three stacks of 25 quantum wells each; the first 25 wells were 3.9 nm wide and were separated by $Al_{0.38}Ga_{0.62}As$ barriers; the second stack consisted of 4.4 nm wide wells. with $Al_{0.30}Ga_{0.70}As$ barriers; the last stack had 5.0 nm wide wells and $Al_{0.24}Ga_{0.76}As$ barriers. All the barriers were 44 nm wide; the wells and the contacts were uniformly doped with Si to $n = 4 \times 10^{18} cm^{-3}$.

The absorption spectrum at zero field and room temperature is shown in Figure 1. The measurement was taken with a Fourier transform infrared spectrometer in the usual 45° multipass geometry [12]; the absorption of the light polarized in compliance with the selection rules was normalized by the absorption of light polarized in the perpendicular direction, to allow for only the intersubband contribution. The absorption peak at 1364 cm^{-1} is due to the 3.9nm wells while the stronger absorption at 964 cm^{-1} is the composite contribution of the two other stacks of quantum wells, which, individually, have absorption peaking at 1080 and 920 cm^{-1}. These results agree with our design values; our calculations, which included band nonparabolicity [13] and a band offset value of 0.60, anticipated absorption peaks at room temperature at 1335, 1052, and 880 cm^{-1}, respectively. We see that in each of the three different types of wells, light is absorbed by electrons excited from the first subband to a second subband which is located close to the top of the well. The blue shift in the experimental values versus the calculated ones can be explained by the omission of the exchange interaction from our calculations [14] ; in these heavily doped samples, the correction supplied by many-body effects is noticeable. The existence of the two absorption peaks that merge into a wide and strong peak was also experimentally verified by analyzing the absorption of a few additional MBE grown control wafers, which were designed to include, each time, only two of the stacks described above.

Devices were processed out of the grown wafer and prepared as etched mesa, $200\mu m$ in diameter. Figure 2(b) displays the smoothed photocurrent spectroscopy of a device at a temperature of 7K, for different values of applied voltage; the polarity is defined here as positive when the higher potential is applied to the cap layer on top of the mesa (see also figure 6). It is seen that, for low applied field, the first stack of 3.9nm wells, closer to the substrate, provides most of the photocurrent at the appropriate excitation energies around the peak of 1411 cm^{-1}. When the bias is increased above a threshold of 6.5 volts, a sharp transition takes place and the responsivity peak switches to 1140 cm^{-1}; it is apparent that the second stack of quantum wells is now responsible for most of the photocurrent, while the contribution from the first stack has sensibly decreased. Note that the small shifts of the photocurrent peaks with regard to the absorption peaks are due to the different experimental temperatures [15] and to the applied electric field [16] (see also [7]). If we apply a negative bias to the detector [figure 2(a)], again, at low voltages, the photocurrent is due mostly to electrons excited in the first stack of the wells; the responsivity increases with the applied voltage, but its magnitude is always less than that corresponding to the same forward bias; in addition, one observes that the photocurrent peak around 1400 cm^{-1} is much broader in the forward bias mode. When the bias is increased to more negative values, the responsivity extends to lower energies, showing increasing contributions from the second stack: the first stack continues to contribute a constant value to the photocurrent, in contrast to the reduction in response experienced in the opposite polarity of the applied electric field. For still more negative voltages (around -13V), it is apparent from our results that the spectral domain of significant response expands to still lower energies, to include contributions from the third stack of wells, around 900 cm^{-1} (not shown in the figure).

From these results it is apparent that the detector can operate in one of a number of modes. At forward and low bias voltages, the response peaks at a single wavelength (~ 1400 cm^{-1}) and the device functions as a standard bound -to-continuum infrared detector [17]. When exceeding a critical applied voltage, the detector's spectral response switches to a different peak wavelength (~ 1140 cm^{-1}), while the detection at the previous peak is significantly reduced. Thus the detector can operate as a two-color, voltage controlled switching device. At a reverse bias the detection is again centered on the higher energy peak (~ 1400 cm^{-1}) up to a specific applied voltage; for moderately higher voltages, two peaks yield a significant photoresponse. At still higher values of the reverse voltage a third response peak appears, which results in

138

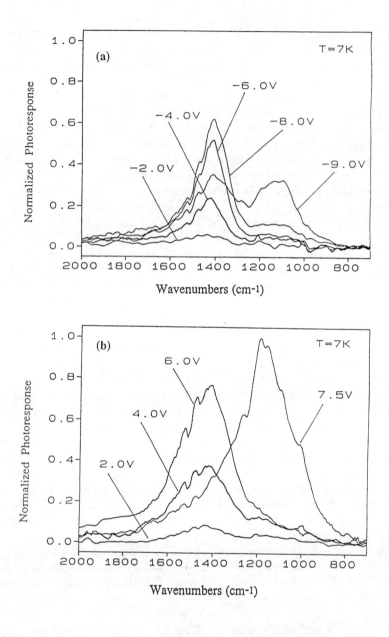

Figure 2: (a) Spectral photoresponse for few values of applied negative voltage for the three-stack quantum well device. Note the broadening in the spectral response below $-8.0V$. (b) Spectral photoresponse for few values of applied positive voltage . Note the switching in peaks at an applied voltage around $6.5V$. The responsivity, at the peak of $1140cm^{-1}$ and the applied voltage of 7.5V, is 0.75 A/W. The units are the same for both (a) and (b).

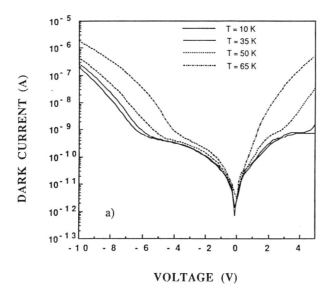

Figure 3: Dark currents at different temperatures.

operation as a wide-band detector. Here the device works as a voltage-controlled, adjustable-spectral-domain detector. These features in the photocurrent, which are observed at a temperature of 7K, persists at higher temperatures. In the reverse bias direction, one observes the same general behavior also at 77K. In the forward polarity, the switching of the photoresponse from the higher energy peak to the lower one is observed up to a temperature of 60K; the critical voltage at which the switching occurs slightly increases with the temperature; at still higher temperatures, the dark current at the required voltages becomes too large for practical detection purposes.

The dark current, measured with a cold shielded window is shown at different temperatures in Figure 3. One can note a strong asymmetry between the two polarities. A fine structure in the plateaux of the I-V curves (not resolved in Figure 3), corresponding to regions of negative differential resistance, was observed and is shown in Figure 4. This measurement gives a most important clue to the origin of the switching phenomena, since these negative differential resistance oscillations are the signature of the formation and expansion of a high field domain along the sample multiquantum well region [28,29,30,31,32].

The data of the photocurrent spectral measurement were taken with a Fourier transform spectrometer, complemented by a setup including a calibrated black-body source and a set of cooled filters at different wavelengths. The noise equivalent

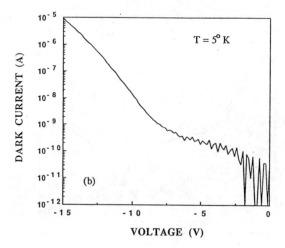

Figure 4: Fine structure in the dark current.

voltage was measured directly with a spectrum analyzer in the cold, shielded window configuration. One should also note the very low values of dark current, which, combined with a responsivity ranging up to 0.75 A/W, ultimately yield high D^* for this detector ($D^*=4 \times 10^{11}$ $cmHz^{1/2}W^{-1}$ at 40K and 1140 cm^{-1}).

3. Analysis of the experimental results

We can interpret the oscillations observed in the I-V characteristics (Figure 4) and the features observed in the photocurrent spectral measurement (Figure 2) as evidence for a rich pattern of high and low electric field domains in the sample [28,29,30,31,32]. The light is absorbed in all 3 stacks of quantum wells but only photoexcited carriers which are in a region with high electric field can be swept out of the quantum well and contribute to the photocurrent. Those in low field region have a high probability of being recaptured by their own well, contributing only negligibly to the photocurrent.

3.1. ELECTRIC FIELD DOMAINS IN BULK SEMICONDUCTORS

The formation of electric field domains (EFD) was first observed in bulk GaAs and InP and is mostly known as the cause of Gunn oscillations [18]. In 1963, when J.B. Gunn was studying the current-voltage characteristics of GaAs and InP devices, he discovered that when the applied electric field was greater than some critical value of several thousand volts per centimeter, spontaneous current oscillation appeared in the circuit. Using probe measurements of the potential distribution across the sample, he established that a propagating high field domain forms in the sample. It nucleates near the cathode, propagates toward the anode with velocity of the order of $10^5 m/s$, and disappears near the anode. Then this process repeats itself. The domain formation leads to a current drop, the domain annihilation results in an increase in the current, and periodic current oscillations exist in the circuit. Later, Kroemer pointed out [19] that all the observed properties of the microwave oscillation were consistent with a theory of negative differential resistance independently proposed by Ridley and Watkins [20] and by Hilsum [21]. The mechanism responsible for the negative differential resistance is a field-induced transfer of conduction-band electrons from a low-energy, high-mobility valley to higher energy, low-mobility satellite valleys. Using the Poisson equation and the equation of current continuity, it can be shown [22,23] that a semiconductor exhibiting bulk negative differential resistivity is inherently unstable, because a random fluctuation of carrier density at any point in the semiconductor produces a momentary space charge that grows exponentially in time. Eventually this leads to the formation of high and low field domains in the sample. The continuity of the current requires in most cases that the high field domain moves from the cathode to the anode.

3.2. ELECTRIC FIELD DOMAINS IN MULTIQUANTUM WELLS

In the case of perpendicular transport in heterojunction superlattices, there are other mechanisms which can generate negative differential conductance, and hence electric field domains. In the miniband conduction regime (i.e. when the applied bias per superlattice period is smaller than the width of miniband), Esaki and Tsu [24] predicted that electrons accelerated perpendicular to the layers may exhibit negative differential velocity due to the negative effective mass they would experience. And this would give rise to negative differential conductivity. Recently, Sibille, Palmier et al. have observed Esaki-Tsu negative differential velocity in GaAs/AlAs superlattices [25]. They also pointed out its fundamental link with Wannier-Stark quantization [26], that these are both manifestations of the same physical phenomenon, i.e. the localization of the superlattice electronic states by Bragg reflection under applied electric field. In the same miniband transport regime, Le Person et al. have observed Gunn oscillations in the growth direction of a nonintentionally doped GaAs/AlAs superlattice [27].

Another negative differential conductance mechanism in superlattice structures was observed by Esaki and Chang [28] in 1974, when the miniband conduction was broken. They studied the transport properties of a GaAs(4.5nm)/AlAs(4.0nm) superlattice. Because of the thin barriers, there is a strong coupling between wells which produces large ground-state bandwidth (5meV) and thus the electron transport is by miniband conduction. By studying the I-V characteristics, they observed a oscillatory negative conductance due to the formation of an expanding high field domain produced by the electric field induced breaking of the miniband conduction. The voltage drop across this domain aligns the ground state of one well with the first excited state of the neighboring well allowing resonant tunneling to occur. A similar phenomenon was observed by Choi et al. in the case of weakly coupled quantum wells [29]. In this case the superlattice consisted of 49 periods of $GaAs(7.6nm)/Al_{0.27}Ga_{0.73}As(8.8nm)$ multiquantum wells. The ground state bandwidth being 0.4meV, the quantum well states are localized (by well width fluctuations and also by applied voltage). Therefore electron transport is dominated by sequential resonant tunneling (SRT). In spite of important differences in transport mechanism (between miniband conduction and SRT), Choi et el. observed similar negative conductance oscillations in the I-V characteristics. SRT was first studied theoretically by Kazarinov and Suris in 1971 [30], who predicted the existence of peaks in the I-V characteristics of weakly coupled multiquantum wells (tight-binding superlattices), which correspond to resonant tunneling between the ground and excited states of adjacent wells. Later, Capasso et al. observed these peaks in a AlInAs/GaInAs superlattice [31]. Recently, in addition to transport experiments which showed high field domain formation, there have been optical studies of these domains in superlattices. Grahn, Schneider et al. used Stark shift in the photoluminescence spectra to identify different electric field domains in a GaAs/AlAs superlattice [32].

3.3.   INTERPRETATION OF EXPERIMENTAL RESULTS

As was mentioned earlier, we can interpret the features observed in the photocurrent spectral measurement (Figure 2) and I-V characteristics (Figure 4) as evidence for formation of high and low electric field domains in the sample (MQWs in our multi-stack detector are weakly coupled, similar to Choi's experiment).

In summary, under an arbitrary applied bias, a uniform distribution of electric field is not *stable* because all of the QWs will be out of resonance i.e. none of the energy levels of pairs of adjacent wells will be aligned. Instead, the system will settle into a different configuration in which the electric field profile includes a high and a low field regions. In the high field region we have ground level to excited level SRT, and in the low electric field region ground level to ground level SRT (see figure 5). Transport within each domain is resonant, while at the boundary between the two regions it is generally non-resonant. This boundary then acts as a bottleneck that limits the current. There should also be some charge accumulation or depletion at this boundary because of the change in the slope of electric fields, as required by Poisson's equation. This also has to be considered when determining the current flow through the structure. An increase in the bias will cause more QWs to enter the high field domain (HFD) region, and this is reflected by the oscillatory behavior in the I-V curve.

In the case of our sample, several electric field distributions across the 3 stacks of QWs are possible (figure 6(a) shows four possible distributions for applied biases -V1 and -V2). Some of these have only high field domain (HFD) in one of the stacks and others have a combination of HFDs and LFDs (low field domains) in all the stacks. The main rule used to determine different field profiles across the structure is that we may use only the electric field values which result in alignment of energy

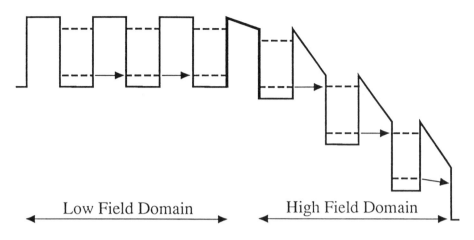

Figure 5: Formation of high and low field domains in the case of a simple Si-doped GaAs/AlGaAs multiquantum well structure.

levels between adjacent wells. In addition the total voltage drop must equal the applied voltage. This results in a large number of possible configurations. The actual electric field distribution should satisfy self consistently Poisson's equation and the equation of current continuity along the superlattice. Because of the complexity of the transport calculations in MQWs a detailed study, which should also include a stability analysis, is very complicated. In this paper, instead, we try to extract some of the parameters which are important for HFD formation, and use them to design samples with a desired electric field distribution.

One of the important parameters is the amount of charge accumulation or depletion at the boundaries, where the slope of the electric field changes. This charge, by altering the tunneling process (resonant or nonresonant) at that boundary, can limit the total current which flows through the structure. If the transport at the boundary is resonant, the LFD will limit the current and its presence should be considered to estimate the current.

For example at low negative biases (figure 6(a), bias -V1), considering that different electric fields are needed in the different stacks for the ground level of a well to be aligned with its neighboring well's excited state, we see that a HFD in stack (a) would lead to a charge accumulation of $2.0 \times 10^{11} electrons/cm^2$ at the boundary between HFD and LFD. While HFD in the stacks (b) or (c) would require 1.6×10^{11} or $1.4 \times 10^{11} electrons/cm^2$ respectively, at the corresponding boundaries. The QWs are doped to $2 \times 10^{12} cm^{-2}$, so providing these amounts of charge would not be a limiting process in EFD formation. The barriers in all 3 stacks being identical, if the current

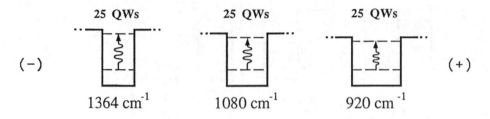

(a) Negative bias:

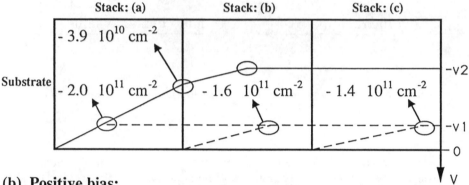

(b) Positive bias:

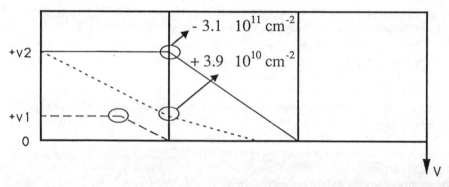

Figure 6: Voltage drop across the three stacks of quantum wells at negative (a) and positive (b) biases. The light is absorbed in all 3 stacks of QWs but only the photoexcited carriers which are in a region with high electric field can be swept out of the quantum well and contribute to the current. Note the amount of charge accumulation or depletion at the boundaries where the slope of electric field changes.

is limited by the domain boundary, one would expect that the configuration in which there is a HFD in stack (a) accommodate more current than other two configurations, and will probably be more stable . At low positive biases (figure 6(b), bias +V1) we see again, and at the first sight surprisingly, HFD formation in stack (a). This shows that the screening effect [29] which would cause the domain formation to start from the anode and then expand toward the cathode is not the dominant effect here.

The high field domain switching at +6.5V and expansion at −8.5V (figure 2) can also be explained in terms of the charges at the boundaries. At high biases, when the HFD would expand to more than one stack, the charge accumulation or depletion at the boundary between two stacks, can limit the current in the structure. This charge is due to the difference between the values of HFD electric fields in different stacks. As can be seen from figure 6(a) (bias -V2), in the case of negative bias the presence of HFD in both stacks (a) and (b) results in a charge accumulation of $3.9 \times 10^{10} cm^{-2}$. In the case of positive bias instead, the same configuration would result in a charge depletion of the same amount at the boundary between the two stacks (see figure 6(b)). This charge depletion can reduce the current through the whole device and thus other configurations become more favorable. Eventually, the configuration observed in the experiment for high positive bias incorporates HFD only in stack (b).

As was mentioned earlier, sequential resonant tunneling is the origin of NDR and subsequent EFD formation in the device. The large switching voltage of 6.5V across 25 periods as observed, corresponds to $260 meV/period$. This is much larger than the excited state-ground state separation $(E_2 - E_1)$ of $169 meV$ as inferred from absorption measurements. This indicates a nonzero voltage drop across the low electric field domain region (in agreement with a nonzero linewidth of the subband) or at space charge regions. More importantly, it also shows that the transport might rely on SRT between the ground state and some excited states which are located higher in the continuum, above the top of the well. The very long barriers of 44nm in this sample and the proximity of the first excited state to the top of the well, can cause the system to "choose" a configuration in which there is SRT through high continuum states (since these states "see" a lower effective barrier). Another interesting point to note is the reduction in the strength of the high energy peak in the photocurrent measurement at the negative bias of −8.5V, when the second peak appears (see figure 2). This can be explained by a rearrangement of the HFD in the stack (a), in such a way that SRT to lower states in the continuum occurs. From an analysis of the peak strengths at different voltages, one can see that this configuration corresponds to a voltage drop across the stack (a) of $\approx 4V$ or $160 meV/period$, which is the separation between the ground state and the first excited subband.

3.4. DESIGN OF MULTI-STACK INFRARED DETECTORS

Even though there are a lot of processes and parameters which can influence the transport in the superlattice, such as impurity or phonon-assisted tunneling, reso-

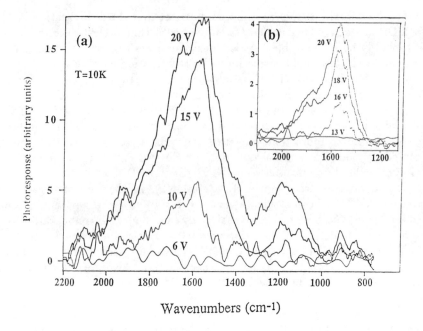

Figure 7: (a) Spectral photoresponse for the two-stack MQW device with $44nm$ barriers in both stacks. At low bias there is a peak at $\approx 1600cm^{-1}$ and as we increase the applied bias another peak at *longer* wavelengths $\approx 1200cm^{-1}$ appears. (b) (Inset) Spectral photoresponse of another two-stack MQW device with similar well characteristics but the barriers in the long wavelength stack are shortened to $20nm$. At low bias there is a peak at $\approx 1550cm^{-1}$ and as we increase the applied bias another peak at *shorter* wavelengths $\approx 1800cm^{-1}$ appears.

nant tunneling through different states in the continuum, relaxation times and space charge effects; it is still possible to design samples with the desired EFD configuration for applications like tunable infrared detectors [11]. This is done by considering the charge accumulation effects as was discussed earlier. Figure 7(a) shows the photocurrent spectroscopy at different applied biases for a two-stack MQW IR-detector with a design similar to that of our original three-stack device (stack (a):4.0 nm $GaAs$ wells separated by 44 nm $Al_{0.36}Ga_{0.64}As$ barriers. Stack (b):5.2 nm $GaAs$ wells separated by 44 nm $Al_{0.24}Ga_{0.76}As$ barriers.). At low bias there is a peak at short wavelengths $\approx 1600cm^{-1}$ and as we increase the applied bias another peak at longer wavelengths $\approx 1200cm^{-1}$ appears. As a test vehicle for our formalism we set out to design a two-stack MQW detector which displays the opposite pattern in the photocurrent, i.e. a detector with long wavelength peak at low bias voltages, with an added short wavelength peak contributing to the photocurrent at higher bias voltages. To achieve this goal, we have to modify the pattern of EFD formation. This was done by reducing the width of the barriers in one of the stacks. This increases the value of electric field in the high field domain of that stack (It takes a larger field to align the $n = 1$ and $n = 2$ levels of neighboring wells); and subsequently increases the accumulated charge at the boundary between the HFD and LFD.

Figure 7(b) displays the photocurrent of a second two-stack MQW detector where the barriers in the stack (b) (having absorption peak at longer wavelength) were shortened to 20 nm (stack (a):4.0 nm $GaAs$ wells separated by 44 nm $Al_{0.38}Ga_{0.62}As$ barriers. Stack (b):4.7 nm $GaAs$ wells separated by 20 nm $Al_{0.30}Ga_{0.70}As$ barriers.). By this mean we can achieve the requirement that the electric field for ground state to excited state SRT be increased in stack (b) and become larger than the corresponding value of the electric field in stack (a). As a result we see that this time the peak at longer wavelength ($\approx 1550cm^{-1}$) appears first, and then, by increasing the bias further, the peak at shorter wavelength ($\approx 1800cm^{-1}$) appears. It is interesting to note that the spectral photoresponse of these two-stack MQW devices has a similar behavior when we reverse the polarity of the applied bias (not shown in the figure); this is in contrast with the switching behavior observed in the three-stack MQW device. This shows the importance of the LFD in the third stack as a current limiting process.

In conclusion, we have discussed some of the important parameters governing the formation, expansion and readjustment of electric field domains in multiquantum well structures and its application to tunable multi-color detectors.We also showed how the pattern of electric field domain formation can be manipulated by careful design of the device.

This work was supported by DARPA and the Office of Naval Research.

References

[1] J.S. Smith, L.C. Chiu, S. Margalit, A. Yariv, and, A.Y. Cho, 'A new infrared

detector using electron emission from multiple quantum wells', *J. Vac. Sci. Technol. B*, **1**,376 (1983).

[2] M.A. Kinch, and A. Yariv, 'Performance limitations of GaAs/AlGaAs infrared superlattices', *Appl. Phys. Lett.*, **55**,2093 (1989).

[3] B.F. Levine, K.K. Choi, C.G. Bethea, J. Walker, and R.J. Malik, 'New $10\mu m$ infrared detector using intersubband absorption in resonant tunneling GaAlAs superlattices', *Appl. Phys. Lett.*, **50**,1092 (1987).

[4] B.F. Levine, A. Zussman, S.D. Gunapala, M.T. Asom, J.M. Kuo, and W.S. Hubson, 'Photoexcited escape probability, optical gain, and noise in quantum well infrared photodetectors', *J. Appl. Phys.*, **72**,4429 (1992).

[5] H.C. Liu, A.G. Steele, M. Buchanan, and Z.R. Wasilewski, 'Dark current in quantum well photodetectors', *J. Appl. Phys.*, **73**,2029 (1993).

[6] H. Schneider, F. Fuchs, B. Dischler, J.D. Ralston, and P. Koidl, 'Intersubband absorption and infrared photodetection at 3.5 and 4.2 μm in GaAs quantum wells', *Appl. Phys. Lett.*, **58**,2234 (1991).

[7] E. Rosencher, E. Martinet, F. Luc, Ph. Bois, and E. Bockenhoff, 'Discrepancies between photocurrent and absorption spectroscopies in intersubband photoionization from GaAs/AlGaAs multiquantum wells', *Appl. Phys. Lett.*, **59**,3255 (1991).

[8] B.F. Levine, C.G. Bethea, V.O. Shen, and R.J. Malik, 'Tunable long-wavelength detectors using graded barrier quantum wells grown by electron beam source molecular beam epitaxy', *Appl. Phys. Lett.*, **57**,383 (1990).

[9] E. Martinet, F. Luc, E. Rosencher, Ph. Bois, and S. Delaitre, 'Electrical tunability of infrared detectors using compositionally asymmetric GaAs/AlGaAs multiquantum wells', *Appl. Phys. Lett.*, **60**,895 (1992).

[10] I. Gravé, A. Shakouri, N. Kuze, and A. Yariv, 'Voltage-controlled tunable GaAs/AlGaAs multistack quantum well infrared detector', *Appl. Phys. Lett.*, **60**,2362 (1992).

[11] A. Shakouri, I. Gravé, Y. Xu, A. Ghaffari, and A. Yariv, 'Control of electric field domain formation in multiquantum well structures', *Appl. Phys. Lett.*, **63**,1101 (1993).

[12] B.F. Levine, R.J. Malik, J. Walker, K.K. Choi, C.G. Bethea, D.A. Kleinman, and J.M. Vandenberg, 'Strong 8.2 μm infrared intersubband absorption in doped GaAs/AlAs quantum well waveguides', *Appl. Phys. Lett.*, **50**,273,(1987).

[13] Z.Y. Xu, V.G. Kreismanis, and C.L. Tang, 'Photoluminescence of GaAs-AlGaAs multiquantum well structures under high excitations', *Appl. Phys. Lett.*, **43**,415 (1983).

[14] J.W. Choe, O. Byungsung, K.M.S.V. Bandara, and D.D. Coon, 'Exchange interaction effects in quantum well infrared detectors and absorbers', *Appl. Phys. Lett.*, **56**,1679 (1990).

[15] G. Hasnain, B.F. Levine, C.G. Bethea, R.R. Abott, and S.J. Hsieh, 'Measurement of intersubband absorption in multiquantum well structures with monolithically integrated photodetectors', *J. Appl. Phys.*, **67**,4361 (1990).

[16] A. Harwit, and J.S. Harris Jr., 'Observation of Stark shifts in quantum well intersubband transitions', *Appl. Phys. Lett.*, **50**,685 (1987).

[17] B.F. Levine, C.G. Bethea, K.K. Choi, J. Walker, and R.J. Malik, 'Bound-to-extended state absorption GaAs superlattice transport infrared detectors', *J. Appl. Phys.*, **64**,1591 (1988).

[18] J.B. Gunn, 'Microwave oscillation of current in III-V semiconductors', *Solid. State. Commun.*, **1**,88,(1963); 'Instabilities of current in III-V semiconductors', *IBM J. Res. Dev.*,**8**, 141 (1964)

[19] H. Kroemer, 'Theory of Gunn effect', *Proc. IEEE*, **52**,1736 (1964).

[20] B.K. Ridley, and T.B. Watkins, 'The possibility of negative resistance in solids', *Proc. Phys. Soc.*, **78**,293 (1961).

[21] C. Hilsum, 'Transferred electron amplifiers and oscillators', *Proc. IRE*, **50**,185 (1962).

[22] S.M. Sze, 'Physics of semiconductor devices', 2nd edition, *Wiley & Sons publishers*, Ch.11 (1981).

[23] M. Shur, 'GaAs devices and circuits', *Plenum Publishing Co.*, Ch.4 (1987).

[24] L. Esaki, and R. Tsu, 'Superlattice and negative differential conductivity in semiconductors', *IBM J. Res. Develop.*, **14**,61 (1970).

[25] A. Sibille, J.F. Palmier, H. Wang, and F. Mollot 'Observation of Esaki-Tsu negative differential velocity in GaAs/AlAs superlattices', *Phys. Rev. Lett.*, **64**,52 (1990); A. Sibille, J.F. Palmier,and F. Mollot, 'Coexistence of Wannier-Stark localization and negative differential velocity in superlattices', *Appl. Phys Lett.*,**60**,457 (1992).

[26] G. Wannier, 'Wavefunctions and effective Hamiltonian for Bloch electrons in an electric field', *Phys. Rev.*, **117**,432 (1960).

[27] H. Le Person, C. Minot, L. Boni, J.F. Palmier, and F. Mollot, 'Gunn oscillations up to 20GHz optically induced in GaAs/AlAs superlattice', *Appl. Phys. Lett.*, **60**,2397 (1992).

[28] L. Esaki, and L.L. Chang, 'New transport phenomenon in a semiconductor superlattice', *Phys. Rev. Lett.*, **33**,495 (1974)

[29] K.K. Choi, B.F. Levine, R.J. Malik, J. Walker and, C.G. Bethea, 'Periodic negative conductance by sequential resonant tunneling through and expanding high-field superlattice domain', *Phys. Rev. B* , **35**,4172 (1987)

[30] R.F. Kazarinov and, R.A. Suris, 'Electric and electromagnetic properties of semiconductors with a superlattice', *Sov. Phys. Semicond.*, **6**, 120 (1972); *ibid*, 'Possibility of the amplification of electromagnetic waves in a semiconductor with superlattice', *Sov. Phys. Semicond.*, **5**,707 (1971)

[31] F. Capasso, K. Mohammad, and A.Y. Cho, 'Sequential resonant tunneling through a multiquantum well superlattice', *Appl. Phys. Lett.*, **48**, 478 (1986)

[32] H.T. Grahn, H. Schneider and K. von Klitzing, 'Optical studies of electric field domains in GaAs/AlGaAs superlattices', *Phys. Rev. B*, **41**, 2890 (1990); H.T. Grahn, R.J. Haug, W. Muller, and K. Ploog, 'Electric field domains in semiconductor superlattices; a novel system for tunneling between 2D systems', *Phys. Rev. Lett.*, **67**,1618 (1991)

INFRARED HOT-ELECTRON TRANSISTOR DESIGN OPTIMIZATION

K. K. CHOI, M. Z. TIDROW, M. TAYSING-LARA, AND W. H. CHANG
U. S. Army Research Laboratory, EPSD
Fort Monmouth, NJ 07703-5601, USA

ABSTRACT. In this work, we discuss the optical and electrical properties of quantum well infrared photodetectors, and issues related to the design of an infrared hot-electron transistor.

1. Introduction

Long wavelength infrared detection using intersubband transitions in quantum well structures has been under intense investigation in recent years. In the following, we present a theoretical calculation on the peak position of the absorption wavelength and the linewidth for a wide range of detector parameters. We will also discuss the photoconductive gain of the detectors with respect to the energies of the final state of the optical transition and the satellite valleys of the detector material.

Recently, thermal imaging using GaAs/AlGaAs MQW structures has been demonstrated.[1,2] However, in all cases, the operating temperature has to be kept around 60 K. Before quantum well infrared photodetector (QWIP) focal plane arrays (FPA) can be operated at or above 77 K, this technology will be less competitive with other infrared technologies that can be operated at higher temperatures. The reason for the lower temperature operation is because of its larger thermally activated dark current, which tends to saturate the readout circuit and prevents background limited infrared photodetection (BLIP) at 77 K.

Since the specific detectivity D^* of a typical QWIP is about 10^{10} cm$\sqrt{\text{Hz}}$/W at 77 K, it is sufficiently high such that the noise equivalent temperature difference (NEΔT) of a QWIP FPA is more likely limited by the fixed pattern noise due to pixel non-uniformity rather than the temporal noise caused by a small D^*, provided that a long integration time can be employed.[3] However, for the current state-of-the-art readout circuit, the maximum total current density J_{tot} must be less than 10 μAcm^{-2} for an integration time of 33ms. Under the normal thermal imaging condition, J_{tot} includes both the dark current density J_d and the 300 K background photocurrent density J_p. If we further require the detection to be BLIP, ie. $J_p > J_d$, then J_d should be much less than 10

151

H. C. Liu et al. (eds.), Quantum Well Intersubband Transition Physics and Devices, 151–165.
© 1994 *Kluwer Academic Publishers. Printed in the Netherlands.*

μAcm^{-2}. On the other hand, for a QWIP with a cutoff wavelength at 10 μm and a doping density N_d of 1 x 10^{18} cm^{-3}, the typical J_d at 77 K is 10 $mAcm^{-2}$, at least three orders of magnitude higher than the limitation. Hence, a long integration time is not feasible at this temperature.

Besides saturating the readout circuit, a large dark current will also reduce the charge injection efficiency and increase the thermal noise of the readout circuit, which will further reduce the sensitivity of the FPA. Therefore, it is critical to reduce the dark current of the detector before it is applicable at 77 K. In order to greatly reduce the dark current, we have designed different electron energy filters placed next to an optimized QWIP to form an infrared hot-electron transistor (IHET). In the following, we will describe several different filter structures.

Although an IHET is useful in reducing the dark current of a QWIP to provide BLIP at 77 K, due to the companion reduction of photocurrent, the dark current limited detectivity D* in some cases may also be reduced. In order to improve the detector performance, it is advantageous to further reduce the noise of an IHET. We found that for certain filter designs, the measured noise level is lower than that predicted by the conventional theory. These findings indicate improvements on the detector performance can further be made using appropriate filter designs.

2. Optical Properties of QWIP

The detectors we studied are the single bound state GaAs/AlGaAs detectors consisting of a number of quantum wells separated by thick barriers. In this calculation, we concentrate on the single electron transition energy and exclude many-body effects such as the exchange energy, the depolarization effect and the excitonlike effect,[4] which turn out to be nearly cancelled by the corrections related to the effective mass of the electrons.[5,6] We found that if one assumes $\Delta E_c = 0.60\Delta E_g$, the result agrees with the existing experimental data.

2.1 BASIC THEORY

In the following, we consider an optical beam incident normal to a MQW material surface, this surface is inclined at an angle θ toward the growth layers. During an optical transition, an electron in the ground state E_1 absorbs a photon $\hbar\omega$ and promotes to an extended state $E_2 = E_1 + \hbar\omega$ above the barriers. The absorption coefficient $\alpha(\hbar\omega)$ can be obtained by the Fermi Golden rule. If we denote the two dimensional electron density in each well to be ρ_s, the period length L, the refractive index of the well n_w and the effective mass of an electron in the well m*, α can then be expressed as

$$\alpha(\hbar\omega) = \frac{\rho_s}{L} \frac{\pi e^2 \hbar}{2n_w \epsilon_o m^* c} \sin^2\theta \, f(E_2) g(E_2) \ , \tag{1}$$

where $f(E_2)$ is the oscillator strength and $g(E_2)$ is the one-dimensional density of the final states. Note that under the present optical illumination condition, the effective period length for light absorption is $L/\cos\theta$.

When the presence of the well and scattering effects are neglected, the density of the final states $g(E_2)$ is simply given by

$$g(E_2) = \frac{L}{2\pi\hbar}\left(\frac{m_b^*}{2}\right)^{1/2}\frac{1}{\sqrt{E_2 - H}} \quad , \tag{2}$$

where m_b^* is the effective mass of an electron in the barrier, and H is the barrier height. $g(E_2)$ here is the density of states for single spin and for states with odd parity only, because the even parity states do not contribute to light absorption. For the dipole transition between two energy states E_1 and E_2, $f(E_2)$ is given by

$$f(E_2) = \frac{2\hbar^2}{m^*(E_2 - E_1)}\left|\langle 2|\frac{\partial}{\partial x}|1\rangle_w + \frac{m^* n_w}{m_b^* n_b}\langle 2|\frac{\partial}{\partial x}|1\rangle_b\right|^2 \quad , \tag{3}$$

where n_b is the refractive index of the barrier. $|1\rangle$ is the ground state localized in the well and $|2\rangle$ is the extended state above the barriers in the present case. In Eq. (3), we have taken different effective mass and refractive indices in the well region and in the barrier region into account.

In order to obtain the position of the absorption peak λ_p, we assume that $g(E_2)$ is relatively slow varying in comparison with $f(E_2)$ due to scattering effects. The value of f is computed as a function of the final state energy E_2. We then determine the value of E_2, denoted by E_m, at which f is a maximum. The transition energy is equal to $E_m - E_1$, and finally, λ_p is equal to

$$\lambda_p = \frac{2\pi\hbar c}{E_m - E_1} \quad . \tag{4}$$

In the following section, we will present the numerical results.

2.2 NUMERICAL RESULTS

Fig. 1 shows E_m as a function of aluminum molar ratio x for different values of well width d. For a given d, E_m increases monotonically with x as expected. Besides setting the transition energy, the absolute value of E_m is also relevant in determining the photoconductive gain and hence the responsivity of the detector. When E_m is

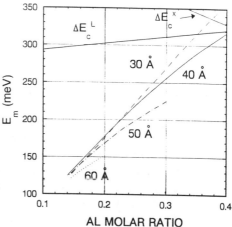

FIG. 1 The energy of the final state E_m, and L and X valleys in the barrier region as a function of x.

154

above the L valley minima or the X valley minima in the well region as well as in the barrier region shown in Fig. 1, the photoelectrons, initially located in the Γ valley, can scatter into these valleys. Once inside these valleys, the momentum of an electron will be randomized rapidly due to the large increase in the density of states,[7] and hence the hot-electron diffusion coefficient and the photoconductive gain will be reduced. In the GaAs well region, the L minima and the X minima are located at 284 meV and 476 meV respectively relative to the Γ conduction band edge, so that the corresponding intervalley scattering occurs within the well region only when E_m is higher than these energies. On the other hand, the L and X valleys associated with the adjacent barrier, being in different spatial locations, are shifted relative to E_m when the MQW structure is subjected to an applied electric field. They can be brought below the E_m state under a strong field even though E_m may be initially below these satellite valleys. Therefore, one expects that the responsivity of a MQW detector will drop beyond a certain applied bias V_{crit}, whose value depends on the initial position of E_m relative to the barrier satellite valleys. This voltage dependence is commonly observed. For instance, from Fig. 1, for a MQW structure with d equal to 50 Å and x equal to 0.25, E_m will align with the L valley of the adjacent barrier when the applied voltage is 110 mV per period. We expect the responsivity will decrease at a higher voltage as the photoconductive gain decreases. Fig. 2 shows the responsivity of four samples containing 30 periods of the stated quantum wells but with different doping densities. V_{crit} is observed ranging from 80 meV to 110 meV, consistent with the estimation. The smaller doping sample acquires a lower V_{crit} since the sample tends to form high field domains in which the potential drop at the light sensitive region is higher than the average value.

After determining E_m, one can calculate the energy E_m relative to the barrier height H. The result is shown in Fig. 3. From the detector design point of view, the thermionic emission current and hence the dark current noise will be minimized if H is chosen to be equal to E_m for the single

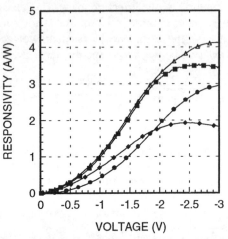

FIG. 2 The responsivity of four QWIP with N_d from 0.5 to 1.2 (x 10^{18}) cm^{-3}.

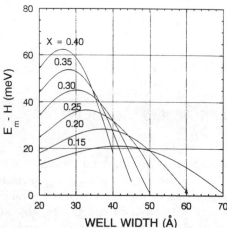

FIG. 3 The energy of the final state E_m relative to the barrier height H as a function of well width.

bound state detectors, since the activation energy of the thermionic emission current is $H - E_1 - E_f$, where E_f is the Fermi energy. However, it may not be the optimum design as the detectivity of a detector depends also on the optical signal. When $H = E_m$, the photoelectrons with energy below the absorption peak need to tunnel out of the well in order to be collected, hence the photocurrent is reduced. Obviously, $E_m - H$ should be around $\gamma/2$ in order to utilize the full absorption peak, where γ is the absorption linewidth. Since a typical γ is around 30 meV, the detector parameters which give $E_m - H$ a value of 15 meV in Fig. 3 are thus preferred.

Fig. 4 shows the calculated λ_p as a function of d. Within the detector parameters shown, λ_p can be varied from 5 μm to over 25 μm. From these tuning curves, the peak wavelength for a given set of detector parameters can be obtained.

Fig. 5 shows the tuning curves as a function x for different d. These tuning curves shows that λ_p are insensitive to different d larger than 30 Å. Actually, λ_p is approximately fixed for a given x regardless of d. In Fig. 5, we also show the experimental λ_p (denoted as crosses) measured by Levine et al.[8] and for other well widths (denoted as circles) from our measurements. Within the experimental error and theoretical uncertainty, the agreement is satisfactory.

Although a given λ_p can be obtained from more than one set of detector parameters, other absorption properties such as the linewidth and linshape may not be the same. In determining γ, one has to consider α over a wide range of energies, thus it is necessary to include the energy dependence of the density of states. At large energies, g(E) given by Eq. (2) is expected to be applicable. Hence, one can obtain the approximate γ by calculating the product of $f(E_2)$ and $g(E_2)$. The resulting γ for three well widths is shown in Fig. 6. It increases as the Al molar ratio decreases. The symbols indicate the experimental results from Levine et al.[8] and from our measurements. Although

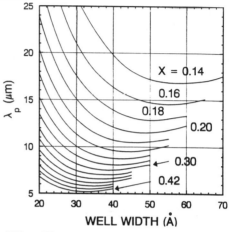

FIG. 4 The absorption peak λ_p.

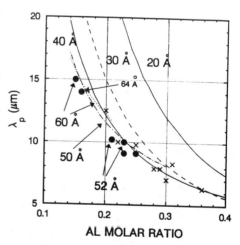

FIG. 5 λ_p and the existing data for d = 40 Å unless indicated otherwise.

there is a substantial scattering of experimental data, presumably due to the differences in absorption background subtraction, impurity broadening and material uniformity, the values of γ and its aluminum molar ratio dependence are quite consistent with the calculation. From this calculation, γ can be varied from 20 % to 50 % of λ_p by changing the well width when λ_p is around 10 μm. Note that although we calculate the linewidth in this work based on the exact eigenfunctions of the structure, the absorption lineshape has not been calculated exactly due to the difficulty in obtaining the exact density of the eigenstates. To calculate the absorption lineshape, it is more convenient to use the perturbation method in which the final states of all transitions are regarded to be originated from a single quantum state broadened by interaction effects.[9] Since the oscillator strength in this formalism is a constant, the absorption lineshape is totally dependent on the spectral function of this quantum state and can be calculated analytically. However, the oscillator strength in this approach may not be obtained from first principles and remains a fitting parameter unless the wave function of the upper state is explicitly defined in the model Hamiltonian.

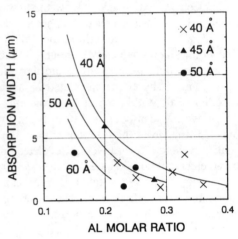

FIG. 6 The absorption width for a given well width as a function of Al molar ratio. The figure also shows existing experimental data.

3. Hot-Electron Spectroscopy

In order to better understand various physical properties of a QWIP, we have performed extensively hot-electron spectroscopy on different types of QWIPs. It is accomplished by placing an electron energy filter next to a QWIP structure. Here, we would like to use it to show that the energy distributions of the photoelectrons ρ_p and the thermally generated electrons ρ_d are different, and that the average energy of the photoelectrons are higher than that of the thermal electrons at temperature equal to 77 K even for a highly doped sample.

The QWIP structure used in this study consists of a 6000-Å n^+ emitter contact layer, a number of periods (N_{per}) of 50-Å GaAs well and 500-Å $Al_{0.25}Ga_{0.75}As$ barrier, and a n^+ base contact layer. The base layer is composed of a 300-Å $In_{0.1}Ga_{0.9}As$ layer and a 200-Å GaAs layer. In the subsequent studies, we usually use the same QWIP structure design. Therefore, the stated QWIP structure will be referred as the *standard structure* in this article.

On the top of the base, a 2000-Å $Al_{0.29}Ga_{0.71}As$ barrier as a high pass filter and a 1.1 μm n^+ GaAs collector are grown. The doping density N_d is 3.5 x 10^{18} cm^{-3} both in the QWIP and in the base. The band diagram of the detector is shown in Fig. 7.

Fig. 8 shows $\beta_d \equiv dI_{dc}/dV_c$ and $\beta_p \equiv dI_{pc}/dV_c$ for a fixed emitter voltage V_e for a QWIP with $N_{per} = 50$. $\beta_d(V_c)$ and $\beta_p(V_c)$ can be shown to be proportional to $\rho_d(E)$ and $\rho_p(E)$ respectively, where E is the electron energy relative to the conduction band edge.[10]

Since a negative V_c raises the potential barrier height of the filter, the more negative V_c represents electrons with higher energies. The photocurrent is generated by a CO_2 laser with photon energy equal to 133 meV modulated at 200 Hz and incident at a 45° light coupling angle. At $V_e = -1V$, a distinctive peak in β_d can be identified. This peak is due to direct thermionic emission (TE) from the resonant state E_2, which is located at 146 meV above the ground state E_1 deduced from the photocurrent spectrum. The rising of β_d at V_c higher than 0.5 V is due to the thermally assisted tunneling current (TAT) occurring below the barrier height.[11] At $V_e = -1V$, the energy of the photocurrent is actually slightly lower than that of the dark current due to the fact that the photon energy used is less than $E_2 - E_1$. At higher V_e, the energy of the photoelectrons is higher than that of the dark electrons as the TAT current increases rapidly with V_e shown in Fig. 8. Note that the present sample is 7 times higher in doping than an optimized QWIP with which the TE peak will not be observable at 77 K.

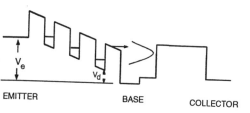

FIG. 7 The energy band structure of detector for hot-electron spectroscopy.

4. Infrared-Hot Electron Transistors

In section 3, the photocurrent peak injected into the base is observed to be relatively sharp and is higher in energy than that of the TAT current. By using the filter to discriminate against the TAT current but allow the photoelectrons be collected in the collector, one can then increase the sensitivity of the QWIP. The new detector is referred as an Infrared Hot-Electron Transistor (IHET). In detector design optimization, there are two figures of merit: one is the specific detectivity D* for individual detector performance, another is the noise equivalent temperature difference (NEΔT) for detector array performance. They are related by the expression

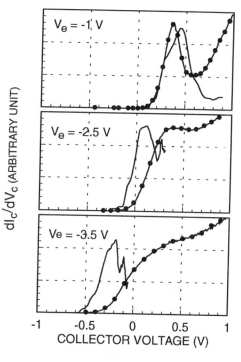

FIG. 8 The dark electron (curves with circles) and the photoelectron energy distributions at different V_e.

$$NE\Delta T = \frac{\sqrt{A}}{D^* \dfrac{dP_B}{dT} \sqrt{2\tau_{in}}} , \qquad (5)$$

where A is the detector area, P_B is the blackbody radiation power, and τ_{in} is the integration time. The dark current limited collector detectivity D^*_c after energy filtering can be shown to be related to the emitter detectivity D^*_e before energy filtering by a factor $\alpha_p/\sqrt{\alpha_d}$, while $(NE\Delta T)_c$ is related to $(NE\Delta T)_e$ approximately by a factor of α_d/α_p, where α_p and α_d are the photocurrent transfer ratio, defined as I_{pc}/I_{pe}, and the dark current transfer ratio, defined as I_{dc}/I_{de}, respectively. For individual detector optimization, one needs to maximize $\alpha_p/\sqrt{\alpha_d}$. On the other hand, α_d/α_p needs to be minimized for detector array operation.

To optimize D*, it turns out that one has to preserve the photocurrent by using a low barrier filter. Fig. 9 and Fig. 10 show the performance of an IHET (referred as detector #1) with a standard QWIP (N_{per} = 30, N_d = 1.2 x 10^{18} cm^{-3}) and a 2000 Å $Al_{0.25}Ga_{0.75}As$ high pass filter (the filter barrier height H_f = 187 meV). Fig. 9 is the maximum D* as a function of V_e at each temperature, and Fig. 10 is the corresponding dark current density at the D* maximum. By eliminating the TAT current, D^*_c is larger than D^*_e at all temperatures. The improvement is larger at lower temperatures due to the larger energy separation between the photocurrent and the dark current. However, the dark current reduction at 77 K in the present case is only a factor of 2. As discussed in the introduction, it is insufficient for focal plane array applications.

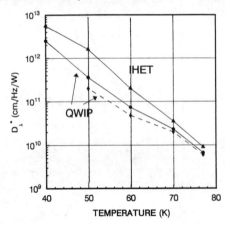

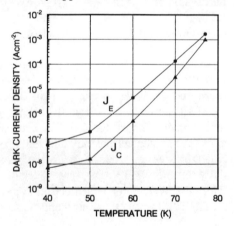

FIG. 9 The D^*_e and D^*_c as a function of temperatures for detector #1. The figure also shows the performance of a QWIP in ref. 8.

FIG. 10 The emitter and collector current density of detector #1 as a function of temperature.

In order to greatly reduce the dark current, we have increased H_f to 217 meV in one of the detectors (referred as detector #2) with the same QWIP as #1 except that N_d = 0.5 x 10^{18} cm^{-3}. The filter consists of a 2000 Å $Al_{0.29}Ga_{0.71}As$ layer with the Al molar ratio of the filter barrier graded from 0.14 to 0.29 for 300 Å on each side of the barrier to reduce the quantum mechanical reflection so that the energy selectivity is improved. Fig.11 shows the 77 K I-V characteristics of the detector. In this figure, the collector voltage V_c relative to the base is always kept at zero volts. J_{ed} is the dark current flowing into the emitter, it represents the dark current level of the corresponding QWIP. The background emitter photocurrent J_{ep} in Fig. 11 is measured by first exposing the detector to room temperature radiation at detector temperature T equal to 10 K, where the dark current is negligible, and then adjusted for a small change in the responsivity at T = 77 K using AC locking techniques. The origin of the background

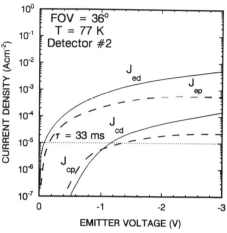

FIG. 11 A plot of the dark current densities (solid curves) and the 300 K background photocurrent densities (dashed curves) of the detector #2.

current is confirmed by checking both the magnitude and the V_e dependence with the AC measurements. Infrared light is coupled to the detectors, which are without an anti-reflection coating, through a 45° edge with a field of view FOV of 36°. The photocurrent level should be equal to that using a one dimensional grating either with an anti-reflection coating or with a AlAs cladding layer. In Fig. 11, we also indicate the saturating current level as the dotted lines when τ = 33 ms. Obviously, J_{ed} is about 2 orders of magnitude higher than this limit, and since $J_{ep} < J_{ed}$, it is not BLIP at 77 K. Note that J_{ep} itself is larger than 10 μAcm^{-2} under a small V_e, and hence it should also be reduced if a long τ is used.

In contrast, the dark current density J_{cd} and the background photocurrent density J_{cp} after filtering are much smaller and fall within the current limit of the readout below certain voltages. For example, at -0.8 V, J_{cd} is 1.68 μAcm^{-2}, 248 times lower than J_{ed} (R_oA = 72 and 6500 Ωcm^2 for the emitter and the collector respectively). The large dark current reduction is the result of employing a high barrier filter in detector, so that only a very small portion of the emitter dark electrons injected into the base has energies that can overcome the barrier.

After filtering, there is also a reduction in the background photocurrent. Nevertheless, J_{cp} is now larger than J_{cd} below certain voltages, the detector is BLIP at 77 K. At V_e = -0.8 V, the detector can provide up to 84 %BLIP when other noise sources such as the Johnson noise and the readout noise are ignored. Due to the large reduction in the total current level, there is a reduction on the calculated D* by a factor of three, and D^*_c at λ = 8.8 μm is 1.1 x 10^{10} cm$\sqrt{}$Hz/W at 77 K with a 9.8 μm cutoff. The spectral response of the IHET is shown in Fig. 12. Note that if J_d and J_p of the present QWIP were suppressed uniformly at the readout, the D* would have been reduced by a factor of 16, demonstrating the advantage of energy filtering.

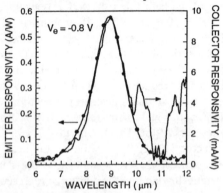

FIG. 12 The spectral response of detector #2.

Although there is a reduction of D* in the present case, the NEΔT of the IHET detector arrays is actually improved because the integration time is inversely proportional to J_{tot} and can be made much longer than a QWIP array. The resultant NEΔT based on the measured spectral response $\Delta\lambda = 1.7$ µm, the specific D*$_c$ and the 33 ms integration time is 14 mK, which is 3.2 smaller than that of the corresponding QWIP array.

In order to further increase the temperature of operation, we construct an IHET (detector #3), with which the well width of the QWIP is 50 Å, and the barrier width is 500 Å equally divided into three layers of different Al molar ratios, 0.28, 0.305 and 0.33. For this QWIP, $N_d = 1.2 \times 10^{18}$ cm^{-3} and $N_{per} = 30$. The filter consists of a double barrier structure with a 50 Å-GaAs well sandwiched between two 40 Å-Al$_{0.3}$Ga$_{0.7}$As barriers followed by a 2000 Å-Al$_{0.25}$Ga$_{0.75}$As barrier. The filter has a passband centered at E = 260 meV. The measured λ_p is shorter for this detector so that the QWIP is BLIP at 77 K, and the IHET is BLIP at 90 K as shown in Fig. 13. At 90 K and -1.0 V, D*$_c$ = 1.5 $\times 10^{10}$ cm$\sqrt{}$Hz/W at λ_p = 7.8 µm, and is 2.5 times higher than D*$_e$. The improvement is partly due to the reduction of the cutoff wavelength from 9.4 µm to 8.8 µm. as shown in Fig. 14, and partly due to the better energy selection of the filter in the elevated operating temperatures.

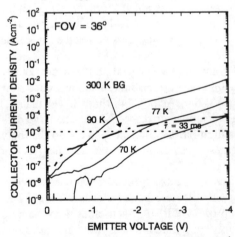
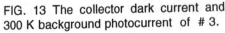

FIG. 13 The collector dark current and 300 K background photocurrent of # 3.

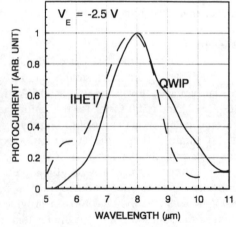

FIG 14. The spectral response of detector #3.

5. Noise properties of IHET

Since the detector noise characteristics play an important role in determining the performance, we will examine the noise characteristics of an IHET in greater detail. For a QWIP structure, the dark current induced generation-recombination (g-r) noise is well described by the standard expression of a photoconductor:[12]

$$i_{ne} = \sqrt{4 e g I_e B} ,$$ (6)

where g is the current gain, and B is the bandwidth. In deriving Eq. 6, it is important to realize that it is the number of electrical pulses N_p received at the contact for a given integration time τ_{in}, each of which contains an electrical charge of ge, but not the number of physical electrons collected (= gN_p), that follows Poisson distribution.

After passing through the energy filter between the base and the collector, the dark current is reduced by a factor of α_d. If the filtering process does not introduce additional noise, the collector noise current i_{nc} will simply be $\alpha_d i_{ne}$, and the filter acts as a noiseless amplifier. The collector detectivity D^*_c in this case is related to the emitter detectivity D^*_e by a factor of α_p/α_d. Since $\alpha_p > \alpha_d$ due to preferential photocurrent collection by the filter, $D^*_c > D^*_e$ regardless of the absolute magnitudes of the transfer ratios.

However, the electrical pulse filtering is a statistical process subject to fluctuations. The associated noise i_{cp} is referred as partition noise.[13] The standard deviation of these fluctuations, σ, obeys binomial statistics and is equal to $\sqrt{\{N_p \alpha_d (1-\alpha_d)\}}$. Thus, i_{cp} is given by

$$i_{cp} = \sqrt{2} \frac{\sigma}{\tau_{in}} e g ,$$

$$= [4 e g \alpha_d (1 - \alpha_d) I_e B]^{1/2} ,$$ (7)

where the extra factor of $\sqrt{2}$ is caused by the statistical fluctuation of g due to recombination process. Including the noise from the emitter, the total collector noise current is

$$i_{nc} = \sqrt{\alpha_d^2 i_{ne}^2 + i_{cp}^2} ,$$

$$= \sqrt{4 e g \alpha_d I_e B} ,$$ (8)

$$= \sqrt{4 e g I_c B} .$$

Eq. 8 happens to be the standard expression for g-r noise. With partition noise, i_{nc} is proportional to $\sqrt{\alpha_d}$ instead of α_d, hence $D^*_c/D^*_e = \alpha_p/\sqrt{\alpha_d}$ and can be less than 1 even when $\alpha_p > \alpha_d$.

It turns out that Eq. 8 is valid only for devices with g and α_d independent of electron energy E, which is not true for an IHET. In an IHET, the thermally activated current has a relatively wide energy distribution of 30 to 50 meV,[14] with an additional 20 meV broadening at the base.[15] This energy spread introduces two effects in the noise consideration. One is a lifetime effect: the hot-electrons with higher E are expected to have a longer lifetime and hence a larger gain, the stated g is actually an energy averaged value. Since the collector only accepts the higher energy electrons, it is possible that the gain measured at the collector g_c is larger than that at the emitter g_e. Another one is a quantum effect: α_d should also be different for electrons with different energies governed by the transmission coefficient of the filter T(E). According to Büttiker and others,[16,17] partition noise should be applied to each separate quantum channels which, in our case, are represented by different energy values. Therefore, σ in Eq. 7 should be more accurately expressed as

$$\sigma^2 = \int_0^{\infty} \rho_{th}(E)\, T(E)\, [1-T(E)]\, dE ,$$

$$\equiv N_p K ,$$

(9)

where ρ_{th} is the hot-electron energy distribution at the front boundary of the filter. The factor K depends on both T(E) and the electron phase coherence. If the resulting T(E) is slow varying, K approaches $\alpha_d(1-\alpha_d)$, and Eq. 8 is recovered. In another extreme, if T(E) only assumes values 0 and 1, ie. the transmission is deterministic, then K = 0, in which case the partition noise is completely suppressed. Incidentally, i_{nc} and the base noise current, i_{nb}, are totally correlated in this regime with $i_{ne} = i_{nc} + i_{nb}$, instead of the usual expression $i_{ne}^2 = i_{nc}^2 + i_{nb}^2$ for uncorrelated noise. Accounting only for the quantum effect, i_{nc} is given by

$$i_{nc}^2 = \alpha^2 4e g I_e B + 4e g K I_e B .$$

(10)

In order to incorporate the lifetime effect, let us consider only those electrons with E > H_f; the lower energy electrons are irrelevant in the present consideration. We denote the portion of the emitter dark current carried by these electrons to be I_{eh} and the averaged transfer ratio to be α_h with $I_c = \alpha_h I_{eh}$. Obviously, the average current gain of these electrons is equal to g_c, and the emitter noise current carried by these electrons is $\sqrt{(4e g_c I_{eh} B)}$. Therefore, including the lifetime effect, i_{nc} is now given by

$$i_{nc} = \sqrt{4e g_c (\alpha_h + K_h / \alpha_h)\, \alpha_h I_{eh} B} ,$$

$$= \sqrt{4e g_c (\alpha_d + K / \alpha_d)_h\, I_c B} ,$$

$$= \sqrt{4e g_c' I_c B} ,$$

(11)

where g_c' is the measured apparent current gain including both effects discussed above, and $(\alpha_d + K/\alpha_d)_h$ is evaluated for E > H_f. Since the two effects oppose each other, g_c' can be larger or smaller than g_e.

Here, we present the noise characteristics of three IHET with different filter structures but with similar emitter structures. The first IHET is detector #2 described in section 4. The T(E) characteristics is depicted in Fig. 15 (a).

The QWIP structure of the second IHET (referred as #4) is that of a standard

QWIP with $N_d = 1.0 \times 10^{18}$ cm^{-3} and $N_{per} = 30$. The filter consists of a 30-Å $Al_{0.3}Ga_{0.7}As$ barrier, a 65-Å GaAs well, another 30-Å $Al_{0.3}Ga_{0.7}As$ barrier, and a 2000-Å $Al_{0.22}Ga_{0.78}As$ barrier. The barriers form a band pass filter with T(E) depicted in Fig. 15 (b).

For the third IHET (referred as #5), the QWIP structure is that of a standard QWIP with $N_d = 3.5 \times 10^{18}$ cm^{-3} and $N_{per} = 5$. The filter consists of 20 periods of 30-Å $Al_{0.4}Ga_{0.6}As$ barrier and 15-Å GaAs well, and a 1000-Å $Al_{0.25}Ga_{0.75}As$ barrier. The barriers form a band pass filter with T(E) depicted in Fig. 15 (c).

In order to know the range of electrons accepted into the collector, we have measured the activation energies E_a of I_e and I_c as a function of emitter voltage V_e with $V_c = 0$ V. The results are shown in Fig. 16, in which E_{ac} is always higher than the corresponding E_{ae}, indicating the collector accepting only the high energy electrons from the injected distribution. From the activation energy difference, we deduced that only about 20 meV of the upper hot-electron distribution is being accepted into the collector even at a relatively high V_e.

Fig. 17 (a) shows the measured g_e and g_c' at temperature equal to 77 K for detector #2. In this plot, g_e increases linearly with V_e below -2 V and saturates at a value ≈ 0.5 as usually observed. For the collector, the measured g_c' is always higher than g_e as expected from the lifetime effect consideration. A computation of the factor $(\alpha_d + K/\alpha_d)$ in Eq. 11 assuming a constant ρ_{th} up to 20 meV above the bottom of the passband is equal to 0.96, showing the quantum effect indeed negligible for this filter structure.

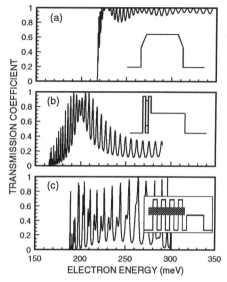

FIG. 15 The T(E) of three different filters. The inserts show the corresponding filter band structures.

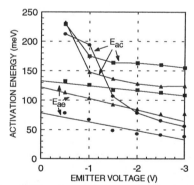

FIG. 16 The E_a of detector #2 (squares), #4 (triangles), and #5 (circles). V_e of #5 has been scaled for the same electric field as #2 and #4.

Fig. 17 (b) shows the noise characteristics for detector #4 which fall into an intermediate case where the energy effect is larger than the quantum effect below -1.6 V and vise versa at a higher V_e. The calculated $(\alpha + K/\alpha)$ factor is 0.82 in this case, bringing the expected g_c from slightly higher than 0.5 to the measured value of 0.4 at large V_e.

Fig. 17 (c) shows the noise characteristics for detector #5, which are mostly domi-

164

nated by the quantum effect. For this detector, the majority of I_e are of high energy electrons relative to H due to the larger E_F, which cause a large $g_e \approx 0.45$ at low voltages (< -1 V). At higher V_e, the apparent g_e increases rapidly due to the rapid increase of the 1/f noise observed in this QWIP. From the results of the previous two detectors, we expect that the true g_e remains to be about 0.45 at higher biases. On the other hand, the 1/f noise is absent in the collector noise measurements, consistent with the fact that 1/f noise is carried by low energy electrons which are effectively blocked by the energy filter. Therefore, g_c' can be measured at higher V_e. When V_e is small, g_c' is about 0.5 because of the lifetime effect induced by a large activation energy difference. When $V_e > -1$ V, the activation energy difference decreases, which leads to a rapid decrease of g_c' to 0.21 due to the quantum effect. The $(\alpha_d + K/\alpha_d)$ factor for this filter is calculated to be 0.66, quite consistent with the expected value of 0.47 given by the ratio of g_c' and g_e.

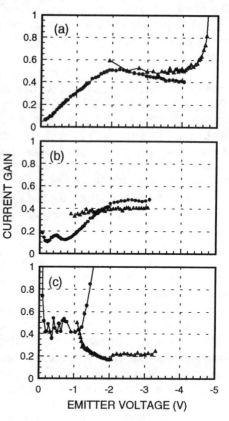

FIG. 17 The emitter current gain g_e (circles), and the collector current gain g_c' (triangles) for (a) detector #2, (b) #4 and (c) #5 as a function of V_e. g_e is defined as $i_{ne}^2/4eI_eB$. g_c' is defined as $i_{nc}^2/4eI_cB$. The V_e of detector #5 has been multiplied by a factor of 5 to account for the smaller number of periods in its QWIP structure.

6. Summary

In summary, we have discussed the electrical and infrared properties of GaAs/AlGaAs multiple quantum well structures. By properly selecting the structural parameters, detectors with different absorption wavelengths, linewidths and lineshapes can all be chosen to satisfy a specific application. By using an additional energy filter, not only the physical properties of a QWIP can be better understood, but also the sensitivity of the detector be increased and the operating temperature range be extended. An IHET is also more flexible for circuit insertion since it separates the light detection process and the electron collection process into two independently controllable operations, so that an arbitrary portion of electrons can be used for signal integration. We have also elucidated the noise theory of an IHET. We found that those filters with transmission coefficient fluctuating between a small value and a large value offer a better noise performance. In fact, a noise factor can be assigned to each filter structure, analogous to a signal amplifier. The sensitivity of an IHET can be enhanced by using a proper filter with a low noise factor. With the versatility of band-gap engineering and a better understanding of the hot-electron physics, the quantum well infrared technology by no means reaches its full potential and is expected to flourish in the future.

Acknowledgement: The authors would like to thank C. H. Kuan, L. Fotiadis, Yuan P. Li, C. W. Farley, C. Chang, D. C. Tsui, J. S. Ahearn and G. J. Iafrate for participating in this project in various stages.

References

1. C. G. Bethea, B. F. Levine, V. O. Shen, R. R. Abbott, and S. J. Hsieh, IEEE Trans. on Elect. Dev. **38**, 1118 (1991).
2. L. J. Kozlowski, G. M. Williams, G. J. Sullivan, C. W. Farley, R. J. Anderson, J. Chen, D. T. Cheung, W. E. Tennant, and R. E. DeWames, IEEE Trans. on Elect. Dev. **38**, 1124 (1991).
3. B. F. Levine, Appl. Phys. lett. **56**, 2354 (1990).
4. M. O. Manasreh, F. Szmulowicz, T. Vaughan, K. R. Evans, C. E. Stutz, and D. W. Fisher, Phys. Rev. B **43**, 9996 (1991).
5. Z. Ikonić, V. Milanović, and D. Tjapkin, Appl. Phys. Lett. **54**, 247 (1989).
6. K. K. Choi, J. of Appl. Phys. **73**, 5230 (1993).
7. S. M. Sze, *Physics of Semiconductor Devices* (John Wiley and Sons, New York, 1981), pp 647.
8. B. F. Levine, C. G. Bethea, G. Hasnain, V. O. Shen, E. Pelve, R. R. Abott, and S. J. Hsieh, Appl. Phys. Lett. **56**, 851 (1990).
9. K. K. Choi, M. Taysing-Lara, P. G. Newman, W. Chang, and G. J. Iafrate, Appl. Phys. Lett. **61**, 1781 (1992).
10. K. K. Choi, L. Fotiadis, M. Taysing-Lara, and W. H. Chang, Appl/ Phys. Lett., **60**, 592 (1992).
11. E. Pelvé, F. Beltram, C. G. Bethea, B. F. Levine, V. O. Shen, S. J. Hsieh, and R. R. Abbott, J. Appl. Phys. **66**, 5656 (1989).
12. G. Hasnain, B. F. Levine, S. Gunapala, and N. Chand, Appl. Phys. Lett., **57**, 608 (1990).
13. W. B. Davenport and W. L. Root, An Intro. to the Theory of Random Signals and Noise, (McGraw-Hill, New York, 1958), p. 140.
14. C. H. Kuan, D. C. Tsui, K. K. Choi, P. G. Newman, and W. H. Chang, submitted to Appl. Phys. Lett.
15. K. K. Choi, L. Fotiadis, M. Taysing-Lara, W. Chang, and G. J. Iafrate, Appl. Phys. Lett., **60**, 592, (1992).
16. M. Büttiker, Phys. Rev. Lett., **65**, 2901 (1990).
17. C. W. J. Beenakker and H. van Houten, Phys. Rev. B. **43**, 12066 (1991).

16 μm INFRARED HOT ELECTRON TRANSISTOR

S. D. Gunapala, J. K. Liu, J. S. Park, and T. L. Lin
Center for Space Microelectronics Technology
Jet Propulsion Laboratory, California Institute of Technology
Pasadena, CA 91109
U.S.A.

ABSTRACT. We have demonstrated a bound to continuum state GaAs/Al$_x$Ga$_{1-x}$As infrared hot electron transistor which has a peak response at $\lambda_p = 16.3$ μm. An excellent photo-current transfer ratio of $\alpha_p = 0.12$ and very low dark current transfer ratio of $\alpha_d = 7.2 \times 10^{-5}$ is achieved at a temperature of T = 60 K.

1. INTRODUCTION

Many advanced NASA satellite missions will require long wavelength infrared (IR) instruments out to 19 μm cutoff wavelength. Examples of these instruments are the Atmospheric IR Sounder (AIRS), the Tropospheric Emission Spectrometer (TES), the High Resolution Dynamic Limb Sounder (HIRDLS), and the Stratospheric Wind IR Limb Sounder (SWIRLS) which are being planned for NASA's Earth Observing System (EOS). These space applications have placed stringent requirements on the performance of the IR detectors and arrays including high detectivity, low dark current, uniformity, radiation hardness and lower power dissipation. In addition, the infrared spectrum is rich in information vital to the understanding of composition, structure and the energy balance of molecular clouds and star forming regions of our galaxy. Therefore, NASA has great interest in infrared detectors both inside and outside the atmospheric windows. This paper will present a study and development of a low dark current very long wavelength intersubband IR hot electron transistor (IHET).

There has been a lot of interest recently in the detection of long wavelength ($\lambda = 8$-12 μm) infrared radiation using multiple quantum wells, due to the fact that these quantum well IR photodetectors [1-13] (QWIPs) and IHETs [8,14-16] can be fabricated using the mature III-V materials growth and processing technologies. This superior materials control results in high uniformity and thus allows fabrication of large staring arrays ($\lambda = 8$-12 μm) with excellent imaging performance [17-19]. One of the problems associated with the very long wavelength QWIPs is the higher dark current which adversely affects detector

H. C. Liu et al. (eds.), Quantum Well Intersubband Transition Physics and Devices, 167–176.

performance. By analyzing the dark current of shallow quantum wells we have realized that the total tunneling current (sequential tunneling + thermionic assisted tunneling) is significantly higher than the thermionic dark current (Fig. 1). The conduction electrons carrying these two tunneling current components are lower in energy than the photoelectrons [8]. Therefore, a 16μm GaAs/Al$_x$Ga$_{1-x}$As IHET which can effectively filter out sequential tunneling and thermionic assisted tunneling currents was fabricated.

2. DARK CURRENT

In this section the dark current of a single quantum well, which has intersubband absorption peak at 16 μm will be analyzed. First effective number of electrons [11,20] n(V) which are thermally excited into the continuum transport states, as a function of bias voltage V were calculated, using the following expression.

$$n(V) = \left(\frac{m^*}{\pi \hbar^2 L_p} \right) \int_{E_0}^{\infty} f(E)T(E,V)dE$$

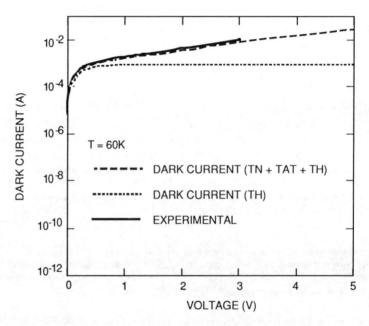

FIG.1 Theoretical and experimental (solid) dark current-voltage curves at T = 60K. Dotted curve shows the dark current (theoretical) due to thermionic emission only. Dashed curve shows the total dark current (thermionic + tunneling + thermionic assisted) versus bias voltage.

The first factor containing the effective mass m* represents the average three dimensional density of states. Where L$_p$ is the superlattice period, f(E) is the Fermi factor

$f(E) = [1+ \exp(E-E_0-E_F)/KT]^{-1}$, E_0 is the bound state energy, E_F is the two-dimensional Fermi energy, E is the energy of the electron, and T(E,V) is the tunneling current transmission factor. This tunneling transmission factor obtained by applying WKB approximation to a biased quantum well is:

$$T(E) = (4\sqrt{E}\sqrt{(V_0 - E)} / V_0)e^{-2\tau}$$

where $\tau = (2L\sqrt{2m^*} / 3\hbar\Delta V)(V_0 - E)^{\frac{3}{2}}$, V_0 is the barrier height, ΔV is the bias voltage per superlattice period, and L is the barrier width. The number of electrons, given by n(V), accounts for thermionic emission above the barrier height when $E>V_0$ and thermionic assisted tunneling and tunneling when $E<V_0$. Then the bias-dependent dark current $I_d(V)$ was calculated, using $I_d(V) = eAn(V)v(V)$, where v(V) is the average transport velocity , A is the device area, and e is the electronic charge. The average transport velocity was calculated using $v(V) = \mu F[1+ (\mu F/v_s)^2]^{-1/2}$, where μ is the mobility, F is the electric field, and v_s is the saturated drift velocity. In order to obtain T = 60K bias-dependent dark current $\mu = 1200$ cm²/Vs and $v_s = 5.5\times10^6$ cm/s was used. Fig. 1 shows the T= 60K dark current due to thermionic emission, total dark current (thermionic + thermionic assisted tunneling + tunneling), and experimental dark current of a QWIP sample which has wavelength cutoff λ_c = 17.8 μm. According to the calculations tunneling through the barriers dominate the dark current at temperatures below 30K, at temperatures between 30-55 K thermionic assisted tunneling might become important, and at temperatures above 55 K thermionic emission into the continuum transport states dominate the dark current.

3. EXPERIMENT

As shown in Fig. 2 the device structure consisted of a multi-quantum well region of 50 periods of 500 Å undoped $Al_{0.11}Ga_{0.89}As$ barrier and 65 Å doped GaAs well. The quantum wells were doped to n = 5×10^{17} cm⁻³, and sandwiched between a heavily doped (n = 1×10^{18} cm⁻³) 1 μm GaAs contact layer at the bottom as the emitter contact and a doped (n = 3×10^{17} cm⁻³) 500 Å GaAs layer on the top as the base contact. On top of the base a 2000 Å undoped $Al_{0.11}Ga_{0.89}As$ layer and a doped (n = 3×10^{17} cm⁻³) 0.5 μm GaAs layer were grown. The 2000 Å undoped $Al_{0.11}Ga_{0.89}As$ As layer acted as a discriminator between the tunnel-electrons and photo-electrons, and the top 0.5 μm GaAs layer served as the collector. This device structure was grown on a semi-insulating GaAs substrate using molecular beam epitaxy.

The intersubband absorption was measured on a 45° polished multipass waveguide [21] as shown in the inset of Fig. 3. As shown in the Fig. 3 the T = 300K absorption coefficient spectra α_p has a peak infrared absorption coefficient $\alpha_p = 534$ cm⁻¹ at $\lambda_p = 17.1$ μm with absorption half heights at 14.2 and 18 μm (i.e., a full width at half maximum of $\Delta\lambda = 3.8$ μm). At low temperature the half width narrows and the peak absorption coefficient increases [22,23] by a factor of 1.3 so that $\alpha_p = 694$ cm⁻¹ at T = 60K corresponding to an unpolarized quantum efficiency $\eta = (1 - e^{-2\alpha l})/2 = 16.5\%$.

To facilitate the application of bias to the quantum well structure, the following processing steps were carried out. First arrays of 200x200 μm² square collectors were chemically etched. In the next processing step the 6.25x10⁻⁴ cm² QWIP mesas which overlap with collector mesas were etched. Finally, Au/Ge ohmic contacts were evaporated onto the emitter, base and collector contact layers. The emitter and collector dark currents versus base-collector bias voltage are shown in Fig. 4. This figure also shows the excellent dark current filtration capability of the quantum filter. The dark current transfer ratio

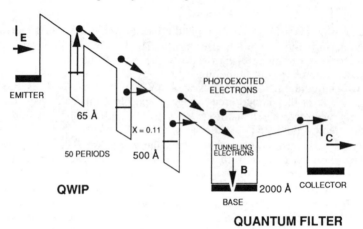

FIG.2 Conduction-band diagram of a infrared hot electron transistor, which utilizes bound to continuum intersubband transition.

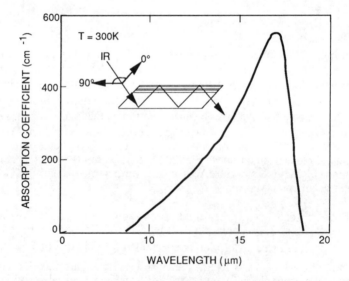

FIG.3 Absorption coefficient spectra $\alpha(\lambda)$ of the long wavelength quantum well infrared detector. This absorption spectra was measured at room temperature using a 45° multipass waveguide geometry as shown in the inset.

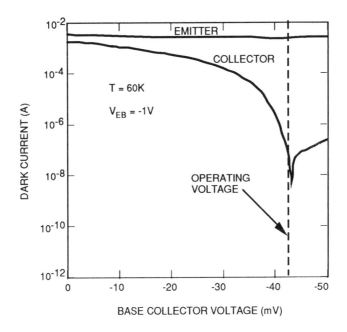

FIG.4 IHET emitter and collector dark currents versus base-collector voltage at T = 60K. Emitter bias was kept at -1V relative to the base potential. This figure also shows the lower energy dark current filtration capability of the quantum filter.

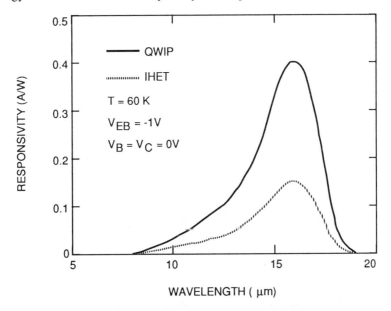

FIG.5 Emitter and collector responsivity spectra at temperature T = 60K. Emitter was kept at -1V bias relative to the base and collector.

$\left(\alpha_{d} = I_{C(\text{dark})} / I_{E(\text{dark})}\right)$ is 7.2×10^{-5} at operating base-collector bias voltage $V_{CB} = -42$ mV (Fig. 7).

These 200×200 μm^2 square detectors were back illuminated through a 45° polished facet as described in detail previously [1] and responsivity spectrums were measured with a tunable source consisting of a 1000K blackbody and a grating monochromator. The emitter and collector responsivity spectrums measured at T = 60K are shown in Fig. 5. These two spectrums are similar in shape and peak at $\lambda_p = 16.3$ μm. The values of the cutoff wavelength λ_c and the spectral width $(\Delta\lambda / \lambda)$ (full width at half maximum) are 17.3 μm and 20% respectively. The absolute responsivity was measured by two different methods; by comparing the detector photo-response with the photo-response of a calibrated pyroelectric detector, and by using a calibrated blackbody source. The peak responsivity Rp of the detector was 400 mA/W. Fig. 6 shows the IHET emitter and collector photo currents versus base-collector voltage at T = 60K. The emitter was kept at -1V bias relative to the base potential. Due to the hot electron relaxation in the wide base region, the photo current at collector is smaller relative to the emitter photo current. Photo current transfer ratio $\left(\alpha_p = I_{C(\text{photo})} / I_{E(\text{photo})}\right)$ is 1.2×10^{-1} at $V_{CB} = -42$ mV (Fig. 7). It is worth noticing that α_d is more than three orders of magnitude smaller than α_p.

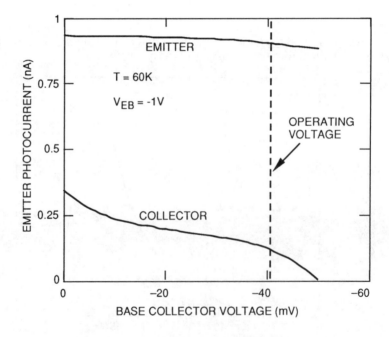

FIG.6 IHET emitter and collector photo currents versus base-collector voltage at T = 60K. Emitter was kept at -1V bias relative to the base potential.

4. RESULTS

The optical gain g of the detector determined from $R = (e / h\nu)\eta g$ is given by $g = 0.2$. The noise current [24] i_n was calculated using $i_n = \sqrt{4eI_d g \Delta f}$, where Δf is the bandwidth. The calculated noise current of the detector is $i_n = 17$ pA at T = 60K. The peak D^* can now be calculated from $D^* = R \sqrt{A\Delta f} / i_n$. The calculated D^* between the emitter and the base (QWIP) at $V_{EB} = -1$V, $V_{CB} = -42$ mV and T = 60K is 5.8×10^8 cm$\sqrt{\text{Hz}}$/W. The detectivity D* at the collector (IHET) is determined[15] from $D^*(\text{IHET}) = (\alpha_p / \sqrt{\alpha_d}) D^*(\text{QWIP})$. Table 1 shows the QWIP and IHET detectivity D^* at temperature T = 60 K for several base-collector bias voltages. As shown in Fig. 8 detectivity D* increases

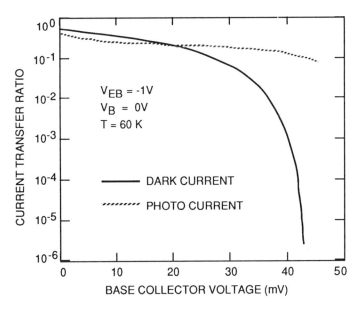

FIG.7 Photo current and dark current transfer ratio of IHET as a function of base-collector bias voltage at temperature T = 60 K.

dramatically with decreasing temperature reaching $D^* = 1 \times 10^{12}$ cm$\sqrt{\text{Hz}}$/W at T = 25 K and is even larger at lower temperatures. In contrast, detectivity D^* of HgCdTe detectors are saturated as the temperature is lowered.

5. CONCLUSIONS

In summary, we have demonstrated a very long wavelength ($\lambda_c = 17.3$ μm) IHET. This device clearly shows the dark current filtration capability of the energy filter. Therefore, the D* of IHET is much higher than the D^* of two terminal multi-quantum well detectors. It is also worth noting that the power dissipation of these detectors are two orders of magnitude smaller than that of HgCdTe detectors. In addition, these detectors show

174

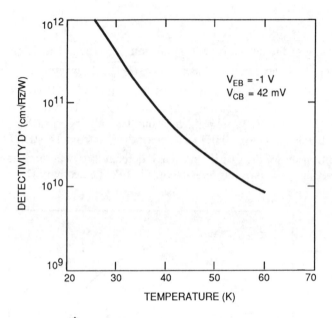

FIG. 8 Peak detectivity D^* for a IHET having a cut-off wavelength of $= 17.3$ μm as a function of temperature T.

TABLE I. Comparison of QWIP and IHET detectivity D^* at temperature T = 60 K for several base-collector bias voltages.

V_{BC} (mV)	$\alpha_p / \sqrt{\alpha_d}$	D* QWIP (cm√Hz/W)	D* IHET (cm√Hz/W)
-30	0.72	6.0×10^8	4.3×10^8
-35	1.17	5.9×10^8	6.9×10^8
-40	3.98	5.9×10^8	2.3×10^9
-42	14.14	5.8×10^8	8.2×10^9

absolutely no change in dark current and responsivity after an exposure of 6.5 Mrad of 1 MeV proton radiation [25] which is equal to 5 years of radiation damage in space. Due to excellent uniformity, radiation hardness, lower power dissipation and lower 1/f noise these GaAs based QWIPs and IHETs are extremely attractive to NASA applications such as EOS missions and IR astronomy.

6. ACKNOWLEDGMENTS

We are grateful to K. K. Choi of the Army Research Laboratory and K. M. S. V. Bandara and B. F. Levine of the AT&T Bell Laboratories for many useful discussions and C. A. Kukkonen, V. Sarohia, S. Khanna, K. M. Koliwad, B. A. Wilson, and P. J. Grunthaner of the Jet Propulsion Laboratory for encouragement and support of this work.

The research described in this paper was performed by the Center for space Microelectronics Technology, Jet Propulsion Laboratory, California Institute of Technology, and was jointly sponsored by the Ballistic Missile Defense Organization/Innovative Science and Technology Office, and the National Aeronautics and Space Administration, Office of Advanced Concepts and Technology.

7. REFERENCES

1. B. F. Levine, C. G. Bethea, G. Hasnain, J. Walker, and R. J. Malik, Appl. Phys. Lett. **53** (1988) 296.
2. B. F. Levine, *Proceedings of the NATO Advanced Research Workshop on Intersubband Transitions in Quantum Wells*, Cargese, France Sept. 9-14, 1991, edited by E. Rosencher, B. Vinter, and B. F. Levine (Plenum, London, 1992).
3. J. Y. Andersson and L. Lundqvist, Appl. Phys. Lett. **59**, 857 (1991).
4. S. D. Gunapala, B. F. Levine, D. Ritter, R. A. Hamm, and M. B. Panish, Appl. Phys. Lett. **58**, 2024 (1991).
5. S. D. Gunapala, B. F. Levine, D. Ritter, R. A. Hamm, and M. B. Panish, Appl. Phys. Lett. **60**, 636 (1992).
6. B. K. Janousek, M. J. Daugherty, W. L. Bloss, M. L. Rosenbluth, M. J. O'Loughlin, H. Kanter, F. J. De Luccia, and L. E. Perry, J. Appl. Phys. **67**, 7608 (1990).
7. S. R. Andrews and B. A. Miller, J. Appl. Phys. **70**, 993 (1991).
8. K. K. Choi, M. Dutta, P. G. Newman, M. L. Saunders, and G. L. Iafrate, Appl. Phys. Lett. **57**, 1348 (1990).
9. L. S. Yu and S. S. Li, Appl. Phys. Lett. **59**, 1332 (1991).
10. A. G. Steele, H. C. Liu, M. Buchanan, and Z. R. Wasilewski, Appl. Phys. Lett. **59**, 3625 (1991).
11. S. D. Gunapala, B. F. Levine, L. Pfeiffer, and K. West, J. Appl. Phys. **69**, 6517 (1990).
12. M. J. Kane S. Millidge, M. T. Emeny, D. Lee, D. R. P. Guy, and C. R. Whitehouse in Ref. 2.
13. C. S. Wu, C. P. Wen, R. N. Sato, M. Hu, C. W, Tu, J. Zhang, I. D. Flesner, L. Pham, and P. S. Nayer, IEEE Tran. Electron Devices **39**, 234 (1992).
14. K. K. Choi, M. Dutta, and R. P. Moerkirk, Appl. Phys. Lett. **58**, 1533 (1991).
15. K. K. Choi, M. Taysing-Lara, L. Fotiadis, and W. Chang, Appl. Phys. Lett. **59**, 1614 (1991).

16. K. K. Choi, L. Fotiadis, M. Taysing-Lara, and W. Chang, Appl. Phys. Lett. **59**, 3303 (1991).

17. B. F. Levine, C. G. Bethea, K. G. Glogovsky, J. W. Stayt, and R. E. Labenguth, Semicon. Sci. Technol. **6** (1991) c114.

18. M. T. Asom et al., Proceedings of the **IRIS** Specialty Group on Infrared Materials, Boulder, CO, August 12-16,1991, Vol. **I**, p. 13.

19. L. J. Kozlowski, G. M. Williams, G. J. Sullivan, C. W. Farley, R. J. Anderson, J. K. Chen, D. T. Cheung, W. E. Tennant, and R. E. DeWames, IEEE Trans. Electron Devices **38**, 1124 (1991).

20. B. F. Levine, C. G. Bethea, G. Hasnain, V. O. Shen, E. Pelve, R. R. Abbott, and S. J. Hsieh, Appl. Phys. Lett. **56**, 851 (1990).

21. B. F. Levine, R. J. Malik, J. Walker, K. K. Choi, C. G. Bethea, D. A. Kleinman, and J. M. Vandenberg, Appl. Phys. Lett. **50**, 273 (1987).

22. G. Hasnain, B. F. Levine, C. G. Bethea, R. R. Abbott, and, S. J. Hsieh, J. Appl. Phys. **67**,4361 (1990).

23. M. O. Manasreh, F. Szmuiowicz, D. W. Fischer, K. R. Evans, and C. E. Stutz, Appl. Phys. Lett. **57**, 1790 (1990).

24. B. F. Levine, A. Zussman, S. D. Gunapala, M. T. Asom, J. M. Kuo, and W. S. Hobson, J. Appl. Phys. **72**, 4429 (1992).

25. S. D. Gunapala, J. S. Park, T. L. Lin, and J. K. Liu (to be published).

Long Wavelength λ_c = 18μm Infrared Hot Electron Transistor

C.Y. Lee[a,c,d,e],M.Z. Tidrow[a,b],K.K. Choi[a],W. Chang[a],L.F. Eastman[e]
a. U.S. Army Research Laboratory, EPSD, Ft. Monmouth, NJ 07703
b. National Research Council Research Associate Program
c. National Science Foundation Grant ECS-9217579
d. Geo Centers Inc., Lake Hopatcong, NJ 07849
e. Cornell University, Ithaca NY 14853

1. Introduction

GaAs/$Al_xGa1_{-x}As$ quantum well infrared photodetectors[1,2] (QWIPs) have advanced considerably in the last several years for infrared detection in the spectral regions λ = 8-12 μm and λ=3-5 μm. Recently the performance of extended-wavelength QWIPs operating over the spectral range λ = 11-15 μm[3] and λ = 14-18 μm[4] have also been demonstrated. These extended-wavelength detectors are of interest for a variety of applications including the detection of "cold" objects in space and for observing the earth's atmosphere and composition. For these applications, it is desirable to achieve very low levels of dark current for two reasons. Firstly, the background photon flux is very low in this wavelength range. Therefore, operation of a QWIP requires a very low dark current level for background limited infrared photodetection (BLIP). Secondly, for thermal imaging purposes, it is desirable to incorporate the extended-wavelength QWIPs into focal plane arrays (FPAs). However, the thermally activated dark current of a QWIP will saturate a typical FPA readout circuit, even at a rather low temperature. Here, we will describe an infrared hot-electron transistor (IHET) structure, whose dark current is two to three orders of magnitude lower than that of a QWIP. This IHET is capable of background-limited thermal imaging at T = 55K at λ_c = 18μm. In this paper, the performance of the extended-wavelength IHET is presented, including the dark current, the responsivity and detectivity as a function of temperature, bias and wavelength.

2. Sample Design and Growth

In order to tailor the intersubband absorption to occur at extended-wavelengths, the barrier height of a GaAs/AlGaAs QWIP has to be relatively low. This results in a large tunneling dark current, even at low temperatures. To reduce the tunelling current, a high energy pass filter is placed next to the QWIP to form an infrared hot-electron transistor (IHET).[5] We investigate an IHET designed to absorb light at λ = 13-18 μm in which the quantum well

H. C. Liu et al. (eds.), Quantum Well Intersubband Transition Physics and Devices, 177–186.

intersubband absorption takes place between the ground state, E1 and the continuum excited states. The QWIP structure consists of a 1μm n⁺ emitter contact layer, and 30 periods of 60-Å GaAs well and 500-Å $Al_{0.15}Ga_{0.85}As$ barrier. Following the QWIP is a thin 500-Å $In_{0.08}Ga_{0.92}As$ base layer. On top of the base, a 1400-Å $Al_{0.17}Ga_{0.83}As$ barrier (high pass filter) and a 1.1μm n⁺ GaAs collector are grown. The Al molar ratio of the filter barrier is graded from 0.07 to 0.17 for an additional 300Å on each side of the barrier to reduce the quantum mechanical reflection of the injected electron. The doping density N_d is 0.8×10^{18} cm⁻³ in the quantum wells and 1.0×10^{18} cm⁻³ in the base. The structure was grown via molecular-beam epitaxy. The band structure of the IHET is shown in Fig. 1.

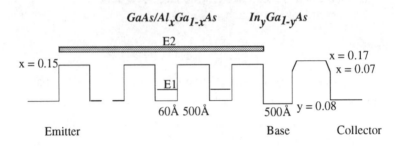

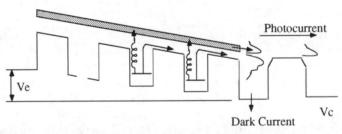

Fig. 1 The energy band structure of the infrared hot electron transistor at zero bias, and with emitter bias Ve. The emitter consists of an infrared sensitive QWIP. The structure between the base and collector is the electron energy filter.

3. Dark Current

The temperature dependent dark current characteristics of the IHET are shown in Fig. 2(a) and (b). A bias V_e is applied to the emitter while holding the base and collector at zero bias. The dark current flowing into the emitter, J_{de}, is the dark current level of the corresponding QWIP. The dark current flowing from the collector is denoted as J_{dc}. Also shown are the 300K emitter and collector background photocurrent, J_{pe} and J_{pc}, obtained by exposing the detector to room temperature radiation at a field of view $\phi = 36°$, while maintaining the detector at T = 10K. Using AC lock-in techniques, the responsivity is

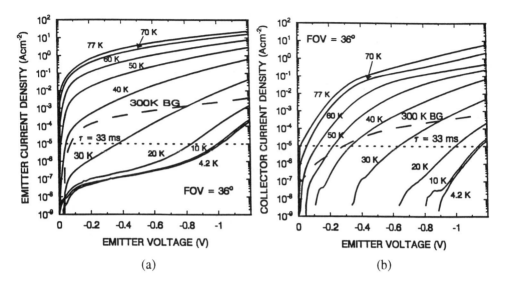

Fig. 2 Emitter and collector dark current density as a function of the emitter bias Ve for different temperatures. The 300K background photocurrent (300K BG) is shown as the dashed line. The saturating current level for a typical readout circuit for an integrtion time τ = 33ms is shown as the dotted line.

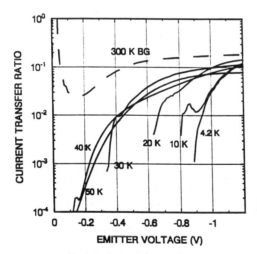

Fig. 3 Current Transfer Ratio for 300K background photocurrent (300K BG) and labelled dark currents.

180

found to be independent of temperature. The saturating current level for a typical readout circuit[6] for an integration time $\tau = 33$ms is shown in Fig. 2 as the dotted line. In this example, Fig 2(a) demonstrates that the QWIP is not BLIP for temperatures larger than T = 35K, and in order to satisfy readout current limits, detector operation is restricted to emitter biases less than 0.08V at T=35K. In comparison to the QWIP, the IHET demonstrates BLIP performance for temperatures up to T = 55K without saturating the readout circuit, as seen in Fig. 2(b). The increase of the BLIP temperature is achieved by blocking the low energy dark current with the electron energy filter while allowing a larger fraction of the photocurrent to be collected at the collector. A plot of the dark current transfer ratio $\alpha_d = J_{dc}/J_{de}$ given in Fig.3 shows that the collector dark current is reduced by up to two to four orders of magnitude compared to that of the emitter.

The dark current is expected to increase exponentially with temperature as the electrons are thermally excited out of the well into the continuum states. This can be modelled via the thermionic emission current equation[7],

$$I_d \, \alpha \, Ce^{-\Delta E/kT} \tag{1}$$

where C is a constant. $\Delta E = E_c - E_f$ is the thermal activation energy, E_c is the barrier height relative to E1, and E_f is the fermi energy in the quantum well. The emitter dark current I_{de} versus 1/T plot shown in Fig. 4(a) demonstrates an excellent linear fit of the

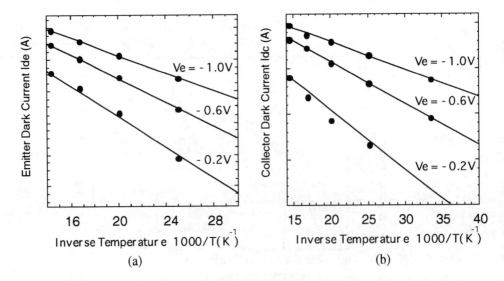

Fig. 4 Emitter and Collector dark current plotted against inverse temperature at different emitter biases (V_e). The straight line slope gives an emitter activation energy $\Delta E = 52$ meV and collector activation energy $\Delta E = 61$ meV at $V_e = -0.2$V

dark current at different emitter biases. From the slope of this line we determine the quantity ΔE, for various biases, which is shown in Fig. 5. At low biases, where thermally assisted tunelling (TAT) is negligible, ΔE is equal to 52 meV. This is in excellent agreement with the expected value of 52.8 meV which is given by the difference between E_c (= 70 meV) and E_f (= 17.2 meV) at zero bias condition. As the bias is increased, ΔE decreases linearly due to an increasing contribution from the thermally assisted tunelling, whose activation energy can be substantially lower than $E_c - E_f$.

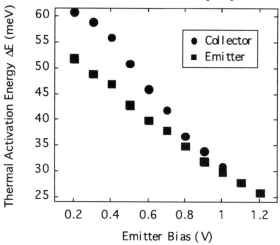

Fig. 5 Thermal activation energy $\Delta E = Ec - E1$ for emitter and collector plotted as a function of emitter bias Ve.

At low biases, due to the energy relaxation in the base, only very high energy electrons injected from the emitter can be collected at the collector. Thus, the collector activation energy is much higher than that of the emitter. As the bias increases, more injected electrons can overcome the filter barrier due to the increase in the depletion voltage in the base. At 0.5 V, the collector activation energy is 52 meV, same as that of the emitter at low biases, indicating that the thermionic emission current can now go through the filter barrier, but not the TAT current which has an activation energy of 43 meV. At even higher biases, most of the injected electrons have energy higher than the filter barrier, which leads to the same activation energy between the emitter and collector. This observation demonstrates the selective energy filtering effect of the collector barrier.

4. Responsivity

The responsivity spectra was measured by fabricating two-level mesa transistor structures with emitter area $A_e = 7.92 \times 10^{-4}$ cm^2 and collector area $A_c = 2.25 \times 10^{-4}$ cm^2, using a polished 45° incident facet. A blackbody source and a monochromator were used as the infrared source. The blackbody source power was calibrated with a HgCdTe detector, using a 10μm cut-on filter to block unwanted contributions from higher-order diffraction of low-wavelength light. A calculation of the theoretical blackbody spectral power was used to verify the measured power as follows. Eqn.(2) gives the theoretical blackbody spectral density, i.e. the power radiated per unit wavelength at wavelength λ by a unit area of blackbody at temperature T_B.

$$W(\lambda) = \frac{2 \pi h c^2}{\lambda^5} \left[\exp\left(\frac{h c}{\lambda k T_B} \right) - 1 \right]^{-1} \tag{2}$$

The theoretical blackbody power incident per unit area can then be given by

$$I(\lambda) = W(\lambda) \, E_u(\lambda) \, F \tag{3}$$

where $E_u(\lambda)$ is the monochrometer grating efficiency for unpolarized light and F is a constant used to fit the measured power spectrum. Using this method, an accurate fit was obtained between experimental and theoretical power density over the entire range $\lambda = 13 - 20 \,\mu m$ as shown in Fig 6. Finally, the spectral power incident upon the device itself is given by

$$P(\lambda) = I(\lambda) \, A \, T \, (1\text{-}r) \, E_p \cos(\phi) \tag{4}$$

where A is the detector area, ϕ is the angle of incidence, T is the transmission of the dewar window, $r = .28$ is the reflectivity of the GaAs surface. $E_p = 0.727$ is a parameter that accounts for the difference in the grating efficiency of s and p polarized light.

The spectral responsivity for unpolarized light for both emitter and collector are shown in Fig. 7(a) and (b) respectively for several values of emitter bias V_e at T = 10K. The values shown are the DC responsivity converted from the measured AC responsivity by multiplying by a factor of 2.22. For the emitter, the peak responsivity at Ve = -0.5V is $R_{pe} = 0.88$ A/W which occurs at a wavelength of $\lambda_p = 15\mu m$. The cutoff wavelength is found to be $\lambda_c = 18\mu m$, yielding a spectral width $\Delta\lambda = 5\mu m$ and $\Delta\lambda/\lambda = 33\%$. As the bias was increased, the responsivity increases in magnitude as expected due to an increase in the mean free path of the hot electron. The collector responsivity lineshape is similar to that of the emitter, but its magnitude is reduced by a factor referred to as the photocurrent transfer ratio $\alpha_p = I_{pc}/I_{pe}$. α_p of the 300K background photocurrent is shown in Fig. 3. For example, at Ve = -0.5V, T = 10K, the collector peak responsivity $R_{pc} = 0.12$ A/W at $\lambda = 15\mu m$. The ratio R_{pc}/R_{pe} then yields a value $\alpha_p = 0.136$, agreeing well with the value of $a_p = 0.130$ obtained in Fig. 3. Both the emitter and collector responsivity were found to be relatively temperature independent up to T = 50K.

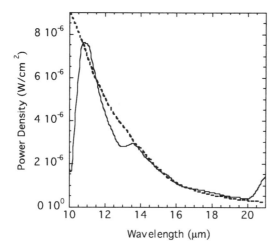

Fig 6 Fitted theoretical blackbody power density (dotted line) and the power density as measured by a Hg Cd Te detector (solid line).

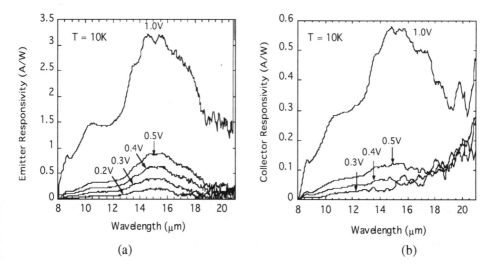

Fig. 7 Emitter and collector responsivity taken for emitter bias |Ve| = 1.0, 0.5, 0.4, 0.3, 0.2V

184

5. Detectivity

The detectivity of the IHET can be determined by,

$$D^* = \frac{R\sqrt{A}}{\sqrt{4\,e\,g\,I_d}} \tag{5}$$

where R is the responsivity, A is the area of the device, g is the gain and I_d is the dark current. Noise measurements performed on the IHET shows that while 1/f noise dominates in the emitter, the filter blocks the frequency-dependent noise component effectively so that the collector current consists of only the g-r noise with g = 0.5 up to Ve = -0.6 V. An analysis of the dark current and photocurrent transfer ratios predicts an enhancement of the collector detectivity of the collector. Combining equation (5) with the definitions,

$$I_{dc} = \alpha_d I_{de} \tag{6a}$$

$$I_{pc} = \alpha_p I_{pe} \tag{6b}$$

it can be shown that

$$D_c^* = \frac{\alpha_p}{\sqrt{\alpha_d}}\, D_e^* \tag{7}$$

Therefore, there is an enhancement of the detectivity at the collector, provided that $\alpha_p > \sqrt{\alpha_d}$. Using equation (7), it can also be shown that the noise equivalent temperature for collector ($NE\Delta T_c$) and emitter ($NE\Delta T_E$) are related as $NE\Delta T_c = (\alpha_d/\alpha_p)NE\Delta T_E$. From Fig. 3, the $NE\Delta T$ of the IHET is improved by a factor at 100 at T = 55K, and Ve = 0.2V.

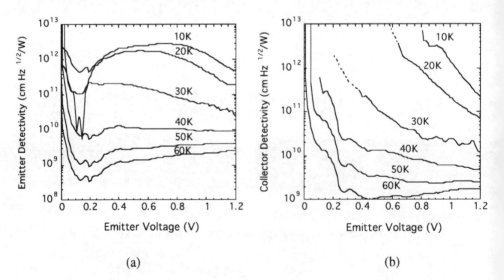

(a)   (b)

Fig. 8 Emitter and collector detectivity vs. emitter voltage |Ve| at T = 10, 20, 30, 40, 50K. The dotted line was extrapolated from the dark current fit given in Fig. 4b.

The detectivity of emitter and collector are shown in Fig. 8(a) and (b). Due to limited instrument sensitivity, the collector dark current levels (I_{dc}) at very low bias could not be accurately measured. Therefore, to calculate the detectivity in the low bias regions, I_{dc} values are extrapolated from the linear fits of the thermionic emission dark current given in Fig. 4. The temperature dependence of the maximum emitter detectivity D_e^* and collector detectivity D_C^* at $V_e = 0.2$ V is shown in Fig.9. There is an enhancement of the detectivity at the collector as predicted by eqn. (7). For example, the emitter detectivity D_e^* (T=30K) $= 2 \times 10^{11}$ cm $Hz^{1/2}$/W at Ve = 0.2V, T = 30K, $\lambda = 14.5$ mm. Using eqn. (7), the collector should be enhanced to a detectivity of D_c^* (T=30K) $= 1.2 \times 10^{12}$ cm $Hz^{1/2}$/W, which agrees well with the measured value shown in Fig. 9. Due to the increase of thermionic dark current with the temperature, the D* value will decrease with

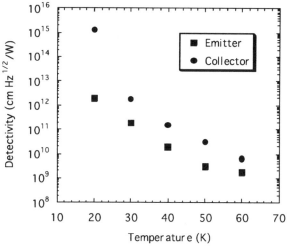

Fig. 9 Emitter and collector detectivity vs. temperature T. The T = 20K and 30K collector detectivity was calculated using extrapolated dark current values from Fig. 4b.

increasing temperatures. However, because $\alpha_p > \sqrt{\alpha_d}$, even at T = 55K, the detectivity is still enhanced from D_e^*(T=50K) $= 9.2 \times 10^8$ cm $Hz^{1/2}$/W to D_c^* (T=50K) $= 3.2 \times 10^{10}$ cm $Hz^{1/2}$/W at Ve = 0.2V.

In conclusion, we have designed and demonstrated an extended-wavelength $\lambda_c = 18\mu$m IHET, which utilizes a high-energy electron filter next to a QWIP to significantly reduce the tunneling current. This structure is background limited and its current level is compatible with a typical readout circuit at T = 55K, making it a viable candidate for use in thermal imaging focal plane array applications.

References

1. S.R. Andrews, B.A. Miller, *"Experimental and theoretical studies of the performance of quantum-well infrared photodetectors"*, J. Appl. Phys. 70, 993 (1991)
2. B.F. Levine, S.D. Gunapala, R.F. Kopf, *"Photovoltaic GaAs quantum well infrared detectors at 4.2 μm using indirect Al_xGa_{1-x} barriers"*, Appl. Phys. Lett. 58, 1551 (1991)
3. A. Zussman, B.F. Levine, J.M. Kuo, J. de Jong, *"Extended long-wavelength $\lambda = 11$-$15\mu m$ GaAs/Al_xGa_{1-x}As quantum-well infrared photodetectors"*, J. Appl. Phys. 70, 5101 (1991)
4. B.F. Levine, A. Zussman, J.M. Kuo, J. de Jong, *"$19\mu m$ cutoff long-wavelength GaAs/Al_xGa_{1-x}As quantum well infrared photodetectors"*, J. Appl. Phys. 71, 5130 (1992)
5. K.K. Choi, L. Fotiadis, M. Taysing-Lara, W. Chang, G.J. Iafrate, *"High detectivity InGaAs base infrared hot-electron transistor"*, Appl. Phys. Lett, 59, 3303 (1991)
6. K.K. Choi, M.Z. Tidrow, M. Taysing-Lara, W. Chang, C.H. Kuan, C.W. Farley, F. Chang, *"Low Dark Current Infrared Hot-Electron Transistor for 77K Operation"*,
7. M.J. Kane, S. Milidge, M.T. Emeny, D. Lee, D.R.P. Guy, C.R. Whitehouse, *"Performance Tradeoffs in the Quantum Well Infrared Detector"*, Intersubband Transitions in Quantum Wells, Ed. E. Rosencher et al., Plenum Press, New York, 1992, pp. 31-42

A NOVEL TRANSPORT MECHANISM FOR PHOTOVOLTAIC QUANTUM WELL INTERSUBBAND INFRARED DETECTORS

Harald Schneider, Stefan Ehret, Eric C. Larkins, John D. Ralston, and Peter Koidl
Fraunhofer-Institut für Angewandte Festkörperphysik
Tullastrasse 72
79108 Freiburg, Federal Republic of Germany

ABSTRACT. We have investigated photovoltaic quantum well (QW) intersubband infrared detector structures in which the photovoltaic behavior is caused by internal electric fields across the barrier regions. The photovoltage arises from a rectification of the photoexcited carrier transport in the barrier regions. The effect is demonstrated experimentally using GaAs/AlAs/AlGaAs double-barrier QW structures operating in the 3-5 μm regime. We discuss theoretically the detectivity limit expected for similar, but optimized QW infrared detector structures. In addition, we present a selfconsistent calculation of the band profile of an optimized photovoltaic detector structure designed for the 8-12 μm-regime.

1. Introduction

Intersubband transitions in quantum well (QW) structures have found promising applications in infrared photodetection and nonlinear optics in the $3 - 5\mu$m and $8 - 12\mu$m spectral ranges [1–3]. Infrared detection studies have mainly concentrated on photoconductive detection mechanisms, including successful implementations in focal-plane array cameras [2,3].

Device concepts for photovoltaic QW infrared detector structures have been developed by Goossen et al. [4] and by Kastalsky et al. [5]. We have previously observed large photovoltaic effects in double barrier quantum well (DBQW) intersubband detector structures, originally realized in order to demonstrate intersubband photodetection in the $3 - 5\mu$m regime using GaAs QW's [6,7]. In these devices, a conduction band edge profile similar to those shown in the inset of Fig. 1 is realized by incorporating thin tunnel barriers (AlAs) between the GaAs QW's and the AlGaAs barriers. Although asymmetric AlAs layer thicknesses [8] or interfaces have some influence on the observed photovoltaic behavior, the main effect is a consequence of space charges, caused by an asymmetric dopant distribution with respect to the well center. The space charges produce a built-in electric field across the barrier layers. This field gives rise to a photovoltaic behavior through a rectification of the photoexcited carrier motion in the barrier region [9].

H. C. Liu et al. (eds.), Quantum Well Intersubband Transition Physics and Devices, 187–196.
© *1994 Kluwer Academic Publishers. Printed in the Netherlands.*

188

The purpose of the present study is to work out the influence of these internal fields on the photovoltaic behavior and on the dark current of DBQW infrared detector structures. We also discuss the theoretical detectivity limit that can be obtained using the underlying transport mechanism. Finally, we propose a different, optimized QW infrared detector structure with which this detectivity limit should be approached.

The transport mechanism is demonstrated and analyzed in section 2. The detectivity limit is discussed in section 3. In section 4, we present an optimized QW detection concept for photovoltaic detection in the 8-12 μm regime.

2. Space Charge Induced Photovoltage in DBQW Infrared Detectors

In order to investigate the influence of internal electric fields in the barrier layers on the photovoltaic behavior, we prepared a series of DBQW infrared detector structures in which the doped region is located differently with respect to the QW layers. This allows us to change the built-in field across the AlGaAs barriers in a well-controlled manner. The samples are 50 period DBQW structures with 5 nm wide Si-doped GaAs QWs, sandwiched between 2 nm thick AlAs tunnel barriers, and further separated by 25 nm $Al_{0.3}Ga_{0.7}As$ layers. The dopant spikes ($2 \cdot 10^{18}$ cm^{-3} Si) of the three samples under study are nominally 4 nm wide and located, respectively, at the lower, center, and upper parts of the 5 nm wide GaAs QW's, as schematically indicated in the inset of Fig. 1. Si-doped GaAs contact layers ($1 \cdot 10^{18}$ cm^{-3}) with thicknesses of 0.5 μm and 1.0 μm were included above and below the active region, respectively. The structures were grown at 580 °C by molecular-beam epitaxy on (100)-oriented semi-insulating GaAs substrates and processed into mesa detectors of 0.04 mm^2 area with ring-shaped ohmic contacts. The photocurrent was excited by narrow-band illumination of a 45-degree polished facet at the peak wavelength of the responsivity using a glowbar and a monochromator. The measurement was performed using a lock-in amplifier. A calibrated thermopile detector was used as a reference.

Fig. 1 shows the voltage dependence of the photocurrent responsivity of the three samples. All three structures show a steady-state photocurrent at 0 V bias, with responsivities of 2.9 ± 0.4 mA/W, and a photovoltage under open-circuit conditions. The polarity of this photovoltage indicates a preferential motion of the photoexcited electrons towards the substrate.

The observed photovoltaic behavior is associated with a potential distribution like the one shown schematically in Fig. 2(a). Such a potential distribution is expected to occur in an asymmetrically doped multiple DBQW in which the dopant atoms are displaced towards the right with respect to the QW center. Here the space charges caused by the partial depletion of the right-hand part of the dopant distribution induce electric fields of opposite signs across the QW layers and across the barrier regions. This potential distribution gives rise to a photoconduction mechanism as indicated by the arrows in Fig. 2(a). Electrons which are excited by an intersubband transition into the upper subband of a QW (vertical arrows) can be emitted from the well before intersubband relaxation back to the ground state occurs. This photoemission occurs towards both sides of the QW layer (unless the tunnel barriers are made inequivalent [8]). However, the motion of the emitted carriers is rectified because of the built-in field across the AlGaAs barrier layers, thus giving rise to an electron current towards the left-hand side of each barrier.

The highest possible current gain g expected for this transport mechanism is $1/2N$ for a structure containing N QW's with symmetric tunnel barriers. This holds for photoemission probability (out of the QW) of 1/2 into each direction and for a back relaxation of the emitted carriers into the QW located at the left-hand side of each barrier. Our 50 period samples result in a gain of about 0.3 % at 0 V, which is three times less than the theoretical

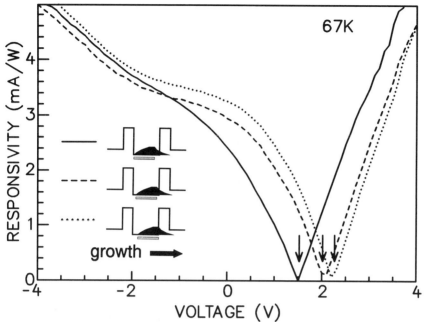

Figure 1: Photocurrent responsivity versus applied voltage of three DBQW detectors with different nominal positions of the dopant spike. Arrows indicate the voltage V_N where the photocurrent changes its sign. The assignment of the data to the individual samples is shown in the inset. The shaded and black areas represent the intended position of the Si doping spike and a realistic dopant distribution, respectively, both with respect to the spatial profile of the conduction bandedge. Growth direction is from left to right.

limit. Preliminary results indicate an increase of the responsivity by a factor of about two if the thickness of the tunnel barriers is further reduced. The precise value of the measured gain therefore allows us to conclude that the rectification due to the space charge fields is almost complete and that the main deviation from the idealized behavior in the present samples is caused by an energy relaxation of the photoexcited carriers back to the ground state of the QW before the emission occurs.

In Fig. 2(b), we have outlined the situation in which the transport asymmetry of the photoexcited electrons is compensated by an external voltage of magnitude V_N. This situation occurs when the built-in field across the $Al_{0.3}Ga_{0.7}As$ layers is approximately compensated by the external field so that the barrier potential is no more rectifying. The voltage V_N at which the photocurrent changes its sign therfore serves to determine the built-in electric field across the AlGaAs barrier. It becomes obvious from the vertical arrows in Fig. 1, that V_N increases systematically when the doping spike is displaced along the growth direction.

The experimental variation of V_N with the nominal spike position is consistent with the behavior expected from a simple analysis of the space charge fields, using the one-

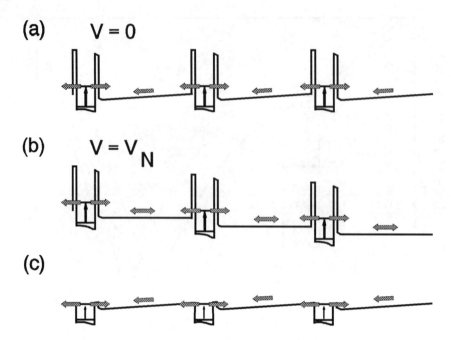

Figure 2: Potential distribution of the conduction bandedge of an asymmetrically doped DBQW structure (a) at vanishing external electric voltage and (b) at an external voltage corresponding to flatband condition in the barrier region; (c) same for an asymmetrically doped single-barrier QW structure. The arrows indicate the processes relevant for detector operation.

dimensional Poisson equation [9]. The absolute value of V_N, however, depends critically on migration of the dopant atoms due to segregation during epitaxial growth. We have recently simulated this dopant segregation numerically [10] by solving a rate equation describing the Si dopant incorporation kinetics in the GaAs/AlGaAs system at the experimental growth temperature. The theoretical results reproduce the experimental values for V_N reasonably well. This agreement indicates that the observed photovoltaic asymmetry at nominally symmetric DBQW potential can be explained by dopant segregation. Unintended asymmetries due to interface roughness scattering only play a minor role. The simulation also shows that the built-in field across the AlGaAs layer is rather inhomogeneous since part of the dopants are segregation-shifted deep into the barrier.

The dopant distribution also has an important influence on the dark current, giving rise to the asymmetric dark current versus voltage characteristics shown in Fig. 3. The dark current at positive bias voltage is higher than the dark current at negative bias. This behavior can be understood from the potential profiles of Figs. 2(a) and 2(b). At positive bias, the electrons are emitted towards the right side of the DBQW. Here the effective barrier height (activation energy E_A) is smaller than at negative bias. Since the dark current j_D at the temperature T scales with $\exp(-E_A/k_B T)$, a smaller barrier height gives rise to a higher dark current. In addition, the asymmetry of the dark current behavior

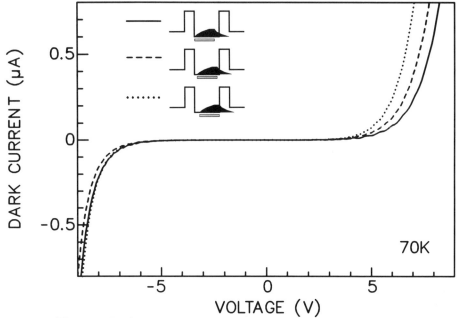

Figure 3: Dark current vs. voltage of the DBQW samples at 70 K.

increases with increasing asymmetry of the dopant distribution as expected. Interestingly, the positive branch of the dark current characteristics reacts more sensitively on the dopant distribution than the negative branch. We attribute this feature to self-consistent changes in the potential distribution, giving rise to a stronger dependence of E_A on the dopant position for positive bias than for negative bias.

3. Detectivity Limit

Photovoltaic detection has the conceptual advantage over photoconductive detection that the *shot noise* $i_s = \sqrt{4eJ_D g_I \Delta f}$ per bandwidth Δf induced by the dark current J_D per unit area observed in a photoconductive QW infrared detector with gain g_I is not present. Instead, only the *Johnson noise* $i_J = \sqrt{4k_B T \frac{\partial J_D}{\partial V} \Delta f}$ is observed.

In order to calculate the noise properties we first determine the dark current density J_D within the tunneling-assisted thermionic emission model [12,13]

$$J_D = \frac{m^* e}{\pi h^2 p} v_D(F) \int_{E_0}^{\infty} f(E) T(E, F) dE. \tag{1}$$

In Eq. (1), $f(E) = \left(1 + \exp(\frac{E - (E_0 + E_F)}{k_B T})\right)^{-1}$ is the Fermi-distribution with the subband energy E_0 and the Fermi energy $E_0 + E_F$. $T(E, F)$ is the probability for transmission of an electron of effective mass m^* with energy E across the barrier. p is the period of the structure. $v_D(F) = \mu F / \sqrt{1 + (\mu F / v_s)^2}$ is the field-dependent drift velocity of the photoemitted carriers, with the mobility μ and the saturation drift velocity v_s. In the context of DBQW structures, this parameter should be re-expressed in terms of a characteristic scattering time [11]. A discussion going into this direction will be the subject of a later publication.

A simple functional dependence of J_D is obtained from the rigid-barrier approximation, $T(E, F) = 0$ for $E \leq E_B$, $T(E, F) = 1$ for $E > E_B$, and from the assumption $E_B - (E_0 + E_F) >> k_B T$, giving rise to

$$J_D = e v_D \frac{m^* k_B T}{\pi \hbar^2 p} \exp(-\frac{E_B(F) - (E_0 + E_F)}{k_B T}). \tag{2}$$

In Eq. (2), the field dependence of the transmission probability has been cast into a field dependent barrier energy $E_B(F)$. In a linear expansion which is valid for small positive fields, $E_B(F)$ is given by

$$E_B(F) = E_B^{(0)} - eF\delta, \tag{3}$$

where we have defined the coefficient δ. In particular, J_D now shows a thermally activated behavior, with the activation energy $E_A = E_B(F) - (E_0 + E_F)$. We note that Eq. (3) is not valid for negative fields at which carrier emission occurs towards the other side of the well. A closed expression for J_D can be obtained in principle by considering thermionic emission out of the two sides of the QW seperately, starting with two expressions of the form of Eq. (1).

Within the present approximations, the ratio between the Johnson noise i_J of a photovoltaic QW detector at $F = 0$ and the shot noise i_s of a photoconductive device at its operating field $F = F_{op}$ is

$$\frac{i_J}{i_s} = \sqrt{\frac{\alpha g v}{g_I}} \exp(-\frac{eF_{op}\delta}{2k_B T}), \tag{4}$$

with $\alpha = k_B T \mu / e p v_s$ and the photoconductive and photovoltaic gain coefficients g_I and g_V.

It is now possible to compare the achievable detectivities of QW infrared detector structures $D_I^* = R_I \sqrt{\Delta f}/i_s$ in the photoconductive mode and $D_V^* = R_V \sqrt{\Delta f}/i_J$ in the photovoltaic mode. Defining the hopping distance $l_I = N p g_I$ of the photoconductive device and using $l_V = p$ for the photovoltaic device, we arrive at

$$\frac{D_V^*}{D_I^*} = \sqrt{\frac{p}{l_I \alpha}} \exp(\frac{eF_{op}\delta}{2k_B T}). \tag{5}$$

From Eq. (5), we obtain $D_V^*/D_I^* > 1$ if $eF_{op}\delta > k_B T \ln(l_I \alpha / p)$. This relation indicates that the detectivities of optimized photovoltaic and photoconductive QW intersubband detectors are of the same order of magnitude.

It is interesting to evaluate these expressions for realistic device parameters. Typical operating fields [2] in a photoconductive QW infrared detector correspond to 50-100 meV per period. This means that $eF_{op}\delta$ =25-50 meV in Eq. (5) for $\delta = p/2$, i. e., that the activation energy of the photoconductive device is reduced correspondingly. On the other hand, the optimized photovoltaic QUID requires an *excess energy* $E_X > k_B T$ to obtain a rectification of the photocarrier motion in the barrier region (see Fig. 2), which, in turn, leads to a reduced barrier height E_B. Consequently, a value of E_X =10-20 meV (at 77K) would lead to the desired photovoltaic detector performance. It should thus be possible to obtain $D_V^*/D_I^* > 1$.

We have calculated Eq. (1) numerically within the two-band approximation [14] for a selection of potential distributions [15]. These potential distributions are represented here for simplicity by compositional gradients in the GaAs/AlGaAs material system and not by

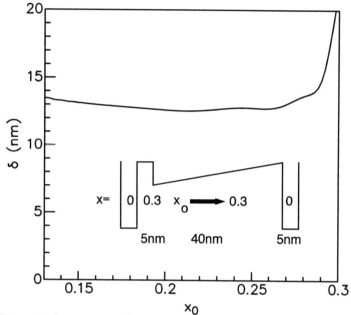

Figure 4: δ versus x_0 of a photovoltaic QW infrared detector structure at 77K. The inset shows the spatial distribution of the Al-content x. x_0 denotes the initial Al-content of the graded layer ($x_0 = 0.25$, $E_F = 15$ meV, $\mu = 1$ m^2/Vs).

space charge fields. The calculations indicate a thermally activated behavior of the dark current over many orders of magnitude, in excellent agreement with the approximation of Eq. (2). Using this equation, the parameter δ, as defined by the series expansion of Eq. (3), can be expressed as

$$\delta = \frac{k_B T}{e} \frac{1}{I} \frac{\partial I}{\partial F}. \tag{6}$$

$I = \int f(E)T(E, F)dE$ is the integral expression in Eq. (1).

In Fig. 4, we have plotted the resulting value of δ as obtained numerically for a particular class of asymmetric potential distributions consisting of a GaAs well, an Al$_{0.3}$Ga$_{0.7}$As barrier and a graded potential barrier with Al-content x from $x = x_0$ to $x = 0.3$, as shown schematically in the inset of the figure. For $0.30 > x_0 > 0.29$, δ decreases strongly with decreasing start-value x_0 of the graded layer since resonant states in the barrier region are being formed. For $x_0 < 0.29$, δ depends only weakly on x_0 because of the presence of these localized states. Similarly to the discussion of Eq. (3), δ only corresponds here to tunneling-assisted thermionic emission towards the right-hand side of the well. For emission towards the left, a corresponding value of $p - \delta$ should be used. In the example of Fig. 4, the activation energy of J_D for $x_0 = 0.25$ is by about 30 meV smaller than for $x_0 = 0.30$.

4. Photovoltaic Long-Wavelength QW Infrared Detector

Potential distributions like the one shown in the inset of Fig. 4 are promising in order to obtain simultaneously a good efficiency of the above-barrier rectification of the photoex-

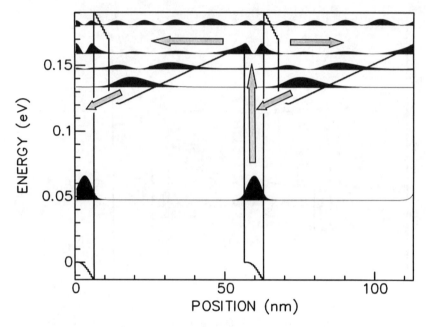

Figure 5: Conduction bandedge profile, subband energies, and probability distributions of a novel photovoltaic 8-12 μm QW intersubband detector structure. The proposed device consists of 6.5 nm GaAs, 5.0 nm $Al_{0.25}Ga_{0.75}As$, and 45.0 nm $Al_{0.20}Ga_{0.8}As$, modulation-doped at about 10 nm with respect to the QW center with a sheet carrier density of $4.0 \cdot 10^{11}$ cm^{-2}. Device operation is indicated by arrows.

cited carriers and a small excess energy between the upper QW subband and the barrier bandedge. Detectors which rely on compositional gradients, however, are difficult to prepare by epitaxial methods.

We therefore propose to use again asymmetric dopant distributions, or modulation doping, in order to realize the internal electric fields. The expected transport mechanism for such a device is indicated in Fig. 2(c). Exploiting this approach quantitatively, we have performed self-consistent quantum mechanical calculations of the potential profile and subband structure of useful layer sequences. Fig. 5 shows results for the case of a photovoltaic infrared detector structure at 8-12 μm detection wavelength. In this example, intersubband excitation is from the first to the fourth subband. The second and third subbands, necessary for the transport mechanism producing the photovoltage, are mainly localized in the barrier region. The expected transport mechanism is indicated by arrows.

The present approach to photovoltaic QW intersubband infrared detectors is particularly useful for applications in the 8-12 μm-regime. Here the small responsivity and the large impedance of this detector allows long integration times in spite of the large background and signal photon fluxes at these wavelengths. These properties should give rise to a very large dynamic range of these detectors.

In the $3-5\mu$m regime, on the other hand, the higher degree of preamplification required

for photovoltaic intersubband detection may turn out critical. This is because the low detector noise requires low-noise amplification of the signal in order to keep a large detectivity.

5. Conclusions

We have presented and analyzed a particular carrier transport mechanism observed in
double-barrier QW intersubband infrared detector structures which produces a large photovoltaic effect. The mechanism is based on internal electric fields across the barrier layers.
These internal fields give rise to a rectified motion of the photoemitted carriers, inducing a
steady-state photocurrent or, under open circuit conditions, photovoltage.

The transport mechanism allows for low-noise infrared detection, with similar detectivities but smaller responsivities and noise currents than in the case of photoconductive
intersubband detector structures. Therefore, photovoltaic QW intersubband devices should
be useful for $8 - 12\mu$m infrared detection, giving rise to a high saturation threshold and
an extremely large dynamical resolution. We have proposed and simulated a layer structure which should result in an optimized photovoltaic QW intersubband infrared detector
structure at these wavelengths.

ACKNOWLEDGEMENTS. The authors are grateful to H. Rupprecht for his encou-
ragement and continuous support of this work. We would also like to thank J. Fleissner,
C. Hoffmann, M. Hoffmann, and K. Räuber for diode processing, and to H. Biebl and B.
Dischler for absorption measurements.

References

1) Rosencher, E., Vinter, B., and Levine, B. (eds.) (1992), 'Intersubband Transitions in Quantum Wells', (Plenum London).

2) Levine, B. F., Bethea, C. G., Glogovsky, K. G., Stayt, J. W., and Leibenguth, R. E. (1991), 'Long-wavelength 128x128 GaAs quantum well infrared photodetector array', Semicond. Sci. Technol. **6**, C114.

3) Kozlowski, L. J., Williams, G. M., Sullivan, G. J., Craig, W. F., Anderson, R. J., Chen, J., Tennant, W. E., and DeWames, R. E. (1991), 'LWIR 128x128 GaAs/AlGaAs multiple quantum well hybrid focal plane array', IEEE Trans. Electron Devices **38**, 1124.

4) Goossen, K. W., Lyon, S. A., and Alavi, K. (1988), 'Photovoltaic quantum well infrared detector', Appl. Phys. Lett. **52**, 1701.

5) Kastalsky, A., Duffield, T., Allen, S. J., and Harbison, J. (1988), 'Photovoltaic detection of infrared light in a GaAs/AlGaAs superlattice', Appl. Phys. Lett. **52**, 1320.

6) Schneider, H., Fuchs, F., Dischler, B., Ralston, J. D., and Koidl, P. (1991), 'Intersubband absorption and infrared photodetection at 3.5 and 4.2 μm in GaAs quantum wells', Appl. Phys. Lett. **58**, 2234.

7) Schneider, H., Koidl, P., Fuchs, F., Dischler, B., Schwarz, K., and Ralston, J. D. (1991), 'Photovoltaic intersubband detectors for 3-5 μm using GaAs wells sandwiched between AlAs tunnel barriers', Semicond. Sci. Technol. **6**, C120.

8) Ralston, J. D., Schneider, H., Gallagher, D. F. G., Kheng, K., Fuchs, F., Bittner, P., Dischler, B., and Koidl, P. (1992), 'Novel MBE grown GaAs/AlGaAs quantum well structures for infrared detection and integrated optics at 3-5 and 8-12 μm', J. Vac. Sci. Technol. **B 10**, 998.

9) Schneider, H., Larkins, E. C., Ralston, J. D., Schwarz, K., Fuchs, F., and Koidl, P. (1993), 'Space-charge effects in photovoltaic double barrier quantum well infrared detectors', Appl. Phys. Lett., August 9.

10) Larkins, E. C., Schneider, H., Ehret, S., Fleissner, J., Dischler, B., Koidl, P., and Ralston, J. D. (1993), 'Influence of MBE Growth Processes on Photovoltaic 3-5 μm Intersubband Photodetectors', IEEE Trans. Electron Devices, submitted.

11) Liu, H. C., Steele, A. G., Wasilewski, Z. R., and Buchanan, M. (1993), 'Dark current mechanism and the cause of the current-voltage asymmetry in quantum well intersubband photodetectors', presented at the 2nd Int. Symp. on 'Physical concepts and materials for novel optoelectronic device applications', May 24-27, Trieste (Italy).

12) Levine, B. F., Bethea, C. G., Hasnain, G., Shen, V. O., Pelve, E., Abbott, R. R., and Hsieh, S. J. (1990), 'High sensitivity low dark current 10 μm GaAs quantum well infrared photodetectors', Appl. Phys. Lett. **56**, 851.

13) Rosencher, E., Luc, F., Bois, Ph., and Delaitre, S. (1992), 'Injection mechanism at contacts in a quantum well intersubband infrared detector', Appl. Phys. Lett. **61**, 468.

14) Bastard, G. (1982), 'Theoretical investigations of superlattice band structure in the envelope-function approximation', Phys. Rev. **B 25**, 7584.

15) Schneider, H. (1993), 'Optimized performance of quantum well intersubband infrared detectors: photovoltaic versus photoconductive operation', J. Appl. Phys., to be published.

INTERSUBBAND STARK-LADDER TRANSITIONS IN MINIBAND-TRANSPORT QUANTUM-WELL INFRARED DETECTORS

J. W. Little and F. J. Towner
Martin Marietta Laboratories
1450 S. Rolling Road
Baltimore, MD 21227
USA

ABSTRACT. We have measured electro-absorption in miniband-transport quantum-well infrared detectors consisting of doped quantum wells surrounded by superlattice barrier layers. We show that a theoretical model of the quantum-well/superlattice-barrier system in an electric field accurately predicts features observed in the photocurrent spectra such as a field-induced blue-shift of the main low-field transition and the development of intersubband Stark-ladder transitions at high fields.

1. Introduction

The miniband-transport (MBT) infrared detector[1,2] consists of doped quantum wells separated by superlattice barrier layers in which a miniband of energy states is formed. The coupling of an excited state of the quantum well to the miniband provides enhanced optical absorption and ensures that photoexcited carriers have a relatively high-mobility channel for transport to an external contact. Under normal operating conditions, the electric fields in the MBT are relatively low, and the electrical and optical properties of the detector are determined, in part, by the strong mixing of the superlattice states (i.e., by the miniband). We have recently fabricated MBT detectors that respond in the mid-wave infrared (MWIR) region of the spectrum (3-5μm wavelength) and use very deep quantum wells with modified superlattice barrier layers. Because of the strong confinement of the dopants in the deep wells, the MWIR detectors can sustain very high electric fields with relative low dark currents.

The properties of a superlattice in a high electric field are dramatically different from those at low field. These changes will lead to equally dramatic modifications to the electrical and optical properties of the detector. By studying the properties of MBT detectors over a large range of fields, we have observed new features in the optical absorption spectra that result from changes in the coupling between various states in the system. We have shown that a theoretical model of the detector system based on fundamental calculations of the energies and wavefunctions associated with the electronic subbands can be used to accurately predict the optical properties of the quantum-well/superlattice system over the full range of electric fields.

H. C. Liu et al. (eds.), Quantum Well Intersubband Transition Physics and Devices, 197–206.
© 1994 *Kluwer Academic Publishers. Printed in the Netherlands.*

2. Sample Description

Figure 1 shows the energy-band diagram for a typical mid-wave MBT quantum-well infrared detector. It consists of a stack of pseudomorphically-strained InGaAs quantum wells surrounded by superlattice barrier layers comprised of alternating layers of $Al_xGa_{1-x}As$ and $Al_yGa_{1-y}As$. The quantum-well and superlattice material parameters (i.e., the layer thicknesses and compositions) are chosen so that the ground-state to excited-state transition energy equals the energy of a photon in the 3-5-μm wavelength range and the excited state is nearly in resonance with the bottom of the miniband (see Fig. 1). This can be accomplished using indium mole fractions in the range of 0.1-0.2, quantum well widths on the order of 40 Å, aluminum mole fractions in the 0.25-0.6 range, and superlattice periods on the order of 50 Å. Carriers that are photoexcited out of the ground state of the well into either the excited state or the miniband states move through the miniband under the influence of an applied electric field and are collected at the contacts as a photocurrent.

The energy width of the miniband is primarily determined by the thickness of the high-bandgap layer in the superlattice, and the number of superlattice layers can be chosen to optimize certain electrical properties of the detector (e.g., the dark current or the internal electric field per volt of bias).

$Al_xGa_{1-x}As/Al_yGa_{1-y}As$ superlattice

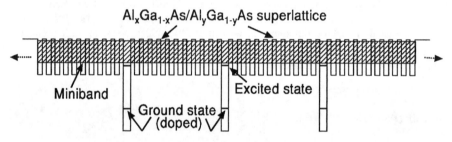

Figure 1. Energy-band for a typical mid-wave MBT detector.

For the samples studied here, the material parameters were chosen to give a ground-state to excited-state transition energy of ~250 meV. A Kronig-Penney calculation of the width of the miniband for an infinite superlattice having the same parameters as the (finite) superlattices used here gives a value of ~115 meV. This would place the top of the miniband very close in energy to the top of the high bandgap superlattice layer (i.e., nearly in the continuum), as illustrated in Fig. 1. One of the samples used in this study (sample A) had barrier layers consisting of 5.5 superlattice layers, and a second sample (sample B) had 10.5 superlattice layers comprising the barriers. Results of calculations of the transition energies and strengths for the InGaAs quantum-well/finite superlattice-barrier system are given in a later section.

Single-crystal x-ray diffraction measurements were made on the samples using both specular ([0,0,4]) and non-specular ([3,3,5]) reflections to determine the average lattice constants of the epilayer parallel and perpendicular to the [1,0,0] growth surface and the period of the multiple quantum-well (MQW) stack (i.e., the thickness of the quantum well and the superlattice barriers). It was found that the quantum wells were essentially pseudomorphically strained (i.e., the lattice constant parallel to the growth surface matched that of the GaAs substrate), and the period of the MQW stack was within 0.5% of the design values for both samples.

3. Sample Geometry and Experimental Setup

One quarter of each of the two-inch wafers (for samples A and B) was processed into sets of 8x8 arrays of 100-μm-square mesas on 200-μm centers. The mesas were defined using standard photolithography and etched using reactive-ion-etching to a depth that just exposed the bottom doped GaAs contact layers. Gold/tin contacts were formed on the mesa tops and on the bottom contact layer (in a grid pattern around the mesas) using liftoff techniques and rapid thermal annealing. The unprocessed quarters of each wafer were reserved for characterization (e.g., x-ray diffraction, room-temperature optical transmission, etc.) and for future infrared detector fabrication.

Due to the polarization selection rules governing intersubband transitions[3], it is necessary to provide a component of the optical polarization along the quantization direction (i.e., perpendicular to the plane of the quantum wells). This is most easily accomplished by polishing a window in the side of the chip at a 45° angle to the plane of the quantum wells[4]. We used mechanical lapping followed by chemical polishing to obtain a specular finish on the window. Figure 2 shows the geometry used for obtaining good thermal contact with the cold finger of a liquid nitrogen cryostat while providing the appropriate optical coupling and easy electrical contact to several devices on a chip. The chip was mounted substrate-side down in a 68-pin ceramic chip carrier using a thermally-conducting epoxy. A 45-degree turning mirror (made by polishing and metalizing a piece of GaAs) was placed opposite the window in the sample chip to direct downwardly-propagating light (as viewed in Fig. 2) into the sample. Light refracted at the sample window strikes the surface of the wafer at a 57.5° angle to the normal, and, therefore, contains a component of the polarization perpendicular to the plane of the quantum wells. Gold wires were bonded onto the top contacts of several mesas and to the bottom (common) contact and connected to the leads of the ceramic chip carrier to allow for the application of electric fields and the for the collection of photocurrents.

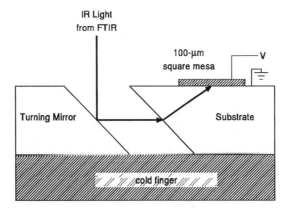

Figure 2. Sample mounting geometry used to couple light into the detector pixel.

The chip carriers were mounted on the cold finger of a liquid-nitrogen pour-filled cryostat and placed in the beam of a Mattson Fourier-transform infrared (FTIR) spectrometer. A low-noise power supply was used to apply biases to the top contact (with the bottom contact as the ground plane), and photocurrents were amplified using a Keithly current-sensitive preamp. The signals were referenced to a background scan taken with the internal detector of the FTIR. Photocurrent spectra were obtained as functions of photon energy over a large bias range with the sample held at 80K. The internal electric field was determined by assuming that the bias potential was dropped uniformly across the entire MQW stack.

4. Photocurrent Spectra

Figure 3 shows photocurrent spectra for sample A (with 5.5 superlattice periods comprising the barriers) for low-to-moderate electric fields (17.4 to 69.7 kV/cm). Note that for low fields the spectra are dominated by a single peak at an energy of ~250 meV and show high-energy tails that extend to around 400 meV. The small feature at ~ 290 meV is due to incomplete removal of a strong atmospheric absorption line, and is not a property of the sample.

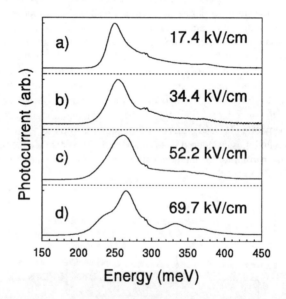

Figure 3. Low-to-moderate-field photocurrent spectra
from 5.5-period superlattice sample.

For moderate fields (e.g., 52.2 and 69.7 kV/cm), structure develops in the high-energy tail of the photocurrent spectra, and a low-energy shoulder appears on the main transition peak. The dominant peak shifts to higher energy with increasing field and exhibits a maximum blue-shift of about 16 meV (relative to the 17.4 kV/cm peak) at a field of 69.7 kV/cm.

In addition to the main transition and the high-energy tail discussed above for the low-field photocurrent spectra, a well-defined (but weak), high-energy peak was observed that may be the result of an increased extraction probability for carriers excited into states near the top of the barrier (which coincides with the top of the miniband, as discussed earlier).

Figure 4 shows photocurrent spectra for sample A under high electric fields. Note that the (originally) dominant peak shifts rapidly to lower energies and decreases in intensity relative to a new feature that appears at ~290 meV in the 87.1 kV/cm curve (Fig. 4b) and grows to become the largest of the three remaining peaks at the highest fields (Fig. 4d). The exchange of oscillator strength and the increased splitting between the two peaks is characteristic of a strong anti-crossing of two energy levels such as was observed in band-to-band spectra of quantum wells embedded in a strongly-coupled superlattice[5].

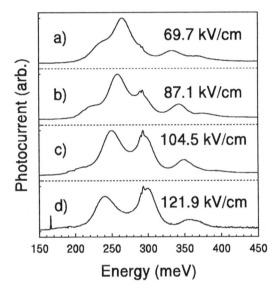

Figure 4. Moderate-to-high-field photocurrent spectra
from 5.5-period superlattice sample.

The blue shift of the dominant transition and the development of additional structure in the moderate-field spectra shown in Fig. 3 are also similar to features that were observed in interband photocurrent spectra of semiconductor superlattices[5] and were shown to be the result of Stark localization of the miniband states. The field dependence of the intersubband transitions reported here is the result of a mechanism that is related to the localization of the superlattice states but is unique to the coupling of the subbands in the quantum-well/superlattice-barrier system.

We found[6] that the field-dependent spectra obtained from sample B (with 10.5 superlattice layers) were essentially the same as those obtained from sample A (with 5.5 superlattice layers). At all measured fields, the energies and relative intensities of the main features in the spectra were the same for both samples. This indicates that the optical properties of the quantum-well/superlattice-barrier system are determined by a few (≤ 5) nearest neighbor superlattice wells.

In the following section, we will identify the features observed in the field-dependent photocurrent spectra using theoretical calculations of the wavefunctions, transition energies, and relative absorption strengths.

5. Theoretical Description

To calculate the subband energies and the wavefunctions for the quantum-well/superlattice system, we use Airy-function solutions to the single-electron Schroedinger equation and match the boundary conditions at each interface in the usual way[7]. Biaxial strain[8] and band nonparabolicity[7] are very important in these systems because they affect the depth of the well and the energy of the excited state, respectively, and are therefore included in the calculation.

The structure used for the calculation includes the quantum well, an n-period superlattice barrier on either side of the well, and a high-bandgap terminating layer that ensures that the wavefunctions decay outside of the structure even at very high electric fields. The terminating layer has negligible effect on the low-field states and on the high-field states that are localized near the quantum well. As discussed above (and substantiated below), the high-field optical properties are dominated by these few near-neighbor states.

Figure 5 shows a comparison of the measured energies (symbols) of the peaks observed in the photocurrent spectra of sample B and the calculated transition energies (solid lines) for a 10.5-period superlattice sample with the appropriate layer thicknesses and compositions. The calculation includes transitions between the ground state of the quantum well and all of the excited states of the system (i.e., the excited state of the well and the twenty miniband states originating from the low bandgap layers in the superlattices). Anomalous states located in the continuum (i.e., above the barriers) were also found but were shown[6] to be due to the terminating layer and not real states of the system. These have been removed from Fig. 5, but their effects can be seen as residual anticrossings in the calculated transitions. A detailed discussion of the results of this calculation will be presented elsewhere[6], and we will only summarize them here.

Data represented by stars in Fig. 5 indicate relatively strong transitions, and those denoted by plus signs indicate relatively weak features.

The electric field region labeled "Miniband region" in Fig. 5 corresponds to a strong mixing of the superlattice states due to the high degree of degeneracy. In this region, the observed spectra were dominated by a single strong peak (see Figs. 3a-c). This is the transition between the ground state and the excited state of the quantum well. The calculated transitions that are dropping linearly with electric field in the miniband region are superlattice states that are strongly localized in layers remote from the quantum well (i.e., in the tenth, ninth, etc. superlattice wells). There is negligible overlap between these wavefunctions and the ground state wavefunction that is localized in the doped quantum well, and therefore, there are no spectral features associated with them.

The high-field regions in Fig. 5 are labeled "Stark localization" and correspond to fields for which the superlattice wavefunctions extend over only a few layers. Here, transitions between the ground state of the quantum well and at least the first two nearest neighbors on either side can be seen (in addition to the excited-state transition) and exhibit the linear shift in energy with electric field that is typical of Stark-ladder transitions[5]. The labels on the high-field transitions (0, ±1, ± 2) refer to the excited state, the first nearest neighbors, and the second nearest neighbors, as shown in Figure 6. At a field of about 100 kV/cm , the states labeled "0" and "-1" are strongly mixed as the nearest neighbor state crosses the excited state.

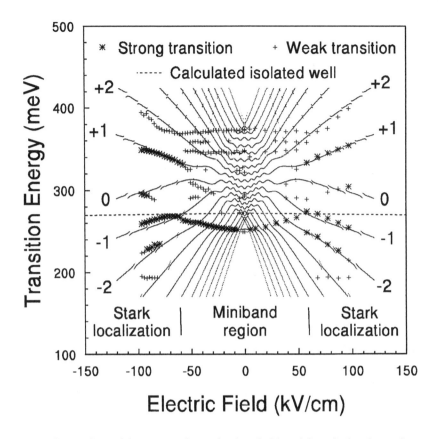

Figure 5. Comparison of the measured energies (symbols) and the calculated energies (solid lines) for a 10.5-period superlattice sample. In the "Miniband region", the dominant feature is the excited state transition that blue shifts and anticrosses with superlattice states as the magnitude of the field increases. In the "Stark localization" region, the linearly-shifting Stark-ladder states are observed.

Note that the main transition in the miniband region increases in energy (i.e., blue shifts) with increasing field up to a field of 60-70 kV/cm. The blue shift is the result of a decreased coupling between the excited state of the quantum well and the miniband states as the miniband states localize with increasing field. The dashed line in Fig. 5 is the calculated transition between the ground state and the excited state of a quantum well with infinitely-thick, uniform barriers composed of the higher-bandgap AlGaAs material instead of the superlattice. At low field, the superlattice-barrier transition is lowered in energy with respect to the uniform-barrier transition due to strong coupling between the excited state and the miniband states. As the miniband states localize with increasing field, the excited state moves up in energy as a result of the decoupling.

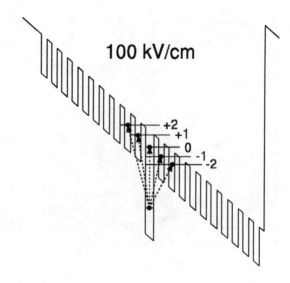

Figure 6. Main excited state transition ("0") and the first two
nearest-neighbor transitions ("±1" and "±2") for a 10.5-
period superlattice at 100 kV/cm electric field.

The relative strengths of the transitions can be determined from the wavefunctions by calculating the cross section for absorption that is proportional to the square of the dipole matrix element between the various excited states and the ground state[9]. Figure 7 shows a comparison between observed photocurrent spectra (a) and the calculated absorption spectra (b) for sample A. The curves in Fig. 7b were obtained by convolving the calculated absorption cross sections for each transitions with a Gaussian of half-width at the 1/e intensity point of 15 meV (chosen to match the observed width of the spectra). All of the salient features of the observed spectra are reproduced in the calculated curves. The blue shift of the main transition between the 17.4 kV/cm curve and the 69.7 kV/cm curve in Fig. 7a is also seen in Fig. 7b for comparable fields. At the highest field, four main transitions (the Stark-ladder transitions) are seen in both curves with similar relative intensities.

Figures 5 and 7 show that the theoretical model used here can accurately predict both the peak energies and the relative intensities of the intersubband transitions in the MBT detector structure.

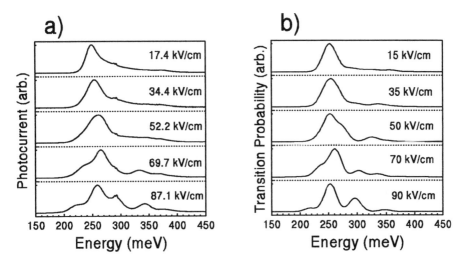

Figure 7. a) Measured photocurrent spectra and b) calculated absorption spectra for a 5.5-period superlattice sample. Note that the main features of the measured spectra are reproduced in the calculated spectra including the blue shift of the dominant transition with increasing field up to about 70 kV/cm and the appearance of several (Stark-ladder) peaks for higher fields.

6. Conclusions

We have shown that the electric field dependent coupling between the excited state and the miniband states in an MBT quantum-well infrared detector can dramatically affect its optical properties. A blue shift in the main transition with increasing field was shown to arise from the decoupling of the states, and transitions between the ground state of the quantum well and the quasi-localized Stark-ladder states of the miniband were observed and identified. A theoretical model of the MBT structure was shown to accurately predict the transition energies and the shape of the absorption spectra over a very large range of electric fields.

Acknowledgments

The authors gratefully acknowledge Make Taylor for processing the arrays used in this study and R. P. Leavitt of Army Research Laboratories for many enlightening discussions concerning the electro-optic properties of semiconductor superlattices.

References

[1] T.S. Faska, J.W. Little, W.A. Beck, K.J. Ritter, and A.C. Goldberg, JPL Innovative Long Wavelength Infrared Detector Workshop, April 7-9, 1992.

[2] The MBT detector is equivalent to the bound-to-miniband design described, for example, in L.S. Yu and S.S. Li, Appl. Phys. Lett. **59**, 1332 (1991).

[3] L.C. West and S.J. Eglash, Appl. Phys. Lett. **46**, 1156 (1985).

[4] A. Zussman, B.F. Levine, J.M. Kuo, and J. De Jong, J. Appl. Phys. **70**, 5101 (1991).

[5] R.P. Leavitt and J.W. Little, Phys. Rev. B **41** (8), 5174 (1990).

[6] J.W. Little, in preparation.

[7] R.P. Leavitt, Phys. Rev. B **44** (20), 11,270 (1991).

[8] H. Kato, N. Iguchi, S. Shika, M. Nakayama, and N. Sano, J. Appl. Phys. **59**, 588 (1989).

[9] E. Merzbacher, *Quantum Mechanics*, John Wiley & Sons, Inc., New York, 1970.

CO₂-LASER HETERODYNE DETECTION WITH GaAs/AlGaAs MQW STRUCTURES

E.R. Brown, K.A. McIntosh, K.B. Nichols, F.W. Smith[†], and
M.J. Manfra
Lincoln Laboratory, Massachusetts Institute of Technology
Lexington, MA 02173-9108

ABSTRACT. Optical heterodyne experiments are carried out on GaAs/AlGaAs multiple-quantum-well (MQW) structures using a CO_2-laser local oscillator operating near 10 μm. In a structure having 100 4.5-nm-wide GaAs quantum wells and 40-nm-wide AlGaAs barriers, the electrical bandwidth B_{IF} is approximately 13 GHz for 77-K operation and the heterodyne quantum efficiency η_{EH} is 20%. This value of η_{EH} is near the theoretical photon-noise limit for photoconductors as applied to the MQW detector. The measured B_{IF} is roughly five times that of the fastest HgCdTe detectors, making the GaAs/AlGaAs MQW structure useful for applications in broadband terrestrial spectroscopy, astrophysics, communications, and radar.

1. Introduction

1.1 GaAs MQW DETECTORS

Since the initial observation of strong intersubband absorption [1] and sensitive direct detection [2] in multiple quantum wells (MQWs), detectors made from n-type GaAs MQWs have been the prototype for a new class of infrared (IR) detectors. The advantages of the GaAs MQW detector over alternative IR detector technologies are well known and discussed at length in other chapters of this text. The primary benefits are: (1) the peak wavelength λ_p of the spectral response curve can be "engineered" through control of the quantum-well width and the barrier height, and (2) GaAs is much more robust than the standard IR detector materials (e.g., HgCdTe) and is, therefore, more amenable to fabrication in large focal-plane arrays. These advantages have been utilized to demonstrate a multiple-color detector [3] and a long-wavelength detector ($\lambda_p \approx 18$ μm) [4] operating where intrinsic (cross-gap) semiconductors are very difficult to obtain.

The disadvantages of the GaAs MQW direct detector are also well known because they have been a major impediment to the implementation of these devices in IR staring arrays. One disadvantage is that the dark-current density at 77 K is roughly 3 to 4 orders of magnitude higher than in good HgCdTe detectors. This causes the specific detectivity D^* of the MQW detectors to be roughly 100 times lower under low-background radiation. The best detectivity at 77 K reported to date at 10-μm or longer wavelength is

---

[†]Presently at Martin Marrietta Electronics Laboratory, Syracuse, New York 13221

H. C. Liu et al. (eds.), Quantum Well Intersubband Transition Physics and Devices, 207–220.
© 1994 Kluwer Academic Publishers. Printed in the Netherlands.

$D^* = 1.6 \times 10^{10}$ cm Hz$^{1/2}$ W^{-1} [5]. A second disadvantage is the polarization selection rule that results in a strong absorption of light only for the component of the incident IR electric field that is oriented perpendicular to the plane of the quantum wells. This requires that two-dimensional grating structures be fabricated at the top interface between free space and the GaAs wafer so that randomly polarized radiation at normal incidence can be transformed into radiation that can be efficiently absorbed within the MQW structure. As will be discussed in Sec. 1.2, neither of these disadvantages limits the performance of GaAs/AlGaAs MQW detectors in the optical-heterodyne, or coherent mode.

In the majority of GaAs/AlGaAs MQW structures studied to date, the detector response has been found to be photoconductive in the sense that the time-averaged photocurrent I_0 is related to the optical power P by the relation

$$I_0 = \frac{\eta_0 egP}{h\nu} \propto \frac{\eta_0 egE^2}{h\nu}, \tag{1}$$

where η_0 is the external quantum efficiency, e is the electron charge, g is the photoconductive gain, E is the magnitude of the optical electric field, h is Planck's constant, and ν is the optical frequency. The gain is defined by τ/t_T, where τ is the photoconductive lifetime (i.e., relaxation time into a quantum well), and t_T is the transit time of electrons across the MQW structure. The detector noise has been found to be dominated by thermal fluctuations in the generation and recombination of electrons associated with the dark current I_D. This so-called G-R noise mechanism is described by a power spectral density,

$$S_I = 4egI_D. \tag{2}$$

The actual detector noise spectrum is white from dc up to a frequency $\leq B_t \equiv (2\pi\tau)^{-1}$, which defines the photoconductive electrical bandwidth.

1.2. OPTICAL-HETERODYNE DETECTION

In the optical-heterodyne mode, the output beam from a laser local oscillator (LO) is combined with radiation from another IR source (signal) and both are coupled into the MQW structure, as depicted in Fig. 1. Since the photoconductivity in Eq. (1) is inherently quadratic in the optical electric field, the absorption of the combined beams results in the generation of photocurrent at the difference or intermediate frequency (IF) between the LO and signal beams [6]. The magnitude of the photocurrent will be the greatest at IFs less than the detector electrical bandwidth B_{IF}. Hence, photocurrent at the sum frequency between the LO and signal, although present in theory, is usually

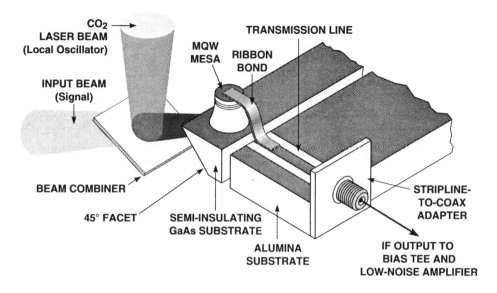

Fig. 1. Experimental configuration using an MQW structure as the heterodyne detector and a CO_2 laser as the local oscillator.

negligible because $B_{IF} \ll \nu_{LO} + \nu_S$ in the vast majority of optical-heterodyne detectors. Another important feature of heterodyne detection is that IF power is produced only for the components of the signal beam which have the same spatial mode(s) and polarization as the laser LO. This is a consequence of the mathematical orthogonality of different output spatial modes from the laser.

The magnitude of the IF photocurrent, i_{IF}, is proportional to the time-averaged LO power P_{LO} and signal power P_S through the relation $i_{IF} = [2(\eta_0 eg)^2 P_{LO} P_S/(h\nu)^2]^{1/2}$, where it has been assumed that $\nu_{LO} \approx \nu_S \equiv \nu$. Along with the IF signal, fluctuations in the rate of LO photon flux, or photon noise, leads to a G-R-like noise in the photocurrent that is also in proportion to P_{LO}. Using the appropriate expressions for the LO-induced and dark-current G-R noise factors, one can write the following equation for the IF signal-to-noise power ratio:

$$\frac{S}{N} \approx \frac{2(\eta_0 eg)^2 P_S P_{LO} R_A / (h\nu)^2}{4eg(I_{LO} + I_D)R_A \Delta f + kT_A \Delta f} = \frac{2(\eta_0 eg)^2 P_S P_{LO} R_A / (h\nu)^2}{4eg(\eta_0 eg P_{LO}/h\nu + I_D)R_A \Delta f + kT_A \Delta f}, (3)$$

where R_A is the amplifier small-signal input resistance.and Δf is the width of the IF passband. The last term in the denominator represents the amplifier noise power with T_A being the equivalent noise temperature. A more useful figure of merit for heterodyne performance is the noise equivalent power, NEP_{HET}. This is defined as the value of P_S in Eq. (3) that yields an IF S/N ratio of unity.

It is now clear why the two disadvantages of MQW detectors in direct detection discussed in Sec. 1.1 do not necessarily limit the heterodyne performance. To overcome the dark-current disadvantage, one simply applies enough LO power so that the photon-induced G-R noise dominates the dark-current noise. According to Eq. (3), the IF S/N ratio then approaches the value $\eta_0 P_S/2h\nu$ and the NEP_{HET} approaches $2h\nu\Delta f/\eta_0$. These values define the photon-noise limit, which represents a maximum sensitivity that can be surpassed only by increasing η_0. To indicate how the magnitude of the NEP_{HET} relates to this limit, a more convenient quantity is the "effective-heterodyne" quantum efficiency η_{EH}, defined by $\eta_{EH} \equiv 2h\nu \, \Delta f/NEP_{HET}$. This quantity will be used henceforth to describe the detection process.

The polarization-rule disadvantage does not limit the heterodyne performance because most laser LOs, particularly CO_2 lasers, emit in only one polarization. In this case, a high η_0 can be obtained simply by coupling the LO and signal through a 45° facet lapped in the substrate. This is the coupling technique used in the present experiments. Because η_0 for a single polarization in MQW detectors can readily approach that of HgCdTe detectors, the photon-noise limit of the two detector types should be comparable. The superior speed of the MQW detector can then be used to advantage.

TABLE I.
MATERIAL CHARACTERISTICS OF MQW SAMPLES

Sample	No. Wells	Well Width (nm)	Barrier Width (nm)	L_{MQW} (nm)
1	50	4.5	20	1245
2	50	5.1	40	2295
3	100	4.5	40	4490

2. GaAs/AlGaAs Coherent Detector Fabrication

2.1. EPITAXIAL GROWTH

Three different MQW structures were used in the present study, and the materials characteristics are summarized in Table I. All three were grown by molecular-beam epitaxy on semi-insulating GaAs substrates at a temperature of 600°C. Samples 1 and 3 had 4.5-nm-wide quantum wells, while sample 2 had 5.1-nm wells. All samples had 21% Al composition in the barriers. The center 2.5 nm of each well was doped n-type with Si to a density of approximately 2.5×10^{18} cm^{-3}, yielding a sheet density of 6×10^{11} cm^{-2}. Samples 1 and 2 had 50 wells, but sample 3 had 100 wells so that the effect of the expected increase in η_0 upon heterodyne detection could be determined. The contact layers below the MQW structure consisted of 750 nm of GaAs doped to 1.0×10^{18} cm^{-3} followed by 250 nm of GaAs doped to 2.5×10^{18} cm^{-3}. The contact layer above the MQW structure consisted of 350 nm of GaAs doped to 2.5×10^{18} cm^{-3}. L_{MQW} defines the total length of each MQW structure.

2.2. FABRICATION AND PACKAGING

After epitaxial growth, detectors were fabricated by the following sequence of steps. First, 150- and 75-μm-diameter Ni/Ge/Au metal contacts were patterned by optical lithography and metal lift-off techniques. Wet-chemical etching was then used to define the MQW mesas using the contact as a self-aligned mask. Next, an ohmic field metal was deposited upon the exposed n$^+$ cladding layer below the MQW structure. After fabrication of the mesas, the wafers were cleaved into rectangular chips, and a 45° facet was lapped into one edge of each chip to couple the IR radiation to the MQW structure.

After fabrication, the chips were soldered onto copper mounting blocks and wire- or ribbon-bonded to a microstrip stripline as shown in Fig. 1. The separation between the mesa and the microstrip line was roughly 250 μm. Both types of bonds were attempted to maximize the high-frequency response of the detectors. The best approach consisted of three parallel bonds made with 12-μm-diameter wire. This appears to be a good compromise between the high capacitance of a ribbon bond and the high inductance of a single-wire bond. For all of the experiments, the mounting blocks were fastened on the cold head of a closed-cycle refrigerator. Power in the IF band was thereafter taken out of the refrigerator through a stripline-to-coaxial adapter on the mounting block and coaxial line .

3. Detector Measurements

3.1. MQW INTERSUBBAND ABSORPTION CHARACTERIZATION

After growth, portions of each wafer were thinned by mechanically lapping the bottom sides, which were then polished with a Clorox solution. This preparation is necessary to reduce nonspecular scattering and free-carrier absorption effects in the samples. Small $(3 \times 10 \times 0.4$ mm) pieces were cleaved from each wafer, and parallel 45° facets were lapped and polished on two opposing ends to allow multiple-pass transmission experiments. IR transmission was measured with a Fourier-transform spectrometer over the spectral range 400 to 4800 cm^{-1}. The linearly polarized radiation was incident normally on one of the 45° facets, and the transmitted radiation came out the opposing facet. The resulting transmission spectra showed Gaussian-like absorption peaks similar to those in previous GaAs MQW structures. The measured peak wavelength λ_P and full width at half maximum ΔE for samples 1, 2, and 3 were $\lambda_P = 10.53$, 9.70, and 10.19 μm, and $\Delta E = 43$, 45, and 42 meV, respectively. Each of these absorptions completely covers

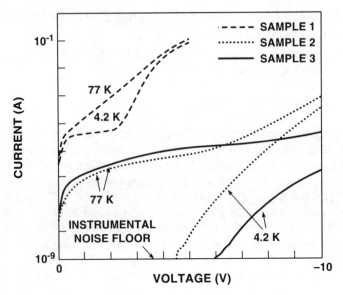

Fig. 2. Current-voltage characteristics of three MQW samples at 4.2- and 77-K operating temperature.

the range of emission wavelengths, 9 to 11 μm, from a generic CO_2 laser.

3.2. CURRENT-VOLTAGE CHARACTERISTICS

After fabrication I_D was measured for several 75-μm-diameter devices from each sample. Fig. 2 shows the 4.2-K and 77-K dark current vs voltage curves for samples 1, 2, and 3. A negative voltage is applied to the top contact because it produces lower I_D than an equal magnitude positive bias. This behavior is commonly seen in MQW detectors and has been attributed to various asymmetries in material composition. Note that I_D in sample 1 drops by only one order of magnitude as the temperature is reduced from 77 to 4.2 K, whereas I_D in samples 2 and 3 drops by several orders as expected for a dark current composed of electrons thermionically emitted from the quantum wells. The excess dark current at 4.2 K in sample 1 is attributed to defect-related transport through the relatively thin (20-nm) barriers.

3.3. EXTERNAL RESPONSIVITY

One indicator of the quality of an MQW detector is its external responsivity r. This is defined as the change in photocurrent per unit change in optical power at a fixed bias voltage V_B. From the fundamental law of photoconductivity, the responsivity is related to the other important detector parameters by $r = \eta_0 eg/h\nu$. The relative wavelength dependence of the responsivity was first measured using a prism spectrometer with a

1000°C blackbody source. The magnitude of r in the three detector samples was measured directly from the current-voltage characteristics under laser illumination. Radiation from a waveguide CO_2 laser was tuned to a wavelength near the peak of the spectral response curves and was coupled through the 45° facet on the substrate. The laser wavelength for each sample, listed in Table III as λ_{LO}, was common to all subsequent measurements.

For each sample, the measurement of r vs V_B yielded a curve having a well-defined peak at a voltage V_B^P. The peak values of r were 0.4, 0.28, and 0.25 mA/mW for samples 1, 2, and 3, respectively. A peaked responsivity was also observed in our previous studies on an MQW sample having 50 4.5-nm-wide GaAs wells and 20-nm-thick barriers, and was correlated to a peaked photoconductive gain vs V_B [7]. The rise in g below the peak is expected from the decrease in t_T with V_B at low bias levels. The fall in g beyond the peak is not well understood. However, it can be argued that the physical mechanism behind the fall is sensitive to the electric field across the MQW structure, E_{MQW}, since V_B^P increases between samples 1 and 3. Because of their greater L_{MQW}, samples 2 and 3 require proportionately more bias voltage to get a given E_{MQW}.

3.4. G-R NOISE AND ELECTRICAL BANDWIDTH

Measurement of the G-R-noise power spectrum is very useful for an MQW detector, or any photoconductor, because it provides B_{IF} and g. B_{IF} is found from the 3-dB rolloff

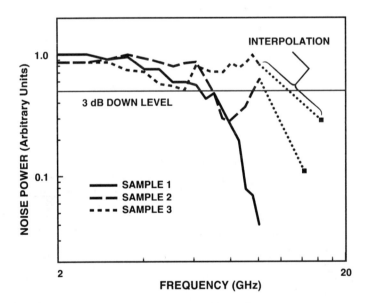

Fig. 3. Normalized G-R noise power spectrum from MQW samples.

214

frequency of the normalized G-R power spectrum. The assumption is that the detector responds dynamically to fluctuations in the steady-state laser power (i.e., the photon noise) in the same way that it responds to a sinusoidal modulation of the laser power. This assumption has been found to be valid for most, if not all, photoconductors [6]. The gain is found by inducing a G-R noise spectrum with a known amount of laser power and calibrating the spectrum at a frequency well below B_{IF}.

G-R-noise power spectra for the three samples are plotted in Fig. 3. The 3-dB rolloff frequencies were $B_{IF} \approx 7$, 9, and 13 GHz for samples 1, 2, and 3, respectively. The normalized power spectra were obtained by chopping the CO_2 laser beam and measuring the IF passband output power using a spectrum analyzer and dual-channel lock-in amplifier as shown in Fig. 4. To correct for the frequency-dependent gain of the amplifiers, white noise from a microwave noise diode was injected into the amplifier chain using a directional coupler. The noise diode was square-wave chopped at a frequency f_d that was approximately ten times higher than the mechanical chopper frequency f_c. The spectrum analyzer was scanned in frequency between 0.5 and 18 GHz,

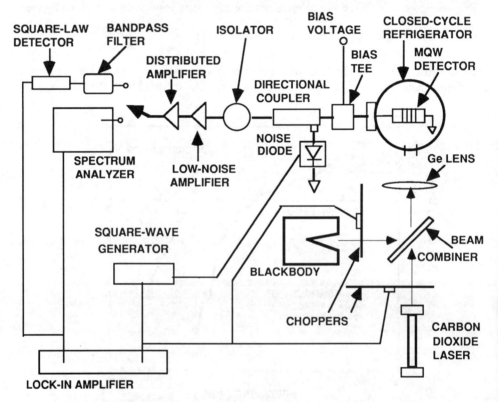

Fig. 4.   Schematic diagram of optical-heterodyne and G-R noise experiments. The directional coupler and noise diode were removed for all measurements except the noise bandwidth.

and its video output was fed into the lock-in amplifier in which one channel was synchronously detected at frequency f_d and the other channel at frequency f_c. The signal at f_c was then divided by the signal at f_d using the normalization option of the lock-in amplifier.

The value of g was determined by chopping the CO_2 laser while it produced a known amount of photocurrent I_0 in the detector. The IF power of the MQW detector was measured using the combination of bandpass filter and square-law detector shown in Fig. 4. Then the MQW detector was replaced with the noise diode and the IF output was remeasured. The power spectral density from the noise diode is given by kT_N, where $T_N = 290 (10^{ENR/10} - 1)$ and ENR is the equivalent noise ratio (15.2 dB). This expression allowed calibration of the G-R noise density, and Eq. (2) yielded g. The values of g are listed in Table II for each sample biased at its respective $V_B{}^P$ The highest g of 0.25 occurs in sample 1. The reason for the inferior g of 0.15 in sample 2 and 0.11 in sample 3 is not clear since one might expect that both τ and t_T would scale in the same way as the barrier thickness is increased with a fixed number of quantum wells. Apparently, the increased barrier thickness increases t_T faster than τ.

Knowing g, one can estimate τ and B_τ using the approximation $t_T \approx L_{MQW}/v_s$, where v_s is the saturation velocity (1×10^7 cm/s) in the AlGaAs barriers. The resulting high values of B_τ in Table II indicate that the B_{IF} of the three samples is probably limited in the present experiments by reactive effects in the load circuit.

TABLE II.
PARAMETERS OF MQW DETECTORS DERIVED FROM G-R NOISE SPECTRUM
AND EXTERNAL RESPONSIVITY

Sample	$V_B{}^P$ (V)	B_{IF} (GHz)	g	τ (ps)	B_τ (GHz)
1	2.0	7	0.25	3.1	51
2	4.0	9	0.15	3.4	47
3	7.5	13	0.11	4.9	33

3.5. EXTERNAL QUANTUM EFFICIENCY

The value of η_0 was calculated from Eq. (1) using the r, g, and hv appropriate to each sample. Samples 1 and 2 displayed comparable values of η_0 consistent with the fact that they had an equal number (50) of quantum wells and electronic sheet density in each well. Sample 3 displayed the highest η_0 consistent with its relatively large number (100) of wells.

These values, listed in Table III, are thought to represent the best measure of η_0 possible from the present set of experiments. Any estimate based on the absorption strength is complicated by the uncertainty in the number of optical passes through the MQW structure. In principle, the relationship between η_0 and the fractional absorption per quantum well, ζ, is given by $\eta_0 = 1 - (1 - \zeta)^{N \cdot P}$, where N (an integer) is the number of wells and P is the number of optical passes through the structure. Since the top of the detector mesa is metallized, one might expect that P = 2. However, better agreement is obtained with the η_0 values in Table III if P = 1 is used in the calculation. This discrepancy is not yet understood.

3.6. HETERODYNE SENSITIVITY

The process of mixing the CO_2 laser with a blackbody generates power over the entire IF band since the blackbody is essentially a very broadband thermal-noise source, the optical analog of a microwave noise diode. The portion of the IR power that appears in the IF band is given by $\nu \pm B_{IF}$, where the plus and minus signs denote the upper and lower sidebands, respectively. For a blackbody of absolute temperature T_{BB} and emissivity ε, the IR spectral density in the single polarization and spatial mode of the laser is given by

$$S_P = \frac{h\nu\varepsilon}{e^{h\nu/kT_{BB}} - 1} \tag{4}$$

The power generated in the IF band is given by Eq. (3) times the optical transmittance between the blackbody and the MQW detector. Eq. (4) provides a means of calibrating the heterodyne sensitivity as quantified by NEP_{HET}. The relationship between NEP_{HET} and the experimental parameters is given by,

$$NEP_{HET} = 2\Delta f \left\{ \frac{h\nu\varepsilon T}{\left(e^{h\nu/kT_{BB}} - 1 \right)} - \frac{h\nu\varepsilon T}{\left(e^{h\nu/kT_{RT}} - 1 \right)} \right\} \frac{N}{S}, \tag{5}$$

where T_{RT} is the ambient temperature.

Table III lists the heterodyne sensitivity for the three samples, quantified by η_{EH}. The bias voltage is kept fixed at V_B^P, and P_{LO} is sampled at 2 and 8 mW. The amplifier noise temperature was first assumed to be zero to study the intrinsic noise mechanisms, and then was set to 100 K (the anticipated performance of a relatively broadband high electron mobility transistor (HEMT) amplifier operating at 300 K). In comparing the results for $T_A = 0$, it is seen that samples 1 and 2 are very similar except at the lowest P_{LO}. The superiority of sample 2 at $P_{LO} = 2$ mW reflects its lower dark current. Hence, less LO power is required to make the photocurrent G-R noise dominate. Sample 3 displays the best performance at all values of P_{LO}, and the η_{EH} values of around 20%

TABLE III

QUANTUM EFFICIENCIES OF MQW SAMPLES

[The values of η_{EH} correspond to $T_A = 0$ K (open) or 100 K (bracketed).]

Sample	λ_{LO} (µm)	$h\nu_{LO}$ (meV)	η_0	η_{EH} ($P_{LO} = 2$ mW)	η_{EH} ($P_{LO} = 8$ mW)
1	10.24	121	0.19	0.07 [0.06]	0.12 [0.11]
2	9.50	131	0.20	0.12 [0.08]	0.13 [0.11]
3	9.56	130	0.34	0.14 [0.09]	0.20 [0.18]

represent the most sensitive heterodyne detection obtained to date with an MQW detector. It should be noted that our results are dependent on the bias voltage, which has been selected for peak responsivity, not peak η_{EH}. For example, sample 2 has an η_{EH} of 18% with $V_B = 1.5$ V, $P_{LO} = 2$ mW, and $T_A = 0$. However, the lower g under these conditions would reduce η_{EH} to 8% for $T_A = 100$ K.

With $T_A = 100$ K, the performance of each sample degrades as expected from Eq. (2). The greatest degradation occurs for sample 3 because of its low g. However, η_{EH} is down by only 10% at $P_{LO} = 8$ mW. This exemplifies the important point stated in Sec. 1.2 that sufficient LO power makes the photocurrent G-R noise dominate the other noise mechanisms. This is particularly useful with GaAs/AlGaAs MQW detectors because very high levels of LO power can be applied before device burnout occurs. For example, a P_{LO} of 100 mW has been applied to sample 2 at 77 K with no degradation in its properties. This opens up the possibility of operating MQW heterodyne detectors at elevated temperatures. To date, the highest temperature at which sensitive heterodyne detection (i.e., near the photon-noise limit) has been demonstrated is 150 K.

4. Systems Applications

4.1. INTEGRATION WITH LOW-NOISE ELECTRONICS

The subunity g of the MQW detector requires the first-stage amplifier shown in Fig. 4 to have a very low noise figure so that the photon-induced G-R noise can dominate with tolerable levels of LO power. At the same time, the amplifier must have a bandwidth of at least 10 GHz to match or exceed B_{IF}. The amplifier technology that is best suited for this application is based on the HEMT. This type of transistor can have a very low noise figure at room temperature and a substantially lower noise figure at cryogenic temperatures. In addition, the bandwidth of state-of-the-art HEMT amplifiers can readily exceed 10 GHz.

A final advantage of the GaAs/AlGaAs MQW detector over its HgCdTe counterparts is that it can be integrated monolithically with high-quality HEMT

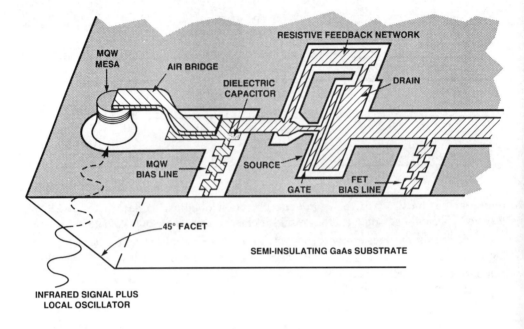

Fig. 5.   Monolithic integration of GaAs MQW detector and low-noise HEMT amplifier on a GaAs substrate.

transistors on a common GaAs substrate. Shown in Fig. 5 is a possible monolithic layout of such a combination. The HEMT is connected with resistive feedback between the drain and gate contacts to provide flat gain over the required bandwidth without spurious oscillation. In addition to the low noise figure and wide bandwidth, the monolithic configuration will present less parasitic reactance to the MQW detector than in the present configuration. This follows from the usage of an air bridge between the detector and amplifier in the monolithic circuit. The air bridge would have roughly 10 times lower capacitance than wire bonds, making it possible that the B_{IF} of sample 3 would exceed 20 GHz.

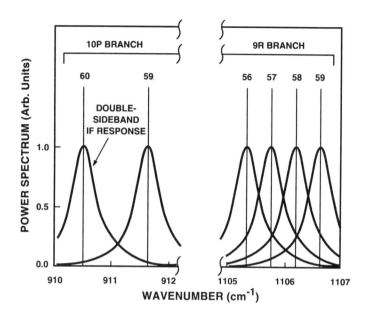

Fig. 6. Concept for a continuous-wavelength spectrometer based on the MQW heterodyne detector and an isotopic CO_2 laser.

4.2. CONTINUOUS-FREQUENCY COHERENT SPECTROMETER

The superior B_{IF} of MQW detectors enables several infrared systems applications beyond the capability of HgCdTe photodiodes. One intriguing possibility is continuous-wavelength remote spectroscopy of atmospheric molecules involved in ozone depletion, global warming, and toxic waste. This is achieved by filling in the frequency space between adjacent lines of a molecular gas laser. For example, the entire wavelength range from approximately 9 to 11 μm can be analyzed with extreme sensitivity by tuning through the lines of a laser containing a CO_2 isotope, such as $^{16}O^{12}C^{18}O$. This concept is presented schematically in Fig. 6, which shows the CO_2 lines at the low- and high-wavenumber extremes of the emission band between 910 cm^{-1} (10.99 μm) and 1107 cm^{-1} (9.03 μm). The separation between adjacent CO_2 lines is roughly 1 cm^{-1} (30 GHz) or less. Each CO_2 line is superimposed on the double-sideband IF response of the MQW detector, assuming that B_{IF} =15 GHz.

In practice the CO_2 lines will be selected by adjusting a grating within the laser cavity. The sum of all the IF response curves constitutes the total IR spectroscopic response. Because of the overlap between adjacent IF response curves, molecular lines practically anywhere in the spectrum between 910 and 1107 cm^{-1} can be detected with great sensitivity.

220

Summary

When operated as an optical-heterodyne detector with a CO_2-laser LO, the GaAs/AlGaAs MQW detector offers three advantages over HgCdTe-based technology. First and foremost is the higher electrical bandwidth, at least five times greater, that can be achieved with these detectors. In this work a B_{IF} of 13 GHz was obtained, but the photoconductive parameters indicate that at least 30 GHz is possible. Second, the MQW detector is very robust and can tolerate high levels of LO power, making elevated temperature (T > 77 K) and other adverse operating conditions feasible. Third, it is monolithically compatible with GaAs HEMT amplifiers, which should facilitate the extension of B_{IF} beyond 20 GHz while maintaining high detection sensitivity.

Acknowledgments

The authors thank C.A. Graves, K.M. Molvar, and D.J. Landers for expert technical assistance with fabrication and packaging, and R.A. Murphy and D.L. Spears for useful comments on the work. This effort was sponsored by the Department of the Air Force.

References

[1]   L.C. West and S.J. Eglash, *Appl. Phys. Lett.* **46**, 1156 (1985).
[2]   B.F. Levine, C.G. Bethea, G. Hasnain, V.O. Shen, E. Pelve, R.R. Abbott, and S.J. Hsieh, *Appl. Phys. Lett.* **56**, 851 (1990).
[3]   H.C. Liu, J.R. Thompson, Z.R. Wasilewski, M. Buchanan, J. Li, and J.G. Simmons, 1993 Device Research Conference Digest, Paper VIB-2.
[4]   B.F. Levine, *Intersubband Transitions in Quantum Wells* (Plenum Press, New York, 1992), p. 43.
[5]   V. Swami Swaminathan, DARPA Infrared Focal Plane Array Program Review, McLean, VA, December 1992.
[6]   R.H. Kingston, *Detection of Optical and Infrared Radiation* (Springer, New York, 1978), Sec. 6.1.
[7]   E.R. Brown, K.A. McIntosh, F.W. Smith, and M.J. Manfra, *Appl. Phys. Lett.* **62**, 1513 (1993).

INTERSUBBAND ABSORPTION IN N - TYPE Si AND Ge QUANTUM WELLS

KANG L. WANG, CHANHO LEE AND S. K. CHUN
Device Research Laboratory
64-127B Engineering IV, Electrical Engineering Department
University of California, Los Angeles
Los Angeles, CA 90024-1594.

ABSTRACT. Intersubband absorption of infrared light of n-type quantum wells was formulated within the framework of effective mass tensors, and normal incidence absorption was observed in the conduction bands of Si and Ge quantum wells on Si (110) and Si (001) substrates, respectively, grown by a Si molecular beam epitaxy (MBE) system. Normal incidence electron transitions are allowed due to the coupling of normally incident light and the off-diagonal terms of the inverse effective mass tensor, which are caused by the tilted energy ellipsoids. The quantum wells were formed by the band offset in the SiGe heterostructure and/or by Sb δ - doping. An absorption coefficient of 17,000 cm^{-1} was obtained as measured by a Fourier transform infrared (FTIR) spectrometer. Absorption peaks ranging 5 - 10 μm were obtained for a variety of samples studied. The detection range can be easily tuned by changing either the doping concentration, the band offset or both. The polarization angle dependencies of the absorption spectra were also obtained from waveguide structures. These results offer the opportunity for the application of focal plane arrays and of integration with Si integrated circuits built on Si substrates.

1. Introduction

Recently, electron intersubband transitions have been successfully demonstrated for an infrared detector application using AlGaAs/GaAs (and other III-V compound semiconductors) and SiGe/Si quantum well structures[1-7]. For GaAs quantum wells, the isotropic (or scalar) effective mass of the Γ conduction band results in the selection rule : the intersubband transitions are allowed only for the incident light with the electric field polarized along the quantum well growth direction (z polarization). For clarity of discussion, we define the xy polarization for which the field is on the quantum well plane, whereas for the z polarization the field is parallel to the quantum well growth direction. For Si quantum wells, the ellipsoids for orientations other than [001] are tilted and there are off-diagonal elements in the effective mass tensor[8-9]. In the investigation of the intersubband spectroscopy of the MOS inversion layer, Nee et al.[9] observed an indication of the transition under both perpendicular and parallel incident lights for Si (110) MOS inversion layers. They used the effective mass tensor formulated by Stern and Howard[10] and showed that the xy-polarized electric field induces the intersubband transition through the off-diagonal terms of the inverse mass tensor. One of the problems of using the inversion layer is the low quantum efficiency. Another problem is that the transition energy usually occurs in the far IR range (10's meV) because of its large effective mass, thus making it impossible for 8 - 12 μm (~ 100 meV) applications. With the recent advent of molecular beam epitaxy (MBE) and low temperature chemical vapor deposition

H. C. Liu et al. (eds.), Quantum Well Intersubband Transition Physics and Devices, 221–235.

(CVD), the strained SiGe layers have been successfully grown, allowing the exploration of using Si/SiGe quantum well heterostructures and/or δ – doped layers. Park *et al.* [4] have demonstrated intersubband transitions near $\lambda = 10 \, \mu m$ for 8 - 12 μm detection applications using p - type δ - doped SiGe/Si quantum wells. Similar results have been reported for n - type multiple quantum well structures on (001) oriented Si[6,11-13]. However, all the n - type quantum well results were for samples grown on (001) and showed a similar polarization dependence as that of GaAs/AlGaAs quantum wells due to the fact that the major axes of the ellipsoids lie on <001> equivalent directions. For samples grown on orientations other than these major ellipsoidal axes, the mass tensor will have off-diagonal elements, resulting in different polarization dependencies. Theoretically, Yang and co-workers[14-15] have calculated the absorption coefficient of conduction intersubband transitions of the SiGe/Si quantum wells grown in the [110] and [111] directions for the xy and z polarizations. The absorption coefficient for the xy polarization was shown to be comparable to that of GaAs/AlGaAs in this case. In their calculation, no strain effect was included and all indirect conduction valleys were assumed to be degenerate. However, we will show that the strain can affect the population of each valley and the occupancy of different valleys, which will in turn determine the polarization behavior of the intersubband transitions of the quantum well. Further, the absorption coefficient changes according to an appropriate sum of the effective masses of the occupied valleys projected onto the growth direction. Thus the strain effect is an important factor to be taken into account for understanding the intersubband transition physics of SiGe quantum wells.

The paper will discuss the electron transitions induced by normally incident light as demonstrated already in the conduction band of Si (110) multiple quantum wells which are composed of Si wells and $Si_{1-x}Ge_x$ barriers[7]. We have also observed the normal incidence detection in Ge multiple quantum wells of $Si_{1-x}Ge_x$/Ge structures grown on Si (001) substrates by Si MBE[16]. The transition energy and absorption strength of the intersubband absorption are measured. The peak absorption positions of the absorption spectra (i.e., transition energy) will be shown to vary with the Ge composition in the barriers and with the doping concentration in the wells. Since Si can be doped to a level two orders higher than GaAs, Si-based structures have a much higher range of wavelength control due to many body effects. Many-body effects will be shown to be important in determining the transition energy. The infrared detectors using intersubband transitions built on Si-based material are attractive as they can be conveniently integrated monolithically with electronic devices for on-chip signal processing.

2. Effective Mass Tensor Formulation

2.1. OSCILLATOR STRENGTH FOR INTERSUBBAND TRANSITION

The physics of conduction band intersubband transitions has been formulated using an effective mass tensor[17]. We briefly outline the results below. Within the one-band effective mass formulation, the Hamiltonian for an electron in the quantum wells shown in Fig. 1(a) can be modified as follows[10]

$$H = \frac{1}{2}P^T W P + V(z) \tag{1}$$

where P is the vector momentum operator and W is a 3 x 3 inverse mass tensor (instead of the mass tensor) to account for the anisotropic mass. For a given quantum well (growth) direction, $\hat{z}$, in the

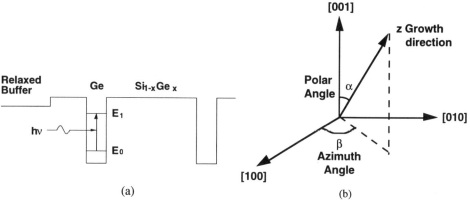

Fig. 1. (a) Typical one dimensional energy band structure of a sample. For this example, the multiple quantum wells were grown along the z direction on relaxed SiGe buffer layer. The L valleys in the conduction band of the strained Ge and $Si_{1-x}Ge_x$ forms quantum wells and barriers, respectively. (b) Coordinate system chosen for the formulation having the growth direction along the z axis.

coordinate system shown in Fig. 1(b), the inverse mass tensor for each indirect conduction valley can be obtained using a coordinate transformation in which Euler's angles are used to describe the relationship between the direction of a conduction ellipsoid and the growth direction. The above expression describes the motion of an electron in the potential well and is valid if the well width is much larger than the atomic dimension. Within this framework of the envelope function approximation, the oscillator strength for any direction of the incident light can be written as[10,17]

$$f_{i \to j} = 2 \frac{m_o}{\hbar \omega} (\varepsilon_x w_{zx} + \varepsilon_y w_{yz} + \varepsilon_z w_{zz})^2 |\langle F_j| P_z |F_i \rangle|^2 \qquad (2)$$

where m_o is the free electron mass, ω is the angular frequency of subband transition, ε_i is a component of the photon polarization vector, and F_i and F_j are the initial and final state envelope functions, respectively.

Unlike the spherically symmetric conduction valley case, the xy-polarized optical electric field can cause the electronic motion in the quantum well direction through the off-diagonal terms of the inverse mass tensor, w_{zx} and w_{yz}, and thus, induce the intersubband transitions. Obviously, for the z-polarization transition, the electric field acts on the envelop through w_{zz} and the behavior is the same as the spherically symmetric Γ band case.

The oscillator strengths of both parabolic and square potential well cases have been evaluated. For the parabolic potential well, the oscillator strength for the transition from n to n+1 in terms of elements of the inverse mass tensor, w_{ij}, is

$$f_{n \to n+1} = \frac{m_o}{w_{zz}} (\varepsilon_x w_{zx} + \varepsilon_y w_{yz} + \varepsilon_z w_{zz})^2 (n+1) . \qquad (3)$$

For the isotropic conduction valley at the Γ point, the off-diagonal elements of the inverse mass ten-

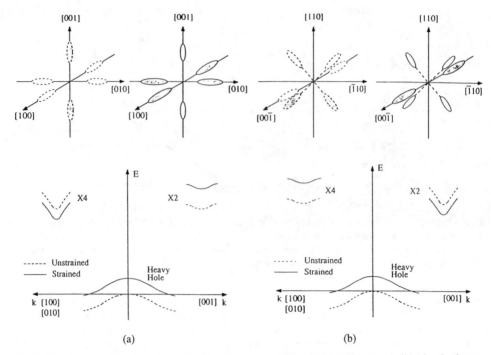

Fig. 2. Change of band structure due to the biaxial compressive strain when the growth direction is chosen along (a) [001] and (b) [110] directions. The dotted lines of the bottom figure shows the positions of the conduction and heavy hole bands for the unstrained case while the solid lines indicate the shifted positions

sor, w_{zx} and w_{yz}, are zero and the above oscillator strength reduces to that for an infinite parabolic potential for the Γ valley[18]. For an infinite square potential well, the oscillator strength can be written as (for comparing with the above expression of the parabolic potential well),

$$f_{n \rightarrow n+1} = \frac{m_o}{w_{zz}} (\varepsilon_x w_{zx} + \varepsilon_y w_{yz} + \varepsilon_z w_{zz})^2 \frac{64}{\pi^2} \frac{(n+1)^2 (n+2)^2}{(2n+3)^2}. \tag{4}$$

For the isotropic conduction band with an infinite square potential well, the above expression gives the same results as Ref. 1. As shown in eqs. (3) and (4), the mass tensor term in the brackets is the same for the infinite parabolic and square potential wells. That is, the oscillator strengths are independent of the type of the potential well.

Since the occupied valleys determine the physics of transition, it is necessary to determine the occupancy of the conduction valleys with and without strain first. For a relaxed SiGe alloy, the energy gap can be approximately obtained by the virtual crystal approximation, i.e., a linear interpolation of the lowest band edge at the X and L points of Si and Ge[19], respectively. In the unstrained bulk material case, the minima of Si- or Ge-like conduction valleys are all degenerate. In the case of quantum wells (for the unstrained Si-like case), the conduction valleys with the largest directional mass give the lowest subband energy level and are occupied as shown in Fig. 2(a) and (b) (dotted lines) for the [001] and [110] growth directions, respectively. Here, for the [001] growth direction as shown in the top part of Fig. 2(a), the two shaded ellipsoids in the [001] and [00$\overline{1}$] directions are occupied due to the higher mass projected to the [001] growth direction. On the other hand, the four equivalent ellipsoids are occupied for the [110] growth direction as shown in top part of Fig. 2(b). In the strained well case, the occupancy changes depending on the strain. When the Si well is compressively strained in plane, the occupancies change as shown in Fig. 2(b). The conduction valleys of the SiGe at the X or L point are no longer degenerate and the determination of occupied valleys cannot be assessed in terms of the directional mass alone. The occupancy of these split sets of the valleys will thus depend on the strain and their effective masses projected to the quantization (or growth) direction. The oscillator strength is evaluated by summing the contributions from each of the occupied valleys according to their occupancies as follows,

$$f_{0 \to 1} = \frac{1}{n^{v}} \sum_{v} (\varepsilon_x w^{v}_{zx} + \varepsilon_y w^{v}_{yz} + \varepsilon_z w^{v}_{zz})^2 \eta^{v}$$ (5)

where n^{v} is number of the occupied valleys and η^{v} is the fractional occupancy of all equivalent valleys. Thus for normal incidence transitions, proper valleys must be occupied with electrons.

2.2. OSCILLAROR STRENGTH IN THE CONDUCTION BANDS OF Si AND Ge

2.2.1. *Si and Ge δ- Doping Quantum Wells*. First, we discuss the unstrained δ - doped quantum well case. The oscillator strength for the intersubband transitions between the ground state and the first excited state is calculated for a parabolic potential well and the results are shown in Figs. 3 through 7. The two angles shown in the figures indicate the growth direction with respect to the [001] direction. For example, the [110] growth direction has a 90° polar angle and an azimuth angle of 45°. Figs. 3(a) and (b) show the oscillator strength of Si δ - doped quantum wells (unstrained) for the xy- and z-polarized electric fields, respectively. For the xy polarization, the coupling of the off-diagonal elements, w_{zx} and w_{yz}, and the optical field is present for the growth directions other than [001] in Si, and thus, can produce the intersubband transitions for the normally incident light. The average oscillator strength for different growth directions can be calculated and the maximum value of normal incidence, 0.8243, is found to be near the [023] growth direction. For the z-polarized electric field, shown in Fig. 3(b), the comparison of the oscillator strength with the GaAs case is given elsewhere[17].

For Ge (unstrained) quantum wells, the oscillator strength for the xy and z polarizations, are shown in Figs. 4(a) and (b), respectively for the L or <111> valleys. For the xy polarization, Fig. 4(a) illustrates that the intersubband transition is forbidden when grown in [111], and the maximum oscillator strength of 3.6755 is obtained when the growth direction is near [144].

2.2.2. *Strained Quantum Wells*. Since the strain will affect the energy band and thus occupancy, the oscillator strengths for several strained quantum wells are also obtained. In the calculation, only the occupied valleys are used, and the inverse mass tensor of the strained Si layers is assumed to be the same to the first order as that of unstrained Si[20].

First, we discuss the result for the strain-symmetrized $Si_{1-x}Ge_{x}$/Si quantum well case[21]. For the case of the $Si_{0.6}Ge_{0.4}$ well under the biaxially compressive strain, grown on $Si_{0.2}Ge_{0.8}$, the calculated results of the oscillator strength for the xy and z polarization are shown in Figs. 5(a) and (b), respectively. In this case, $Si_{0.6}Ge_{0.4}$ experiences a biaxially compressive strain and the valleys are Si-like. When compared with that of relaxed Si shown in Fig. 3, the oscillator strength is smaller

226

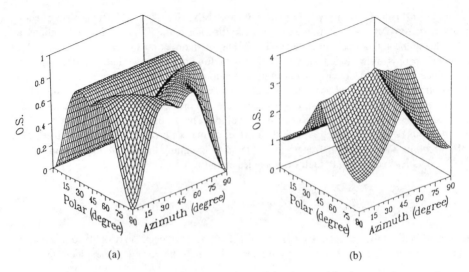

(a)   (b)

Fig. 3. Oscillator strength of the relaxed Si parabolic well for the transition between the ground state and the first excited state: (a) xy polarization and (b) z polarization.

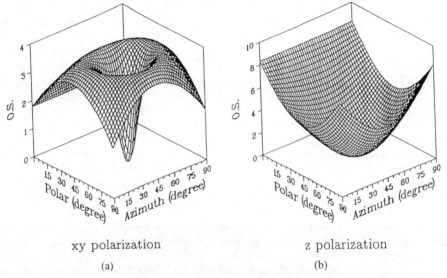

xy polarization   z polarization

(a)   (b)

Fig. 4. Oscillator strength of the relaxed Ge parabolic well for the transition between the ground state and the first excited state: (a) xy polarization and (b) z polarization.

for the xy polarization and larger for the z polarization. This is due to the fact that the occupied valleys under the biaxially compressive strain are different from those of relaxed Si and the different mass tensor components for the newly occupied valleys change the oscillator strength.

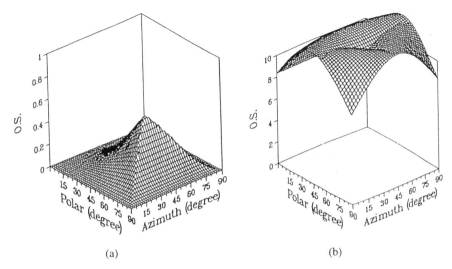

(a)   (b)

Fig. 5. Oscillator strength for the transition between the ground state and the first excited state in the parabolic quantum well: (a) xy polarization and (b) z polarization. The biaxial-compressively strained $Si_{0.6}Ge_{0.4}$ layer grown on $Si_{0.8}Ge_{0.2}$ is the quantum well, where the Si-like conduction valleys are occupied.

Next, the structure of $Ge/Si_{0.6}Ge_{0.4}$ layers grown on $Si_{0.2}Ge_{0.8}$ (for strain symmetrization) is discussed. In this case, the strained Ge layers are the quantum wells and the lowest valleys are Ge-like. Figs. 6(a) and (b) show the oscillator strength for the xy and z polarizations, respectively.

2.2.3. *Waveguide Structure.* In actual experiments, a waveguide structure is often employed in order to increase the absorption strength through multiple internal reflections. For convenience of comparison with the experimental data, we also calculate the oscillator strength for a multipass waveguide. Therefore, the polarization angle dependence of the measured absorption strength can be compared with the calculation.

First, we will discuss the Si (110) case. The waveguide structures shown in the insets of Fig. 7 is usually used for increasing the absorption with multiple reflections in experiments. As shown in the insets, the samples have bevel angles of θ, and the x axes are chosen to be along the [$\bar{1}$10] and [00$\bar{1}$] directions, respectively, in Figs. 7(a) and (b). In the structure, for the incident light normal to the polished plane, the polarization angle ϕ is defined as $\phi = 0°$ for the zx polarization having the optical field components along the x axis (on the plane) and the growth direction. For $\phi = 90°$, however, the field is along the y direction only. The results for general Si-like valleys are shown in Figs. 7(a) and (b), and the contributions for both X2 and X4 valleys are given separately, as functions of the bevel angle θ and the polarization angle ϕ, assuming a parabolic potential. For the case when the $\hat{x}$ is along [$\bar{1}$10], Fig. 7(a) shows the polarization angle dependence, identical to that of the [001] growth case. Fig. 7(b) shows the oscillator strength for the case when the x axis is along the [00$\bar{1}$] direction. In the figure, the oscillator strength (dotted line) from the two X2 valleys in the [001] and [00$\bar{1}$] directions decreases as the polarization angle increases, showing the same dependence as in the Γ valley case. On the other hand, the dependence on the polarization angle of the oscillator strength for the other four X4 valleys (shown in solid lines) is quite different: at the 90° polarization angle, there is a finite contribution and it yields transitions by the normally incident light.

Next, we will discuss the Ge (001) case. Most of the transitions occur between the ground state

228

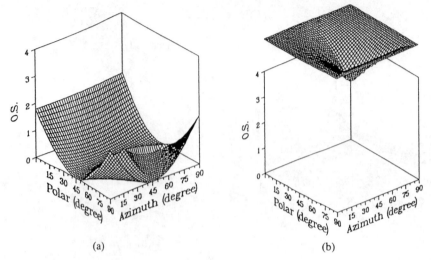

(a)   (b)

Fig. 6. Oscillator strength for the transition between the ground state and the first excited state in the parabolic quantum well: (a) xy polarization and (b) z polarization. The biaxial-compressively strained Ge layer grown on $Si_{0.8}Ge_{0.2}$ is the quantum well, where the Ge-like conduction valleys are occupied.

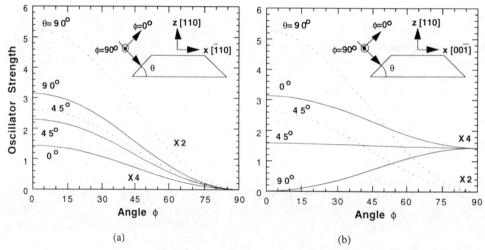

(a)   (b)

Fig. 7. Oscillator strength for the transition between the ground state and the first excited state when the minimum conduction valley is Si-like. The growth direction is along [110] direction and the x axis is chosen along (a) [$\bar{1}$10] and (b) [00$\bar{1}$] directions.

and the first excited state for a narrow quantum well. The total oscillator strength is summed for all of the L valleys from eq. (5) as follows:

$$f_{0 \to 1} = 0.12\eta \sum_{i,j,k=1}^{2} (\varepsilon_x(-1)^i w_{zx} + \varepsilon_y(-1)^j w_{yz} + \varepsilon_z w_{zz})^2 \; , \tag{6}$$

assuming each valley is equally occupied. For Ge, $w_{zx} = w_{yz} = 3.86/m_o$ and $w_{zz} = 8.33/m_o$.

When the polarization angle, ϕ, is defined as in Fig. 7 with $\theta = 45°$, the polarization angle dependence of the oscillator strength becomes

$$f_{0 \to 1} = 0.48\eta m_o \left(\frac{(w_{zx})^2}{w_{zz}} \cos^2\phi + \frac{2(w_{yz})^2}{w_{zz}} \sin^2\phi + w_{zz} \cos^2\phi \right). \tag{7}$$

The first two terms are due to the coupling of the off-diagonal terms of the inverse mass tensor and the optical field, and the third term is due to the diagonal terms. The optical field can be decomposed into two components: the z-polarized and the xy-polarized fields. The z-polarized field couples with the diagonal terms, inducing the intersubband transition. This intersubband transition occurs for all carriers occupying any of the valleys. The xy-polarized field couples with the off-diagonal terms of the inverse mass tensor and a normal incidence intersubband transition is also produced. This intersubband transition is possible only for the carriers in the energy ellipsoids which are not on the principal axes such as Si (110) and Ge (001). In other words, the off-diagonal terms of the inverse mass tensor should be non-zeros. For the normal incidence case, $\phi = 90°$ and the oscillator strength is $0.96\eta m_o(w_{yz})^2/w_{zz}$.

Eq. (7) represents the polarization angle dependence of the oscillator strength and absorption strength. This polarization angle dependence has the same form regardless of the quantum well shapes as long as the waveguide structures are the same. Because it is difficult to obtain the exact shape of the quantum well for the grown samples, it is more prudent to compare the polarization angle dependence of the measured data with the *normalized* absorption strength for the calculated values from eq. (7). The measured data are obtained from the absorption integral as a function of the polarization angle ϕ of the incident light. Here, an absorption integral is defined as the integrated area of an absorption spectrum.

3. Experiment

Three types of structures as shown in Figs. 8(a) - (c) were used for the study. Samples were grown in a Si-MBE system. The growth and doping procedures were described elsewhere[6]. The first structure (A) consists of a 3000 Å layer of an undoped Si buffer layer followed by 5 periods of 50 Å n^+ Si layers doped with Sb and 300 Å spacer layers of undoped Si, and then an undoped Si cap layer with a thickness of 3000 Å grown on a Si (001) substrate. The second structure (B) consists of an undoped $Si_{1-y}Ge_y$ buffer layer followed by 10 periods of a 50 Å *Si* well and a 300 Å $Si_{1-x}Ge_x$ barrier, and a 2500 Å $Si_{1-y}Ge_y$ undoped cap layer. The Ge composition x of the barrier is varied from 0 to 50%. Si well layers are doped with Sb in the center 40 Å region and the doping concentration of the Si layers is about 1.3×10^{20} cm^{-3} according to secondary ion mass spectrometry (SIMS) profiles. The Ge content (y) in the buffer layer is chosen so as to maintain the symmetric strain condition in the multiple quantum wells[20]. More importantly, this is done in order to provide a favorable condition for normal incidence detection. The last structure (C) is composed of 10

230

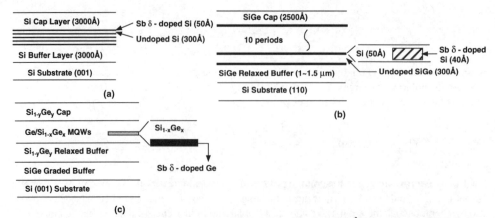

Fig. 8. Sample structures. (a) The structure A on Si (001) consist of a 3000 Å layer of undoped Si buffer layer followed by 5 periods of 50 Å n^+ Si layers doped with Sb and 300 Å spacer layers of undoped Si, and then an undoped Si cap layer with a thickness of 3000 Å. (b) The structure B on Si (110) consists of a 1 ~ 1.5 μm undoped $Si_{1-y}Ge_y$/Si buffer layer followed by 10 periods of a 50 Å Si layer doped with Sb in the 40 Å center region, an undoped $Si_{1-x}Ge_x$/Si barrier of 300 Å thick, and finally a 3000 Å undoped $Si_{1-y}Ge_y$/Si cap layer. (c) Structure C on Si (001). The active region consists of ten periods of a 250 ~ 320 Å $Si_{1-x}Ge_x$ barrier and a 42 Å thick Sb δ - doped *Ge* well. An undoped $Si_{1-y}Ge_y$ cap layer followed the active layers.

Sb doped *Ge* quantum wells and undoped $Si_{1-x}Ge_x$ barriers grown on a Si (001) substrate. The Ge composition (x) of the barriers ranges from 40% to 80%. The Ge content (y) in the buffer layer is chosen similarly to maintain the symmetric strain condition. The doping is done similar to the case B and the doping concentrations of the Ge layers are varied from 10^{17} to 10^{20} cm^{-3}, as verified by SRP and SIMS profiles. The Ge (001) well is chosen for this study because it has several advantages. Firstly, normal incidence detection is possible as discussed before. Secondly, Ge (001) has a large inverse effective mass: $8.33/m_0$ in the growth direction and $3.86/m_0$ in the quantum well plane direction. If a waveguide structure is used, the inverse effective mass is $10.12/m_0$ for $\phi = 0^\circ$. The inverse mass for GaAs is $14.93/m_0$, and thus, $f_{Ge}/f_{GaAs} = 0.68$. However, Ge can be doped up to 20 times higher and, therefore, the absorption strength of Ge can be higher than that of GaAs by one order of magnitude. Finally, because Ge (001) can be grown on Si (001) substrate, it can potentially be integrated with the conventional Si electronic integrated circuits.

Waveguide structures are prepared by polishing a 45° angle wedge on both edges. The absorption spectra of the samples are measured at room temperature using a Nicolet 740 Fourier transform infrared (FTIR) spectrometer.

4. Result and Discussion

Fig. 9 shows the absorption spectra of a sample of structure C with varying polarization angles. The sample has 10 Ge/$Si_{0.6}Ge_{0.4}$ multiple quantum wells and the 2D carrier density in each Ge well is 6.2×10^{13} cm^{-2}. The maximum absorption coefficients at the 0° and 90° polarization angles are 14,500 cm^{-1} and 11,000 cm^{-1}, respectively. The absorption coefficient is smaller for the normal in-

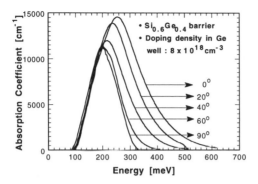

Fig. 9. Absorption Spectra of a Ge well sample for various polarization angles. The sample has ten Ge/Si$_{0.6}$Ge$_{0.4}$ multiple quantum wells and the 2D carrier density in each Ge well is 6.2 $\times$ 10^{13} cm^{-2}.

cidence case due to the smaller inverse effective mass. The peak absorption wavelengths at 0° and 90° are 5.0 μm (249 meV) and 6.3 μm (196 meV), respectively. The peak absorption position shift of 53 meV at 90° is due to the depolarization effect[9] and is close to the calculated value of 59 meV. The depolarization shift becomes larger as the carrier density increases.

The polarization angle dependence of the absorption strength (absorption integral obtained from the spectra) is plotted in Figs. 10(a) and (b). Fig. 10(a) shows the polarization angle dependence of the absorption strength for structures A and B (Si (001) and Si(110) wells). Normal incidence detection is observed when the light is incident along the [110] direction. However, when the light is incident to the [001] direction, it is identical to Si (001), and thus, there is no absorption at the 90° polarization angle. For the [110] incidence case, experimental values are smaller than the calculated values because the X2 valleys, which do not contribute to the normal incidence detection, are partially occupied by electrons. The experimental values of the [001] incidence case at 90° is not zero due to the difficulty in locating the exact [001] direction in the FTIR absorption measurement. For the normal incidence detection, it is important that the X4 valleys are occupied. The latter condition can be accomplished with a proper strain condition as discussed above. Using the Si$_{1-y}$Ge$_y$ buffer layer, the Si layers suffer an in-plane tensile strain and the X2 and X4 energy bands in Si layers split, making the X4 valleys lower than the X2 valleys. The energy level separation of the X2 and X4 valleys are 85 meV for the Si$_{0.7}$Ge$_{0.3}$/Si multiple quantum well structures if the doping effect is not considered.

Fig. 10(b) shows the polarization angle dependence of structure C (*Ge(001) well*). The solid circles are the average values of many samples having various Ge compositions in the Si$_{1-x}$Ge$_x$ layers (40% ~ 80%) and with various doping concentrations in the Ge layers (4 $\times$ 10^{17} cm^{-3} ~ 1.5 $\times$ 10^{19} cm^{-3}). The error bars represent the standard deviations. The solid line represents the calculated values of absorption strength of the strained Ge. The calculated absorption strength is normalized. The transition width is assumed to be a δ - function in the calculation, however, the actual energy subbands are broadened due to phonon, impurity and other scatterings. The discrepancy between the experimental data and the calculated data may be due to the non-ideal nature of the quantum wells and in part due to the use of the absorption integral in comparison with the calculated oscillator strength. Although the conduction band minima lie in the L valleys in the latter case, there is a possibility that the X valleys are partially filled with carriers because of the high doping used and the

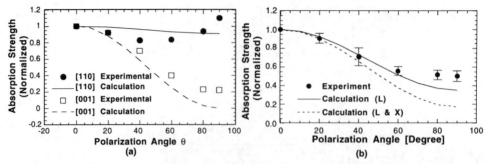

Fig. 10. (a) Absorption strength vs. polarization angle for a Si (110) sample. For the optical field along the [110] direction, the absorption strength decreases first and increases again as the polarization angle is increased. For the optical field along the [001] direction, the absorption strength decreases in accordance with the $\cos^2\theta$ rule. (b) Polarization angle dependence of absorption strength for Ge (100). The solid line and the dashed line represents the calculated oscillator strength of Ge L valleys only and of Ge L and X valleys, respectively, and the solid circles represent the average value of many samples of various Ge compositions in $Si_{1-x}Ge_x$ layers and of different doping concentrations in Ge layers. Error bars represent the standard deviation.

lowering of the X4 valleys by the compressive strain in the Ge wells. However, the experimental results of the polarization angle dependence shown in Fig. 10(b) indicates that the absorption comes mainly from the electrons in the L valleys. The oscillator strength of electrons in the X valleys is

$$f_{0 \rightarrow 1} = 0.48 \eta m_o w_{zz} \cos^2\phi \tag{8}$$

where $w_{zz} = 5.56/m_o$ and is equal to zero at $\phi = 90°$ as expected. The dashed line in Fig. 10(b) represents calculated values of absorption strength when the L and X valleys are equally occupied by electrons. If the absorption had a significant contribution from the X valleys, the experimental absorption strength data would be smaller than the calculated values for the 90° polarization angle as shown in the figure. Therefore, the measured absorption comes mainly from the electrons in L valleys.

Figs. 11 (a) - (c) show the effects of strain and doping. Fig. 11 (a) shows the strain effect on the transition energy and the absorption strength of *structure B* for different Ge composition in $Si_{1-x}Ge_x$ barriers. The increase of the peak transition energy is due to the larger band offset of X valleys in the $Si/Si_{1-x}Ge_x$ heterostructure as the strain increases (or the Ge composition in $Si_{1-x}Ge_x$ barriers increases). The transition energy difference between the 0° and 90° of polarization angle is due to the depolarization effects[9]. The reason for the decreases in absorption strength as the strain increases is not yet clear. The absorption strength for the polarization angle of 90° is smaller than that for 0° because the off diagonal terms of the inverse mass tensor of Si (110) are smaller than those of the diagonal terms.

Fig. 11 (b) shows the strain effect on the peak transition energy and the absorption strength for *structure C*, i.e., Ge wells. The intended doping concentration in the Ge wells is 8×10^{18} cm^{-3} for each sample. The strain increases as the Ge composition (x) in the $Si_{1-x}Ge_x$ barrier decreases. The peak transition energy and the absorption strength are observed to increase as the strain increases and to have different values at the polarization angle of 0° and 90°. The transition energy increase

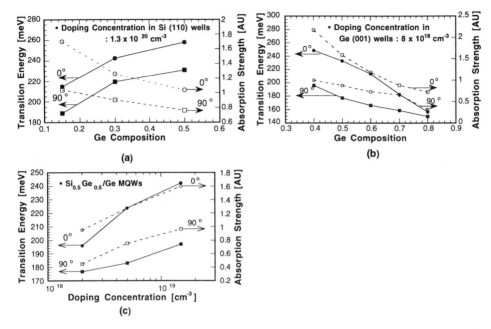

Fig. 11. (a) Ge composition dependence of transition energy and normalized absorption strength for Si (110). The Ge compositions of the barriers are varied. The peak energy position shifts to higher energies as the Ge composition is increased. (b) Ge composition dependence of transition energy and absorption strength for Ge wells grown on Si (001). The maximum transition energies and absorption integrals decrease as the Ge composition increases. The intended doping levels are same for all Ge compositions. (c) The doping concentration dependence of transition energy and absorption strength for structure C. The doping concentrations of samples are varied from 4×10^{17} to 8×10^{18} cm^{-3}.

comes from a deeper potential and higher carrier density. As the strain increases, the conduction band offset of L valleys increases, and thus, the quantum well potential becomes deeper. The higher carrier concentration gives rise to the larger many-body effects[22], which also result in the increased transition energy (to be discussed later). With higher Ge contents, Sb dopants tend to segregate more readily, and consequently, the incorporated Sb doping concentration decreases. The loss of dopants will result in a lower Hartree-Fock potential and a lower depolarization shift. In the Si well case, Sb segregation is minimal and the doping density remains as the intended density for all the barrier height values. As can be seen in Fig. 11(b), the difference of the transition energies of the 0° and 90° polarization cases becomes very small as the Ge content approaches to 0.8, indicating a lower carrier density. In contrast, the Si well case shows a constant difference for two polarization angle cases in Fig. 11(a) and there is no doping loss in this case. The FWHM's of the absorption spectra, which are not shown in the figure, are also observed to increase as the strain increases. A similar trend is observed for the amount of the depolarization shift, which increases as the carrier density increases (to be discussed later). Indeed, the absorption strength decreases as the Ge content increases due to the loss of dopants, supporting this argument. The difference of the absorption strengths for the 0° and 90° polarization angle cases is due to the difference of the diagonal and the off-diagonal terms in the inverse mass tensor, respectively.

The doping effects on the transition energy and absorption strength are described next. The absorption spectra in Fig. 11(c) show that the transition energy shifts to a lower energy for structure C due to the depolarization effect as the polarization angle increases. Each sample has $Si_{0.5}Ge_{0.5}$ barriers. The peak transition energy and the absorption strength are observed to increase as the doping concentration increases and to have different values at the polarization angle of $0°$ and $90°$. Since the absorption strength is proportional to the density of electrons which can make transitions in the quantum wells, it is obvious that the absorption strength increases as the doping concentration increases. The transition energy increase can be explained by two factors. One is the Hartree-Fock potential. As the doping concentration increases, the Hartree-Fock potential increases in the doped layers and the quantum well potential becomes deeper, leading to a higher the transition energy. The other is another many-body effect, the depolarization[22]. The difference of the transition energy at $\phi = 0°$ and $\phi = 90°$ can be attributed to the depolarization effect. It is also observed that the transition energy difference becomes larger with the increase of the doping concentration because the depolarization effect becomes larger[9,23]. For example, the depolarization shift for a doping concentration of 2×10^{18} cm^{-3} is 19 meV and that for a concentration of 1.5×10^{19} cm^{-3} is 45 meV. The absorption strength at $\phi = 0°$ is larger than that at $\phi = 90°$ because the diagonal terms of the inverse effective mass tensor are larger than the off-diagonal terms. The ratio of the oscillator strength at $\phi = 0°$ to that at $\phi = 90°$ from eq. (7) is 2.86. The ratio obtained from the measurement is 2.20.

5. Conclusion

We have calculated the oscillator strength of electrons in the conduction band of Si and Ge and have observed normal incidence absorption by electron intersubband transition in the conduction bands of Si (110) and Ge (001) quantum wells. These observations were made on multiple quantum well structures grown on Si (110) and Si (001) substrates, respectively, in a Si MBE system. δ - doping was obtained using thermally evaporated Sb with a low substrate temperature. A high absorption coefficient has been obtained due to the large inverse effective mass and the high doping density. The absorption wavelength can be easily controlled in the range of 5 - 12 μm by changing the Ge composition in the $Si_{1-x}Ge_x$ layers and the doping concentration in the Si and Ge layers. These structures have potential application for focal plane arrays and for integration with Si electronic VLSI circuits.

Acknowledgment

The authors would like to thank Dr. Karunasiri for helpful discussion. This work was in part supported by ARO, AFSOR, and SRC.

References

1. West L. C. and Eglash S. J. (1985) 'First observation of an extremely large-dipole infrared transition within the conduction band of a GaAs quantum well', Appl. Phys. Lett. **46**, 1156.
2. Levine B. F., Choi K. K., Bethea C. G., Walker J. and Malik R. J. (1987) 'New 10μm infrared

detector using intersubband absorption in resonant tunneling GaAlAs superlattices', Appl. Phys. Lett. **50**, 1092.

3. Karunasiri R. P. G., Park J. S. and Wang K. L. (1991) 'Si$_{1-x}$Ge$_x$/Si multiple quantum well infrared detector', Appl. Phys. Lett. **59**, 2588.

4. Park, J. S., Karunasiri R. P. G., Mii Y. J. and Wang K. L. (1991) 'Hole Intersubband absorption in δ-doped multiple Si layers', Appl. Phys. Lett. **58**, 1083.

5. Hertle H., Schuberth G., Gornik E. and Abstreiter G. (1991) 'Intersubband absorption in the conduction band of Si/Si$_{1-x}$Ge$_x$ multiple quantum wells', Mat. Res. Soc. Symp. Proc. **220**, 379.

6. Lee C. and Wang K. L. (1992) 'Intersubband absorption in Sb δ-doped molecular beam epitaxy Si quantum well structures', J. Vac. Sci. Technol. **B 10**, 992.

7. Lee C. and Wang K. L. (1992) 'Intersubband absorption in Sb δ-doped Si/Si$_{1-x}$Ge$_x$ quantum well structures grown on Si (110)', Appl. Phys. Lett. **60**, 2264.

8. Yi K. S. and Quinn J. J. (1983) 'Optical absorption and collective modes of surface space-charge layers on (110) and (111) silicon', Phys. Rev. **B 27(4)**, 2936.

9. Nee S. M., Claessen U. and Koch F., (1984) 'Subband resonance of electrons on Si (110)', Phys. Rev. **B 29**, 3449.

10. Stern F. and Howard W. E. (1967) 'Properties of Semiconductor Surface Inversion Layers in the Electric Quantum Limit', Phys. Rev. **B 163**, 3449.

11. Ziendel H. P., Wegehaupt T., Eisele I., Oppolzer H., Reisinger H., Tempel G. and Koch F. (1987) 'Growth and characterization of a delta-function doping layer in Si', Appl. Phys. Lett. **50**, 1164.

12. Tempel G., Schwarz N., Müller F. and Koch F. (1990) 'Infrared resonance excitation of δ-layers - A silicon-based infrared quantum-well detector', Thin Solid Films **184**, 171.

13. Müller F., Tempel G. and Koch F. (1990) 'Infrared Excitation of (100) Si δ - layer Detected by Surface Barrier Tunneling', *Proc. ICPS II*, 1254.

14. Yang C. I. and Pan D. S. (1988) 'Intersubband absorption of silicon-based quantum wells for infrared imaging', J. Appl. Phys. **64**, 1573.

15. Yang C. I., Pan D. S. and Somoano R. (1989) 'Advantages of an indirect semiconductor quantum well system for infrared detection', J. Appl. Phys. **65**, 3253.

16. Lee C., Chun S. K. and Wang K. L. (1993) 'Electron transition by normally incident light in Si$_{1-x}$Ge$_x$/Ge multiple quantum wells grown on Si (001)', SSDM, Tokyo, Japan.

17. Chun S. K. and Wang K. L. (1992) 'Oscillator strength for intersubband transitions in strained n-type SiGe quantum wells', Phys. Rev. **B 46**, 7682.

18. Karunasiri R. P. G. and Wang K. L. (1988) 'Infrared absorption in parabolic multiquantum well structures', Superlatt. Microstruct. **4**, 661.

19. Braunstein R., Moore A. R. and Herman F. (1958) 'Intrinsic Optical Absorption in Germanium-Silicon Alloys', Phys. Rev. **109**, 695.

20. Balslev I. (1966) 'Influence of Uniaxial Stress on the Indirect Absorption Edge in Silicon and Germanium', Phys. Rev. **143**, 636.

21. Kasper E., Herzog H.-J., Jorke H. and Abstreiter G. (1987) 'Strained layer Si/SiGe superlattices', Superlatt. Microstruct. **3**, 141.

22. Karunasiri R. P. G., Wang K. L. and Park J. S. (1992) 'Intersubband Transitions in SiGe/Si Quantum Structures', in Z. C. Feng, (eds.) *Semiconductor Interfaces and Microstructures*, World Scientific, Singapore, p. 252.

23. Ando T. (1975) 'Subband Structure of an Accumulation Layer under Strong Magnetic Fields', J. Phys. Soc. Jpn. **39**, 411.

INTERSUBBAND TRANSITIONS IN P-TYPE SiGe/Si QUANTUM WELLS FOR NORMAL INCIDENCE INFRARED DETECTION

R. P. G. KARUNASIRI

Department of Electrical Engineering
66-127G Engineering IV
University of California Los Angeles
Los Angeles, CA 90024

ABSTRACT. The study of intersubband transitions in both quantum wells and superlattices have rapidly developed to the point where intersubband photodetectors are becoming attractive for a number of applications. This is particularly significant for Si-based heterostructures due to its advantage of monolithic integration with the conventional silicon signal processing electronics. In this paper, experimental observations of intersubband and intervalence band transitions in SiGe/Si quantum wells will be described. The use of polarization dependence measurement for the identification of different transitions will be illustrated. Finally, the application of SiGe/Si multiple quantum well structures for the fabrication of infrared detectors operating in both the 2-5 μm and 8-12 μm ranges will be described.

1. Introduction

Because of the potential for monolithic integration and the well developed processing technology, Si-based optoelectronic devices have attracted considerable interest in recent years. Photodetectors operating in the 1.3-1.5 μm range have been fabricated using interband transitions of SiGe/Si multiple quantum wells [1]. The recent development of visible light emission from porous-Si [2] has also shed new light on the optoelectronic application of Si-based structures. In this paper, infrared transitions between the subbands of SiGe/Si quantum wells will be reviewed and then the potential applications of these transitions in infrared detection will be discussed. In SiGe/Si system, hole intersubband transitions were first observed between heavy hole subbands using p-type multiple quantum wells [3]. This is primarily due to the large band offset at the valence band. Recently, in addition to intersubband transitions, transitions between different hole bands were also reported [4]. The origin of this intervalence band transition can be traced back to the coupling of the conduction band with the valence band [5, 4]. Figure 1 illustrates the possible subband transitions in the valence band of a SiGe/Si quantum well structure. One of the key advantages of the intervalence band transition is the

237

H. C. Liu et al. (eds.), Quantum Well Intersubband Transition Physics and Devices, 237–250.
© 1994 *Kluwer Academic Publishers. Printed in the Netherlands.*

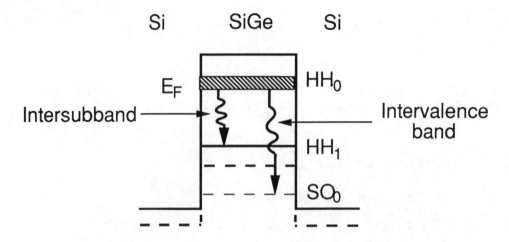

Figure 1: Band diagram of the quantum well structures showing possible subband transitions.

possibility of normal incidence detection which is forbidden in the case of the intersubband transition [6]. More recently, intersubband transitions in the conduction band have been observed [7] which includes normal incidence absorption due to the off- diagonal effective masses for structures grown on specific orientations [8]. The advantage of subband transitions is that the characteristic properties are independent of the host materials rather controlled by the band offsets and layer dimensions. This is particularly important in the Si/Ge system where both constituents have indirect bandgaps which limits its application in optoelectronic devices. In the following, we will begin with a discussion on the material and the design issues that are important in the optimization of detector response for a desired wavelength range.

2. Material Considerations

The important parameters in the design of p-type SiGe/Si multiple infrared detectors are: the critical thickness, the band-offset, the relative positions of the three hole bands (i.e., light, heavy, and split-off), and the respective effective masses. The critical thickness of strained SiGe layer on Si is strongly dependent on the growth parameters, especially the substrate temperature. For a typical growth temperature ($\sim$ 500-600 °C) used in the heteroepitaxy of SiGe on Si, the values of the critical thickness as a function of the Ge composition is given by People and Bean [9]. For example, a SiGe layer with a 50% Ge composition has a critical thickness of about 100 Å. In the case of multiple layer growth, the critical thickness is obtained using the average Ge composition of a single period of the structure (i.e, the average Ge content of a period, $x_{Ge} = (x_1 d_1 + x_2 d_2)/(d_1 + d_2)$, where the x's and d's are the Ge content and thickness of each constituent layer, respectively) [10]. In order to avoid the generation of misfit dislocations as a result of strain relaxation, the individual

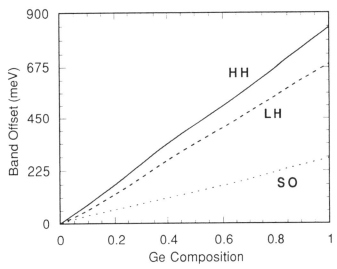

Figure 2: Band offsets of the three hole band as a function of Ge composition for a strained SiGe layer grown on Si.

layer thicknesses as well as the multiple layer thickness must be kept below their respective critical thicknesses. One of the other important parameters in determining detector response is the band-offset. This is obtained by interpolation using the band offsets of Si and Ge which were originally obtained by Van de Walle and Martin [11]. Figure 2 shows the valence band offset for the three hole bands of a SiGe layer grown on Si substrate. For most practical purposes the conduction band offset is negligible if the structure is grown on a Si substrate. For the detectors based on internal photoemission (2D free carrier absorption) from quantum wells, the heavy hole band-offset mainly determines the cut-off of the photoresponse. On the other hand, for the detectors based on inter-valence band absorption, the detector response is strongly dependent on the strain induced splitting of the three valence bands. Such splittings can be calculated as a function of Ge composition using deformation potentials [12]. Figure 3 shows the positions of three valence bands as a function of the Ge composition of a SiGe layer grown pseudomorphically on a Si substrate. Another important parameter in designing a detector with a given spectral response is the effective masses of the three valence bands. The values of the effective masses, including strain effects have been calculated using the k·p approximation by Chun and Wang [13].

3. Optimization of Detector Structure

For attaining optimum sensitivity, a quantum well structure with a single bound state and an excited extended state close to the barrier is desirable [6, 14]. The photon energy required for the transition is usually higher than the energy difference between the bound and the extended states. This is due to well known plasma shifts [15]. In the case of heavy doping in the well, the exchange-correlation interactions

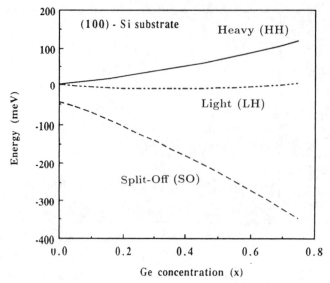

Figure 3: Relative positions of the three hole band as a function of Ge composition for a strained SiGe layer grown on Si.

between holes can also affect the positions of the energy levels [16]. Such effects are incorporated by solving the Schrodinger's and Poisson's equations self-consistently using the exchange- correlation interaction potential in a density-functional approximation [15, 16]. The peak position, including the shifts due to these two effects, is given by

$$\tilde{E}_{10}^2 = E_{10}^2(1 + \alpha - \beta) \tag{1}$$

where $\tilde{E}_{10}$ is the shifted energy, E_{10} is the difference $(E_1 - E_0)$ between the subband energy levels calculated self- consistently, and α and β are the depolarization and exciton- like effects, respectively. The details of the calculation of the quantities α and β are given by Ando [15].

The doping is typically done in the well in order to minimize the scattering in the barriers where the photoexcited carriers spend most of their time before reaching the contacts. However, doping in the well can reduce the lifetime of the excited carriers, which can be captured back into the well without contributing to the photocurrent. This can have a strong effect on the response of the detector. Therefore, when considering the amount of dopants in the well, a compromise must be made between quantum efficiency and the leakage current (mainly due to thermionic emission) that can be tolerated for a given application. In the case of p- type SiGe/Si detectors, an order of magnitude larger doping density is required to obtain comparable quantum efficiencies to that of n-type GaAs/AlGaAs detectors. This is because the hole effective mass is larger than the electron mass in GaAs.

4. Infrared Transitions in p-type SiGe/Si Quantum Wells

Two types of subband transitions are possible in the valence band of SiGe/Si quantum well structures: (a) an intersubband transition which is between quantized states of a same hole band and (b) an intervalence band transition which involves the participation of two different hole bands (for example, between the light and heavy hole subbands). In addition to these transitions, strong two dimensional (2D) free carrier absorption occurs for relatively long wavelengths. The structures used in the experiment were grown in a molecular beam epitaxy (Si-MBE) system on high resistivity (100 Ω-cm) (100)-Si wafers. The growth temperature is kept as high as possible (in the 500-600 °C range, depending on the Ge composition) in order to improve the quality of the Si barrier where most of the photoexcited carrier transport occurs. Several $Si_{1-x}Ge_x$/Si multiple quantum well structures with different Ge compositions, x = 0.15, 0.3, 0.4 and 0.6 in the quantum well, were employed to study the strain dependence of the subband transitions. A period of the SiGe/Si multiple quantum well structure consists of a 40 $\mathring{A}$ $Si_{1-x}Ge_x$ well and a 300 $\mathring{A}$ Si barrier. The center 30 $\mathring{A}$ of the $Si_{1-x}Ge_x$ wells is boron doped to about 5×10^{19} cm^{-3} while the Si barriers are undoped. For the $Si_{0.85}Ge_{0.15}$/Si, $Si_{0.7}Ge_{0.3}$/Si and $Si_{0.6}Ge_{0.4}$/Si samples, 10 periods of multiple quantum wells are grown, while for the $Si_{0.4}Ge_{0.6}$/Si sample, only five periods are grown because of the critical thickness limitation of the SiGe strained layers. The absorption spectra of the samples were taken at room temperature using a Fourier transform infrared (FTIR) spectrometer. The different transitions were identified using the characteristic polarization dependence of the absorption [4].

Figure 4 shows the measured absorption spectra of the $Si_{0.4}Ge_{0.6}$/Si quantum well sample at two different polarization angles using a waveguide structure as shown in the inset. At the 0° polarization angle (electric field having a component in the growth direction as depicted in the inset), an absorption peak occurs at 5.3 μm. At the 90° angle (beam polarized parallel to the plane), this peak vanishes, but another peak appears at a shorter wavelength, 2.3 μm. The peak at the 0° polarization is due to the intersubband transition between two heavy hole states [17]. This is confirmed by the polarization dependence of the peak shown in the Fig. 5 (a). The peak at 2.3 μm shows a polarization dependence opposite to that of the intersubband transition as illustrated Fig. 5 (b). At normal incidence (90° polarization), the absorption reaches the maximum and it drops as the angle decreases. Two other samples, $Si_{0.7}Ge_{0.3}$/Si and $Si_{0.6}Ge_{0.4}$/Si, show an absorption spectra similar to that of the $Si_{0.4}Ge_{0.6}$/Si sample discussed above [18]. For the sample with 15 % Ge concentration, the absorption spectra revealed a clear peak at 0°, due to the heavy hole intersubband transition; however, no obvious peaks are observed at 90°.

The origin of the absorption peak at the 90° (or normal incidence) can be understood qualitatively by considering the matrix element of the optical transition including band mixing. The matrix element of the transition when the light is incidence normal to the layer, is given by [19]

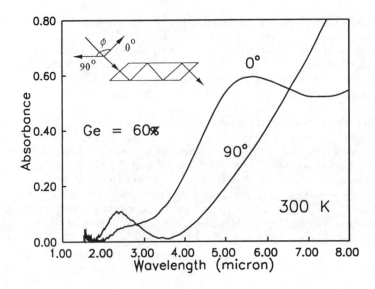

Figure 4: Measured room temperature absorption spectra of a $Si_{0.6}Ge_{0.4}/Si$ multiple quantum well sample at 0 and 90° polarization angles. At 0°, a clear peak due to intersubband transition is shown at near 5.3 μm. However, at 90° this peak vanishes, and another peak appears at 2.3 μm which is due to intervalence band transition. The inset shows the waveguide structure used for the measurement.

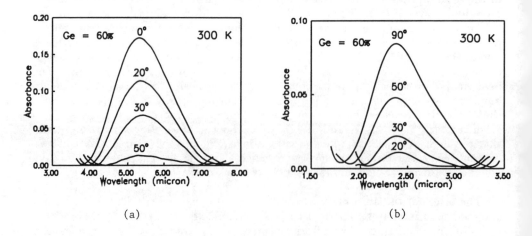

Figure 5: Polarization dependence absorption spectra of the two peaks shown in the Fig. 4.

$$M \approx\ <U_f\ |\ \hat{\epsilon}\cdot\vec{p}\ |\ U_i\ ><F_{fs}\ |\ F_{is}\ > \tag{2}$$

where U_i and U_f are the initial and final Bloch functions, respectively; F_{is} and F_{fs} are the s-like component of the initial and final envelope functions, respectively; $\hat{\epsilon}$ is the photon polarization vector; $\vec{p}$ is the momentum of the hole; and z indicates the growth direction. The matrix element is zero to the first order if U_f and U_i are both p-like at the Γ-point assuming no band mixing [5]. If, however, the light and spin-orbit split-off Bloch functions have significant band mixing from the s-like conduction band, a substantially large matrix element can result. The band mixing can be significant if the (Γ) bandgap is small enough and if the light hole and spin-orbit split-off bands involved in the transition have a large $k_\parallel$ [20]. For Si, the coupling is weak between the conduction and valence bands due to the large direct band gap at the Γ point (4.2 eV). On the other hand, for Ge, a significantly large coupling is expected since the bandgap at the Γ point is small (0.89 eV). Due to this coupling, the valence band wave functions can have a component of the s-like conduction band state, particularly for large $k_\parallel$. For $Si_{1-x}Ge_x$, the bandgap at the Γ-point decreases as the Ge content increases, and therefore, with a sufficient Ge content, the intervalence band transition will become stronger.

For estimating the absorption strength qualitatively, the coupling of the valence band Bloch states with the conduction band can be expressed by using the first order perturbation theory [5],

$$U_i\ =\ U_i^0\ +\ \frac{\hbar}{m_0}\frac{<U_c^0\ |\ \vec{k}\cdot\vec{p}\ |\ U_i^0\ >}{(E_g\ +\ |\ E_i\ |)}U_c^0 \tag{3}$$

where the index i $(= h,\ l,\ s)$ refers to the heavy, light and spin-orbit split-off holes, respectively, and U_c is the Bloch state of the conduction band. E_g is the Γ energy gap and E_i is the energy of the light and split-off hole band edges measured from the heavy hole band edge. It can be seen that as the energy gap reduces, the coupling with the conduction band increases. This explains the fact that the intervalence band absorption was observed only for high Ge concentrations. From Eq. 3, it is expected that the absorption strength will be proportional to $(E_g\ +\ |\ E_i\ |)^{-2}$. Indeed such a dependence is observed experimentally as illustrated in Fig. 6.

The observed polarization dependence shown in Fig. 5 (b) can be understood by considering the optical matrix element between Bloch functions. The matrix element involved in the present case is proportional to $<U_c^0\ |\ \hat{\epsilon}\cdot\vec{p}\ |\ U_{HH}>$ where the Bloch function U_{HH} is only a function of the (x,y) plane [5]. Thus, the largest absorption is expected when the field is polarized along the (x,y) plane. Next, we will discuss the use of the subband transitions in SiGe/Si quantum wells described above for infrared detector applications.

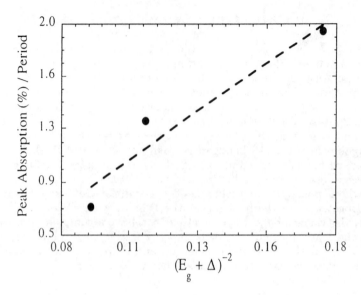

Figure 6: Measured absorption strength as a function of $(E_g + \mid E_i \mid)^2$, where E_g is the Γ energy gap and E_i is the band edge energy of the light or spin-orbit split-off hole bands measured from the heavy hole band edge.

5. Photoresponse

For the photoresponse measurement, mesa diodes of 200 μm in diameter were fabricated using the above structures. The spectral dependence of the photocurrent was measured using a Glowbar source and a grating monochromator with lock-in detection. Infrared light was illuminated on the mesa at either an angle or from the backside of the wafer depending on the transition process to be probed. The measured leakage current for the detectors with 15% and 60% Ge composition at 77 K as a function of bias across the device is shown in Fig. 7. It can be seen that there is a drastic reduction of the leakage current as the Ge composition is increased. This is mainly due to the increase of the barrier height at higher Ge compositions which suppresses the leakage current.

Figure 8 shows the photoresponse of the detector with 15% Ge composition at $0°$ and $90°$ polarizations [21]. The data was taken at 77 K using a 7 kΩ load resistor with a 2 V bias across the detector. In the $0°$ polarization case, a peak is found near 8.6 μm which is in agreement with the FTIR absorption spectra [21]. The full width at half maximum (FWHM) is about 80 meV while for the $90°$ polarization, the peak near 7.2 μm with a similar FWHM is observed. For the latter, the shift of the peak to a shorter wavelength may be due to the sharing of phonon energy with momentum conserving processes such as impurity, alloy and phonon scatterings. The photoresponsivity for both cases is about the same, 0.3 A/W. In the case of unpolarized light incident on the structure, a peak is found near 7.5 μm and the photoresponsivity is about 0.6 A/W, which is approximately the sum of the two polarization cases. The photoresponse is also measured by illuminating the light

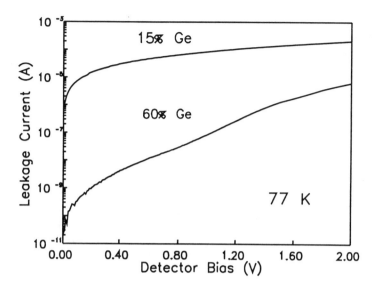

Figure 7: Current-voltage and differential resistance measured at 77 K for the 15% and 60% Ge composition devices. Positive bias means the top contact positive.

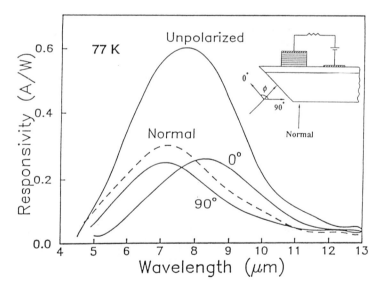

Figure 8: Photoresponse at 77 K for two polarizations angles with a 2 V bias applied across the detector. Infrared is illuminated on the facet at the normal such that the incident angle on the multiple quantum well structure is 45° as shown in the inset. Dashed curve shows the normal incidence photoresponse, when the light is illuminated normally on the backside of the detector.

normally on the backside of the device. The photoresponse with a similar peak responsivity and width to the 90° polarization case is observed, as shown by the dashed curve in Fig. 8.

It is clear that for the 0° polarization case, the photoresponse is due to intersubband transitions between two heavy hole subbands [3] with some contributions from the internal photoemission of holes which are excited via free carrier absorption. For the 90° polarization case, the intersubband transition is forbidden but the free carrier absorption is stronger than that of the 0° polarization. This is because the entire photon electric field lies in the xy-plane. The photoresponse in this case is due to the internal photoemission of free carriers, since the mixing of the hole bands due to the s-like conduction band or/and nonparabolicity is too small to have significant absorption. The following explains the normal incident photoconduction process. If the energy of the incoming photon is large enough to create holes having an energy larger than the barrier height, the photoemitted holes can travel above the barriers and be collected under an applied electric field (internal photoemission). This gives rise to the photocurrent. As the photon energy decreases (wavelength increases), the photocurrent increases due to the large free carrier absorption at longer wavelengths as shown in Fig. 4. When the photon energy is further decreased such that the energy for generating holes is lower than the barrier height, the photocurrent vanishes since the flow of holes is blocked by the potential barriers. This results in the cutoff of the photocurrent. This type of photoemission is similar to that observed by Lin et al. for a SiGe/Si heterojunction [22]. The difference is that for their case the free carrier absorption is three dimensional. In our case, all multiple layers contribute to the photoemission and thus a larger quantum efficiency is expected. The internal quantum efficiency (η) of the detector for either polarization is given by $\eta = (1 - e^{-2\alpha l})$ [23]. For the present detector, a quantum efficiency of $\eta \sim 14$ % is obtained.

The photoresponse of the structures which showed intervalence band transition have been measured using mesa diodes similar to above except that the infrared light is illuminated from the backside of the substrate as schematically shown in the inset of Fig. 9. Responsivity spectra as a function of wavelength at 77 K for the samples with 30 % and 60 % Ge compositions are shown in Fig. 9. It can be clearly seen from the measured result that as the Ge composition is increased the peak photoresponse moves towards a shorter wavelength due to the large splitting of the heavy hole and split-off hole bands. In comparison with the absorption data at room temperature, the photoresponse shows several peaks while the absorption spectrum has only one broad peak. The existence of several peaks is due to the transitions to both bound and continuum states which can be discriminated in the photocurrent measurement [16].

In order to identify the origin of these transitions, we have estimated the bound state energies for the three hole bands using a multiband self-consistent calculation including the splitting of the hole bands due to strain. Many-body effects due to hole-hole Coulomb interaction have also been taken into account in the calculation using a density functional approximation [15]. The potential well, calculated

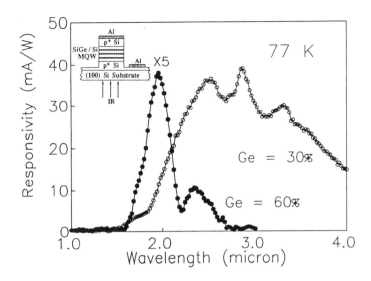

Figure 9: Measured responsivity at 77 K as a function of wavelength for 30 % and 60 % samples. Infrared is incident from the backside of the wafer (normal to the substrate plane) as shown in the inset.

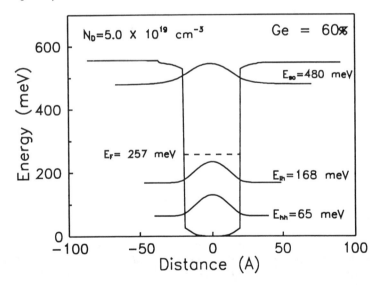

Figure 10: Calculated wave functions and energies for the three hole bands for 60% Ge sample. In this plot the hole energy is taken to be positive. For the doping used in this sample, only the ground states of the light and heavy holes are occupied as indicated by the Fermi energy. Many-body effects are also included in the calculation.

ground state energies, and wavefunctions of the three valence bands for 60 % Ge sample, are shown in Fig. 10. For the doping concentration used, only the ground states of the heavy and light holes are populated as indicated by the Fermi level of Fig. 10. This indicates that for normal incidence, light transitions can occur between the heavy hole ground state and the ground states of the light and split-off bands and the continuum states [4]. The excited bound states mainly participate in the intersubband transition which is forbidden for normal incident light [4]. The observed peak at 2.0 μm is most likely due to the transition from the heavy hole ground states to the continuum states (i.e., a mixture of the light, heavy and split-off subbands) while the peak at 2.4 μm may be attributed to the transition from the heavy hole ground state to the ground state of the split-off band. In the latter case, photoexcited carriers must tunnel through the barrier to reach the contact and should give a strong dependence on the external bias, which is in agreement with the measured bias dependent data. The calculated positions for these transitions are 438 meV and 523 meV, respectively. These energies have been corrected for the well known depolarization shift associated with subband transition [24]. From our data, the experimentally observed energies for the same transitions as above are 516 meV and 620 meV, respectively.

The detectivity of the detectors with different Ge compositions was estimated using the measured dark current and the quantum efficiency [25]. For the detector with 15 % Ge, the estimated detectivity at 77 K is about $D^*(9 \ \mu m) = 1 \times 10^9 \ cm\sqrt{Hz}/W$ while for the one with 60 % Ge is about $D^*(3 \ \mu m) = 4 \times 10^{10} \ cm\sqrt{Hz}/W$. These values are an order of magnitude lower than that of comparable GaAs/AlGaAs detectors. This is mainly due to the larger doping density required for attaining appreciable quantum efficiency.

6. Summary

In summary, infrared absorption between different subbands of $Si_{1-x}Ge_x/Si$ multiple quantum well structures are described. The origin of different transitions were identified using the polarization dependence measurements. Infrared detectors based-on intersubband and intervalence band transitions have been demonstrated. The photoresponse was shown to cover both the 3-5 μm and the 8-14 μm atmospheric windows. The continued progress in the study of SiGe intersubband transitions indicates that monolithic integration of SiGe/Si multiple quantum well infrared detectors with Si signal processing circuits should be realizable for focal plane applications.

Acknowledgements: The author would like to thank J. S. Park, K. L. Wang Y. J. Mii and S. K. Chun for collaboration on this research. This work was supported in part by Army Research Office (Dr John Zavada) and the Air Force Office of Scientific Research (Dr. Gerald Witt).

References

[1] H. Temkin, T. P. Pearsall, J. C. Bean, R. A. Logan and S. Luryi. *Appl. Phys. Lett.*, 48, 963, 1986.

[2] L. T. Canham. *Appl. Phys. Lett.*, 57(10), 1046, 1990.

[3] R. P. G. Karunasiri, J. S. Park, and K. L. Wang. *Appl. Phys. Lett.*, 59, 2588, 1991.

[4] J. S. Park, R. P. G. Karunasiri, and K. L. Wang. *Appl. Phys. Lett.*, 61, 681, 1992.

[5] E. O. Kane. *J. Phys. Chem. Solids.*, 1, 82, 1956.

[6] D. D. Coon, and R. P. G. Karunasiri. *Appl. Phys. Lett.*, 45, 649, 1984.

[7] H. Hertle, G. Schuberth, E. Gornik, G. Abstreiter, and F. Schaffler. *Appl. Phys. Lett.*, 59, 2977, 1991.

[8] Chanho Lee and K. L. Wang. *Appl. Phys. Lett.*, 60, 2264, 1992.

[9] R. People, and J. C. Bean. *Appl. Phys. Lett*, 39, 538, 1986.

[10] R. Hull, J. C. Bean, F. Cerdeira, A. T. Fiory, J. M. Gibson. *Appl. Phys. Lett.*, 48, 56, 1986.

[11] C. G. Van de Walle and R. Martin. *J. Vac Sci. Technol.*, B3, 1257, 1985.

[12] R. People. *Phys. Rev. B*, 32(2), 1405, 1985.

[13] S. K. Chun and K. L. Wang. *IEEE Trans. Electron Devices*, 39, 2153, 1992.

[14] A. G. Steele, H. C. Liu, M. Buchanan, and Z. R.Wasilewski. *Appl. Phys. Lett.*, 59, 3625, 1991.

[15] T. Ando. *Z. Phys. B*, 26, 263, 1977.

[16] R. P. G. Karunasiri, J. S. Park, and K. L. Wang. *Appl. Phys. Lett.*, 61, 2434, 1992.

[17] R. P. G. Karunasiri, J. S. Park, Y. J. Mii, and K. L. Wang. *Appl. Phys. Lett.*, 57, 2585, 1990.

[18] J. S. Park, R. P. G. Karunasiri, and K. L. Wang. *Appl. Phys. Lett.*, 58, 1083, 1991.

[19] K. L. Wang and R. P. G. Karunasiri. Infrared Detectors using SiGe/Si Quantum Well Structures, In M. O. Manasreh, editor, *Semiconductor Quantum Wells and Superlattices for Long-Wavelength Infrared Detectors*, page 139. Artech House, Norwood, 1993.

[20] P. Man and D. S. Pan. *To be published*, 1993.

[21] J. S. Park, R. P. G. Karunasiri, and K. L. Wang. *Appl. Phys. Lett.*, 60(1), 103, 1992.

[22] T. L. Lin and J. Maserjian. *Appl. Phys. Lett.*, 57, 1142, 1990.

[23] B. F. Levine, C. G. Bethea, G. Hasnain, V. O. Shen, E. Pelve and P. R. Abbott. *Appl. Phys. Lett.*, 56(9), 851, 1990.

[24] S. J. Allen, D. C. Tsui, and B. Vinter. *Solid St. Commun*, 20, 425, 1976.

[25] M. A. Kinch, and A. Yariv. *Appl. Phys. Lett.*, 55(20), 2093, 1989.

LARGE ENERGY INTERSUBBAND TRANSITIONS IN HIGH INDIUM CONTENT InGaAs / AlGaAs QUANTUM WELLS

H. C. Chui, E. L. Martinet, M. M. Fejer, and J. S. Harris, Jr.
Solid State Laboratory
Stanford University
Stanford, CA 94305-4055
U.S.A.

ABSTRACT. Large energy intersubband transitions are necessary for extending intersubband applications to the near-infrared ($\leq 2\mu m$ wavelength) where compact diode laser based sources are available. By growing high indium content InGaAs / AlGaAs quantum wells (QWs) on GaAs substrates with linearly graded InGaAs buffers, we have demonstrated peak intersubband absorption energies as high as 580meV ($2.1\mu m$ wavelength). We have also demonstrated intersubband absorption at 580meV in asymmetric coupled InGaAs / AlAs QWs. These are the largest bound-to-bound intersubband transition energies to date. Experimental studies of buffer and QW growth parameters for optimized intersubband absorption have been performed. The well width dependence of intersubband transition energies in both $In_{0.5}Ga_{0.5}As / Al_{0.45}Ga_{0.55}As$ QWs and $In_{0.5}Ga_{0.5}As / AlAs$ QWs have been measured and theoretically modelled.

1. Introduction

Intersubband transitions have been used for applications such as photodetectors, modulators, and nonlinear optics.[1,2,3] These intersubband applications, however, are restricted in wavelength of operation by the range of intersubband transition energies available. Intersubband transition energies in GaAs / AlGaAs QWs and InGaAs / InAlAs QWs lattice-matched to InP have been typically limited to less than 300meV ($\geq 4.1\mu m$ wavelength), and only recently have intersubband energies of greater than 400meV ($\leq 3.1\mu m$) been achieved.[4,5] By using InGaAs and AlGaAs for the well and barrier materials, respectively, QWs can have large conduction band offsets ΔE_c and thus, large intersubband transition energies. However, these InGaAs / AlGaAs QWs are highly strained and are difficult to grow. Lord, et. al.[6] demonstrated that by the initial growth of a linearly graded InGaAs buffer, high quality $In_{0.5}Ga_{0.5}As / AlGaAs$ QWs could be grown on a GaAs substrate by molecular beam epitaxy (MBE). This graded buffer acts as an effective substrate with a lattice constant corresponding to the InGaAs at the top of the buffer. By grading the buffer up to a lattice constant intermediate between the QW barrier and well materials, the wells and barriers can be strain balanced. Using this growth technique, we had previously demonstrated intersubband transitions in InGaAs / AlGaAs QWs with indium compositions of up to 50%.[7] We present here the following: (1) MBE growth studies performed on $In_{0.5}Ga_{0.5}As$ / AlGaAs QWs for optimization of intersubband absorption, (2) measurements of the well width dependence of intersubband absorption in both $In_{0.5}Ga_{0.5}As / Al_{0.45}Ga_{0.55}As$ QWs and $In_{0.5}Ga_{0.5}As / AlAs$ QWs, (3) a theoretical model for predicting the intersubband energies, and (4) results on large energy intersubband transitions in both square and asymmetric coupled QWs.

H. C. Liu et al. (eds.), Quantum Well Intersubband Transition Physics and Devices, 251–259.

2. Sample Description

The layer structure for the QW samples is shown in Figure 1. All of the InGaAs / AlGaAs QW samples were grown by molecular beam epitaxy (MBE) in a Varian Gen-II system on semi-insulating (SI) GaAs substrates. These QWs were grown atop linearly graded InGaAs buffers which were graded at a rate of 16% indium composition per μm from GaAs up to a final buffer indium composition, usually near the average indium concentration of the QWs. In addition, all of the QW samples were uniformly silicon doped n-type over the well regions only. The indium concentrations of the InGaAs wells and buffers were verified to be within 3% indium composition of the target values by high resolution x-ray diffraction (HRXRD) measurements on thick, relaxed InGaAs samples grown with the same conditions and in the same run as the QW samples. The intersubband absorption spectra from these samples were measured using a Fourier transform infrared spectrometer (FTIR) with the samples mounted at Brewster's angle to the linearly polarized light.

3. Growth Optimization

Growth studies of $In_{0.5}Ga_{0.5}As$ / $Al_{0.45}Ga_{0.55}As$ QWs with 40Å wells were performed to minimize the intersubband absorption linewidth and maximize the integrated absorption fraction (IAF).[8] The optimal substrate temperature during the growth of the QWs was found to be between 350°C and 400°C for both As_2 and As_4 species. At higher temperatures, the absorption linewidths were broader, indicative of poor interface quality,[9] while at lower temperatures the IAF and measured Hall sheet charge density dropped, indicative of an increased trap density and a reduced material quality. By adding 1 monolayer GaAs smoothing layers at the QW interfaces and ramping the substrate temperature to 600°C for growth of the AlGaAs barriers, we observed that the linewidths and IAFs were significantly improved. However, the added temperature ramps and growth interrupts greatly increased the growth time.

The final indium composition y_b of the $In_yGa_{1-y}As$ buffer was also determined to be a critical growth parameter; FTIR results from samples with varying y_b are shown in Figure 2. The linewidth monotonically decreased with increasing y_b up to $y_b = 0.3$. For samples grown with $y_b = 0.4$ and 0.5, the wafer had an extremely rough surface and was optically very lossy, and the intersubband absorption was weak, indicative of relaxed material. This optimal effective substrate lattice constant corresponded to that of an $In_{0.3}Ga_{0.7}As$ buffer; if a GaAs substrate or an InP substrate (corresponding to an $In_{0.53}Ga_{0.47}As$ buffer) was used without a graded InGaAs buffer, significantly poorer QW material quality would result. This optimal $y_b = 0.3$ is larger than the average indium concentration of the multi-quantum well, $y_{avg} = 0.17$, but lower than the indium composition of the InGaAs well, $y_{well} = 0.5$. That is, the optimum buffer composition appears to minimize the strain in the well without exceeding a critical thickness due to strain for an individual barrier layer or due to cumulative strain from the MQW structure. With different QW geometries, the optimal y_b is likely to be different; however, optimal material quality should have y_b higher than y_{avg} but lower than y_{well}. For instance, if a thinner barrier was used, a higher y_{avg} would result, and a $y_b > 0.3$ would likely be optimal.[6]

Using near optimized growth conditions (As_2, $y_b = 0.3$, substrate temperature = 400°C), we were able to obtain full-width at half-maximum (FWHM) linewidths of 34meV for 1 to 2 transition energies of 350meV. This linewidth corresponds to a relative linewidth (FWHM divided by the transition energy) of less than 10%; similar relative linewidths were observed in well doped GaAs / AlGaAs QWs. This FWHM is also comparable to that obtained in InGaAs / InAlAs QWs grown on InP with similar transition energies (29meV for unstrained and 60meV for strained).[10]

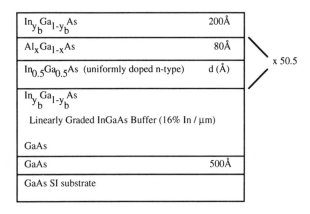

Figure 1. Layer diagram of quantum well structures.

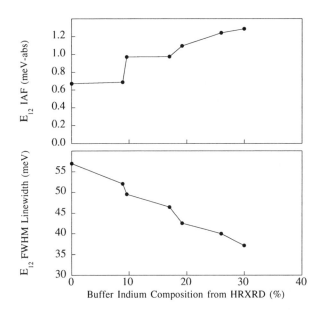

Figure 2. Dependence of intersubband absorption linewidth and IAF on final buffer indium composition y_b as determined by HRXRD. For samples with y_b = 0.4 or 0.5, the QWs were relaxed, and the intersubband absorption was extremely weak.

4. Well Width Dependence of Intersubband Absorption

The dependence of the intersubband absorption on well width for both $In_{0.5}Ga_{0.5}As$ / $Al_{0.45}Ga_{0.55}As$ and $In_{0.5}Ga_{0.5}As$ / AlAs QWs was also determined. These samples each had 50 QWs and were doped at a sheet charge density per QW of 2.8×10^{12} cm^{-2} and 4.0×10^{12} cm^{-2} for the $Al_{0.45}Ga_{0.55}As$ and AlAs barrier samples, respectively. The samples were grown at a substrate temperature of 400°C with As$_4$ for the $Al_{0.45}Ga_{0.55}As$ samples and with As$_2$ for the AlAs samples. The final buffer indium composition was targeted to be near the average indium concentration of the QWs for the AlGaAs samples and slightly higher than the average for the AlAs samples. As determined by our growth parameter studies, using a final buffer indium composition slightly greater than the average lattice constant of the quantum wells should help improve the QW material quality. The intersubband absorption peak energies for these QWs are plotted in Figure 3. The 1 to 2 transition energies for the narrow AlAs QWs are among the largest observed to date for intersubband transitions. For the $Al_{0.45}Ga_{0.55}As$ barrier samples, the 23Å well had a 1 to 2 transition energy of 384meV, lower than that for the 30Å well. This indicates that the second energy level was near the top of the well so that narrowing the well width from 30Å to 23Å caused the ground state subband energy to increase more rapidly than the second subband energy.

The FWHM linewidths and IAFs for the $In_{0.5}Ga_{0.5}As$ / $Al_{0.45}Ga_{0.55}As$ and $In_{0.5}Ga_{0.5}As$ / AlAs QWs samples are shown in Figure 4. The measured IAFs are approximately half of the theoretically calculated values (model is described in Section 5) for all of the samples except the narrowest well samples for both series for which the IAF dropped significantly below 50% of the theoretical value. The IAFs are calculated by using dipole moments obtained from our theoretical model and by using the targeted dopant concentration for the carrier concentration.[11] We have observed this 50% experimental to theoretical IAF ratio, even for GaAs / AlGaAs QW samples. The reduced IAFs in the narrowest wells may be due in part to the high dopant concentrations in these narrow wells required to achieve the same sheet charge density as the wider wells. The linewidths follow a generally increasing trend with narrower well width. Many linewidth broadening mechanisms contribute to this well width dependent linewidth, including increased impurity and interface scattering with decreased well widths and increased transition energies. The broad linewidth for the narrowest well may be due to Γ-X scattering in the AlAs barrier, since according to our model, the second energy level is above the AlAs X-valley.

5. Theoretical Modelling

The transition energies and dipole moments for these quantum wells were modelled with a single band effective mass model, and band nonparabolicity was taken into account by using an energy dependent effective mass m*(E) given by[12,13]

$$m^{*}(E) = \frac{m_0^{*}}{1 + \dfrac{\alpha}{E_g}\left(E - E_{c0}\right)} \qquad (1)$$

where E is the electron energy, E_{c0} is the conduction band edge energy, α is the nonparabolicity factor,[14] and m_0^{*} is the effective mass at the bottom of the conduction band.[15] For energies less than 400meV above the conduction band edge, the nonparabolic band parameterization given by Equation 1 is used. However, for energies of more than 400meV above the band edge, the fourth order approximation to the band dispersion (energy E versus wavenumber k) becomes inaccurate for $In_{0.5}Ga_{0.5}As$ and in fact, approaches a maximum in E vs. k so that the effective mass

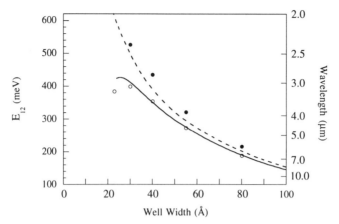

Figure 3. Well width dependence of 1 to 2 intersubband absorption peak energies for In$_{0.5}$Ga$_{0.5}$As / Al$_{0.45}$Ga$_{0.55}$As (hollow circles) and In$_{0.5}$Ga$_{0.5}$As / AlAs QWs (filled circles). Solid and dashed lines are calculated energies corresponding to the Al$_{0.45}$Ga$_{0.55}$As and AlAs QW samples, respectively.

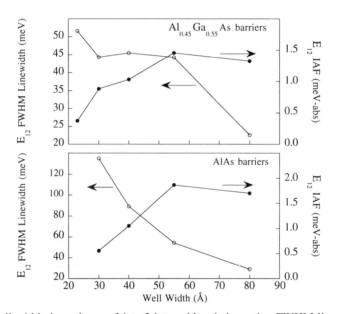

Figure. 4. Well width dependence of 1 to 2 intersubband absorption FWHM linewidth (hollow circles) and IAF (filled circles) for In$_{0.5}$Ga$_{0.5}$As / Al$_{0.45}$Ga$_{0.55}$As (top graph) and In$_{0.5}$Ga$_{0.5}$As / AlAs QWs (bottom graph).

becomes infinite. In order to resolve this problem, we noted that the E vs. k curves for InAs and GaAs, calculated using a pseudopotential method,[16] are nearly linear for energies high above the band edge around the Γ-point. Thus, we assumed a linearized E vs. k for energies of more than 400meV above the conduction band edge with the E vs. k curve and slope matched at 400meV. The calculated energies using this method for both the $In_{0.5}Ga_{0.5}As$ / $Al_{0.45}Ga_{0.55}As$ and $In_{0.5}Ga_{0.5}As$ / AlAs QWs are also plotted in Figure 3. Good agreement between the calculated and measured energies is observed even for energies of more than 750meV above the conduction band edge. This energy dependent effective mass formulation may be a good guide for intersubband and hot electron transistor modelling where electron energies are extremely high above the conduction band edge.

6. Large Energy Intersubband Absorption Results

To obtain even larger transition energies, we have investigated samples with higher indium concentrations. Three $In_{0.6}Ga_{0.4}As$ / $Al_xGa_{1-x}As$ QW samples with 50 QWs and aluminum compositions of x = 0.45, 0.67, and 1.0 were grown with a final buffer indium composition of 0.3 using As_2 at a substrate temperature of 375°C. The samples were uniformly Si doped in the well regions at 5.7×10^{12} cm^{-2}/QW. The intersubband absorption spectra for these three samples are shown in Figure 5. The AlAs barrier sample has a peak transition energy of 580meV (2.1μm wavelength). To our knowledge, this is the largest bound-to-bound intersubband transition energy reported to date. The relative linewidths for these samples are 14%, 17%, and 25% for the x = 0.45, 0.67, and 1.0 samples, respectively.

Asymmetric coupled double $In_{0.6}Ga_{0.4}As$ / $Al_xGa_{1-x}As$ QWs with large transition energies were also grown for future nonlinear optics experiments.[17] The double QW structure is shown in the insert of Figure 6. 200 double QWs were grown with Si doping in the well regions at a sheet charge density of 3.0×10^{12} cm^{-2} per double QW. The QWs were grown atop a graded InGaAs buffer with a final buffer indium composition of 0.3 using As_4 at a substrate temperature of 375°C. The intersubband absorption spectrum for the QWs is shown in Figure 6. Two absorption peaks are seen at 295meV (4.2μm wavelength) and 580meV (2.1μm) with relative linewidths of 24% and 18% corresponding to the E_{12} and E_{13} transitions, respectively. This E_{13} = 580meV is the largest 1 to 3 intersubband transition to date. This coupled QW sample should exhibit a strong second order nonlinear susceptibility for 4μm to 2μm wavelength second harmonic generation since it is doubly resonant.

7. Conclusions

We have demonstrated intersubband transitions in high indium content InGaAs / AlGaAs QWs grown on GaAs using a linearly graded InGaAs buffer. Growth studies were performed to optimize growth conditions for minimum intersubband absorption linewidth with maximum IAF. The well width dependence of intersubband absorption for both $In_{0.5}Ga_{0.5}As$ / $Al_{0.45}Ga_{0.55}As$ and $In_{0.5}Ga_{0.5}As$ / AlAs QWs was determined, and good agreement of the intersubband energies to a simple model was obtained. A peak 1 to 2 intersubband transition energy as high as 580meV (2.1μm wavelength) was observed in $In_{0.6}Ga_{0.4}As$ / AlAs QWs. A E_{13} also at 580meV (2.1μm) was observed in asymmetric coupled $In_{0.6}Ga_{0.4}As$ / AlAs QWs. These are the largest bound-to-bound intersubband transition energies to date. The large transition energies observed in these InGaAs / AlGaAs QWs should be useful for short wavelength intersubband applications in the near-infrared ($\leq 2\mu$m wavelength) where compact diode laser based sources are available.

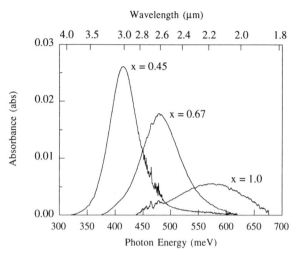

Figure. 5. Intersubband absorption spectra for $In_{0.6}Ga_{0.4}As$ / $Al_xGa_{1-x}As$ QWs with 30Å wells for x = 0.45, 0.67, and 1.0.

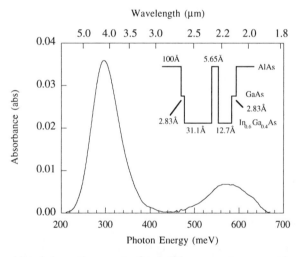

Figure. 6. Intersubband absorption spectra for doubly resonant asymmetric coupled $In_{0.6}Ga_{0.4}As$ / AlAs QWs with conduction band diagram (insert).

258

Acknowledgements

H. C. Chui acknowledges fellowship support from the Office of Naval Research (ONR), and E. L. Martinet, from ONR and Lockheed. This work was supported by ONR under contract numbers N00014-91-J-0170 and N00014-92-J-1903 and by ARPA under contract number N00014-90-J-4056. FTIR measurements were performed on a Bruker FTIR at the Stanford Free Electron Laser facility.

References

1. *Intersubband Transitions in Quantum Wells*, edited by E. Rosencher, B. Vinter, and B. Levine (Plenum, New York, 1992).
2. C. G. Bethea, B. F. Levine, V. O. Shen, R. R. Abbott, and S. J. Hseih, `10-μm GaAs/AlGaAs multiquantum well scanned array infrared imaging camera', *IEEE Electron. Devices* **38**, 1118 (1991).
3. M. M. Fejer, S. J. B. Yoo, R. L. Byer, A. Harwit, and J. S. Harris, Jr., `Observation of extremely large quadratic susceptibility at 9.6-10.8 μm in electric-field-biased AlGaAs quantum wells', *Phys. Rev. Lett.* **62**, 1041 (1989).
4. L. H. Peng, J. H. Smet, T. P. E. Broekaert, C. G. Fonstad, `Transverse electric and transverse magnetic polarization active intersubband transitions in narrow InGaAs quantum wells', *Appl. Phys. Lett.* **61**, 2078 (1992).
5. H. Asai and Y. Kawamura, `2.4μm intersubband absorption in $In_{1-x}Ga_xAs/AlAs_{1-y}Sb_y$ multiple quantum wells', *Fourth International Conference on InP and Related Compounds* (1992).
6. S. M. Lord, B. Pezeshki, and J. S. Harris, Jr., `Investigation of high In content InGaAs quantum wells grown on GaAs by molecular beam epitaxy', *Electron. Lett.* **28**, 1193 (1992).
7. H. C. Chui, S. M. Lord, E. Martinet, M. M. Fejer, and J. S. Harris, Jr., `Intersubband transitions in high indium content $In_{0.5}Ga_{0.5}As/AlGaAs$ quantum wells', *Appl. Phys. Lett.* **63**, 364 (1993).
8. More details of these growth studies will be presented at the 1993 North American Conference on MBE with proceedings submitted as H. C. Chui and J. S. Harris, Jr., `Growth studies of InGaAs/AlGaAs quantum wells grown on GaAs with a linearly graded InGaAs buffer', *J. Vac. Sci. Technol. B*.
9. M. J. Ekenstedt, S. M. Wang, and T. G. Andersson, `Temperature-dependent critical layer thickness for $In_{0.36}Ga_{0.64}As/GaAs$ single quantum wells', *Appl. Phys. Lett.* **58**, 854 (1991).
10. H. Asai and Y. Kawamura, `Intersubband absorption in highly strained InGaAs/InAlAs multiquantum wells', *Appl. Phys. Lett.* **56**, 1149 (1990).
11. L. C. West and S. J. Eglash, `First observation of an extremely large dipole infrared transition within the conduction band of a GaAs quantum well', *Appl. Phys. Lett.* **46**, 1156 (1985).
12. G. Bastard, *Wave Mechanics Applied to Semiconductor Heterostructures* (les éditions de physique, France, 1988), Chapt. 2.
13. S. J. B. Yoo, *Linear and Nonlinear Spectroscopy of Quantum Well Intersubband Transitions*, Stanford Univ. Thesis, 1991.
14. J. S. Blakemore, `Semiconducting and other major properties of gallium arsenide', *J. Appl. Phys.* **53**, R123 (1982).

15. The material parameters used for this calculation were a Γ-point band gap energy E_g of 0.755eV, 1.424eV, 1.985eV, and 3.018eV for $In_{0.5}Ga_{0.5}As$, GaAs, $Al_{0.45}Ga_{0.55}As$ and AlAs, a conduction band offset ratio $\Delta E_c/E_g$ of 0.5 for $In_{0.5}Ga_{0.5}As$ to GaAs and 0.6 for GaAs to $Al_xGa_{1-x}As$, a m_0^* of 0.034 m_0, 0.098 m_0, and 0.141 m_0 for $In_{0.5}Ga_{0.5}As$, $Al_{0.45}Ga_{0.55}As$ and AlAs, respectively, and an $\alpha = -0.86$ for $In_{0.5}Ga_{0.5}As$.

16. J. R. Chelikowsky and M. L. Cohen, `Nonlocal pseudopotential calculations for the electronic structure of eleven diamond and zinc-blende semiconductors', *Phys. Rev. B* **14**, 556 (1976).

17. Initial results are presented here at the 1993 *NATO Advanced Research Workshop on Intersubband Transition Physics and Devices* by E. L. Martinet, B. J. Vartanian, G. L. Woods, H. C. Chui, J. S. Harris, Jr., M. M. Fejer, B. I. Richman, and C. A. Rella , `Applications of high indium content InGaAs/AlGaAs quantum wells in the 2-7µm regime'.

APPLICATIONS OF HIGH INDIUM CONTENT InGaAs/AlGaAs QUANTUM WELLS IN THE 2–7 μm REGIME

E.L. Martinet, B.J. Vartanian, G.L. Woods, H.C. Chui, J.S. Harris, Jr., M.M. Fejer.
Center for Nonlinear Optical Materials, Stanford University, McCullough 226, Stanford CA 94305-4055

B.A. Richman, C.A. Rella.
W.W. Hansen Laboratory of Physics, Stanford University, Stanford CA 94305-4085

ABSTRACT: We report on some applications of high indium content quantum wells for mid–infrared (2–7μm) applications. Studies have been done to explore intra–valence band absorption in the mid–infrared between valence band hole states. Second harmonic generation spectroscopy is demonstrated in the conduction band near 5.5 μm.

1. Introduction

Until recently, intersubband phenomena in the mid–IR (2–7μm) regime have remained largely unexplored because of both the lack of semiconductor materials presenting sufficient bandgap engineering possibilities to create useful resonances and the lack of tunable high power sources in this wavelength region for characterization. The introduction of high indium content (HIC) InGaAs/AlGaAs quantum wells (QW)[1], together with the development of free-electron laser (FEL) facilities[2] uncover a new field of experiments in the mid–IR. We have recently demonstrated HIC–QWs with intersubband transition resonances out to 2.1μm, due to large band offsets in the conduction band (up to 1.4eV) and low effective mass.[3-5] HIC structures can be realized by reducing epilayer strain with a graded buffer[1]. This enables one to increase the range of bandgap engineering to highly strained systems in order to design a well geometry that optimizes the linear or nonlinear reponse of the system at a given transition energy. We demonstrate normal incidence absorption from heavy- to light- holes in *p*–doped HIC–QWs at 4.13μm. We also report the measurement of the second order nonlinear susceptibility ($\chi^{(2)}$) of asymmetric QWs as a function of detuning from the near double resonance (5.5μm to 3μm) using a FEL.

2. Intra–Valence Band Transitions

To date, most work on intersubband absorption in QWs has concentrated on the conduction band (CB) using *n*–doped QWs. However, due to the symmetries involved with electron states in the conduction band, a component of the polarization of the incident radiation normally must lie in the growth direction of the QWs in order to effect a transition. Most often, this is

H. C. Liu et al. (eds.), Quantum Well Intersubband Transition Physics and Devices, 261–273.

done by orienting the sample at Brewster's angle to maximize the z component of light in the QWs. Methods have been found to surmount this difficulty by, for example, grating coupling into a waveguide. Normal incidence absorption has been observed in n–type X or L valley QWs, but these transitions are limited by the shallow wells depths to the far-IR[30]. For holes, however, the symmetry selection rules are different. Due to the mixing of the heavy– and light– hole states away from the Brillouin zone center[6] interactions between mostly heavy– and mostly light– hole like mixed states (henceforth referred to as heavy– and light– hole states) are allowed for light propagating in the growth direction as previously shown by Levine[7]. This enables one to achieve normal incidence absorption without the use of gratings or other techniques[8-12]. Since HIC–QWs have large valence band offsets ΔE_V, transitions in the mid–IR are possible. With the use of p–doped HIC–QWs, we demonstrate normal incidence absorption in the 3–5µm atmospheric window for normal incidence photodetector and modulator applications.

Since these HIC–QWs are highly strained, there is some difficulty in accurately predicting the band edge profile and energy levels. The problem is especially acute in the valence band, where the strain results in heavy hole–light hole band edge splitting and electronic state mixing. As a result, prior to making devices, one must first do a fair amount of characterization to determine the energies of the levels, the dipole matrix elements, and the effect of the residual strain on the heavy hole–light hole band edge splitting. To this end, we have designed and grown several sets of QWs to accurately determine the effects of such variables as: barrier thickness, well thickness, doping concentration, and dopant location (well or barrier).

The structures we used were 50 HIC–QWs all grown on (100) semi-insulating GaAs substrates by molecular beam epitaxy. The InGaAs/AlGaAs HIC–QWs had a 50% indium concentration in the wells[13] and a 45% aluminum concentration in the barriers. The absorption measurements were made on a Bruker Fourier transform infrared spectrometer using three different geometries: Brewster's angle (73° for GaAs) with TM incident, Brewster's angle with TE incident, and normal incidence (NI). The TM geometry has a component of the incident electric field in the direction of growth and is commonly used for intra–conduction band absorption, whereas the TE and NI geometries confine the direction of the electric field to the plane of the QW.

For the TE and NI measurements, a reference sample with a structure identical to the sample but without carriers was necessary to compensate for the multiple reflection effects from the QW epilayers. The difference between using such a reference sample rather than a GaAs substrate is marked. Without the epilayers included in the reference sample, etalon fringes completely mask the intra–valence band absorption. For our reference sample, we used an undoped structure, grown immediately preceding the doped samples. Measurements were also made on a QW sample, where the reference sample was a proton bombarded portion of the doped QW sample. It has been shown previously[14] that proton bombardment creates deep levels which trap the carriers in n–doped QWs making them unavailable for intra–conduction band excitation. We compared a proton bombarded p–doped sample to an undoped reference and saw that the absorption peak also disappeared.

The QW samples were grown on a graded InGaAs buffer layer with the same type of optimized structures we found for HIC conduction band QW structures[15]. The well was Be doped to $p = 1 \times 10^{19}$ cm^{-3}, and the InGaAs buffer was graded to a maximum indium concentration of 17%. In figure 1, we show the absorption spectra from a 40Å In$_{0.5}$Ga$_{0.5}$As well with 80Å Al$_{0.45}$Ga$_{0.55}$As barriers for all three geometries, TM, TE and NI. As can be seen from the figure, there is a transition in all three geometries at very nearly the same energy. Because there is absorption at normal incidence, the transition is between heavy and light hole states. The magnitude of the transition is smaller than for similarly doped n–type QWs, but we can compensate for this by doping the valence band higher than is possible in the conduction band. This will be discussed in section 2.3.

2.1. BARRIER WIDTH VARIATION

We performed a study to determine the effect of barrier thickness on the energy and the strength of the transition. Assuming the QWs and barriers had some effective combined strain relative to the buffer layer, then changing the proportion of the barrier to well thickness should change the strain in the well. This would result in a different band profile due to the resultant strain splitting.

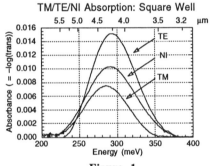

Figure 1

Absorption spectra of square well samples at Brewster's angle TM and TE polarizations and at normal incidence. Well: 40Å In$_{0.5}$Ga$_{0.5}$As; $p = 1 \times 10^{19}$ cm^{-3} throughout the well. Barrier: 80Å Al$_{0.45}$Ga$_{0.55}$As.

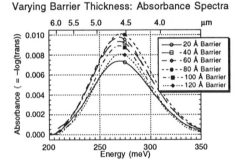

Figure 2

Absorption spectra of 40Å quantum wells doped $p = 1 \times 10^{19}$ cm^{-3} throughout the well. Buffer graded from In$_{0.015}$Ga$_{0.985}$As to In$_{0.3}$Ga$_{0.7}$As. Barrier thickness varies from 20 to 120 Å Al$_{0.45}$Ga$_{0.55}$As.

We grew a series of 40Å In$_{0.5}$Ga$_{0.5}$As QWs varying only the Al$_{0.45}$Ga$_{0.55}$As barrier thickness. In each sample, the QW was p–doped with Be to $p = 1 \times 10^{19}$ cm^{-3}. Also, the buffer was graded from 1.5 to 30% indium. The absorption spectra are shown in figure 2 and were obtained by making Brewster's angle measurements with TM radiation. A GaAs substrate was used as a reference. As can be seen from the plot, there is almost no shift in the position of the peak at 270 meV. However, the strength of the absorption is affected. The integrated absorption fraction (IAF) is a maximum with 80Å barriers: 5.82 Abs-cm^{-1}. This is 11% larger than the average IAF seen for these samples. These results indicate that if the strain were changing as the barrier thickness is varied, the change has little or no discernible effect on the position of the hole energy levels relative to each other, implying that the band edge profile also remains unchanged.

2.2.   WELL WIDTH DEPENDENCE

We examined the well width dependence of the intra–valence band absorption energy by growing several $In_{0.5}Ga_{0.5}As$ QW samples of varying well width. In each, the barrier was 50Å $Al_{0.45}Ga_{0.55}As$, and the buffer was graded from 1.5 to 30% indium composition. As before, the entire QW was doped with Be to $p = 1 \times 10^{19}$ cm^{-3}. However, unlike the varying barrier thickness study, in this case, the total sheet charge density changes since, with uniform doping, it is directly proportional to the well width.

We see in figure 3a that the transition energy reaches a peak in the 27Å well, and then begins to drop again as the wells get narrower. This indicates that the upper level rises more rapidly than the lower level for the wider wells, as is expected for a lighter mass hole. For even narrower wells, however, the upper level is no longer rising as rapidly and is effectively "pinned" at the top of the well. This is because the upper state is now more delocalized and is mostly in the barriers, making it virtually unaffected by well width variations. Figure 3b shows how the absorption rises with well width while the full width half maximum (FWHM), decreases. The increase in the FWHM with decreasing well width is probably due to interface scattering. It may also be due to the change in the nonparabolicity of the bands in k–space as the well width varies[16, 17]. In addition, the total absorption rises, in general, as the well width increases. This would be affected by two factors: the number of carriers available for absorption and the dipole matrix element between the two states involved since the magnitude of the absorption is directly related to both of these.

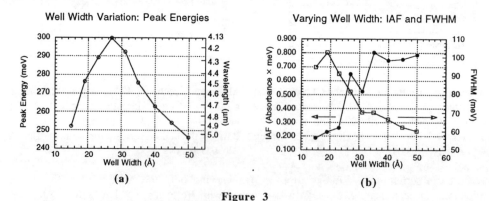

(a)   (b)

Figure 3
Well widths vary from 15 to 50Å. The QWs are doped $p = 1 \times 10^{19}$ cm^{-3} throughout the well. The barrier is 50Å $Al_{0.45}GaAs_{0.55}As$ and the buffer is graded from 1.5 to 30% indium. a) Peak energies for QWs as a function of the well width. b) IAF and FWHM vs. well width.

2.3. DOPING DEPENDENCE

Figures 4 (a) and (b) show the results of a study on the effects of doping on the peak position and IAF for the three geometries used. We use two different doping configurations: one in which the well is doped and the other with the barrier doped. The QWs consist of 45Å $In_{0.50}Ga_{0.50}As$ with 50Å $Al_{0.45}Ga_{0.55}As$ barriers. The maximum indium concentration in the graded buffer is 34%. For the well doped case, the entire well is doped, whereas for the barrier doped case, we have 10Å undoped, 10Å doped, 10Å undoped, 10Å doped, and 10Å undoped. The barriers were doped such that the sheet charge density would be the same as in the lowest three well doped cases. For the barrier doped samples, we see that with increased doping, band bending results in a decrease in the absorption energy.

It is interesting to note that the IAF, rises steadily with the doping for the lower doping concentrations. Also, all three geometries, TM, TE and NI, behave the same. However, for the higher doping concentrations, the TE and NI geometries, in which the electric field is contained entirely in the plane of the QWs, the IAF begins to decrease. On the other hand, in the TM geometry, where we also have a component of the electric field in the direction of growth, the IAF continues to rise. The difference in behavior between the TM and TE/NI cases may be due to a difference in the joint density of states. In all cases, we found the FWHM to be on the order of 60–85 meV, where, in general, it was narrower for the barrier doped samples.

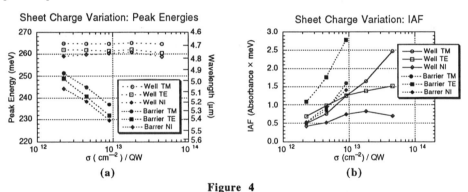

Figure 4

a) Peak energies and b) IAF for various doping concentrations (per QW) for TM, TE and NI orientations. The doping is either throughout the wells or in the barriers. The wells are 45Å $In_{0.50}Ga_{0.50}As$ with 50Å $Al_{0.45}Ga_{0.55}As$ barriers.

3. Second Harmonic Generation in Asymmetric Quantum Wells

QW intersubband nonlinear effects have been widely demonstrated in the 8–12μm regime[18] showing much stronger susceptibilities than their bulk counterparts. Resonant HIC–QW structures could develop into mid–infrared nonlinear devices if they show sufficient susceptibilities at the wavelength of an available source of radiation (solid state laser sources for example). However, as the QW system presents resonant absorption close to the nonlinear susceptibility maximum, both linear and nonlinear spectra have to be known accurately in order to calculate the figure of merit of any device[19]. Optical spectroscopy of optimized QW structures can provide a test for the simple models of the susceptibilities. We report second harmonic generation (SHG) of light at 2.5 μm from a nearly doubly resonant (5.5–3μm) $Al_{0.45}Ga_{0.55}As/In_{0.50}Ga_{0.50}As$ step multi–QW structure. The spectral behavior of the susceptibility $\chi^{(2)}$ of the QW epilayer is measured as a function of detuning, by an interference method, using the second harmonic electric field generated from the GaAs substrate [20, 21].

3.1. DOUBLY RESONANT QUANTUM WELL STRUCTURE

SHG in n–type QWs have been reported in the 10μm region[18, 19, 22], in which non inversion symmetry is achieved by applying an electrical bias on a square well[23] or growing compositionally asymmetric QWs (step wells[22, 24] or coupled QWs[25] for example). The SHG susceptibility can be expressed in that case as[26]:

$$\chi_{QW}^{(2)} \propto \sum_{m,n} \frac{z_{1n} \times z_{nm} \times z_{m1}}{\left(\hbar\omega - E_{1n} - i\,\Gamma_{1n}\right) \times \left(2\,\hbar\omega - E_{1m} - i\,\Gamma_{1m}\right)} \tag{1}$$

where z_{mn} are dipole matrix elements of the transition between the m and n QW envelope states, E_{mn} the intersubband energy and Γ_{mn} is the dephasing energy. Doubly resonant quantum wells were reported to present the highest SHG susceptibility, for which the dominant term of (1) corresponds to $n = 2$ and $m = 3$. The design of a quantum well showing maximum susceptibility consists of achieving both maximum dipole matrix element product $z_{12} \times z_{23} \times z_{31}$ by designing the well shape, together with a minimum in the denominator. The latter consideration consists of achieving a double resonance and minimum dephasing terms for a given targeted transition energy. This can be done by adjusting the growth parameters[15].

3.1.1 Sample optimization

Our sample was grown by molecular beam epitaxy on a (100) GaAs wafer, using an InGaAs graded buffer[1]. The asymmetric step quantum well (ASQW) structure consists of 50 periods of a 39Å $In_{0.54}Ga_{0.46}As$ and 31Å $In_{0.24}Ga_{0.76}As$ QW separated by 80Å $Al_{0.45}Ga_{0.55}As$ barriers, as shown on figure 5.

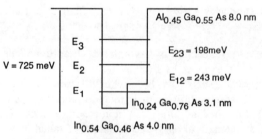

Figure 5.
Conduction band diagram of the asymmetric step quantum wells used in this study.

The quantum well is n–doped to 2.9×10^{12} cm^{-2} in the high indium section of the well. Maximum conduction band discontinuity was 725meV allowing in principle, resonances as high as 3μm. The well was designed to be doubly resonant around 6μm.

3.1.2 Linear spectrum and Calculation of $\chi^{(2)}$ of the QW structure

The absorption spectrum of the structure was measured with a FTIR spectrometer, the sample being at Brewster's incidence angle and the light TM polarized. We used a proton bombarded sample as a reference[14]. The normalized spectrum of absorption per quantum well is plotted on Figure 6a. A strong intersubband absorption is observed at 227meV (5.5μm) with a FWHM of 25meV corresponding to the 1–2 transition. The measured IAF is 27 mAbs-meV-per-QW. A smaller feature corresponding to the 1–3 transition is observed at 412meV (3.0μm) with a FWHM of 42meV and a IAF of 2.9 mAbs-meV / QW.

From the absorption spectrum we can extract the intersubband energies (E_{12} = 227meV, E_{13} = 412meV) and linewidths (Γ_{12} = 12.5meV, Γ_{13} = 21meV) for the transitions 1–2 and 1–3. One can note that the 1–2 transition is resonant at 5.5μm and the 1–3 at 3μm which means that the double resonance criterion is only approximate for this structure. We will assume an effective doping of 1.45×10^{12} cm^{-2} per QW in subsequent calculations of nonlinear susceptibilities, corresponding to the measured IAFs. We then assume the dipole matrix elements as calculated with a single band model[27], taking into account band non-parabolicity[4, 5]: z_{12}= 16.4 , z_{23} = 19.3, z_{13} = 2.52 Å. The $\chi^{(2)}$ deduced from (1) is shown on figure 6b, together with the two terms of the denominator of the dominant component, in absorption units.

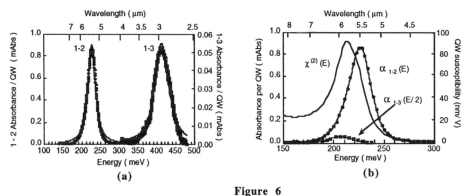

Figure 6

Absorption spectrum of the step quantum well sample (a). Simulated susceptibility and linear resonances leading to it in absorbance units (b)

The resonance of $\chi^{(2)}$ is close to 6μm with a value of 130nm / V as compared to the 5.5μm resonant absorption because the energy of the 1–3 transition is less than twice that of the 1–2 transition. In$_{0.60}$Ga$_{0.40}$As/AlAs asymmetric coupled QWs have also been optimized for maximum $z_{12}\times z_{23}\times z_{31}$ and minimum linewidth. The absorption spectrum demonstrates exact 4–2μm double resonance[5]. The simulated $\chi^{(2)}$ was about 150nm/V, about 100 times larger than the $\chi^{(2)}$ of bulk GaAs (180pm/V)[28], already a good nonlinear material.

3.2. $\chi^{(2)}$ MEASUREMENTS

3.2.1 Interference measurements

On the HIC–QW structure grown on GaAs, both the epilayer ($\chi^{(2)}{}_{QW}$) and the substrate ($\chi^{(2)}{}_{GaAs}$) are sources of SHG. The extraction of $\chi^{(2)}{}_{QW}$ makes use of the interference between the GaAs and multilayer generated second harmonic electric fields [20, 21]:

$$E(2\omega) = \left(\chi^{(2)}_{GaAs} + \chi^{(2)}_{QW}(\omega) \right) E(\omega) E(\omega) \tag{2}$$

Figure 7 shows the geometry of the experiment. The light enters the sample with an incidence angle, ϑ (internal angle r), and an angle φ between the (110) direction of the GaAs substrate and the plane determined by the direction (001) and the polarization of the incident electric field, as shown. A harmonic electric field builds up from the epilayer QW over a length L_{QW} (= 0.6μm). Then a harmonic electric field is generated from the GaAs substrate. The substrate of length L being many coherence length long (l_c is a few tens of microns in the mid IR). Thus energy flows back and forth between the fundamental and harmonic waves[26]. Our SHG signal measured after the sample is then a coherent superposition of those two effects.

For comparable values of the susceptibilities $\chi^{(2)}{}_{GaAs}$ and $\chi^{(2)}{}_{QW}$, the SHG signal from the bulk overwhelms that of the epilayer, by a factor given by the ratio of the lengths in which the radiation is generated. However, the symmetry and electronic structures of the bulk and epilayer region are different: $\chi^{(2)}{}_{GaAs}$ and $\chi^{(2)}{}_{QW}$, have different non-zero elements *(zzz)* and *(xyz)* respectively, so that polarization selection leads to measure either a coherently superimposed signal or only the GaAs generated signal. For the parallel polarization configuration, the SHG conversion can be expressed as[19]:

$$\frac{I_{2\omega}}{I_\omega^2} = T(\vartheta) \left| \chi^{(2)}_{GaAs} L G_{xyz}(r, \varphi) \cdot \left(1 - e^{-2i\zeta} \right) + \chi^{(2)}_{QW}(\omega) L_{QW} F_{zzz}(r) \right|^2 \tag{3}$$

where $I_{2\omega}$, and I_ω are the harmonic and fundamental power densities. $T(\vartheta)$ is the Fresnel transmission of the ω and 2ω electric fields at the sample interfaces and G_{xyz} and F_{zzz}, stand for the projection of electric fields on the principal axes of the crystal. The phase difference, ζ, between the ω and 2ω electric fields after the sample, can be expressed as:

$$\zeta = \frac{\pi}{2} \frac{L}{2 l_c \cos r} \tag{4}$$

The extraction of $\chi^{(2)}{}_{QW}$ consists of fitting ζ and the complex $\chi^{(2)}{}_{QW}$ from Eq. (3) for the SHG measured for different polarization configurations. This provides both amplitude and phase information if the phase difference ζ were known with sufficient accuracy. The value of the coherence length is calculated[26] from the measured value of the GaAs indices in the mid–infrared[29].

3.2.2 Experimental arrangement

The experiment was performed with the Free Electron Laser in the Stanford FEL Center[2], which emits light micropulses of about 2 picoseconds (energy 0.5µJ) at a repetition rate of 11.8MHz and at a wavelength continuously tunable between 3.5 and 6.5µm. The laser beam arriving in the experimental area, was split between a sample channel and reference channel (Figure 8) and focused to a beam diameter of 100µm with parabolic reflectors on both the QW sample and a AgGaSe₂ crystal. The SHG signal from the AgGaSe₂ crystal was used to monitor micropulse power fluctuations.

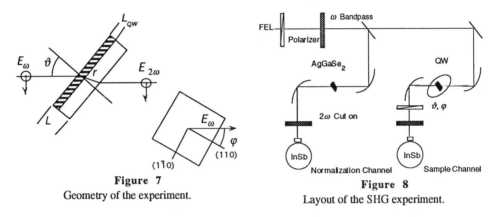

Figure 7
Geometry of the experiment.

Figure 8
Layout of the SHG experiment.

Linear polarization of the incident beam is selected, and the laser harmonics are removed by an interferometric filter. The sample consists of either the QW sample or a proton bombarded version of it, where no carriers are available for excitation, so that $\chi^{(2)}_{QW}$ is null. The latter sample provides a normalization of the susceptibilities to that of GaAs.

3.2.3 Results

Figure 9 shows the scans obtained when φ is varied for three different wavelengths and for a SHG polarization parallel to that of the fundamental light beam. The 6.3, 5.4 and 5.1µm wavelengths have been chosen so that the SHG signal generated from the GaAs is a maximum. Coherence lengths of 77, 60, and 49µm are obtained from the best Sellmeier fits of measured refractive indices known with 10^{-3} accuracy[29]. The substrate thickness is measured to be 514µm both directly and deduced from Fabry Perot oscillations measured with a FTIR spectrometer. This thickness is then 7, 9 or 11 times l_c at the respective wavelengths.

According to Eq. (3), as the angle φ is varied, the SHG from a proton bombarded sample yields 4 even peaks following a $\cos^2 2\varphi$ dependence. This arises from the pure contribution of GaAs, as seen in figure 9a, at 6.3µm. A best fit of the experimental data is given as well, in which the only fitting parameter is the ratio of measured SHG to susceptibility of GaAs, taken as 180pm/V[28].

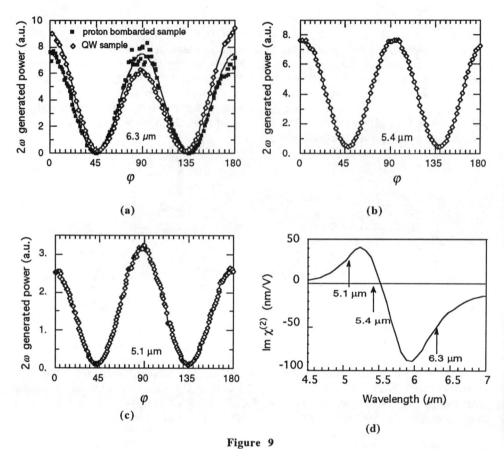

Figure 9

Variation of the SHG power as a function of φ angle for original and proton bombarded samples at 6.3μm.(a) Variation of the SHG power as a function of φ angle for original sample at 5.4 (b) and 5.1μm. (c). The spectrum of the imaginary part of $\chi^{(2)}{}_{QW}$ using a simple perturbative model (parameters from absorption spectrum) is presented (d).

Figure 9a also shows the variation of the SHG power with angle φ arising from the unprocessed sample, at an incidence angle of 65°. Two features are noted: (1) an unevenness between the 0°, 180° peaks and 90°, 270° peaks, and (2) a DC baseline of the SHG. According to Eq. (3), the unevenness of the peaks comes primarily from the imaginary part of $\chi^{(2)}{}_{QW}$ if we assume that the wafer thickness is close to an odd multiple of coherence lengths at this wavelength. With the same assumption, the DC offset arises primarily from the real part of the $\chi^{(2)}{}_{QW}$. A best fit of the φ scan is given assuming the value of 77μm for l_c at 6.3μm. The only two fit parameters are the imaginary and real part of $\chi^{(2)}{}_{QW}$. The accuracy of the result is strongly dependent on the assumed value of l_c. An accuracy of 10^{-3} for the indices does not allow accurate deduction of the amplitude of the real and imaginary parts of

$\chi^{(2)}_{QW}$. Direct measurements of the coherence length are also performed using a wedged[20] GaAs sample to increase the accuracy of l_c. In our present knowledge of the coherence length at that wavelength, the imaginary part of $\chi^{(2)}_{QW} = -30$nm/V ($\pm 50\%$) is deduced.

The same measurements were also performed at 5.4 and 5.1μm, as shown in figure 9b and 9c. The non uniformity of the peaks changes from higher 0° and 180° peaks, to even peaks, to higher 90° and 270° peaks for 6.3, 5.4, and 5.0μm respectively. Assuming a phase mismatch ζ of respectively $\pi/2$, $-\pi/2$ and $\pi/2$ as deduced from the values of the coherence length l_c, we can infer that the imaginary part of the $\chi^{(2)}_{QW}$ goes from negative to null and then to positive respectively. This feature in the variation of Im $\chi^{(2)}_{QW}$ is consistent with the spectrum calculated from the perturbation development (resonant and non resonant terms)[26] as shown in Figure 9d, in which the values of intersubband energies and dephasing terms were fit from the absorption spectrum of the sample.

We notice on figure 9b that the DC offset is maximum at 5.4μm where the imaginary part of $\chi^{(2)}_{QW}$ is null. We can deduce a magnitude of 70nm/V for the $\chi^{(2)}_{QW}$ at this wavelength. Better accuracy of l_c is needed in order to extract both the real part and the magnitude of the susceptibility. Incidence angle, ϑ, variations of the SHG are consistent with the values extracted from the φ scans. Spectrally resolved measurements of $\chi^{(2)}_{QW}$ are in progress on other optimized structures.

4. Conclusion:

We have shown normal incidence absorption from heavy hole to light hole states. Variation of barrier thickness was found not to effect the transition energy, but only the strength of the absorption. Width and doping variation studies were also conducted. The maximum intravalence band transition energy was found to be 300meV (= 4.13μm) in the 27Å well. Higher IAFs were found in barrier doped samples, though the transition energy decreased due to band bending. In the conduction band, SHG of 2.5μm light was demonstrated using a HIC asymmetric step QW with intersubband resonances corresponding to photons of wavelength 5.5 and 3.1μm. $\chi^{(2)}_{QW}$ measurements were performed at 6.3, 5.4 and 5.1μm. The largest susceptibility we measured, was 70nm/V at 5.4μm. A sign reversal of a $\chi^{(2)}$ is demonstrated in a QW system for the first time. By demonstrating normal incidence absorption at 4.1μm in the valence band and second harmonic generation of 2.5μm light in the conduction band, we have shown that the high Indium content quantum well (HIC-QW) system is promising for mid-IR applications.

Acknowledgements:
E. Martinet acknowledges fellowship support from the Office of Naval Research (ONR) and Lockheed, and H.C. Chui from an ONR Fellowship. This work was supported by ONR under contract N00014-91-J-0170 and N00014-92-J-1903 and by ARPA under contract N00014-90-J-4056. FTIR measurements were performed on a Bruker FTIR at the Stanford Free Electron Laser facility.

References:

1. Lord S M, Pezeshki B, Harris J S Jr. *Investigation of High In Content InGaAs Quantum Wells Grown on GaAs by Molecular Beam Epitaxy.* Electronics Letters 1992;**28**(13):1193-1195.
2. Smith T I, Schwettman H A, Berryman K W, Swent R L. *Facilities at the Stanford Picosecond FEL Center.* In: Schwettman HA, ed. *Free Electron Laser spectroscopy in Biology, Medicine, and Material Science.* Los Angeles, CA: , 1993: 23.
3. Chui H C, Lord S M, Martinet E, Fejer M M, Harris J S Jr. *Intersubband transitions in high indium content InGaAs/AlGaAs quantum wells.* Appl. Phys. Lett. 1993;**63**:364.
4. Chui H C, Martinet E, Fejer M M, Harris J S Jr. *2.1µm wavelength intersubband transitions in InGaAs / AlGaAs quantum wells.* Appl. Phys. Lett. 1993;.
5. Chui H C, Martinet E L, Fejer M M, Harris J S Jr. *Large energy intersubband transitions in high Indium content InGaAs / AlGaAs quantum wells,.* In this volume.
6. Kane E O. *Band structure of indium antimonide.* J. Phys. Chem. Solids 1956;**1**:249.
7. Levine B F, Gunapala S D, Kuo J M, Pei S S, Hui S. *Normal incidence hole intersubband absorption long wavelength $GaAs/Al_xGa_{1-x}As$ quantum well infrared photodetectors.* Appl. Phys. Lett 1991;**59**(15):1864.
8. Park J S, Karunasiri R P G, Wang K L. *Intervalence-subband transition in SiGe/Si multiple quantum wells—normal incident detection.* Appl. Phys. Lett. 1992;**61**(6):681-683.
9. Katz J, Zhang Y, Wang W I. *Normal incidence intervalence subband absorption in GaSb quantum well enhanced by coupling to InAs conduction band.* Appl. Phys. Lett. 1993;**62**(6):609-611.
10. Karunasiri R P G, Park J S, Wang K L. *Normal incidence infrared detector using intervalence-subband transitions in $Si_{1-x}Ge_x/Si$ quantum wells.* Appl. Phys. Lett. 1992;**61**(20):2434-2436.
11. Xie H, Katz J, Wang W I. *Infrared absorption enhancement in light- and heavy-hole inverted $Ga_{1-x}In_xAs/Al_{1-y}In_yAs$ quantum wells.* Appl. Phys. Lett. 1991;**59**(27):3601.
12. Gunapala S D. *InGaAs/InP hole intersubband normal incidence quantum well infrared photodetector.* J. Appl. Phys. 1992;**71**(5):2458.
13. Indium concentrations were determined by high resolution X-ray diffraction measurements on a separate sample grown at the time of the other growths.
14. Yoo S J B, Fejer M M, Byer R L, Harris J S Jr. *Second Order Susceptibility in Asymmetric Quantum Wells and its Control by Proton Bombardment.* Appl. Phys. Lett. 1991;**58**(16):1724-1726.
15. Chui H C, Harris J S Jr. *Growth Studies of $In_{0.5}Ga_{0.5}As/AlGaAs$ Quantum Wells Grown on GaAs with a Linearly Graded InGaAs Buffer.* submitted to: J. Vac. Sci. Tech. B .
16. Andreani L C, Pasquarello A, Bassani F. *Hole subbands in strained $GaAs-Ga_{1-x}Al_xAs$ quantum wells: Exact solution of the effective-mass equation.* Phys. Rev. B 1987;**36**(11):5887.
17. O'Reilly E P, Witchlow G P. *Theory of the hole subband dispersion in strained and unstrained quantum wells.* Phys. Rev. B 1986;**34**(8):6030.
18. Bibliography Rosencher E, Vinter B, Levine B, ed. *Intersubband Transitions in Quantum Wells.* London: Plenum, 1992:345. ; vol 288).
19. Yoo S B J. *Linear and Nonlinear Spectroscopy of Quantum Well Intersubband Transitions* [Stanford University Thesis]. Stanford, 1991.

20. Wynne J J, Bloembergen N. *Measurement of the lowest-order nonlinear susceptibility in III-V semiconductors by second-harmonic generation with CO2 laser.* Phys. Rev. 1969;**188**(3):1211.

21. Hollering R W J. *Bulk and surface second-harmonic generation in noncentrosymmetric semiconductors.* Optics Comm. 1992;**90**:147.

22. Rosencher E, Bois P, Nagle J, Delaître S. *Second harmonic generation by intersubband transitions in compositionally asymmetrical MQWs.* Electron. Lett. 1989;**25**:1063.

23. Fejer M M, Yoo S J B, Byer R L, Harwit A, Harris J S Jr. *Observation of extremely large quadratic susceptibility at 9.6 - 10.8 μm in electric-field-biased quantum wells.* Phys. Rev. Lett. 1989;**62**:1041.

24. Boucaud P, Julien F H, Yang D D, et al. *A detailed analysis of second harmonic generation near 10.6 μm in GaAs/AlGaAs asymmetric quantum wells.* App. Phys. Lett. 1990;:(in the press).

25. Capasso F, Sirtori C, Sivco D, Cho A Y. *Nonlinear Optics of Intersubband Transitions in AlInAs/GaInAs Coupled Quantum Wells: Second Harmonic Generation and Resonant Stark Tuning of $\chi_{2\omega}^{(2)}$.* In: Rosencher E, Vinter B, Levine B, ed. Intersubband Transitions in Quantum Wells. New York: Plenum, 1992: 141-9.

26. Shen Y R.*The Principles of Nonlinear Optics.*New York: Wiley, 1984

27. Bastard G, Delalande C, Ferreira R, Liu H W. *Assisted relaxation and vertical transport of electrons,holes and excitons in semiconductor heterostructures.* J. Lumin. 1989;**44**:247.

28. Levine B F, Bethea C G. *Non linear susceptibility of GaP: Relative Measurement and Use of Measured Values to Determine a Better Absolute Value.* Appl. Phys. Lett. 1972;**20**:272.

29. Pikhtin A N, Yas'kov A D. *Dispersion of the refractive index of semiconductors with diamond and zinc-blende structures.* Sov. Phys. Semicond. 1978;**12**(6):622.

30. Zhang Y, Baruch N, Wang W I. *Normal incidence infrared photodetectors using intersubband transitions in GaSb L-valley quantum wells.* Appl. Phys. Lett. 1993;**63**(8):1068

SPECTROSCOPY OF NARROW MINIBANDS IN THE CONTINUUM OF MULTI QUANTUM WELLS

D. GERSHONI, R. DUER, J. OIKNINE-SCHLESINGER,
E. EHRENFREUND and D. RITTER
Solid State Institute, Technion–Israel Institute of Technology, Haifa
32000, Israel

R.A. HAMM, J.M. VANDENBERG and S-N.G. CHU
AT&T Bell Laboratories, Murray Hill, NJ 07974, USA

ABSTRACT.   We report on the optical studies of intersubband transitions in In-GaAs/InP multi-quantum-well superlattices. Absorption, photoinduced absorption and time-resolved photoinduced absorption are used in a complementary way in our studies. By comparing with a multi-band $\vec{k} \cdot \vec{p}$ theoretical model we spectroscopically identify all the observed transitions, explain their relative strength, and spectral shape. We show that for energies below the InP barrier's conduction band, the spectrum is composed of inhomogeneously broadened transition between the first to the second confined electronic level. At low temperature the broadening is due to monolayer fluctuations in the well width. We measure characteristic time of a few nano-seconds for carriers to reach thermal equilibrium in the broadened first electronic level. Above the InP conduction band the spectrum is composed of well defined narrow minibands with minigaps between them. The optical transitions to these newly observed minibands, their polarization selection rules, temperature dependence, and characteristic lifetime are measured and discussed below.

1. Introduction

A discrete set of energy levels characterizes the energy spectrum of a quantum-well (QW) bellow its barrier's energy. The wavefunctions associated with these levels are confined to the well. Above the barrier, the energy spectrum evolves into a continuum of levels with extended wavefunctions which are no longer confined to the well. Right above the barriers, however, the continuum is not uniform and the variations in the density of states still resemble the discrete spectrum at lower energies. In a multi

275

H. C. Liu et al. (eds.), Quantum Well Intersubband Transition Physics and Devices, 275–289.
© 1994 *Kluwer Academic Publishers. Printed in the Netherlands.*

quantum-well (MQW) structure, at energies lower than the energy of the barriers, the spectrum is discrete and is similar to that of a single QW, since wavefunctions of neighboring wells do not interact. However, right above the barriers, minibands of finite width, separated by forbidden minigaps, where the density of states vanishes, are formed. The wavefunctions associated with these minibands, are not confined to an individual well, and they extend over the whole structure (similar to the case of a superlattice (SL)). At yet higher energies above the barrier, the forbidden gaps between the minibands diminish, and a continuum of states is restored. Theoretically, the number of minibands and minigaps below the onset of the continuum and their bandwidths depend solely on the details of the periodic potential structure. In reality, however, the miniband structure reflects also the phase coherence length of carriers in these states, since wavefunctions must maintain their coherence over more than one period in order for the MQW degeneracy to be removed.

In this work we study the optical intersubband (ISB) transitions between the first confined conduction subband of a MQW structure, to higher subband and sub-minibands above it. The transitions are studied using three different experimental techniques: absorption, and near steady state (NSS) and time resolved (TR) photoinduced absorption (PIA).

In the PIA experiments, one pumps electrons into the lowest n=1 sublevel, using interband excitation, and probes the changes in the intersubband absorption using tuned infrared (IR) radiation. In the NSSPIA method the pump beam is modulated at frequencies much lower than the inverse lifetime of the photoexcited carriers and steady state conditions are nearly achieved.[1] In the TRPIA technique, both beams are pulsed, and the infrared probing pulse can be delayed relative to the visible pumping pulse. The temporal resolution of such a technique is limited by the pulses widths only, and it is 3 psec in our case. This novel method is reported here for the first time.

Using these techniques, we have observed transitions between the n=1 and n=2 confined levels,[2] and between the n=1 confined level and the continuum minibands.

2. Experimental

The optical measurements reported here were performed on a sample of InGaAs/InP MQW structure lattice matched to InP. The sample was grown by metalorganic molecular beam epitaxy (MOMBE).[3] It contained 15 periods of a 65 Å

InGaAs well followed by a 367 Å InP barrier. The nominally undoped sample contained a background n-type doping concentration of $< 1 \times 10^{16} cm^{-3}$ both in the ternary QWs and in the binary barriers. The dimensions of the sample were determined by high resolution x-ray diffraction[4] and transmission electron microscopy.[3] The carrier concentration was estimated from Hall measurements performed on thick ($> 1\mu m$) ternary and binary layers grown under similar conditions.

All measurements were done in the waveguide configuration.[1] The absorption and NSSPIA methods are described in detail by Garini et al.[5]

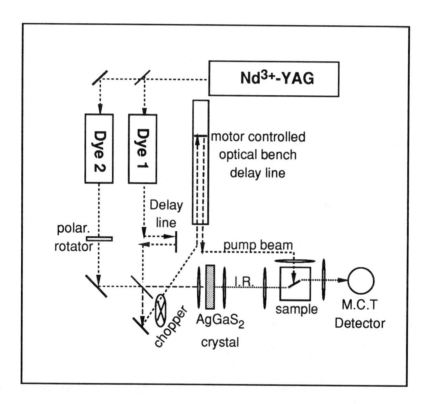

Fig. 1: (a) A schematic description of the experimental set up used for the time resolved photoinduced absorption (TRPIA) measurements.

The TRPIA setup consists of two synchronously pumped picosecond dye lasers simultaneously pumped by a frequency doubled CW mode-locked Nd:YAG laser. Half of the dye lasers visible light (670-730 nm) is used as the pump source. The second half is variably delayed and difference frequency mixed within a silver thiogallate to produce the tunable infrared probe pulse. The change in the IR transmission due to the visible excitation is detected by a liquid nitrogen cooled mercury cadmium telluride detector and a conventional lock-in technique. The experimental setup for the TRPIA is schematically described in Fig. 1.

3. Theoretical model

Our experimental results are analyzed by means of an eight band $\vec{k} \cdot \vec{p}$ theoretical model,[6] which is used to calculate the electronic band structure and wavefunctions of the MQWs. The conduction subband structure is calculated using the material parameters of our sample,[7] and is displayed in Fig. 2a as energy $vs.$ the in-plane wavevector $\vec{k}_\perp$. In addition to the two discrete confined levels e1 and e2, there are also very narrow minibands at energies above the barrier. The widths of these minibands result from the finite overlap between wavefunctions of neighboring periods. This effect is accommodated into the model scheme by performing the calculations at several points within the first SL Brillouin zone (BZ).[6] The overlap becomes more significant for higher subbands and at high enough energies a continuum of states is formed. We count up to 8 minibands of increasing width, well separated by forbidden minigaps, before the onset of the MQW continuum takes place at $\simeq 150$ meV above the bottom of the InP barrier conduction band energy. Thus, it is only the spectral region in the close vicinity to the barrier which distinguishes between the MQW and the single QW case.

The distribution of carriers within the one-electron band structure is calculated using the estimated carrier concentration, and assuming thermal equilibrium.[7] Fig. 2b shows the calculated population of electrons in thermal equilibrium. It is seen that only the first electronic level (e1) is occupied, and that as the temperature increases, a larger portion of the in-plane BZ is occupied by electrons.

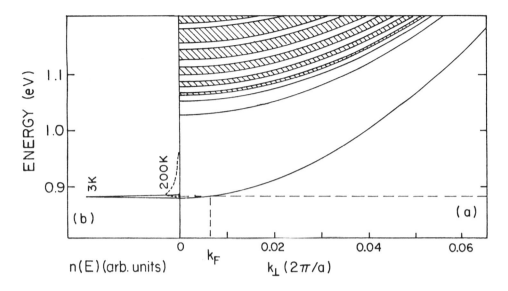

Fig. 2:(a) Calculated MQW conduction band dispersion relations plotted as energy vs. in-plane wavevector, $k_\perp$. The energy is measured from the top of the $InGaAs$ QW valence band. The InP barrier is at 1.047 eV which is 0.17 eV above the e1 level (at $k=0$). The electronic Fermi energy is marked as a horizontal dashed line. (b) The calculated electron occupation density $n(E)$, in thermal equilibrium, for T=3 and 200 K.

The matrix elements for optical transitions between the various states are calculated by the dipole approximation using the calculated wavefunctions.[7] It is important to note that this calculation takes into account contributions to the matrix element from both the dipole moment between the envelope wavefunctions and the dipole moment between the Bloch wavefunctions. Counter intuitively and unlike what is stated in most of the literature, following the pioneering work of West and Eglash,[8] the Bloch wavefunction contribution is the most important one. Indeed, as first noted by Zawadzki,[9] the envelope wavefunction contribution to the matrix elements can be completely neglected, since it is smaller by roughly a factor of the electron effective mass squared than the contribution of the Bloch wavefunctions.

280

Thus, it is the a priori introduction of the electron effective mass into the radiation-matter interaction Hamiltonian, which makes the common model[10] useful. It fails, however, for the inter-valence subband calculations,[11] and likewise, it does not yield reliable results for optical transitions to expanding or continuum conduction mini-bands, which is the subject of the present work.

The absorption or the PIA is finally calculated by summing over all the allowed (crystal-momentum conserving) optical transitions from occupied to unoccupied states.

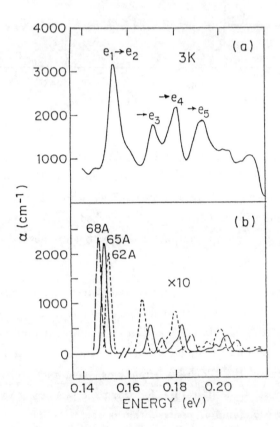

Fig. 3: Measured (a) and calculated (b) intersubband absorption. The dashed line in (b) show the calculated spectra for the nominal MQW dimensions and the solid line is an average over ± one monolayer fluctuations, keeping the same periodicity.

4. Absorption: experimental results and analysis

The absorption measurements are displayed in Fig. 3. Fig. 3a describes the measured absorption spectrum for unpolarized infrared radiation. In Fig. 3b we display the calculated intersubband absorption spectrum of this sample. In the calculations the solid line represents calculations with the nominal dimensions of the structure. The broken lines represent calculations for structures with the same period length and ± 1 monolayer QW thickness difference. Uniform electron density of $1.0 \times 10^{16} cm^{-3}$ and an ambient temperature of 3 K were used for the calculations. We marked the peaks in the figures by the electronic transition between initial and final subbands, which contributes most to their oscillator strength.

Comparing the measured (Fig. 3a) and calculated (Fig. 3b) spectra we see that our model describes well the energy position, and the spectral shape of the observed transitions. The absolute value of the e1-e2 calculated transition agrees within less than a factor of two with the measured one. This is well within the error associated with the estimated density of background dopants. The calculated intensity of the optical transitions to the continuum minibands are almost an order of magnitude smaller than the measured values. This discrepancy will be further discussed below.

We note that, unlike the symmetric single QW situation where only the e1–en (n=2,4,6...) transitions are allowed, here the transitions e1–e3 and e1–e5 are also observed. This is the result of the finite overlap of wavefunctions of electrons of adjacent wells, when these electrons have energy above the barriers.

In the calculated spectrum shown in Fig. 3b (solid line) we have approximated the integration over $k_{||}$ (along the MQW direction), by an average over the contributions of three equally spaced points within the first MQW BZ. Using five or more $k_{||}$ points yielded identical spectra. In order to learn on the relevance of the number of points used for the calculation of the absorption spectrum and the interaction between neighboring wells, we have calculated the absorption due only to the central QW in a non-periodic structure containing only 3 QWs. The spectrum obtained is strikingly similar to the one shown in Fig. 3b (dashed line). This result leads us to believe that the number of points used for the integration over the SL axis indicates the length over which the states maintain their coherence. From agreement between the calculated and the measured spectra, we conclude that electron states in the continuum minibands of this sample must maintain their coherence over distance comparable to at least three MQW periods. This amounts to a few tenths of a micron parallel to the MQW axis. This is possible, since in our InGaAs/InP sample, this direction is composed mainly of binary barrier material of good crystallographic quality. Thus, scattering by potential fluctuations is limited in this direction.

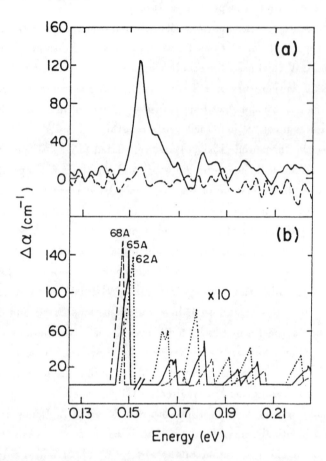

Fig. 4: (a) The photoinduced absorption, $\Delta\alpha$, per well for IR light polarized along the growth direction (solid line) and perpendicular to it (dashed line). (b) The calculated $\Delta\alpha$, for light polarized along the growth direction. Solid line – nominal well width thickness of 22 monolayers. Dashed (doted) line – well width of 23 (21) monolayers.

5. Near steady state photoinduced absorption (NSSPIA)

The NSSPIA results are presented as $\Delta\alpha$, the average change in the absorption coefficient per period. $\Delta\alpha$ is obtained from the measured relative differential transmission, $-\Delta T/T$, by taking into account the number of passes the IR beam makes inside the waveguide, the number of wells, the direction of the IR beam and the period of the MQW. The solid (dashed) line in Fig. 4a is the photoinduced absorption spectrum, $\Delta\alpha$, for infrared light polarized parallel (perpendicular) to the growth direction. In general the NSSPIA spectrum resembles the absorption spectrum of Fig. 3a. The lowest energy optical band centered at 0.155 eV, is the e1–e2 transition. It is polarized along the growth direction as it is in the absorption spectrum. Its shape is distinguishably asymmetric, and somewhat broader than the one observed in absorption. It can be described as composed of two components: a relatively narrow line of 7 meV full width at half maximum, characteristic of intersubband absorption between two confined states, and a broad shoulder in the high energy side. The optical transitions to the continuum minibands are also observed here, centered around energies of 0.18, 0.20 and 0.22 eV, respectively. By comparison with our model calculations as displayed in Fig. 4b, we assigned these bands to the optical transitions from the first electronic level e1 to the fourth (e4), fifth (e5) and sixth (e6) continuum minibands respectively. The first continuum miniband (e3) forms the broad shoulder in the high energy side of the e1-e2 transition.

We have calculated the changes in the absorption due to the laser beam, $\Delta\alpha$, assuming a photoexcited density of $\simeq 6 \times 10^{8} cm^{-2}$ per period. The result is displayed in Fig. 4b. The solid line is the calculated changes in the absorption for the nominal width of the well, 65 Å. The PIA spectra for identical samples having 62 and 68 Å well width and the same periodicity are represented by the broken lines in Fig. 4b. The typical asymmetric shape of the measured PIA lines can be understood by monolayer fluctuations in the QW widths. The sharp rise at low energies and the slower fall at the high energy side, can be approximately reproduced by averaging over the three spectra shown in Fig. 4b.

The spectral positions, the relative intensity and the polarization selection rules of the transitions in the PIA measurements are in better agreement with the model predictions than the absorption measurements. We will return to this point below.

284

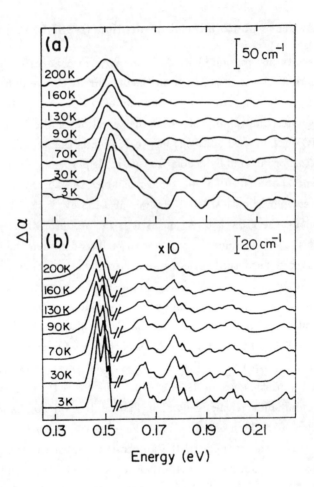

Fig. 5: Measured (a) and calculated (b) $\Delta\alpha$ for various temperatures. The spectra are vertically displaced for clarity. The calculated spectra are averaged over ± 1 monolayer fluctuations.

In Fig. 5a we show the measured PIA spectrum at various temperatures in the range 3–200 K. Two features are apparent in the evolution of the PIA spectra as the temperature increases. First, there is an additional absorption in the low energy side of the e1–e2 transition, which makes this line more symmetric in shape. Second, the "continuum" minibands become weaker relative to the e1–e2 band and are barely observable above 130 K. In Fig. 5b the PIA absorption spectra are calculated for various sample temperatures. In the calculations we average over the monolayer fluctuations as explained above. In agreement with the observations, our calculations explain the evolution of the low energy tail of the e1-e2 transition as the temperature increases. This evolution is the result of the change in the electrons distribution as displayed in Fig. 2b. This, in turn, gives rise to lower energy optical e1–e2 transitions due to the difference in curvature (effective mass) between the two subbands.

The relative decrease of the observed e1 to minibands absorption as a function of temperature, can only be partially accounted for by our model. In the temperature range 3–90 K, the reduction in the relative intensity of the measured minibands absorption is somewhat larger than what we calculate. We believe that this discrepancy is due to important extrinsic contributions to the PIA signal. These contributions explain also the saturation of the PIA at high laser intensities.[12] Since our model does not take any extrinsic effects into account, it cannot explain all the experimental details.

6. Time resolved photoinduced absorption (TRPIA)

In Fig. 6 the dots represent the relative decrease in the transmission of the IR probe beam through the sample as a function of the delay time between this beam and the visible pump beam. We display in Fig. 6 five typical decay curves taken at various energies along the intersubband absorption spectrum of the sample. In these measurements the average power of the pump beam was 150 mW at repetition rate of 76 MHz. The beam was focused on the sample in normal incidence to a spot size of 50 μm, so that maximum modulation of the IR probe transmission was obtained. The beam energy was varied between 1.68 to 1.85 eV and it was composed of the two colors, which were necessary to obtain the required probe energy. We estimate the density of carriers in each quantum well under these conditions to be of the order of or smaller than $1 \times 10^{12} cm^{-2}$. From this estimate and the measured absorption at the end of the pulse (t=0), one obtains oscillator strength of order one. Thus, the oscillator strength for intersubband absorption of photo-electrons is the same as that of the extrinsic donor related electrons.

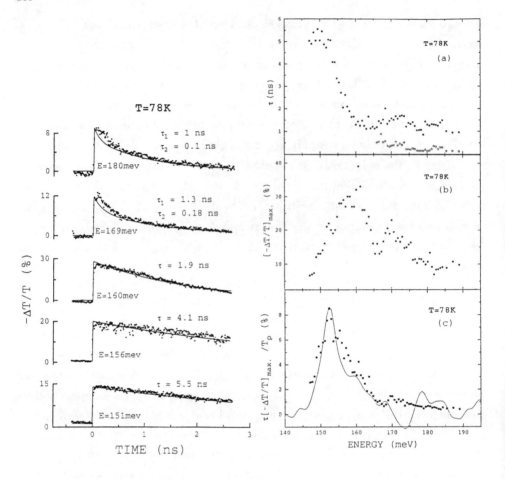

Fig. 6: Decay curves of the photoinduced absorption for various spectral position. The solid lines represent best fitted exponential decay model to the data.

Fig. 7: Fitted decay times (a), maximum absorption (b) and time averaged absorption of the photoinduced absorption as a function of the IR probe energy. Circles in (a) represent second exponential decay lifetime. The solid line in (c) is the NSSPIA measurement.

The solid lines in Fig. 6 represent best fit to an exponential decay of the PIA. Since the measured characteristic decay times of the photogenerated excess carriers are not much shorter than the time, T_p between two successive pulses (about 13 ns) our fitting procedure takes into account non zero background as a result of a pulse to pulse carrier accumulation. As can be seen in Fig. 6 (lowest three curves), the data below $\simeq 165$ meV could be nicely fitted by a single exponential decay. Above this energy, where a single exponential decay could not fit successfully the measured data, a good fit to a sum of two exponentials was obtained (upper two curves). Processes such as carrier cooling and diffusion into the wells, which take place during the first 100 psec after the excitation pulse ends, are evident in the decay curves of Fig. 6. In the following, however, we will discuss only the evolution of the PIA at longer times.

Fig. 7 summarizes the TRPIA measurements at 78K. We display the fitted PIA decay times (τ, Fig. 7a), the maximum PIA, $[-\Delta T/T]_{max}$ (at t=0, Fig. 7b) and the time averaged PIA (defined as $[-\Delta T/T]_{max}\tau/T_p$, Fig. 7c) as a function of the IR probe energy. For comparison, the measured NSSPIA is given by the continues line in Fig. 7c.

Clear and very pronounced evolution in the decay times of the PIA is seen in Fig. 7a. The characteristic lifetime of the induced absorption at the lowest part of the e1-e2 intersubband transition is longer than 5 ns. This is a factor of 5 longer than the life time of the induced absorption at energy of 160 meV which corresponds to the high energy side of this optical band. At higher energies, the PIA decay has two components. A fast one, of the order of a few hundred picoseconds, which becomes even faster with increasing energy above the e1-e2 transition, and a slow one, of order of one nanosecond, which remains essentially unchanged up to 190 meV, well into the region of transitions to the continuum minibands.

More information on the dynamics of the photoexcited electrons can be obtained from Fig. 7b, which describes the photoinduced absorption almost immediately after the pump pulse ends. It is clear that the e1-e2 transition, as seen in the absorption spectrum of Fig. 3a and in the NSSPIA spectrum (solid line Fig. 7c below), is actually composed of at least four narrower optical lines at energies of 152, 154, 158 and 163 meV, respectively. At yet higher energies two additional peaks centered at energies of 172 and 180 meV, respectively, are observed. The PIA spectrum at t=0 is very different from the time averaged one as given in Fig. 7c. The latter is similar to the NSSPIA spectrum which is also displayed in the figure (continues line.)

We interpret our time resolved measurements in the following way. Electrons and holes, which are photogenerated by our above bandgap energy pump pulse, quickly thermalize to separate electrons and holes and excitons at the lowest electronic level at their vicinity. The measured PIA signal is due to excitation of an electron from such a level to a higher electronic subband or miniband by absorbing IR photon from the probe pulse. Thus the TRPIA spectroscopy is a direct technique of measuring the electronic or excitonic population lifetime at the lowest level. As a result of well width fluctuations, the first electronic level (e1) is inhomogeneously broadened, such that regions of wider well width have lower energies. In turn, the optical band e1-e2 is broadened, and its lowest energy side is due to transitions from the lowest energy tail of the e1 level (see Fig. 3b). These tail states have the longest lifetime since they cannot transfer their energies to lower energy regions. Since their density is relatively small, their contribution to the TRPIA signal at short time is small. However, since their lifetime is long, their contribution to the time averaged PIA and to the NSSPIA is dominant.

The well width fluctuations can account also for the longer component of τ above 165 meV (Fig. 7a) and the miniband structure at 170 meV (Fig. 7b). As can be seen in Fig. 3b, for narrower wells the transitions to the minibands (confined to extended) are relatively stronger than in wider wells. Thus, narrow wells contribute more to the transitions to the minibands, resulting in decay times of order of 1 nsec (similar to that of the high energy side of the e1–e2 transition).

The second faster component of the decay which is observed in the PIA at energies of 165 meV and above (see Fig. 7a), clearly indicates that there is an additional process which contributes to the absorption in this energy range. We suggest that this contribution comes from electrons in short lived deep nonradiative centers below the e1 subband. Electrons in these centers have finite probability to be excited within their lifetime to the e2 states by absorbing an IR photon.

Acknowledgments   The work was carried out in the Center for Advanced Optoelectronics, and was supported by the Fund for the Promotion of Research at the Technion and by the Basic Research Foundation Administered by the Israel Academy of Sciences and Humanities, Jerusalem, Israel.

References

1. M. Olszakier, E. Ehrenfreund, E. Cohen, J. Bajaj and G.J. Sullivan, Phys. Rev. Lett. **62**, 2997 (1989).

2. J. Oiknine–Schlesinger, E. Ehrenfreund, D. Gershoni, D. Ritter, M.B. Panish and R.A. Hamm, Appl. Phys. Lett. **59**, 970 (1991).

3. D. Ritter, R.A. Hamm, M.B. Panish, J.M. Vandenberg, D. Gershoni, S.D. Gunapala and B.F. Levine, Appl. Phys. Lett. **59**, 552 (1991).

4. J.M. Vandenberg, D. Gershoni, R.A. Hamm, M.B. Panish and H. Temkin, J. Appl. Phys **66**, 3635 (1989).

5. Y. Garini, E. Cohen, E. Ehrenfreund, D. Gershoni, Arza Ron, E. Linder, L.N. Pfeiffer, K-K. Law, J.L. Merz and A.C. Gossard, (this proceedings).

6. G.A. Baraff and D. Gershoni, Phys. Rev. **B 43**, 4011 (1991).

7. D. Gershoni, C.H. Henry and G.A. Baraff, IEEE Journal of Quantum Electronics (in press, 1993).

8. L.C. West and S.J. Eglash, Appl. Phys. Lett. **46**, 1156 (1985).

9. W. Zawadzki, J. Phys. C: Solid State Phys. **16**, 229 (1983).

10. G. Bastard, "Wave mechanics applied to semiconductor heterostructures", Les editions de physique, Les Ulis, France (1988).

11. Y-C. Chang and R.B. James, Phys. Rev. **B39**, 12672 (1989).

12. E. Ehrenfreund, J. Oiknine–Schlesinger, D. Gershoni, D. Ritter, M.B. Panish and R.A. Hamm, Surface Science **267**, 461 (1992).

INTERSUBBAND ABSORPTION IN STRONGLY COUPLED SUPERLATTICES: MINIBAND DISPERSION, CRITICAL POINTS, AND OSCILLATOR STRENGTHS

M. HELM, W. HILBER, and T. FROMHERZ
Institut für Halbleiterphysik, Universität Linz, A-4040 Linz, Austria

F. M. PEETERS
Physics Department, University of Antwerp (UIA),
B-2610 Antwerpen, Belgium

K. ALAVI and R. N. PATHAK
Center for Advanced Electron Devices and Systems, Department of Electrical
Engineering, The University of Texas at Arlington, Arlington, TX 76019, USA

ABSTRACT. The intersubband absorption is studied in strongly coupled GaAs/AlGaAs superlattices, which exhibit a significant dispersion along the growth direction. The miniband dispersion is directly revealed through two absorption maxima, related to the singularities of the joint density of states at the center and the edge of the superlattice Brillouin zone. The total line shape is strongly asymmetric and can be quantitatively explained, if the k_z-dependent transition matrix elements are taken into account. It is argued that the asymmetry is a fundamental consequence of the oscillator sum rule. When the doping is low, so that the Fermi energy lies in the first miniband, the $k_z=\pi/d$ peak can be suppressed at low temperature due to thermal depopulation, which reflects the curvature of the lowest miniband. At low temperature another absorption line is observed, identified as $1s$-$2p_z$ impurity transition. The existence of this transition implies that the impurity band has not merged with the conduction band despite the metallic behavior.

1. Introduction

Optical absorption between subbands in confined semiconductor structures belongs to the most important techniques for the characterization of such systems. Already in 1974 intersubband absorption was observed in Si accumulation layers [1]. However, it was not until the observation of mid-infrared intersubband absorption in GaAs quantum wells in 1985 [2], that this subject caught wide interest. The reason for this has mainly been its great potential for infrared detector technology. Indeed, rapid progress in research has led us to a point, where GaAs or even SiGe quantum well detectors could substitute HgCdTe interband detectors. Typically, state-of-the-art structures exhibit rather thick barriers to suppress the dark current, and their excited state is located slightly above the top of the barriers in order to facilitate vertical transport of the photexcited carriers [3]. Reviews on these topics can be found elsewhere in this volume.

Apart from the importance for infrared technology, intersubband spectroscopy remains a valuable tool for the investigation of the bandstructure of semiconductor structures. In the

291

H. C. Liu et al. (eds.), Quantum Well Intersubband Transition Physics and Devices, 291–300.
© 1994 *Kluwer Academic Publishers. Printed in the Netherlands.*

present paper we will discuss the intersubband absorption in strongly coupled superlattices, where, due to the wavefunction overlap in neighboring quantum wells, there is a significant energy dispersion along the growth axis. Although these are not the structures of choice for optimized detector applications due to their large dark current, some work on superlattice detectors has been reported [4,5]. More important, however, is the insight into the physics of superlattices which can be gained through intersubband spectroscopy. Whereas it is already more than twenty years that Esaki and Tsu proposed the semiconductor superlattice as a model system for electrons in a periodic potential [6], only recently came long-sought breakthroughs, like the observation of the Wannier-Stark ladder [7], of negative conductivity [8-10] and finally of Bloch oscillations [11]. All these observations rely fundamentally on a well-developed miniband curvature along the growth axis. Here we will demonstrate that intersubband absorption can reveal the dispersion of the minibands directly through the maxima in the joint density of states at the edges of the mini-Brillouin zone. Since all minibands within the conduction band have the same in-plane curvature (apart from nonparabolic effects) the joint density of states has one-dimensional character and shows $1/\sqrt{x}$ -like singularities [12, 13]. This is in contrast to interband transitions, where the joint density of states contains electron and hole states and is much less divergent due to their different curvatures. Interband experiments on superlattices mainly showed exciton features resulting from the zone-edge critical points ("saddle-point excitons")[14].

The main results of the present investigation [15] are the following. (1) The two singularities in the joint density of states at the center and the edge of the mini-Brillouin are clearly observed, thus directly reflecting the miniband dispersion. (2) The line shape is strongly asymmetric, with the low-frequency peak ($k_z=\pi/d$) dominating. (3) This asymmetry can be *quantitatively* reproduced by considering the proper transition matrix elements. It is mainly due to the variation of the oscillator strength across the mini-Brillouin zone. (4) We argue that this variation, i.e., the increase of the oscillator strength towards the edge of the superlattice Brillouin zone, is a fundamental consequence of the f-sum rule. (5) Furthermore, using the 1s-2p$_z$ donor transition, the individual widths of the two lowest minibands (and not only the sum of both) can be determined, and the donor transition can be employed to study the metal-insulator transition in superlattices.

The paper is organized as follows: First we will describe the experimental procedure and the GaAs/AlGaAs superlattices used for the experiment. In the then following section we will present a calculation of the inter-miniband absorption coefficient. Next, in section 4, we will summarize some basic facts about impurity states in superlattices, since they play a significant role in the present investigations. In section 5 the experimental results are presented and compared to the calculations. Finally, in section 6 we want to discuss in more detail the oscillator strength of the miniband absorption and relate it to the basic f-sum rule.

2. Experimental Details

It was decided to employ superlattices of the GaAs/AlGaAs system for our study, since this material system still offers the highest crystalline perfection. Two GaAs/Al$_{0.3}$Ga$_{0.7}$As superlattices with different doping concentrations, but otherwise identical structural parameters were investigated. They were grown by molecular beam epitaxy on semi-insulating GaAs (100) substrates and a 2000 Å wide, undoped GaAs buffer. In both cases the nominal width of the GaAs quantum wells was 75 Å and the width of the AlGaAs barriers 25Å, corresponding to a superlattice period of d=100 Å. Superlattice No. 1 was homogeneously doped n-type, to give an electron concentration of 6×10^{17} cm^{-3}, with a total thickness of 2 μm

(200 periods), superlattice No. 2 was doped to n = 6x10^{16} cm^{-3} and had a thickness of 5 μm (500 periods). Due to the homogeneous doping, band bending can be assumed to be rather small. Sample characterization was performed by X-ray diffraction and Hall measurements, yielding a period of 96.5 Å for both superlattices and a peak mobility of 3000 cm^2/Vs (at 200 K) for sample No. 1 and 6000 cm^2/Vs (at 100 K) for No. 2, respectively.

Infrared absorption measurements were performed with a Bruker IFS 113v rapid-scan Fourier-transform spectrometer. The samples were mounted in a liquid-helium flow cryostat, where the sample temperature could be varied between 5 K and 300 K. In order to achieve an active polarization for the intersubband absorption (electric field perpendicular to the layers), the samples were prepared in a multi-pass waveguide geometry [16]. Then the ratio of active and inactive polarization was measured for the superlattice and a reference substrate in the same geometry. In order to obtain data which were completely free of any system artifacts, a metallic gate (Ti/Au) was deposited on the waveguide samples, and was used to modulate the electron density in the top section of the sample by applying a proper reverse bias.

3. Theoretical Absorption Coefficient

The band structure of the superlattices was calculated within the one-band effective-mass approximation (Kronig-Penney model). Fig. 1 shows the two lowest minibands resulting from this calculation. The first miniband extends from 37 to 55 meV, which is considerably wider than typical broadening induced by scattering processes and layer thickness fluctuations. Hence the miniband dispersion should be well developed. The Fermi energy at T = 5 K is calculated to be at 67 meV for sample No. 1 and at 43 meV for sample No. 2, respectively. Thus, in sample No.1 the first miniband is completely filled, i.e. the Fermi level lies in the minigap (which is of course no real gap, since it contains states with finite in-plane wave vectors), whereas in sample No. 2 it is approximately half-filled. Also shown in Fig. 1 are the theoretical positions of the 1s and 2p$_z$ donor energy levels, which will be discussed in the following section. Note that the Fermi energy has been calculated neglecting all impurity effects.

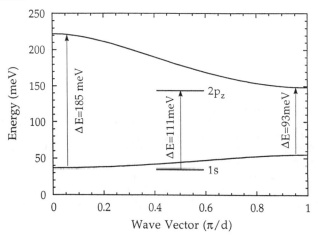

Fig. 1: The calculated dispersion of the two lowest minibands. The 1s and 2p$_z$ impurity states are included schematically. The transition energies at the center and the edge of the mini-Brillouin zone as well as the impurity transition energy are indicated.

Within a single-particle approximation, the absorption coefficient for transitions between the two lowest minibands is given by

$$\alpha = \frac{2}{(2\pi)^3} \frac{\hbar\omega}{I} \int d^3k \frac{2\pi}{\hbar} \left| \langle 1 | (e/m^*) \mathbf{A} \cdot \mathbf{p} | 2 \rangle \right|^2 [f(E_1) - f(E_2)] \cdot \delta(E_2 - E_1 - \hbar\omega). \tag{1}$$

I is the incoming light intensity, $\mathbf{A}$ is the vector potential of the electromagnetic field, and $\mathbf{p}$ is the momentum operator. $f(E_1)$ and $f(E_2)$ are the Fermi-Dirac distribution functions for the ground and excited miniband, respectively. $E_{1(2)}$ and the wavefunctions are a function of k_z. Note that it is mandatory to use the $\mathbf{A}.\mathbf{p}$ interaction for the interaction with the radiation field. The $e\mathbf{E}.\mathbf{r}$ interaction leads to wrong results for an unbounded, periodic system, when the wavefunction does not vanish in the barriers [17].

When a parabolic in-plane dispersion is assumed, the integration over the two in-plane degrees of freedom can be performed analytically and we obtain

$$\alpha = \frac{e^2 kT}{\varepsilon_0 c \hbar^2 \pi m^* \omega} \int_0^{\pi/d} dk_z \left| \langle 1 | p_z | 2 \rangle \right|^2 \ln \left[\frac{1 + \exp([E_F - E_1]/kT)}{1 + \exp([E_F - E_2]/kT)} \right] \left(\frac{\Gamma/\pi}{(E_2 - E_1 - \hbar\omega)^2 + \Gamma^2} \right). \tag{2}$$

Here the energy-conserving δ-function has ben replaced by a Lorentzian, assuming that each transition at wave vector k_z is broadenened in a Lorentzian fashion with the broading parameter Γ. In combination with the k_z-integration, this term represents nothing else than the joint density of states (JDOS). The Fermi energy, E_F, has to be determined numerically from the electron concentration for each temperature. k is Boltzmann´s constant, and η is the refractive index. The other quantities have their usual meaning.

Apart from various constants, there are three factors determining the shape of the absorption coefficient, which are the squared momentum matrix elements, the Fermi-Dirac distribution (already integrated over k_x and k_y) and the broadened JDOS. All three factors depend on k_z. The JDOS for the present superlattices is shown in Fig. 2 (dahed line). It is slightly asymmetric, which stems from the fact that Kronig-Penney bands have a weaker curvature (and so a higher density of states) at their bottom edges than at their top edges. Since the main contribution to the JDOS comes from the higher, broader band, this results in a stronger singularity az $k_z=\pi/d$. (Note that within the tight-binding model, the JDOS can be expressed analytically [12,13] and is exactly symmetric.) The second contribution to α, the thermal occupation factor, has the opposite effect: since there are more occupied states with finite in-plane wave vectors at $k_z=0$ than at $k_z=\pi/d$, the high-frequency peak is enhanced. Finally, the third contribution to α is the squared momentum matrix element, which can be expressed through the oscillator strength, f_{12}, by $f_{12} = (2/m^*\hbar\omega_{21}) | \langle 1 | p_z | 2 \rangle |^2$. For the present system we calculate $f_{12} = 0.3$ at the zone center ($k_z=0$) and $f_{12} = 2.3$ at the zone edge ($k_z = \pi/d$), which is a variation by a factor of 7.5 across the mini-Brillouin zone. (For comparison, an infinite, single quantum well has $f_{12} = 0.96$.) This behavior strongly enhances the low-frequency peak. A detailed discussion of this will be presented in a later section. The resulting absorption coefficient (for sample No. 1 at T = 5 K) is also shown in Fig. 2 (solid line). Surprisingly, this characteristic, strongly asymmetric lineshape has never been explicitly discussed in the literature, even though similar calculations of α have been performed in connection with superlattice infrared detectors [5].

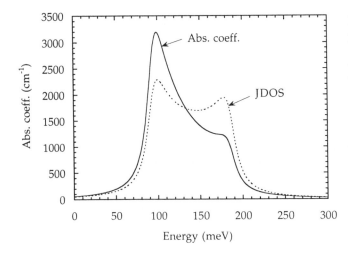

Fig. 2: One-dimensional joint density of states (dashed line) and absorption coefficient according to Eq. (2) for $\Gamma=10$ meV, $n=6\times10^{17}$cm^{-3} and T=5K (full line) for the minibands of Fig. 1. The units of the JDOS are arbitrary.

4. Impurity States

There is no experimental method which allows to completely separate the pure electron miniband structure from other effects. Whereas in *inter*band spectroscopy one has to deal with holes in addition to the electrons and also with excitons, in *intra*band (or *inter-miniband*) spectroscopy one needs proper dopants to provide the electrons in the superlattice. Therefore one inevitably has to consider the donor energy levels. It is well known that due to the broken translational invariance along the growth axis, the normally threefold degenerate 2p state is split, and the $2p_z$ state is pushed up close to the bottom of the second miniband [13,18]. The broken translational symmetry is also known to cause a position-dependent binding energy of the impurities [19,20]. We have calculated the energetic position of the 1s and $2p_z$ states with a variational method [13], and found that the 1s state lies 6.7 meV (5.4 meV) below the first miniband for donors in the center of the wells (barriers), and the $2p_z$ state lies 5.5 meV (7.4 meV) below the second miniband. This results in a transition energy 1s-$2p_z$ of 112.6 meV and 109.4 meV for donors in the wells and barriers, respectively. Due to the relatively broad minibands, the position dependence of the binding energy is small. The spatial energy spreading is only 3.2 meV and can be neglected on the present energy scale. It should be noted that the binding energies of the 1s and $2p_z$ states to the first and second miniband, respectively, are close to the 1s binding energy for bulk GaAs (5.83 meV), again because of the broad minibands. In Fig. 1, the energetic positions of the 1s and $2p_z$ states are indicated schematically. The polarization selection rule for the 1s-$2p_z$ transition is the same as for the intersubband transition. Depending on the doping density and the temperature, the impurity transition and the inter-miniband transition will compete in strength. For low doping, the electrons will freeze out at low temperatures, and only the impurity transition will survive. At higher doping, the donor states will begin to form impurity bands and finally merge with the conduction band. At a certain critical density, n_c, the superlattice will undergo an insulator-metal transition. According to the Mott-criterion the critical density is given by $n_c^{1/3}a_B = 0.25$, (a_B is the donor effective Bohr radius), which gives $n_c = 1.5\times10^{16}$cm^{-3} for bulk GaAs. Both of our samples are in the metallic regime, which has been checked by low-temperature transport measurements. It is evident that the relative strength of impurity and miniband transition can be used as an additional tool to analyze the metal-insulator transition, similarly as has been done for bulk GaAs [21] and GaAs quantum wells [22] using the 1s-2p shallow donor transition and the cyclotron resonance absorption.

5. Experimental Results and Discussion

In Fig. 3 the measured absorption spectrum for the highly doped sample (No. 1) is shown for T = 5 K and T = 300 K. The lineshape remains essentially the same at both temperatures, since the thermal smearing of the distribution function occurs well above the top of the first miniband. Contributions from both critical points in the Brillouin zone are clearly resolved, a peak at 106 meV (at 5K; 103 meV at 300K) and a shoulder at about 180 meV. Comparison with the calculated absorption (Fig. 2) shows that the shape of the absorption, in particular the relative strength of the two peaks is predicted surprisingly well *without adjustable parameters*. In a previous publication a similar lineshape has been reported by some of the present authors for long-period superlattices [13], but could not unambiguously identified due to the weakness of the absorption signal.

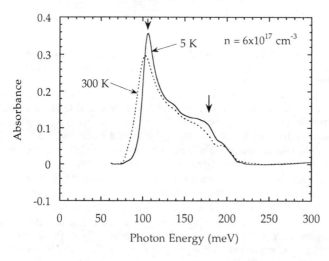

Fig. 3: Experimental absorption spectrum of superlattice No. 1 ($n=6 \times 10^{17} cm^{-3}$) at T=5K and T=300K. The peaks resulting from the critical points at $k_z=0$ and $k_z=\pi/d$ are indicated. (Absorbance = $-\log_{10}$ (Transmisssion)).

In the lower-doped sample the situation is more complex. Absorption spectra for five different temperatures between 5 and 300 K are shown in Fig. 4. At high temperatures, when kT is larger than the width of the lowest miniband, the absorption is similar to the one of the high-doped sample. On decreasing the temperature, the peak at 106 meV disappears, because the top of the miniband at $k_z=\pi/d$ becomes thermally depopulated. *This depopulation demonstrates clearly the well-developed curvature of the first miniband.* In addition, a new peak at 126 meV occurs at low temperature, which can readily be identified as the 1s-2p_z donor transition [13]. Since the 1s and 2p_z level are pinned to the bottom of the first and second miniband, respectively, the spectral position of the 1s-2p_z transition can be used to determine the *individual* widths of the first and second miniband (and not only the sum of both) in the following way. $\Delta_1 = E_1(top)-E_1(bottom) \approx E(1s-2p_z)-E_{21}(k_z=\pi/d)$, and similarly $\Delta_2 = E_2(top)-$

E_2(bottom) $\approx$ $E_{21}(k_z=0)$-$E(1s$-$2p_z)$. Some care has to be taken in relating the calculated energy differences to the peak positions, since the finite broadening effectively shifts the peaks closer together (compare numbers in Fig. 1 and peaks in Fig. 2). One obtains $\Delta_1 = 20$ meV and $\Delta_2 = 60$ meV (both ± 3 meV). Whereas Δ_1 agrees well with the theoretical value (18 meV), Δ_2 is predicted to be 74 meV. This is however not surprising, since our above calculation does not include nonparabolic effects, which are definitely important at energies of ≈ 200 meV. By introducing an energy-dependent effective mass and fitting the Al-content in the barrier (fitted Al-content $\approx 35\%$) we can get a good agreement of both the miniband widths and the absolute energetic positions. Considering the geometrical dimensions of the two samples, even the absolute values of the absorption agree within 15% with the calculation.

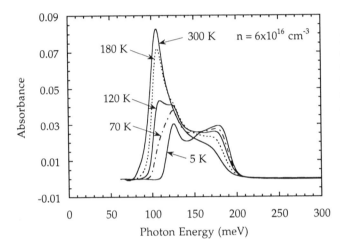

Fig. 4: Experimental absorption spectrum of superlattice No. 2 ($n=6\times10^{16}cm^{-3}$) at different temperatures as indicated.

The occurence of the impurity transition at low temperature is remarkable in itself, since our sample is doped four times higher than the critical density for the Mott metal-insulator transition. Our observations imply that (1) the Fermi energy is located in the first miniband (since the $k_z=0$ miniband peak remains visible), but (2) the impurity band has not merged completely with the conduction band (since the impurity peak is visible). Related work in bulk GaAs [21] agreed with the latter result, however the Fermi energy was found to be in the impurity band.

In an attempt to quantitatively understand the absorption spectra of the low-doped sample, we have carried out two different procedures. In the first procedure we simply calculate the absorption coefficient according to Eq. (1), ignoring the impurity line. The result is shown in Fig. 5(a). The gradual occupation of the miniband zone edge with increasing temperature is clearly seen through the increase of the low-frequency peak. In another fitting procedure, we have added to Eq. (1) a Lorentz oscillator, which describes the impurity transition. The proper occupancy statistics are taken into account. In such a simple model the problem arises that at low temperature the calculated Fermi energy always lies between the impurity level and the band edge. This would result in a complete disappearance of the miniband absorption peak at low temperature, with the impurity line as only remainder. Experimentally, however, the miniband absorption survives even at the lowest temperature, which implies that the Fermi energy is located in the miniband. The only way this can be achieved in this simple model is to assume the number of donors to be smaller than the number of electrons (as it would be the case in modulation doped quantum wells). This forces the Fermi energy into the miniband. Fig. 5(b) shows such a fit using $N_d = n/2$, where N_d is the number of donors, and an oscillator strength for the impurity transition of 0.18. As can be seen, the experimental set of data can be remarkably well reproduced with the same set of parameters over the whole temperature range.

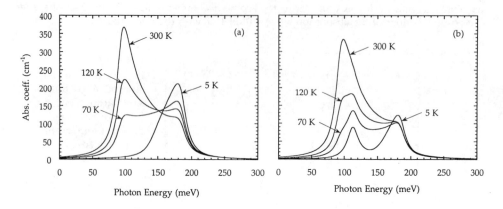

Fig. 5: Theoretical absorption spectrum for superlattice No. 2 ($n=6\times10^{16}cm^{-3}$) at different temperatures as indicated. In (a) the absorption is calculated according to Eq. (2), which means that the impurity transition is ignored. In (b) a Lorentz oscillator is added to Eq. (2) to describe the impurity transition (see text).

6. Oscillator Strengths and Sum Rules

Finally we want to make a few remarks on the variation of the oscillator strength across the superlattice Brillouin zone. Fig. 6 shows the oscillator strength f_{12} as a function of k_z for the present superlattice (solid line). f_{12} increases monotonically across the Brillouin zone, reaching a value of 2.3 at the edge. For a weakly coupled superlattice (thick barriers), $f_{12}(k_z)$ would essentially be constant at around 0.9. The dashed line indicates the case of an infinite single quantum well, where $f_{12}=0.96$. Note that the 1-3 transition is also allowed in a superlattice (in contrast to a single quantum well), although f_{13} is a factor of 1000 smaller than f_{12} in the present case. For specially designed superlattices however, f_{13} can actually be made significantly larger [17].

It is well known that the oscillator strength obeys the so-called f-sum rule (Thomas-Reiche-Kuhn sum rule, or oscillator sum rule), which for a superlattice can be expressed as [17]

$$\sum_{n'} \sum_{k_z'} f_{nn'}(k_z, k_z') = 1 \qquad (3)$$

Here, n (n') and k_z (k_z') are the initial (final) miniband index and wavevector along the superlattice axis, respectively. For absorptive transitions f is positive, for emission it is negative. The sum has to be performed over *all* possible final states, even if this involves non-direct transitions ($k_z' \neq k_z$), where the momentum transfer is provided by phonons or impurities. If $n'=n$, such transitions correspond to *intra*-miniband absorption (Drude-like one-dimensional free-carrier absorption [23]). Now $|n=1, k_z=0\rangle$ is the ground state of the system from which only absorptive processes are possible. Therefore, in order to fulfill Eq. (3), all individual terms of the sum have to be *smaller* than unity, including the most important, direct term $f_{21}(k_z=k_z'=0)$. In contrast, at $k_z=\pi/d$ intra-miniband *emission* processes are possible, which count negative in Eq. (3). Therefore the *inter*-miniband oscillator strength at the *zone edge* has to be *greater* than unity, $f_{12}(k_z=k_z'=\pi/d) > 1$. This tendency becomes more substantial for larger miniband widths. Besides the asymmetry of the absorption spectrum there is second manifestation of this effect. As can be seen in Fig. 4 and Fig. 5 (theoretical and experimental curve!) the area under the absorption curve is not constant, but decreases at low temperature. Following the above arguments, the "missing" area is contained in the intra-miniband absorption at very low frequency (< 10 meV, not measured in the present experiment). At high temperature, when the whole miniband is populated, this Drude absorption vanishes as has been demonstrated recently [23]. Thus the asymmetry of the miniband absorption spectrum is a generic feature which follows from fundamental physical reasons.

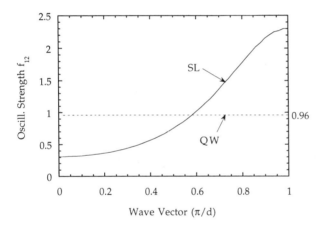

Fig. 6: The oscillator strength f_{12} for the present superlattices (solid line) and for an infinite single quantum well (dashed line).

Inspecting Fig. 6, one might suspect that the average of f_{12} over the mini-Brillouin zone gives something of the order of unity, similar to the situation in quantum wells. Such a behavior can indeed be proven analytically, and a new sum rule for the averaged oscillator strengths can be derived [17].

7. Summary

We have shown that intersubband absorption can reveal the dispersion of superlattice minibands, which manifests itself through the maxima in the joint density of states. The detailed asymmetric shape of the absorption can be understood when considering the proper optical matrix elements and the oscillator sum rule. Impurity transitions, inseparably tied to the miniband absorption, allow investigation of the metal-insulator transition.

300

Acknowledgment: This work has been supported by the "Fonds zur Förderung der wissenschaftlichen Forschung" (FWF), Vienna, by the Texas Advanced Research Program and by the Texas Advanced Technology Program. One of us (F.M.P.) acknowledges support by the Belgian National Science Foundation.

References

1. A. Kamgar, P. Kneschaurek, G. Dorda, and J. F. Koch, Phys. Rev. Lett. **32**, 1251 (1974).
2. L. C. West and S. J. Eglash, Appl. Phys. Lett. **46**, 1156 (1985).
3. B. F. Levine, Semicond. Sci. Technol. **8**, S400 (1993).
4. A. Kastalsky, T. Duffield, S. J. Allen, and J. P. Harbison, Appl. Phys. Lett. **52**, 1320 (1988); L. S. Yu and S. S. Li, Appl. Phys. Lett. **59**, 1332 (1991).
5. Byungsung O, J.-W. Choe, M. H. Francombe, K. M. S. V. Bandara, D. D. Coon, Y. F. Li, and W. J. Takei, Appl. Phys. Lett. **57**, 503 (1990); S. D. Gunapala, B. F. Levine, and N. Chand, J. Appl. Phys. **70**, 305 (1991).
6. L. Esaki and R. Tsu, IBM J. Res. Dev. **14**, 61 (1970).
7. E. E. Mendez, F. Agullo-Rueda, and J. M. Hong, Phys. Rev. Lett. **60**, 2426 (1988); P. Voisin, J. Bleuse, C. Bouche, S. Gaillard, C. Alibert, and A. Regreny, Phys. Rev. Lett. **61**, 1639 (1988).
8. A. Sibille, J. F. Palmier, H. Wang, and F. Mollot, Phys. Rev. Lett. **64**, 52 (1990); A. Sibille, J. F. Palmier, and F. Mollot, Appl. Phys. Lett. **60**, 457 (1992).
9. F. Beltram, F. Capasso, D. L. Sivco, A. L. Hutchinson, S.-N. G. Chu, and A. Y. Cho, Phys. Rev. Lett. **64**, 3167 (1990).
10. H. T. Grahn, K. von Klitzing, K. Ploog, and G. H. Döhler, Phys. Rev. B **43**, 12094 (1991).
11. Ch. Waschke, H. G. Roskos, R. Schwedler, K. Leo, H. Kurz, and K. Köhler, Phys. Rev. Lett. **70**, 3319 (1993); K. Leo, P. Haring Bolivar, F. Brüggemann, R. Schwedler, and K. Köhler, Solid State Commun. **84**, 943 (1992); J. Feldmann et al., Phys. Rev. B **46**, 7252 (1992).
12. R. P. G. Karunasiri and K. L. Wang, Superlattices Microstruct. **4**, 661 (1988).
13. M. Helm, F. M. Peeters, F. DeRosa, E. Colas, J. P. Harbison, and L. T. Florez, Phys. Rev. B **43**, 13983 (1991).
14. J. J. Song et al., Phys. Rev B **39**, 5562 (1989); B. Deveaud et al., ibid. **40**, 5802 (1989); K. Moore, G. Duggan, A. Raukema, and K. Woodbridge, ibid. **42**, 1326 (1990).
15. M. Helm, W. Hilber, T. Fromherz, F. M. Peeters, K. Alavi, and R. N. Pathak, Phys. Rev. B **48**, 1601 (1993).
16. H. Hertle, G. Schuberth, E. Gornik, G. Abstreiter, and F. Schäffler, Appl. Phys. Lett. **59**, 2977 (1991).
17. F. M. Peeters, A. Matulis, M. Helm, T. Fromherz, and W. Hilber, Phys. Rev. B **48**, 15 Oct. 1993.
18. R. L. Greene and K. K. Bajaj, Phys. Rev. B **31**, 4006 (1985)
19. P. Lane and R. L. Greene, Phys. Rev. B **33**, 5871 (1986).
20. M. Helm, F. M. Peeters, F. DeRosa, E. Colas, J. P. Harbison, and L. T. Florez, Surface Sci. **263**, 518 (1992).
21. M.-W. Lee, D. Romero, H. D. Drew, M. Shayegan, and B. S. Elman, Solid State Commun. **66**, 23 (1988); D. Romero, S. Liu, H. D. Drew, and K. Ploog, Phys. Rev. B **42**, 3179 (1990).
22. Y.-J. Wang, B. D. McCombe, and W. Schaff, in"Localization and Confinement of Electrons in Semiconductors", Eds. F. Kuchar, H. Heinrich, and G. Bauer, Springer Series in Solid State Sciences Vol. **97**, Springer, Berlin (1990), p. 314.
23. G. Brozak, M. Helm, F. DeRosa, C. H. Perry, M. Koza, R. Bhat, and S. J. Allen, Jr., Phys. Rev. Lett. **64**, 3163 (1990).

ELECTRONIC QUARTER-WAVE STACKS AND BRAGG REFLECTORS: PHYSICS OF LOCALIZED CONTINUUM STATES IN QUANTUM SEMICONDUCTOR STRUCTURES

C. SIRTORI, F. CAPASSO, J. FAIST, D. SIVCO AND A. Y. CHO
AT&T Bell Laboratories
600 Mountain Avenue
Murray Hill, NJ 07974

ABSTRACT. New semiconductor heterostructures with high localized continuum states and related phenomena are discussed. In quantum wells with quarter-wave stacks as barriers a continuum resonance is highly localized above the well by the near unity reflectivity of the latter. This is observed experimentally in an AlInAs/GaInAs heterostructure as a dramatic narrowing of the bound-to-continuum absorption spectrum. In the superlattice limit the quarter-wave stacks behave as Bragg reflectors and the localized resonance becomes a bound state, a phenomenon observed at cryogenic temperatures. In another set of experiments a heterostructure consisting of doped GaInAs quantum wells cladded by AlInAs/GaInAs λ/4 stacks and by AlInAs barriers is investigated. In a strong electric field of the appropriate polarity the first continuum resonance is strongly confined above the wells, while the second continuum resonance becomes localized in the latter. This manifests itself in a strong narrowing of the intersubband absorption spectrum as a function of the electric field. The same heterostructure exhibits a new type of photoconductivity. For small applied bias this phenomenon manifests itself in directional charge transfer against the electric field as electrons in the wells are photoexcited to resonances localized above the barriers; as the field is increased the photocurrent changes sign. A simple model to describe these features is presented.

1. Introduction

Although widespread attention has been devoted during the last two decades to the investigation of the quantum size effect in semiconductor quantum wells, most of the work has concentrated on quantum confined states at energies below the barrier height. It has not been widely appreciated, however, that under appropriated conditions also the states of the continuum (i.e. energies greater than the barrier height) can become highly localized and eventually bound. In this work we show how quantum wells supporting such states can be designed and discuss recent experimental results. In section 2 we report the equivalent of a Fabry-Perot electron filter in which an electron continuum resonance is spatially localized by high reflectivity quarter-wave stacks. This

H. C. Liu et al. (eds.), Quantum Well Intersubband Transition Physics and Devices, 301–311.
© 1994 *Kluwer Academic Publishers. Printed in the Netherlands.*

produces a sharp narrowing of the bound-to-continuum absorption spectrum. In section 2, we present experimental results on the electric field induced localization of continuum resonances in suitably designed heterostructures consisting of a GaInAs quantum well sandwiched between an AlInAs barrier and an AlInAs/GaInAs $\lambda/4$ stack. In the final section we show how directional charge transfer of photoexcited electrons in this structure gives rise to a striking photocurrent reversal phenomenon.

2. Fabry-Perot Electron Filters and Bound States Above Quantum Wells

In this section we discuss an optimum strategy for localization of electron resonant states above a quantum well [1]. We show that it is also possible to create bound states in this energy range (i.e. above the barrier height) [2], a problem dating back to the late 1920s.

Consider first a conventional rectangular well (Fig. 1(a)). At energies greater than the barrier height one has a continuum of scattering states. For discrete energies corresponding to a semi-integer number of electron wavelengths across the well one finds transmission resonances. Although at these energies the electron amplitude in the well layer is enhanced, the wave functions do not decay exponentially in the barrier, unlike the confined states of the well. These states can be localized above the well using as barriers stacks of layers of thickness $\lambda/4$ each where λ is the de Broglie wavelength in the layer (at the energy of the selected transmission resonance) (Figs. 1(b)-(c)). Constructive interference between the waves partially reflected by the heterointerfaces of the $\lambda/4$ stacks leads to the formation of a quasi-bound state above the center well (Fig. 1(b)). This strongly narrows the transmission resonance in analogy with a Fabry-Perot optical filter where sharp optical resonances are produced using as high reflectivity mirrors quarter-wave stacks. The degree of localization increases with the number of periods due to the increased reflectivity of the $\lambda/4$ stacks; in the structure with just two-period stacks, the wave function is already highly confined (Fig. 1(b)). In the superlattice limit and at low temperatures the stacks become Bragg reflectors; a minigap opens up (Fig. 1(c)) and the localized state becomes a bound state at energies greater than the barrier height. The prediction that certain oscillatory potentials support bound states in the continuum, due to quantum interference, was first put forth by von Neumann and Wigner in 1929 [3]. The formation of such states confined to the barriers of suitably engineered superlattices was proposed in 1977 [4].

To design our structure we have used the envelope function approximation [5] and the following conditions

$$k_W L_{W,C} = \pi, \tag{1}$$

$$k_W L_W = \frac{\pi}{2}, \tag{2}$$

$$k_B L_B = \frac{\pi}{2}, \tag{3}$$

$L_{W,C}$ is the central well thickness, L_W and L_B are the thicknesses of the wells and barriers in the cladding regions; k_W and k_B are the wave numbers in the well and barrier materials. Eq. (1) represents the transmission resonance condition, while Eqs. (2) and (3) are the $\lambda/4$ stacks conditions. The starting point is a reference structure (Fig. 1(a)) consisting of a 32 Å $Ga_{0.47}In_{0.53}As$ QW bound by thick $Al_{0.48}In_{0.52}As$ barriers. The calculations find a single bound

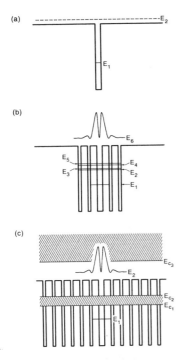

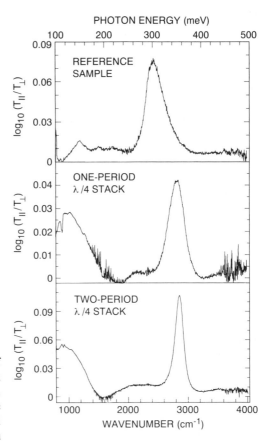

Fig. 1. Conduction band diagrams of AlInAs/GaInAs heterostructures used in the study of localized continuum states. (a) Reference sample. Shown are the ground state of the well (E_1 = 204 meV) and the position (dashed line) of the first transmission resonance in the continuum (E_2 = 560 meV); (b) Quantum well cladded by two-period quarter-wave stacks (Fabry-Perot electronic filter). Shown is $|\psi|^2$ of the localized quasi-bound state (E_6 = 560 meV) formed in correspondence to the transmission resonance and the positions of new states created at lower energies; (c) in the superlattice limit the $\lambda/4$ stacks behave as Bragg reflectors. The state above the well now becomes a bound state localized by the superlattice minigap.

Fig. 2. Absorption spectra at room temperature for the structures with quarter-wave stacks (Fig. 1). The absorbance for polarization parallel to the plane of the layers ($-\log_{10} T_{\parallel}$) is subtracted from the absorbance normal to the plane of the layers ($-\log_{10} T_{\perp}$), to remove the contribution of free-carrier absorption in the buffer layers. The transitions to the confined state above the well corresponds to the peaks at $\simeq 350$ meV. The bound-to-continuum transition in the sample without $\lambda/4$ stacks (reference) corresponds to the broad absorption peak at 300 meV.

state at E_1 = 204 meV and continuum resonances. The first of these (E_2 = 560 meV) satisfies Eq. (1) (Fig. 1(a)). AlInAs/GaInAs of varying number of periods (1, 2, 6 and infinite) are then introduced in the barrier layers. The thickness of the well and barrier layers in the stacks (16 and 39 Å, respectively) satisfy Eqs. (2) and (3), respectively, with k_W and k_B calculated at the energy of the resonance E_2.

The reference sample had twenty 32 Å InGaAs quantum wells n-type doped to $\approx 1 \times 10^{18}$ cm^{-3}, separated by 150 Å undoped AlInAs barriers. In the other three structures the 32 Å wells, doped to the same level, were cladded, respectively, by 1, 2 and 6 undoped periods, each comprising 39 Å AlInAs barriers and 16 Å GaInAs wells, as described above. The phase coherence length is estimated to be ~300 Å at 10 K.

The absorption spectra of the reference structure is broad with a long wavelength cutoff determined by the height of the barrier (Fig. 2(a)). In the structure with one $\lambda/4$ period the peak is considerably narrower and centered at an energy corresponding to the transition between the ground state of the well and the localized resonant state above the well (Fig. 2(b)). As the number of quarter-wave stacks is doubled the absorption peak does not shift and considerably narrows, precisely the behavior expected for a Fabry-Perot (Fig. 2(c)). In fact the observed narrowing (16 meV) can be quantitatively explained in terms of the reflectivity increase of the $\lambda/4$ stacks. In the structure with six periods at low temperatures, the highly localized state becomes effectively a bound state confined by Bragg reflections from the superlattice (Fig. 2(c)) [2]. The absorption spectrum (Fig. 3) shows an isolated peak at 360 meV of width ~10 meV corresponding to the transition from the state E_1 to the state E_2 in Fig. 1(c). It is worth noting that the width of the transition to the confined state above the well in the two- and six-period structure is practically identical to that of the bound-to-bound state transition measured in a conventional 55 Å thick GaInAs well with 300 Å thick AlInAs barriers.

3. Electric Field Induced Narrowing of Bound-to-Continuum Intersubband Transitions

The 40 period structures discussed in this section [7,8] are grown on a semi-insulating <100> InP substrate and are sandwiched between n$^+$ 4000 Å thick Ga$_{0.47}$In$_{0.53}$As contact layers. Each period comprises a Ga$_{0.47}$In$_{0.53}$As quantum well, n-type doped with Si to $\approx 1.5 \times 10^{18}$ cm^{-3}, cladded on one side by an Al$_{0.48}$In$_{0.52}$As barrier and on the other by two Al$_{0.48}$In$_{0.52}$As/Ga$_{0.47}$In$_{0.53}$As quantum wells. The samples were processed similarly to the ones discussed in section 2 and Au-Ge alloyed ohmic contacts were used for the n$^+$ cladding regions. The absorption and short-circuit photocurrent spectra were measured using a Nicolet System 800 Fourier transform infrared spectrometer, and the sample was cooled in a flow dewar equipped with a bias stage.

Fig. 4(a) shows the conduction band diagram of a portion of our structure at zero bias. The wavefunctions and energy levels were calculated in the envelope function approximation. From the shape and spatial extension of the $|\psi|^2$'s one sees that the continuum resonances represent, from a physical point of view, quasi-bound states at energies above the barrier of the doped well plus thick barrier combined. They are spatially confined to this region by the AlInAs/GaInAs ultrathin layers; these act essentially as high reflectivity (≥ 0.90) quarter-wave electron stacks in the energy range from 0.51 eV (top of the barrier) to 0.64 eV. The energy is measured from the bottom of wells. The absorbance spectrum of the structures at zero bias is shown in Fig. 5(a). The features in the absorption spectrum, indicated by arrows, are in good agreement with

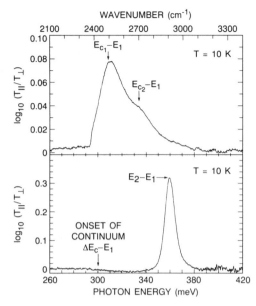

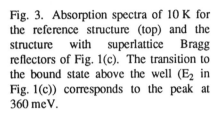

Fig. 3. Absorption spectra of 10 K for the reference structure (top) and the structure with superlattice Bragg reflectors of Fig. 1(c). The transition to the bound state above the well (E$_2$ in Fig. 1(c)) corresponds to the peak at 360 meV.

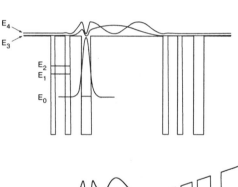

Fig. 4(a). Energy band diagram of another structure showing the calculated energy levels, measured with respect to the bottom of the well and the modulus squared of some of the corresponding wavefunctions (E$_0$ = 192 meV, E$_1$ = 310 meV, E$_2$ = 351 meV, E$_3$ = 513 meV, E$_4$ = 523 meV). The thickness of the widest GaInAs well is 34 Å. The λ/4 stack to the left of this well comprises two 17 Å thick undoped GaInAs wells and two undoped AlInAs barriers, 36 Å and 42 Å thick, respectively. The thickness of the undoped AlInAs barrier to the right of the 34 Å well is 270 Å.

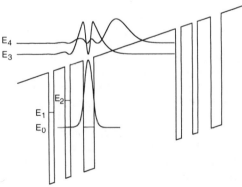

Fig. 4(b). Energy band diagram of the structure in a strong electric field (4.8×10^4 V/cm, positive bias polarity). Note that the first continuum resonance (E$_3$) has been strongly localized above the well while the second one (E$_4$) has been mostly confined to the barrier.

calculations and correspond to transitions from the ground state of the well to the first three resonances in the continuum (states 3, 4 and 5 in our notation). Because of the close spacings of the latter the absorption spectrum is broadened into a band. Note that the absorption peak corresponds to the transition from the bound state of the well to the second continuum resonance. The latter state has a larger spatial overlap with the bound state of the doped well, than the first continuum resonance (Fig. 4(a)). This results in a larger transition matrix element ($|z_{04}| \approx 8$ Å) as compared to the $0 \rightarrow 3$ transition ($|z_{03}| \approx 6.3$ Å).

The multilayer barrier to the left of the doped well in Fig. 4(a) was designed as a quarter wave stack for an electron de Broglie wavelength corresponding to an energy equal to that of the first resonance of the structure in an electric field $\approx 4.8 \times 10^4$ V/cm. At this field the energy of this resonance is 526 meV, measured with respect to the center of the well. The quarter wave stack conditions read

$$\int_0^{L_W} k_W\, dx = \frac{\pi}{2}, \quad \int_0^{L_B} k_B\, dx = \frac{\pi}{2} \tag{4}$$

where k_W and k_B are the wave numbers in the well and barrier material of the stack, calculated at the energy of the first transmission resonance at the above electric field. The thicknesses L_W and L_B of the wells and barriers of the stack are chosen to satisfy Eq. (4). This choice optimizes the spatial confinement of the first continuum resonance (Fig. 4(b)) by maximizing the reflectivity of the multilayer electronic mirror ($R \approx 0.90$). Our calculations show that for smaller electric fields ($0 \leq F \leq 4.8 \times 10^4$ V/cm) the reflectivity increases monotonically to a value ≈ 0.99 at zero bias. This ensures that the first continuum resonance does not appreciably penetrate in the region to the left of the doped well in Fig. 4, for a broad range of electric fields.

As the field is increased, for positive bias polarity, the confining effect of the AlInAs barrier (to the right of the doped well in Fig. 4(a)) is enhanced, leading to increased localization of the first resonance above the center well. This has the effect of enhancing the matrix element of the $0 \rightarrow 3$ transition and hence the corresponding oscillator strength f_{03} and absorption coefficient. An opposite effect occurs for the second continuum resonance which becomes increasingly localized in the barrier, thus reducing the spatial overlap with the ground state of the doped well (Fig. 4(b)) and therefore also the oscillator strength of the $0 \rightarrow 4$ transition. This is similar to the behavior of the ground and first excited states of a rectangular well in an electric field. The centroids of the probability distributions for the two states are shifted in opposite directions by the electric field. The third continuum resonance, whose oscillator strength at zero bias is ~20%, is also squeezed above the rectangular potential of the doped well, and its matrix element with the ground state is strongly reduced ($z_{05} \leq 1$ Å). In summary, an increase of the electric field in the positive polarity should lead to greatly enhanced absorption for the $0 \rightarrow 3$ transition and strongly decreased absorption for the $0 \rightarrow 4$ and $0 \rightarrow 5$ transitions, producing an overall narrowing of the spectrum. In the opposite bias polarity instead there is negligible overlap between the resonances, now localized in the AlInAs barrier and the ground state. The absorption spectrum is then controlled by transitions to extended states at energies above the barrier and is therefore expected to be broad.

The absorption spectra of Fig. 5(b), shown for different electric fields, confirms the above physical picture. At negative bias the spectrum is very broad, while for positive bias the spectrum narrows with increasing field. A detailed study of the spectra shows that the area under the absorption curve is independent of the electric field within the experimental error ($\pm 5\%$). This conservation is a general sum rule previously shown to be valid in the case of interband transitions in quantum wells [9]. One can, in fact, show that the area under the absorption curve

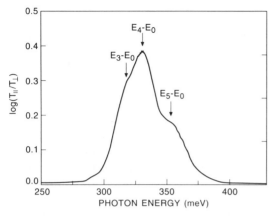

Fig. 5(a). Absorption spectra at 10 K of the structures of Fig. 4 at zero bias. The position of the features in the spectrum indicated by arrows are in good agreement with the calculated bound-to-continuum resonance transitions ($E_3 - E_0 = 317$ meV; $E_4 - E_0 = 328$ meV; $E_5 - E_0 = 348$ meV).

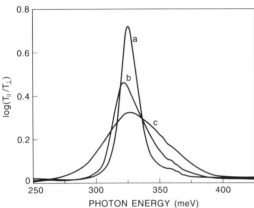

Fig. 5(b). Absorption spectra at 10 K taken for various bias conditions a: +8 V, b: +2 V, c: = −6 V. One volt corresponds to approximately an average electric field of 6 kV/cm across the structure. The position of the peaks in positive polarity is in good agreement with the calculated energy differences ($E_3 - E_0 = 316$ meV at +2 V; $E_3 - E_0 = 332$ meV at +8 V). In opposite polarity the absorption is peaked near the onset of the continuum ($\Delta E_c - E_0$) and is broadened towards lower energies by tunneling effects.

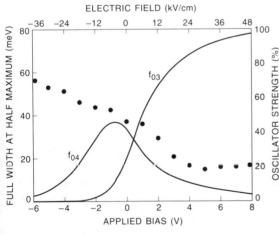

Fig. 6. Full width at half maximum of the measured absorption spectrum as a function of bias for the structure of Fig. 4. Shown are also the calculated modulus squared of the matrix elements of the transitions between the bound state of the doped well to the first two continuum resonances, $|z_{03}|^2$ and $|z_{04}|^2$.

is independent of the electric field in the limit of negligible Stark shift of the ground state. This is a good approximation for our structure in the range of electric fields used in the experiment. There are also other transitions, those from the ground state of the doped well to the confined states of the wells of the $\lambda/4$ stack, i.e. the energy levels E_1 and E_2 (Fig. 4(a) and 4(b)), which are not included in the measured integrated absorption strength obtained from the area under the peaks of Fig. 5(b). These transitions, however, have a very small dipole matrix element (≈ 1 Å) because of their diagonal nature in real space and therefore the contribution to the total integrated absorption strength is negligible. In fact, these transitions are not observed.

Further insight can be gained by plotting the full width at half maximum of the absorption curve as a function of electric field (Fig. 6). In the negative bias polarity the linewidth increases smoothly with increasing field while for positive polarity the linewidth experiences a pronounced drop above $+5\times10^3$ V/cm bias and then saturates. This rapid decrease in linewidth is the result of the increased fraction of the total oscillator strength being concentrated into the ground state-to-first resonance ($0 \rightarrow 3$) as the field is increased, while the $0 \rightarrow 4$ and $0 \rightarrow 5$ transitions lose strength. In fact, the calculated modulus squared of the matrix element of the $0 \rightarrow 3$ transition, $|z_{03}|^2$ rapidly increases with decreasing linewidth (Fig. 6), while $|z_{04}|^2$ is reduced. Note that at zero bias $|z_{04}|^2 > |z_{03}|^2$, in agreement with the absorption measurements (Fig. 5(a)). As the field is increased the centroids of the $|\psi|^2$'s of states 3 and 4 are displaced in opposite direction leading to the crossover between $|z_{03}|^2$ and $|z_{04}|^2$. We estimate that for an electric field $= 4.8\times10^4$ V/cm $\approx 95\%$ of the oscillator strength is concentrated in the $0 \rightarrow 3$ transition. From the area under the absorption peak, the calculated values of the matrix element $<z_{03}> \approx 12$ Å and the transition energy ($E_3 - E_0 \approx 332$ meV at the above electric field) one finds a carrier concentration in the well close to the nominal value.

4. Photocurrent Reversal by Localization of Continuum Resonances

The photocurrent spectrum (Fig. 7) of the structures at zero bias (Fig. 4) reveals features, whose position (arrows) is in good agreement with the absorption data (Fig. 5(a)) and with the calculations. Beyond 600 meV the photocurrent is negligible, in agreement with the calculations which show that the matrix elements to higher lying states become very small (≤ 1 Å). The low energy cutoff is of course determined by the difference between the conduction band discontinuity and the ground state energy.

The photocurrent can be expressed as the difference between fluxes associated with electron transfer to the left and right of the doped well in Fig. 4(a)

$$J_{ph} = J_L - J_R \tag{5}$$

The sign of the photocurrent at zero bias corresponds to photoexcited electrons moving preferentially to the right in Fig. 4(a). This directional charge transfer is easily understood from the energy diagram since the center of charge $<\psi_f|z|\psi_f>$ of the localized continuum resonances is strongly displaced to the right of the doped well.

Fig. 8 shows the photocurrent spectrum for various biases at 5 K. The increased separation of the spectral features corresponding to the continuum resonances, compared to the zero bias case, is a result of the Stark effect. As the bias is increased up to ≈ 1.7 V, corresponding to a field of ≈ 11 kV/cm, the photocurrent flows against the direction of the electric field, i.e. photoelectrons move preferentially to the right in Fig. 4(a). Beyond 1.7 V the photocurrent changes sign and

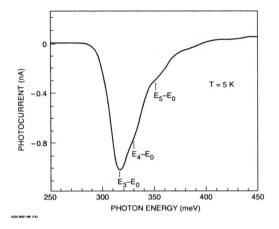

Fig. 7. Photocurrent spectra at 5 K of the structure at zero bias. The negative sign of the photocurrent corresponds to photoexcited electrons moving to the right in Fig. 4(a).

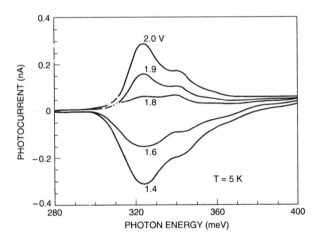

Fig. 8. Photocurrent spectra at various biases and 5 K temperature. The sign of the applied field corresponds to the ground state F_0 in the thick well (Fig. 4(a)) being lowered with respect to the wells of the adjacent period to the right.

electrons move in the usual direction, i.e. "downhill" in the band diagram.

The rapid reversal of the photocurrent with increasing bias can be explained in a straightforward way by the strong decrease of the component of the photocurrent due to electrons flowing against the field, i.e. J_R. In order to contribute to the latter a photoexcited electron must cross the boundary between the thick barrier to the right of the doped well and the quarter-wave stack, so that it can be collected by one of the wells of the neighboring period. Note that the extension of the wavefunctions of the localized continuum resonances beyond this region is completely negligible. The photocurrent J_R is, therefore, proportional to

$$\sum_i L_{oi}(\omega) |z_{oi}|^2 \int |\psi_i|^2 dx \tag{6}$$

where the index i denotes the continuum resonances, ω is the angular frequency of the incident photon, z_{oi} and $L_{oi}(\omega)$ the transition matrix element and the lineshape of the o $\rightarrow$ i transition, respectively. Several transitions are involved for a given photon energy due to the strong overlap and broadening of the continuum resonances. The integral $\int |\psi_i|^2 dx$ represents the probability of finding the photoexcited electrons in the well region to the right of the thick barrier in Fig. 4(a). Evaluation of Eq. (6) shows that it decreases exponentially with increasing bias with a "decay constant" ≈ 1.5 V. This value is comparable to the measured crossover voltage (1.2 V at 300 K and 1.7 V at 6 K) of the photocurrent.

There is, of course, also a concomitant increase of the other component of the photocurrent (J_L) associated with the transfer rate of photoexcited electrons into the thick AlInAs barrier to the left of the doped well in Fig. 1(a). This phonon-assisted process increases with the electric field due to the increase in the density of final states caused by the lowering of the conduction band in the barrier.

The modeling of the photocurrent in these structures represents a difficult problem of quantum transport. Such a theory, although beyond the scope of this letter, would be necessary for an accurate calculation of the crossover voltage and of the bias dependence of the photocurrent.

References

1. Sirtori, C., Capasso, F., Faist, J., Sivco, D. L., Chu, S. N. G., and Cho, A. Y. (1992) 'Quantum wells with localized states at energies above the barrier height: a Fabry-Perot electron filter', Appl. Phys. Lett. 61, 898-900.

2. Capasso, F., Sirtori, C., Faist, J., Sivco, D. L., Chu, S. N. G., and Cho, A. Y. (1992) 'Observation of an electronic bound state above a potential well', Nature 358, 565-567.

3. A detailed discussion of this work is given in Ballentine, E. (1990) Quantum Mechanics, Prentice Hall, Englewood Cliffs, NJ, p. 205.

4. Stillinger, F. H. (1977) 'Potentials supporting positive-energy eigenstates and their application to semiconductor heterostructures', Physica 85B, 270-276.

5. For the parameters of our $Al_{0.48}In_{0.52}As/Ga_{0.47}In_{0.53}As$ heterostructures we used $\Delta E_c = 0.51$ eV, $m_e^*(GaInAs) = 0.043\ m_0$, $m_e^*(AlInAs) = 0.07\ m_0$ and $\gamma = 1.03 \times 10^{-18}\ m^2$ for the nonparabolicity. The latter was taken into account using the method discussed in Nelson, D. F., Miller, R. C., and Kleinman, D. A. (1987) 'Band nonparabolicity effects in quantum wells', Phys. Rev. B35, 7770-7773.

6. Zahler, M., Brener, I., Lenz, G., Salzman, J., Cohen, E., and Pfeiffer, L. (1992) 'Experimental evidence of Bragg confinement of carriers in a quantum barrier', Appl. Phys. Lett. 61, 949-951.

7. Sirtori, C., Faist, J., Capasso, F., Sivco, D. L., and Cho, A. Y. (1993) 'Narrowing of the intersubband absorption spectrum by localization of continuum resonances in a strong electric field', Appl. Phys. Lett. 62, 1931-1933.

8. Sirtori, C., Faist, J., Capasso, F., Sivco, D. L., and Cho, A. Y. (1993) 'Photocurrent reversal induced by localized continuum resonances in asymmetric quantum semiconductor structures', Appl. Phys. Lett. 63, 2670-2672.

9. Miller, D. A. B., Weiner, J. S., and Chemla, D. S. (1986) 'Electric field dependence of linear optical properties in quantum well structures: waveguide electroabsorption and sum rules', IEEE J. Quantum Electron. QE-22, 1816-1830.

Modulation of the Optical Absorption by Electric-Field-Induced Quantum Interference in Coupled Quantum Wells

JÉRÔME FAIST, FEDERICO CAPASSO, ALBERT L.
HUTCHINSON, LOREN PFEIFFER, KEN W. WEST, DEBORAH
L. SIVCO AND ALFRED Y. CHO
AT&T Bell Laboratories,
600 Mountain av, Murray Hill, NJ 07974   USA

ABSTRACT. A coupled-well semiconductor heterostructure which exhibits an extremely large modulation of the intersubband dipole matrix element as a function of the applied electric field is designed and demonstrated. This effect is shown to arise from quantum interference between two spatially separated transitions. Application of these structures for optical modulation is discussed.

1. Introduction

Quantum interference effects in the absorption of atoms and molecules have been investigated since the seminal work of Fano on continuum resonances[1]. Recent examples include interference between multiple absorption pathways in two-photon absorption[2] and in molecular dissociation[3], absorption cancellation in dressed three level atomic systems which may lead to lasing without inversion[4], and the production of a large index of refraction with vanishing absorption by atoms prepared in a coherent superposition of an excited state doublet[5].

The development of Molecular Beam Epitaxy[6] and Bandgap engineering[7] has made possible the design and realization of quantum semiconductor structures with new and unusual optical properties[8]. Of particular interest in this respect are quantum dots[9] and coupled quantum well structures[10-13] which behave respectively as giant artificial atoms or "quasi-molecules" with dipoles matrix elements between electronic states many orders of magnitude larger than atomic and molecular ones.

In this paper we report the design and demonstration of a coupled quantum well semiconductor exhibiting a new quantum interference effect in absorption between electronic states. This phenomenon manifests itself in a striking electric field induced suppression of intersubband absorption, accompanied by a negligible Stark shift of the corresponding transition. As such the large modulation of the dipole matrix element in our effect does not originate from a reduced overlap between the wavefunctions as in the case of the quantum confined stark effect[14] but has its subtle origin in the quantum interference between the two coupled wells.

313

H. C. Liu et al. (eds.), Quantum Well Intersubband Transition Physics and Devices, 313–319.
© 1994 *Kluwer Academic Publishers. Printed in the Netherlands.*

2. Sample structure

The sample, grown by MBE on a semi-insulating GaAs substrate, comprises fifty modulation-doped coupled-quantum-wells. Each period consists of two GaAs wells, respectively 62 Å and 72 Å thick, separated by a 20 Å $Al_{0.33}Ga_{0.67}As$ barrier. The coupled-well periods are separated by a 1450 Å $Al_{0.33}Ga_{0.67}As$ spacer layer. To supply the electron charge in the wells, a δ-doped Si layer (1×10^{12} /cm^2) is inserted in the spacer layers as to ensure a symmetric charge transfer[15]. The growth sequence starts and ends with a GaAs contact layer, doped with Si to n=1×10^{18} cm^{-3}.

3. Theory

Figure 1 (a) shows the conduction band diagram of a period of the sample at zero bias. Indicated are the energy levels and the modulus squared of the wave functions. In contrast to previous theoretical[12] and experimental[13] work on coupled wells structures, where the excited state of the large well was coupled to the ground state of the narrow well, our structure exhibits coupling between states having same quantum number.

The energy levels and wavefunctions are computed by solving Schrödinger's and Poisson's equation in the envelope-function formalism[16,17]. The exchange interaction (Slater term) is included in the local-density approximation.

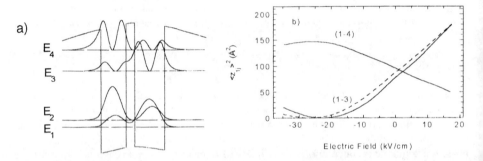

Fig. 1a) The energy-band diagram of a single period of the GaAs/$Al_{0.33}Ga_{0.67}As$ modulation doped structure. Shown are the positions of the calculated energy subbands and the corresponding modulus squared of the wavefunctions. We computed E_1 = 63 meV, E_2 = 80 meV, E_3 = 198 meV, E_4 = 250 meV. b) Square if the intersubband dipole matrix element $(z_{13})^2$ and $(z_{14})^2$ (as indicated) computed as a function of the electric field. Dotted line: tight-binding model for $(z_{13})^2$.

To get a better insight in the behavior of the coupled well system as a function of the applied electric field, let us first consider the two quantum wells, denoted here as well a) and b), coupled by the barrier in a tight-binding approach[16]. In such a model, the calculated wavefunctions ψ_i (i = 1..4) of this system are expanded in terms of the eigenfunctions $\varphi_{1,2}^{a,b}$ of the first two bound states 1,2 of the two isolated wells. In the tight binding approximation, the dipole matrix element $z_{1i} = <\psi_1|z|\psi_i>$ (i = 3,4) between the first and the third or fourth state of the coupled well system can now be written as the sum of the contribution from the two wells a and b

$$z_{1i} = <\psi_1|\varphi_1^a><\psi_i|\varphi_2^a>z_{12}^a + <\psi_1|\varphi_1^b><\psi_i|\varphi_2^b>z_{12}^b \tag{1}$$

where z_{12}^a and z_{12}^b are the transition matrix element computed inside each individual well. As ψ_1 is the ground state of the system, $<\psi_1|\varphi_1^a>$ and $<\psi_1|\varphi_1^b>$ have the same sign. On the contrary, since the second excited state ψ_3 crosses zero twice and is constructed from the antisymmetric wavefunctions $\varphi_2^{a,b}$, $<\psi_3|\varphi_2^a>$ and $<\psi_3|\varphi_2^a>$ have opposite signs. Therefore, if we consider a transition between the first and third state of the coupled well system, the two terms of Eq. (1) have opposite sign. One thus expect large values of z_{13} for large absolute values of the electric field where both wavefunctions are localized either in well a or b (first or last term of Eq. (1) dominates) and a null for some intermediate value of the electric field. At this field the absorption will be suppressed. This behavior is clearly apparent in Fig.1, where we display $(z_{13})^2$ as given by Eq. (1) along with the exact calculation, using the parameters of the GaAs sample. As expected, z_{13} decrease with the applied field and has a null for an electric field of - 22kV/cm. The tight-binding result is in good agreement with the exact computation in which z_{13} is directly computed from the wavefunctions $\psi_{1,3}$, supporting our interpretation of this effect as an interference phenomenon. The self-consistent potential takes into account the screening of the applied electric field by the charges inside the coupled wells. The interference effect still occurs even if the self-consistent potential is neglected, although at a different applied field.

The interference phenomenon occurs for the (1-3) transition because the two terms in Eq. (1) have opposite sign. On the contrary, as $<\psi_4|\varphi_2^a>$ and $<\psi_4|\varphi_2^b>$ have the same sign, the two terms of Eq. (1) for the (1-4) transition add constructively. For large values of the electric field, the ground state wavefunction ψ_1 will be localized in one well and ψ_4 in the other. The two wavefunctions ψ_1 and ψ_4 having now a small overlap, the resulting dipole matrix element z_{14} will be small. This behavior appears clearly on Fig. 1 where $(z_{14})^2$ is displayed as a function of the applied field. z_{14} is maximum at about - 22kV/cm and decrease for positive fields or larger negative fields.

4. Experiments

For the experiments, the samples were processed into square mesas (800µm side) and ohmic contacts were provided to the n+ contact layers. They were then cleaved in strips and the cleaved edges were polished at 45° to provide a two-pass waveguide. The absorption spectra were measured with a Nicolet 800 Fourier transform infrared spectrometer (FTIR) using this wedge waveguide geometry. In Fig. 2 the low -temperature absorption spectra are shown for different applied biases. Positive bias refers to the band diagram configuration in which the thick well is lowered below the thin well. The height of the absorption peak at hv = 136meV corresponding to the 1-3 transition decreases regularly when the applied bias U_b is changed from 30V to -20V, has a minimum at this value and rises again if the bias is decreased further to -30V. At -20V, the integrated absorption of the peak has decreased to 1.6% of its value at 30V. Meanwhile, as shown also on Fig. 2, the (1-3) transition energy has changed by less than 1meV in the same bias range. As expected from the computations, the (1-4) transition has the opposite behavior, its oscillator strength is maximum for U_b = -20V and not observable anymore at 30V.

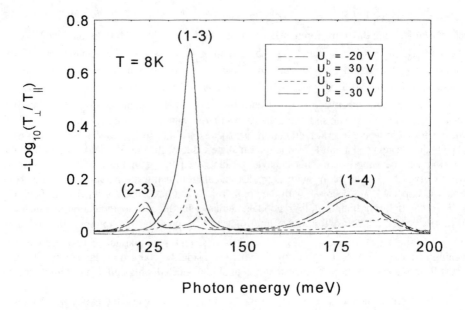

Fig. 2 Intersubband absorption spectra in a two-pass waveguide at T =8K for different applied biases, as indicated. The absorbance for polarization parallel to the plane of the layers (-log$_{10}$T$_{\parallel}$) is subtracted from the absorbance normal to the plane of the layer (-log$_{10}$T$_{\perp}$), to remove instrumental, substrate and free carriers absorption contributions. A strong modulation of the absorption of the (1-3) and (1-4) transitions is observed.

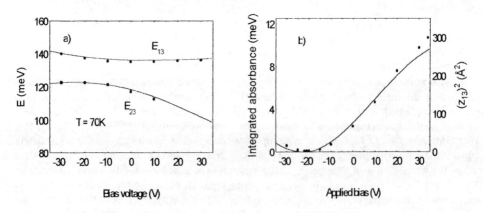

Fig. 3 a) Comparison between the measured (marks) and calculated (solid lines) E$_{13}$ (squares) and E$_{23}$ (dots) transition energies. In order to track E$_{23}$ for a larger range of applied fields, the temperature was raised to 70K to induce a thermal population of the second level. b) Square of the transition matrix element (z$_{13}$)^2 (right axis), as derived experimentally from the integrated absorbance below the (1-3) peak (left axis) at T=8K. The solid line is the calculated value.

Although there is a clear exchange of oscillator strengths between the (1-4) and (1-3) transition as a function of bias, the sum of the oscillator strength of these two transitions is not constant because the (1-2) transition also exchange oscillator strength with the (1-3) and (1-4) transitions. The energy of the (1-2) transition has its minimum near $U_b = -20V$. At this bias, the second state drops slightly below the Fermi energy, allowing the observation of a third peak at hv = 124 meV corresponding to the (2-3) transition.

However, by evaluating the electron sheet density $n_s^{(2)}$ in the second state from the integrated absorbance below the (2-3) peak we find $n_s^{(2)} = 3 \times 10^{10} cm^{-2}$, less than 10% of the ground state density. Therefore, the population of the second state has a negligible impact on the modulation of the (1-3) absorption The characteristic anticrossing of the first two levels as a function of the applied bias is clearly apparent from Fig. 3a) where the measured transition energies E_{13} and E_{23} are reported along with the theoretical predictions using the full self-consistent model. As shown in Fig. 3, the experimental points are in excellent agreement with the theory. On Fig. 3 b), $(z_{13})^2$ is derived from the measured integrated absorption by using a well known relationship[18] and the nominal electron sheet density $n_s = 5 \times 10^{11} cm^{-2}$, as obtained from prior calibration of the MBE system[15] on similar samples. The agreement with the theoretical prediction is excellent. As expected, $(z_{13})^2$ decreases first as the field is increased, has a minimum for $U_b = -20V$, and then rises again. This is a further proof that we are not observing a mere transfer of electrons from one well with one transition energy to an other well with a different transition energy. In the latter case, a monotonic behavior of the absorption strength is expected.

5. Discussion

As expected from a interference phenomenon, the amplitude of the wavefunctions in each well enter in Eq. (1), and these wavefunctions must remain coherent across the coupled-well system. A sufficient amount of disorder will localize the wavefunctions in either well and destroy completely the interference effect. Indeed, this localization effect has been recently observed[19]. In this work, the absorption of a coupled well system with a thicker tunnel barrier (40A) is reported as a function of the applied field. The wavefunctions are localized in the two wells and no modulation of the dipole transition element z_{13} is observed[19]. The modulation of the absorption is entirely due to a transfer of electrons from the ground state of one well to the ground state of the other well having a different transition energy.

6. Modulator applications

This interference effect, able to provide large modulation of the absorption coefficient, is very interesting for optical modulation purposes. As one is not limited by the incoherent transfer of electrons from one well to the other, the modulation is quasi-instantaneous and the ultimate speed limit of a modulator based on such interference effect is its RC time constant. Moreover, at the wavelength of resonance, the modulation is essentially chirp-free.

High temperature operation of the GaAs sample is limited to about 120K due to the current flow caused by thermionic emission above the $Al_{033}Ga_{67}As$ barrier. For this purpose, a similar sample using the large conduction band discontinuity (0.51eV) of $Al_{0.48}In_{0.52}As$ and $Ga_{0.47}In_{0.53}As$ lattice-matched to InP was grown. It comprises of forty modulation-doped coupled-quantum-wells. Each period consists of two $Ga_{0.47}In_{0.53}As$ wells, respectively $100\overset{o}{A}$ and $75\overset{o}{A}$ thick, separated by a $15\overset{o}{A}$ $Al_{0.48}In_{0.52}As$ barrier. The coupled-well periods are separated by a $330\overset{o}{A}$ $Al_{0.48}In_{0.52}As$ spacer layer. The $30\overset{o}{A}$ in the middle of the $Al_{0.48}In_{0.52}As$ barrier is doped with Si to $n=1\times10^{18}cm^{-3}$, providing the charge for the wells. The two growths start and end by a $Ga_{0.47}In_{0.53}As$ contact layer, doped with Si to $n=1\times10^{18}cm^{-3}$.

The sample preparation and the experimental procedures are the same as described in the previous section. The absorption spectra for the InGaAs sample at T=225K and for different applied biases are reported on figure 4.

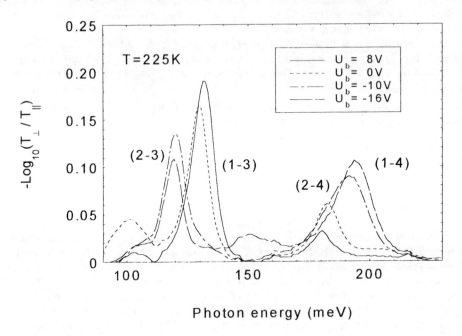

Fig. 4 Intersubband absorption spectra of the InGaAs sample in a two-pass waveguide at T=225K for different applied biases, as indicated. A strong modulation of the (1-3) and (1-4) transition shows that the interference effect is still present close to room temperature.

The (1-3) transition shows a strong modulation due to the interference effect. The absorption is maximum at a bias of 8V, decreases regularly with decreasing voltage, is minimum at -10V and rises again with a strong Stark shift when the voltage is further decreased to -16V. The (1-4) transition has an opposite behavior. Its maximum strength is near -16V, decreases with increasing voltage and has completely disappeared at 8V. In summary, the features of the low-temperature spectra of the GaAs sample essentially remained with little change. Of course, due to the higher temperature, the second state is thermally populated for a larger range of applied biases and the (2-3) and (2-4) transitions have a larger strength.

Because phase-breaking mechanisms increased with temperature, it is remarkable that this interference effect survives at T=225K. We attribute the robustness of this effect to the large transmission probability of the barrier and the small (190Å) overall size of the structure which yields a energy splitting at resonance of about 10meV, still a sizable fraction of kT=20meV.

7. Conclusion

We have demonstrated a new quantum interference effect in optical absorption which results in the electric field induced suppression of the latter. The large modulation of the transition matrix element was shown to survive even at T = 225K and could be useful in modulator applications.

It is a pleasure to acknowledge many stimulating discussions with C. Sirtori.

REFERENCES

1. U. Fano, Phys. Rev. **124**, 1866 (1961)
2. R. B. Stewart and G. J. Diebold, Phys. Rev. A **34**, 2547 (1986)
3. M. Class-Manjean, H. Frohlich and J. A. Beskick, Phys. Rev. Lett. **61**, 157 (1988)
4. S. E. Harris, Phys. Rev. Lett. **62**, 1033 (1989)
5. M. O. Scully, Phys. Rev. Lett. **67**, 1855 (1991)
6. A. Y. Cho, J. Cryst. Growth **111**, 1 (1991)
7. F. Capasso, MRS Bull. **16**, 23 (1991)
8. For a recent review see Physics Today **46**, No. 6 (1993)
9. D. Heitmann and J.P. Kotthaus, Physics Today **46** (No. 6), 56 (1993)
10. C. Sirtori, F. Capasso, D. L. Sivco, and A. Y. Cho, Phys. Rev. Lett. **68**, 1010 (1992)
11. C. Sirtori, F. Capasso, D. L. Sivco, and A. Y. Cho, Appl. Phys. Lett. **60**, 2678 (1992)
12. P. Yuh and K. L. Wang, Phys. Rev. B **38**, 8377 (1988)
13. K. K. Choi, B. F. Levine, C. G. Bethea, J. Walker, and R. J. Malik, Phys. Rev. B **39**, 8029 (1989)
14. W. Chen and T. G. Andersson, Semicond. Sci. Technol. **7** 828 (1992)
15. L. Pfeiffer, E. F. Schubert, and K. W. West, Appl. Phys. Lett. **58**, 2258 (1991)
16. G. Bastard, *Wave Mechanics Applied to Semiconductor Heterostructures* (les Editions de Physiques, Paris, 1990).
17. Nonparabolicities were taken into account using the method of D. F. Nelson, R. C. Miller, and D. A. Kleinmann, Phys. Rev. Rev. B **35**, 7770 (1987).
18. C. Sirtori, F. Capasso, D. L. Sivco, S. N. G. Chu, and A. Y. Cho, Appl. Phys. Lett. **59**, 2302 (1991)
19. N. Vodjdani, B. Vinter, V. Berger, E. Bockenhoff, and E. Costar, Appl. Phys. Lett. **59**, 555 (1991)

Phase Retardation and Induced Birefringence Related to Intersubband Transitions in Multiple Quantum Well Structures

A. Sa'ar and D. Kaufman

Division of Applied Physics
The Fredi and Nadin Herman School of Applied Science
The Hebrew University of Jerusalem
Givat-Ram, Jerusalem 91904, Israel

ABSTRACT

It is well known that the selection rules for intersubband transitions in quantum well structures require that the infrared light should be polarized parallel to the growth direction. As a result the induced intersubband susceptibility tensor becomes highly anisotropic and the crystal becomes birefringent. We have studied the effect of induced birefringence at the mid-infrared range of the spectrum using a number of experimental techniques, including FTIR absorption spectroscopy and optical phase retardation measurements using a tunable CO_2 laser and a cross polarizer set-up. We have observed that a linearly polarized light become almost circularely polarized due to optical phase retardation between the ordinary and the extraordinary components of the optical field over a short path-length of the order of 20 µm near the resonance (but not at the resonance). The real and the imaginary parts of the induced extra-ordinary refractive index were measured and have been found to be of the same order of magnitude. We also show that a solution of the Fresnel equation modified to take into account both the imaginary and the real parts of the susceptibility tensor, is in good agreement with our experimental results.

321

H. C. Liu et al. (eds.), Quantum Well Intersubband Transition Physics and Devices, 321–330.
© 1994 *Kluwer Academic Publishers. Printed in the Netherlands.*

1. INTRODUCTION

Intersubband transitions (ISBT) in semiconductor quantum well (QW) structures have recently attracted much attention due to the novel optical properties [1-2] and the potential applications of these structures for new class of infrared (IR) optical devices. Based on these transitions a variety of applications that make use of these transitions have been proposed and demonstrated such as long wavelength infrared detectors [3], optical rectifiers [4], harmonic generators [5-7], Kerr modulators [8], and others.

Most studies of the optical properties associated with intersubband transitions in QW's have ignored the anisotropy induced in the crystal by these transitions. The ISBT selection rules [1,9] require that the infrared light should have a component of the polarization vector parallel to the growth direction (the principal axis of the crystal). As a result the ISBT susceptibility tensor becomes anisotropic and the crystal becomes birefringent. The induced anisotropy is related to both the imaginary part of the susceptibility tensor, that is proportional to the absorption coefficient, as well as to the real part of the susceptibility tensor that gives rise to induced phase retardation between the ordinary (which is polarized perpendicular to the growth direction) and the extraordinary components of the light .

The purpose of the present work is to study this *linear* optical effect of induced birefringence by ISBT. As will be clarified below the large dipole matrix elements and the resonant nature of the ISBT give rise to a very large birefringence effect which we have measured to be of the order of $\Delta n / n \approx 4 \cdot 10^{-2}$. For comparison, the index anisotropy due to the inter-band quantum confined Stark effect in GaAs QW's has been measured to be of the order of $\Delta n / n \approx 10^{-3}$ [10]. Furthermore, since one can control by external means (i.e. using external bias voltage) the energy separation between the subbands of the well, it is possible to develop an efficient IR optical phase modulator [11], which is based on this effect, that is predicted to have an extremely high modulation rate.

2. LIGHT PROPAGATION IN AN ANISOTROPIC ABSORBING MEDIUM

We begin our analysis by noting that the principal axes of our *uniaxial crystal* are defined by the growth direction of the epilayers. Hence, in the principal axes coordinate system the only non-vanishing component of the susceptibility tensor due to linear ISBT effects (with respect to the optical field) is given by:

$$\chi_{xx}^{ISBT}(\omega) = \frac{Ne^2}{\varepsilon_0} \frac{\langle 1|x|2\rangle\langle 2|x|1\rangle}{(\hbar\omega - \Delta E_{12} - i\hbar\Gamma_{12})} = \chi' + i\chi'' \tag{1}$$

323

where x is the growth direction of the epilayers (see Fig. 1), $\Delta E_{ij} = E_i - E_j$ is the energy separation between subbands i and j of the conduction band QW, $\langle j|x|i \rangle$ is the ISBT exchange dipole matrix element of the $i \to j$ transition, $\hbar\Gamma_{ij}$ is the full width at half maximum associated with the $i \to j$ transition, and χ' and χ'' are the real and the imaginary parts of χ_{xx}^{ISBT} respectively. In eq. (1) we have assumed that only the ground subband is populated (where N is the electron density per unit volume in the QW), and, that the only contribution to the susceptibility comes from the $1 \to 2$ transition which is close to resonance with our laser frequency ω.

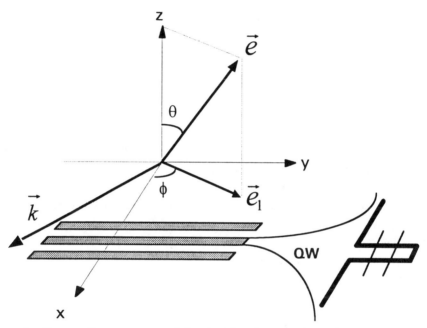

Figure 1. The coordinate system used for obtaining the Fresnel equation and the ordinary and the extra-ordinary solutions. $\vec{k}$ is the wave vector of the IR beam and $\vec{e}$ is a polarization unit vector. $\vec{e}_1$ is the extra-ordinary component of the polarization vector.

If one neglects intra-sub-band and free carrier absorption, only diagonal terms contribute to the dielectric tensor of the structure that is given by:

$$\hat{\varepsilon} = \begin{bmatrix} \varepsilon_{xx} & 0 & 0 \\ 0 & \varepsilon_b & 0 \\ 0 & 0 & \varepsilon_b \end{bmatrix} \tag{2}$$

with $\varepsilon_{xx} = \varepsilon_b + \varepsilon_0(\chi' + i\chi'') = \varepsilon' + i\varepsilon''$ and ε_b is the isotropic dielectric constant of the bulk semiconductor. The generalized refractive index of the structure (which is a complex number for an anisotropic absorbing medium) can be found now for given propagation and polarization directions by solving the Fresnel equation [12]. Two types of solutions can be easily found: (a) ordinary solution for which the light is polarized in the z-direction (i.e. $e_x = e_y = 0$, $e_z \neq 0$, where ej are the projections of the polarization vector in the principal coordinate system) and $n_0^2 = \varepsilon_b$ where n_o is the ordinary refractive index of the bulk crystal; (b) extraordinary solution for light polarized in the x-y plan (see Fig.1) that satisfies the following equation:

$$\frac{1}{\tilde{n}_{ex}^{\,2}(\phi)} = \frac{\cos^2\phi}{\varepsilon_{xx}} + \frac{\sin^2\phi}{\varepsilon_b} \tag{3}$$

where ϕ is the angle between the extra-ordinary polarization wave vector ($\vec{e}_1$) and the x-coordinate (the QW growth direction). Notice that the only difference between eq. (3) and the usual solution of the Fresnel equation for a non-absorbing medium is that the *generalized* extra-ordinary refractive index, $\tilde{n}_{ex}(\phi)$, has real (n_{ex}) and imaginary (κ_{ex}) parts that are related to the extra-ordinary (real) refractive index and the extinction coefficient of the medium respectively (i.e. $\tilde{n}_{ex} = n_{ex} + i\kappa_{ex}$). A comparison of the real and the imaginary parts of eq. (3) yields the following:

$$\frac{n_{ex}^2 - \kappa_{ex}^2}{(n_{ex}^2 + \kappa_{ex}^2)^2} = \frac{\varepsilon'\cos^2\phi}{\varepsilon'^2 + \varepsilon''^2} + \frac{\sin^2\phi}{\varepsilon_b} \tag{4a}$$

$$\frac{2n_{ex}\kappa_{ex}}{(n_{ex}^2 + \kappa_{ex}^2)^2} = \frac{\varepsilon''\cos^2\phi}{\varepsilon'^2 + \varepsilon''^2} \tag{4b}$$

where $\varepsilon' = \varepsilon_b + \varepsilon_0\chi'$ and $\varepsilon'' = \varepsilon_0\chi''$ are the real and the imaginary parts of ε_{xx}. Generally these two equations should be solved self-consistently where n_{ex} and κ_{ex} are complex functions of χ' and χ''. However, in most practical cases we have $\Delta\varepsilon', \Delta\varepsilon'' << \varepsilon_b$ and $\Delta n_{ex}, \kappa_{ex} << n_o$, where we have defined $n_{ex}(\phi) = n_o + \Delta n_{ex}(\phi)$. Hence, one can use first order perturbation theory to derive the following expressions for the extra-ordinary refractive index change and the extinction coefficient of the medium:

$$2\frac{\Delta n_{ex}}{n_o} = \frac{\varepsilon_0}{\varepsilon_b}\chi'\cos^2\phi \tag{5a}$$

$$2\frac{\kappa_{ex}}{n_o} = \frac{\varepsilon_0}{\varepsilon_b}\chi'' \cos^2\phi \qquad (5b)$$

It should be pointed out that both the extinction coefficient and the refractive index change due to ISBT vary as $\cos^2\phi$. In some of the previously published works on ISBT this additional $\cos^2\phi$ factor in the extinction coefficient has been ignored. For example, in many experimental situations the absorption coefficient due to ISBT has been measured in a waveguide geometry [3] where $\phi = \pi/4$. Hence, the measured value of χ'' should be larger by a factor of 2 as compared to the previously reported values.

3. EXPERIMENTS

The sample used in this study for the measurement of the ISBT phase retardation was grown on a semi-insulating GaAs substrate by molecular beam epitaxy. It consists of 25 periods of 90 $\overset{\circ}{A}$ GaAs QW's sandwiched between 275 $\overset{\circ}{A}$ $Al_{0.4}Ga_{0.6}As$ barriers. The central 25 $\overset{\circ}{A}$ of the barriers were Si-doped to $5 \cdot 10^{18}$ cm^{-3}, and the MQW section was clad between two 0.5 µm thick GaAs layers. The sample was designed to have an energy separation of about 115 meV between the first (ε_1) and the second (ε_2) quantized levels so that the transition wavelength of about 10.8 µm will fit to the spectral range of our tunable CO_2 laser. The third confined level in our QW does not play any role in our experiment since the $\varepsilon_1 \rightarrow \varepsilon_3$ transition is not allowed (in the dipole approximation [9]) in our symmetrical QW. The wafer was cleaved to a 4 mm long sample and the sample facets were polished at a 45^0 angle. Results from absorption measurements with a Fourier Transform Infrared Spectrometer (FTIR, Perkin-Elmer 2000) using a multipass waveguide geometry are shown in Fig. 2 (solid line). The absorption peak at 10.6 µm is related to $\Delta E_{12} \cong 117$ meV with a full width at half maximum of 13 meV at room temperature. This value is in reasonable agreement with the calculated energy levels of our QW structure.

326

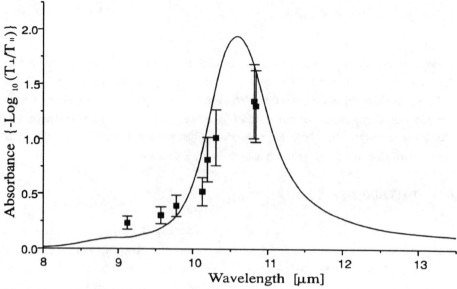

Figure 2.   FTIR intersubband absorbance spectrum of the sample at room temperature (solid line). The square points are the measured values by the tunable CO_2 laser.

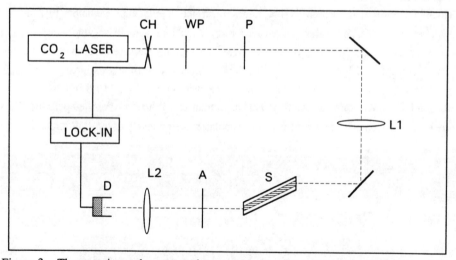

Figure 3:   The experimental setup used to measure the induced phase retardation and the absorbance of the samples with a tunable CO_2 laser. WP: quarter wave plate, CH: chopper, P: polarizer, L1, L2: lenses, A: analyzer, D: pyroelectric detector.

For the measurement of the phase retardation we have used the following experimental set-up, schematically shown in Fig. 3. A linearly polarized IR radiation from a tunable CO_2 laser (Edinburgh, model PL4) passes through a $\lambda/4$ wave plate that converts the radiation into a circularely polarized beam. The circularely polarized beam passes through a high-extinction polarizer to give a variable incident polarization angle, θ, between the z-principal axis of the structure and the polarization vector. The CO_2 laser beam enters the sample at normal incidence through the polished edge and experiences multiple total internal reflections from the boundaries that increase the total optical path length through the MQW section. The optical field at the output edge of the sample can be described as a superposition of an ordinary component that experiences neither attenuation nor phase retardation and an extra-ordinary component that experiences both attenuation $(A(\phi)/2)$ and phase retardation (relative to the ordinary component) $\Gamma(\phi)$ given by:

$$A(\phi)/2 = \kappa_{ex}(\phi)k_0 fL / \cos\phi = f \frac{k_0 n_0 L}{2} \frac{\varepsilon_0}{\varepsilon_b} \chi'' \cos\phi \qquad (6a)$$

$$\Gamma(\phi) = \Delta n_{ex}(\phi)k_0 fL / \cos\phi = f \frac{k_0 n_0 L}{2} \frac{\varepsilon_0}{\varepsilon_b} \chi' \cos\phi \qquad (6b)$$

where f and L are the filling factor and the length of the sample respectively, and k_0 is the vacuum wave-vector of the IR beam. Hence, the output field is ellipticaly polarized due to the optical retardation in our sample with an additional effect of attenuation of the extra-ordinary component of the optical field that changes the ratio between the principal axes of the polarization ellipse. The analyzer at the output of our experimental set-up is used to map the exact profile of the polarization ellipse so that the phase retardation (given by Eq. 6b) can be experimentally determined. For the analysis of the experimental data it is convenient to use the *degree of polarization* parameter [14] that is defined as follows:

$$P(\theta) = \frac{I_{max} - I_{min}}{I_{max} + I_{min}} \qquad (7)$$

where I_{max} (I_{min}) is the maximum (minimum) output optical power for a given incident polarization angle. Notice that $P=1$ for a linearly polarized light and $P=0$ for a circularly polarized light. A simple analysis shows that $P^2(\theta)$ is directly proportional to $sin^2\Gamma$. Hence, by the measurement of the degree of polarization and the absorbance for a given laser wavelength and for various incident polarization angles, θ, one can deduce the retardation $\Gamma(\phi = \pi/4)$ and $\Delta n_{ex}(\phi = \pi/4)$. In Fig. 4, we plot Γ for a certain wavelengths of our CO_2 laser (shown by the square points). The solid line in this Figure shows the calculated phase retardation. For this calculation we have used the Kramers-Kronig

328

relations [13] to transform the absorbance data of Fig. 2 to induced phase retardation using the following relation (which can be easily derived from the Kramers-Kronig relations):

$$\Gamma(\lambda) = \frac{2\lambda}{\pi} P.V. \int_0^\infty \frac{A(\lambda')}{\lambda^2 - \lambda'^2} d\lambda' \qquad (8)$$

where P.V. stands for the principal value of the integral. The maximum retardation and refractive index change in our experiment has been achieved for $\lambda = 9.8$ μm with $\Gamma(\phi = \pi/4) = 1.5$ Radians. Notice that this large value of retardation has been achieved for an optical path length of the order of 20 μm. Hence, the refractive index change at this wavelength is about $\Delta n_{ex}(\phi = \pi/4) \cong 0.1$. Therefore, according to eq. (3a), it is expected that the maximum refractive index change at $\phi = 0$ will be of the order of 0.2. It should be pointed out that fairly large retardation and refractive index changes are also expected at shorter wavelengths (8-9.5 μm) where the absorption is negligible.

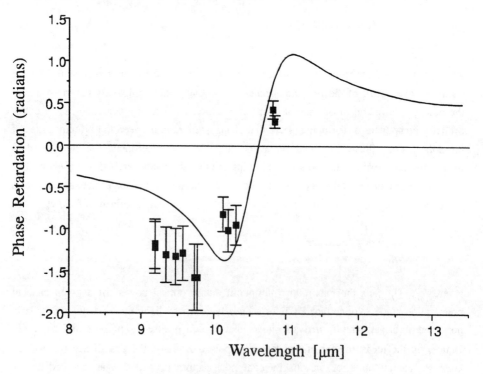

Figure 4. The measured phase retardation and the refractive index change versus laser wavelengths (square points). The solid line shows the calculated phase retardation.

4.   CONCLUSIONS

In conclusion, we have shown that intersubband transitions in multiple quantum well structures induce large anisotropy in the structure that becomes a uniaxial crystal. In addition to the anisotropic ISBT absorption, the birefringence effect causes to a change in the polarization state of the IR beam that, in general, converts a linearly polarized light into an ellipticaly polarized light. The anisotropy induces a phase retardation between the ordinary and the extra-ordinary components of the IR beam which we have measured to be of the order of $\Gamma \approx \pi/2$ for an optical path length as small as 20 μm at $\lambda=9.8$ μm. This effect can be used to make an efficient and small IR phase modulator by using the resonant Stark effect in a properly design structure [11] which is predicted to be very fast.

5.   ACKNOWLEDGMENTS

The MQW samples used in this study were grown by N. Kuze at Caltech, Pasadena CA. We would like to thank Prof. Amnon Yariv for providing us the facilities to grow these samples. This work has been partially supported by grants from the Earnest David Bergmann Science fund, the Zevi Hermann Schapira Research Fund, and by the Intel (Israel) grant for research.

REFERENCES

1.   L.C. West, and S.J. Eglash, *Appl. Phys. Lett.* **46,** 1156 (1985).

2.   A. Harwit and J. S. Harris, Jr., *Appl. Phys. Lett.* **50,** 685 (1987).

3.   B. F. Levine, R. J. Malik, K. K. Choi, C. G. Bethea, D. A. Kleinman, and J. M. Vandenberg, *Appl. Phys. Lett.* **50,** 273 (1987).

4.   E. Rosencher, P. Bois, J. Nagle, E. Costard and S. Delaitre, *Appl. Phys. Lett.* **55,** 1597 (1989).

5.   M .M. Fejer, S.J. Yoo, R.L. Byer, A.Harwit, and J.S. Harris, *Phys. Rev. Lett.* **62,** 1041 (1989)

6.   A. Sa'ar, N. Kuze, J. Feng, I. Grave and A. Yariv, in Proceeding of the NATO Advanced Research Workshop on "Intersubband Transitions in Quantum Wells", Cargese, France, 1991 (NATO ASI Series B: Physics Vol. **288,** Plenum, N.Y., 1992).

7.   E. Rosencher, P. Bois, J. NAgle, and S. Delaitre, *Electron. lett.* **25,** 1063 (1989).

330

8. A. Sa'ar, N. Kuze, J. Feng, I. Grave and A. Yariv, *Appl. Phys. Lett.* **61,** 1263 (1992).

9. A. Sa'ar, *J. Appl. Phys.,.* October 15, 1993, (to be published).

10. J. E. Zucker, I. Bar-Joseph, B. I. Miller, U. Koren, and D. S. Chemla, *Appl. Phys. Lett.* **54,** 10 (1989).

11. E. Dupont, D. Delacourt, V. Berger, N. Vodjdani, and M. Papuchon, *Appl. Phys. Lett.* **62,** 1907 (1993).

12. See for example, A. Yariv, *Quantum Electronics,* 3rd edition, John Wiley, New York, 1989.

13. J. D. Jackson, *Classical Electrodynamics*, 2nd edition, John Wiley, New York, 1975.

14. Born and Wolf, *Priciples of Optics*, 4th edition, Pergamon Press, Oxford, 1970.

THE INTERACTION OF PHOTOEXCITED e–h PAIRS WITH A TWO DIMENSIONAL ELECTRON GAS STUDIED BY INTERSUBBAND SPECTROSCOPY

Y. GARINI, E. COHEN, E. EHRENFREUND, D. GERSHONI,
ARZA RON and E. LINDER
Solid State Institute, Technion-Israel Institute of Technology, Haifa
32000, Israel

L.N. PFEIFFER
AT&T Bell Laboratories, Murray Hill, NJ 07974, USA

K-K. LAW, J.L. MERZ and A.C. GOSSARD
Center for Quantized Electronic Structures (QUEST), University
of California, Santa Barbara, CA 93106, U.S.A.

ABSTRACT. We present a study of the e1-e2 intersubband absorption induced by interband excitation (PIA) in two types of quantum-wells: a. Modulation doped $GaAs/AlGaAs$ MQW's with a density of the two dimensional electron gas (2DEG) in the range of $n_\square = 10^{10} - 10^{12} cm^{-2}$. b. Mixed type I–type II $GaAs/AlAs$ superlattices (MTSL). For the modulation doped MQW's, the PIA is studied as a function of $n_\square$, laser intensity and its chopping frequency. It is observed that when no 2DEG is present, the e1–e2 PIA is due to short lived excitons. For $n_\square \geq 1 \times 10^{10} cm^{-2}$, the PIA decreases with increasing $n_\square$, and for $n_\square \geq 7 \times 10^{10} cm^{-2}$ it vanishes. We interpret the PIA in this density range as due to long lived photoexcited electrons, that in turn result from holes localized at interface fluctuations and have a reduced radiative recombination rate.

In the case of the MTSL's, excitons or separately confined 2DEG and 2DHG are created, depending on the interband excitation energy. This allows us to directly compare the PIA due to excitons and that due to electrons in the same quantum-well. We thus estimate the ratio of the oscillator strengths to be $f_{exciton}/f_{electron} \simeq 10$.

H. C. Liu et al. (eds.), Quantum Well Intersubband Transition Physics and Devices, 331–343.

1. Introduction

The optical properties of doped quantum wells (QW's) strongly depend on the interaction of the photoexcited electron-hole pairs with the two dimensional electron gas (2DEG)[1,2]. This is mainly due to phase space filling and screening[3,4]. Most of the experimental work aimed at studying photoexcitation of doped QW's is based on interband transitions[4,5]. However, the intensity, spectral shape and decay time of the interband transitions vary gradually with $n_\square$ (the electron areal density) in the low 2DEG density range, and therefore, they are not sensitive enough to small variations in $n_\square$.

In this work we studied the photoinduced (e1–e2) intersubband absorption (PIA) in two types of QW's: (a) Modulation doped $GaAs/Al_{0.3}Ga_{0.7}As$ MQW's. (b) Undoped, mixed type I–type II $GaAs/AlAs$ superlattice (MTSL). In the latter case, separately confined two dimensional electron and hole gases (2DEG, 2DHG) are created by photoexcitation. We show that the PIA is strongly dependent on both $n_\square$ and the density of photoexcited e–h pairs, and on the lifetime of the absorbing particles (electrons or excitons). The results are analyzed in terms of a model based on the existence of long lived, photoexcited electrons in a 2DEG, provided that $n_\square \leq 5 \times 10^{10} cm^{-2}$. A comparison between the PIA in doped MQW's and in photoexcited MTSL's allows us to analyze the intersubband transitions in the excitonic and electronic cases, in the very same QW. We thus demonstrate the great sensitivity of the intersubband PIA method to the state of the photoexcited e–h pairs (relative to that observed in interband transitions).

2. Experimental Method

Several $GaAs/Al_{0.3}Ga_{0.7}As$ modulation doped MQW's with different $n_\square$ were studied. They were grown by molecular beam epitaxy on (001)- oriented GaAs substrates. Each MQW consists of 50 GaAs wells of 50 Å nominal thickness, separated by $\simeq 100$ Å thick barriers. Each barrier is Si doped in the middle 40 Å and the densities were in the range of $n_\square = 1 \times 10^{10} - 1 \times 10^{12}$ cm^{-2} (in each well). A schematic description is shown in Fig. 1. The mixed type I–type II superlattice studied was grown in a similar way. Its structure is shown schematically in Fig. 2. It consists of narrow $(L_N = 26\text{Å})$ and wide $(L_W = 68\text{Å})$ $GaAs$ wells that are cladded by $AlAs$ barriers $(L_B = 102\text{Å})$. This structure is symmetric, i.e. every wide well has a neighboring narrow well on each side.

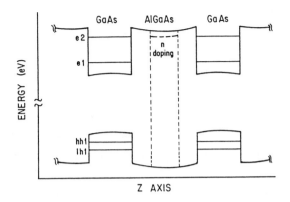

Fig. 1 : The structure of the modulation doped MQW's.

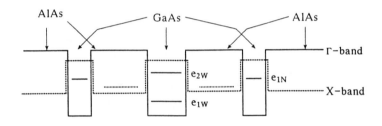

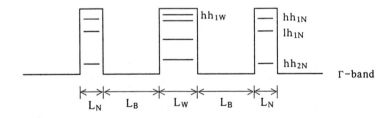

Fig. 2 : The structure of the undoped mixed type I–type II superlattices.

334

For the intersubband spectroscopy, the samples were cut and polished at 45° to the z-axis ($z \parallel [001]$). In this 'waveguide configuration' the infrared radiation had an electric field component along the z-axis. In these experiments, the samples were placed in an immersion type dewar, with an ambient temperature in the range of 2-100K. The laser beam impinged on the sample surface perpendicular to the z-axis. It was chopped by an acousto-optic modulator, in the frequency range of $10^2 < \omega < 3 \times 10^6$ Hz. The photomodulated spectra were obtained using a Fourier transform infrared spectrometer, equipped with a mirror step motor which allows the use of step-and-scan double modulation measurements in this wide dynamic range. A schematic description of the experimental set-up is shown in Fig. 3. The interband excitation was done by irradiating the samples with an Ar^+ laser or with a dye laser tuned resonantly with the e1–hh1 transition.

FTIR BRUKER IFS-66V, RAPID/STEP SCAN

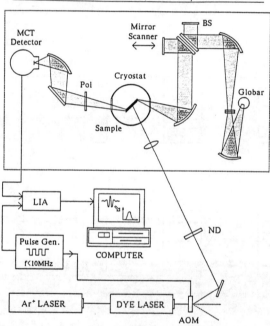

Fig. 3 : A schematic description of the experimental set-up used for photoinduced intersubband absorption. Legend: AOM–acousto-optic modulator, ND–neutral density filter, BS–beam splitter (infrared), Pol–infrared polarizer, LIA–lock-in amplifier.

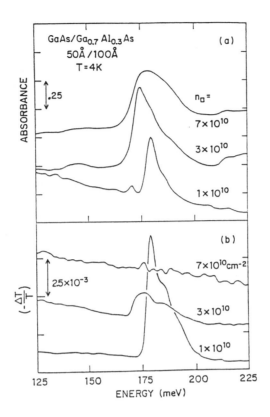

Fig. 4 : *a.* The absorbance of the three MQW's with the lowest doping levels. *b.* The photoinduced absorption of the same samples, excited with laser light: E_L=2.41 eV, I_L=0.64 Wcm^{-2}.

3. Photoinduced Absorption in n-doped MQW's

Fig. 4a shows the absorbance, defined as $-ln(T/T_0)$, for the three MQW's with the lowest doping levels. (T_0 is the transmission through the measuring system). The observed absorption band is due to the (e1–e2) intersubband transition[6], and its integrated intensity is proportional to $n_\square$. More intense (and broader) bands were observed for the MQW's with $n_\square \simeq 10^{11}$ and 10^{12} cm^{-2} (not shown in Fig. 4). The large width of the absorption band (8 mcV for the sample with $n_\square - 1 \times 10^{10}$ cm^{-2}) is probably due to well width fluctuations and/or interaction with Si^+ ions in the barrier. Fig. 4b shows the PIA spectra of the same samples. These were all taken under the same experimental conditions with an excitation (laser) energy of E_L=2.41

eV and intensity of $I_L=0.64$ W/cm^2. It is clearly seen, that the PIA is relatively strong ($[-\Delta T/T]_{max} = 8 \times 10^{-3}$) for $n_\Box = 1 \times 10^{10}$ cm^{-2} and is unobservable for $n_\Box = 7 \times 10^{10}$ cm^{-2}. No PIA signal could be observed at any higher doping level. It should be noted that the direct absorption and PIA of the same MQW have the same spectrum, and are similarly polarized along the growth direction.

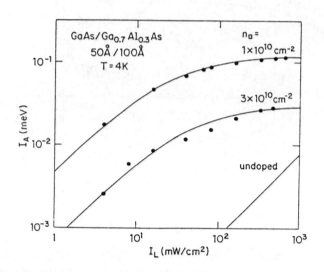

Fig. 5 : The dependence of the integrated photoinduced absorption on the interband laser excitation intensity for the two MQW's with the lowest doping level. The solid curves are the model fitting.

Fig. 5 shows the dependence of the integrated PIA intensity, $I_A = \int(-\Delta T/T)d\varepsilon$, on the exciting laser intensity, I_L, for the MQW's with $n_\Box = 1$ and 3×10^{10} cm^{-2}. The results shown in this figure were obtained with laser excitation at $E_L=2.41$ eV. Similar results were obtained under resonant excitation in the (e1–hh1) band. All the spectra of Fig. 4b and the data points in Fig. 5 were taken with the exciting laser beam modulated at 1.7 kHz. The dependence of I_A on ω, the laser modulation frequency, that was varied over 5 orders of magnitude, is shown in Fig. 6 for the sample with doping density of 1×10^{10} cm^{-2}. (A similar dependence is observed for the $n_\Box=3\times10^{10}cm^{-2}$ MQW).

Before we turn to a quantitative analysis of the PIA, some additional observations should be pointed out. First, we note that the PIA spectrum (observed only for $n_\square < 7 \times 10^{10} cm^{-2}$) has the same shape as that of the direct (e1–e2) absorption spectrum, strongly suggesting that the PIA is due to electrons added to the 2DEG and having the same energy distribution. Secondly, we measured the interband (e1–hh1) photoluminescence and found that its integrated intensity is linearly dependent on the exciting laser intensity, in the same range where the PIA shows a strong saturation (Fig. 5). From these observations we conclude that the PIA is due to a subgroup of the photoexcited electrons, while all of them contribute to the photoluminescence. This idea can be tested by following the phenomenological approach used by Schmitt-Rink et al.[1]. In the case of exciton interband transitions in doped MQW's, they proposed that the exciton generation rate diminishes proportionally to $g(1 - N/N_0)$. Here, g is the photoexcitation generation rate ($g \propto I_L$), N is the exciton density and N_0 is a material dependent "saturation density". We similarly assume that the generation rate for the subgroup of electrons which contribute to the PIA is given by $g(1 - N_{PIA}/N_0)$. We now write the following rate equation for N_{PIA},

$$\frac{dN_{PIA}}{dt} = g(1 - \frac{N_{PIA}}{N_0}) - \frac{N_{PIA}}{\tau} . \tag{1}$$

Its steady state solution is :

$$N_{PIA}^{ss} = \frac{g\tau N_0}{g\tau + N_0} , \tag{2}$$

and this gives $N_{sat} = N_0$. The experimentally measured I_A is directly proportional to N_{PIA}^{ss}: $I_A = I^0 N_{PIA}^{ss}$ where I^0 contains various constants. In our thick samples, cut in the waveguide configuration, we took into account the number of passes of the infrared beam and the gradual decrease of the interband excitation along the MQW structure. Then the solid lines in Fig. 5 are the calculated I_A vs I_L. The parameters obtained by fitting this model to the data points are: $N_0 = 1.2$ and $0.36 \times 10^{10} cm^{-2}$ and $\tau = 12$ and 1.8 μsec for the $n_\square = 1 \times$ and $3 \times 10^{10} cm^{-2}$ MQW's, respectively. We note that the saturation density and electron lifetime decrease with $n_\square$. For comparison we show in Fig. 5 the linear dependence of I_A on the laser intensity, as measured for undoped MQW's.[7] This shows that the PIA in the doped MQW's is clearly due to a different mechanism than in undoped MQW's (long lived electrons vs. short lived excitons). The long lifetime τ of the subgroup of electrons that give rise to the PIA can be obtained from the measured dependence on laser chopping frequency (Fig. 6). We use the same rate equation, eq. 1, but with a modulated generation rate:

$$g(t) = g_0(1 + cos(\omega t))/2 \; . \tag{3}$$

Eq. (1) is then solved by expanding $N_{PIA}(t)$ into a Fourier series and retaining only the ω-component, $N_{PIA}(\omega)$. The solid curve in Fig. 6 is the fit to the experimental points with N_0 and τ taken as the values obtained above. It is seen that this fit describes fairly well the experimental data, and, in particular, it yields the abrupt drop in $N_{PIA}(\omega)$ in the range of $\omega \sim \tau^{-1}$. The deviations of the model from the observed $I_A(\omega)$ indicate that there is a distribution of lifetimes (τ) which determines the frequency dependence.

We have thus shown that the intersubband PIA of a low density 2DEG can be explained by the presence of long lived photoexcited electrons. The question now is, how can such electrons exist, when photoexcited holes are available for radiative recombination ? We propose the following model: a subgroup of the photoexcited hole population consists of localized holes, whose wavefunctions weakly overlap with the electron wavefunctions. It is implicitly assumed that the hole localizing centers do not localize electrons. A finite density of such localized holes results in a saturable density of long lived electrons. Moreover, as $n_\square$ increases, the inter-electron distance decreases to such a degree that the e-h recombination rate is greatly increased. Then the PIA is too weak to be observed.

We have tested this idea of hole localization by using Bastard's model[8] of particle localization by interface potential fluctuations. Assuming that the localization center has a lateral radius of ~ 200 Å and depth of 170 meV (equal to the well-barrier valence band offset), we obtain a single bound state with binding energy of 3.5 meV and a wavefunction that decreases exponentially with an in-plane radius of ~ 100 Å. Then we calculate the radiative recombination rate of such a bound hole relative to that of free hole. We find that the bound hole recombines at a rate $\sim 10^3$ times slower then the free hole.

In summary of this section, we have experimentally demonstrated that the (e1-e2) intersubband transition is a very sensitive means to study the properties of a photoexcited e-h pair in a 2DEG. While the majority of these pairs recombine radiatively at a fast rate ($\sim 10^{10}$ sec^{-1}), a subgroup of the photoexcited electrons are very long lived (μsec range). This, we propose, is due to a subgroup of holes that are localized on interface fluctuations and have a reduced wavefunction overlap with the electrons. The observed saturation of the PIA intensity with increasing interband excitation intensity, is due to the finite density of hole localizing sites.

4. Photoinduced Absorption in Mixed Type I–Type II Superlattices

The electronic properties of $GaAs/AlAs$ superlattices (SL) depend on whether the Γ conduction band (cb) minimum in the $AlAs$ is lowest [9]. As the width of the $GaAs$ layer is reduced, the direct (e1–hh1) bandgap is pushed higher in energy, and for widths smaller than 35 Å the (X–hh1) bandgap becomes lower. This bandgap is indirect in both real and reciprocal spaces. Galbraith et al[10] have recently introduced the mixed type I–type II $GaAs/AlAs$ superlattice. In a MTSL such as shown in Fig. 2, narrow $GaAs$ layers with $L_N < 35$ Å are separated by $AlAs$ barriers from wide $GaAs$ layers, with $L_W > 35$ Å. Galbraith et al[10] studied the absorption spectrum of such a MTSL, as a function of photoexcitation intensity, when the laser energy is tuned above the narrow well bandgap. They showed that a two-dimensional electron gas (2DEG) is formed in the wide well, and the absorption spectrum is strongly modified as its density increases. The lifetime of the 2DEG is determined by the rate of hole tunneling from the narrow wells, and this can be very long ($\simeq 10^{-6}$ sec for $L_B \simeq 100$ Å) in comparison with the recombination time of the 2DEG with holes generated in the wide well ($\simeq 10^{-9}$ sec)[11].

The MTSL provides us with a system in which the (e1–e2) intersubband transition can be studied under two photoexcitation conditions. (a) For interband excitation with a laser energy $E(e1W) - E(hh1W) < E_L^{(1)} < E(e1N) - E(hh1N)$, only excitons are created in the wide wells. Then, the PIA is due to the excitonic intersubband transition (e1:hh1)$_W$–(e2:hh1)$_W$, similar to that observed in undoped MQW's[7]. Fig. 7a shows this transition in the MTSL described in Sec. 2. (b) For interband excitation with $E_L^{(2)} > E(e1N) - E(hh1N)$ a 2DEG is formed in the wide well and a 2DHG in the narrow well. The PIA spectrum obtained under such excitation conditions is shown in Fig. 7b. The lowest energy band of Fig. 7b is very similar in energy and spectral shape to the only band of Fig. 7a. We studied the PIA intensity of these bands as a function of the exciting laser chopping frequency ω (similar to that shown in Fig. 6). For the case of excitation energy $E_L^{(1)}$, the PIA intensity is independent of ω up to the limit of our measuring system (3 MHz). For $E_L^{(2)}$, however, the two bands of Fig. 7b have the same dependence on ω, and the PIA intensity drops sharply at $\omega \simeq 80$ kHz.

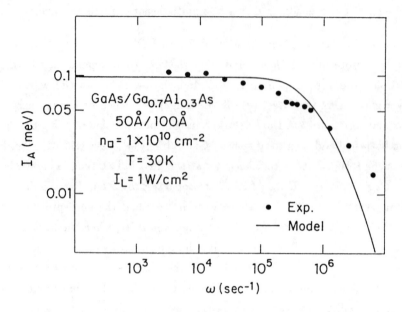

Fig. 6 : The dependence of the integrated photoinduced absorption on the chopping frequency of the exciting laser beam. The solid curve is the model fitting.

In order to determine the origin of the observed PIA bands, we have calculated the dispersion curves of the confined conduction and valence subbands using the method of Baraff and Gershoni[12]. The calculated transition energy $E(e2W) - E(e1W) = 220$ meV is exactly equal to the peak energy of the lowest PIA bands in Figs. 7a,b. Therefore, these bands are due to the e1–e2 intersubband transition within the wide well. However, the distinct dependence on ω when the PIA is excited with $E_L^{(1)}$ or $E_L^{(2)}$ indicates that the particles undergoing the intersubband transitions are different. For excitation at $E_L^{(1)}$ these are <u>excitons</u>, namely the PIA is due to the $(e1:hh1)_W$– $(e2:hh1)_W$ transition. For excitation at $E_L^{(2)}$ the particles are long lived electrons and the PIA is due to the $(e1–e2)_W$ intersubband transition. An important deduction can be made by comparing the integrated intersubband absorbance for the two excitation energies. It is given by[7]:

$$I_A = CGg\tau f \ , \tag{4}$$

where C is a factor lumping together several constants and G is a geometrical factor. These two factors are identical for both excitation energies. g is the generation rate (e–h pairs per unit time and unit area of a quantum-well), τ is the lifetime of the excitons or the electrons and f is the respective oscillator strength.

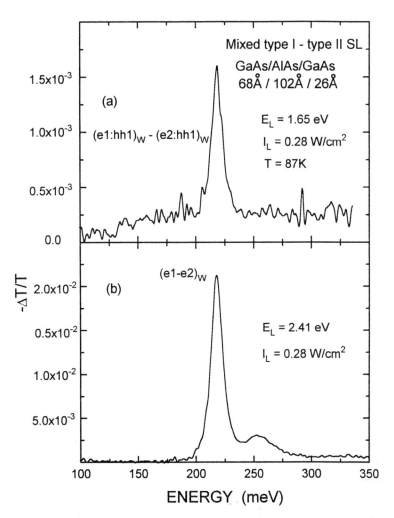

Fig. 7 : Photoinduced absorption spectra of the mixed type I–type II $GaAs/AlAs$ superlattice. a. Excitation energy $E_L^{(1)} < E(e1N) - E(hh1N)$, namely, e–h pairs are generated only in the wide wells. b. $E_L^{(2)} < E(e1N) - E(hh1N)$. A 2DEG is formed in the wide well and a 2DHG in the narrow well.

The integrated PIA is measured from data such as that shown in Figs. 7a,b. We estimate the generation rate by assuming a quantum efficiency of unity and the fact that at both $E_L^{(1)}$ and $E_L^{(2)}$ only the $GaAs$ layers of the MTSL absorb the light. The lifetime of the photoexcited electrons (for $E_L^{(2)}$) is measured to be $\tau_e \simeq 2 \times 10^{-6}$ sec. For the case of the excitons, our ω dependence shows that the lifetime must be shorter than 25 nsec. Using the published values of the photoluminescence decay [13] we have $\tau_{ex} \simeq 3 \times 10^{-9}$ sec (at 90 K). From these we obtain the ratio of the oscillator strengths for the excitonic and electronic e1–e2 intersubband transition:

$$\frac{f_{exciton}}{f_{electron}} \simeq 10 . \tag{5}$$

This result is consistent with our previous finding [7] of a larger oscillator strength of the excitonic intersubband transition than that of the electronic transition. The present study is important since the comparison between the two cases is done in the very same quantum structure.

5. Summary

In this study we demonstrated that the photoinduced (e1–e2) intersubband transition is a very sensitive probe of the e–h state in the presence of a 2DEG. In the case of modulation doped MQW's, the e1–e2 PIA is observed only when the 2DEG density is in the range of $1 \times 10^{10} \leq n_\square \leq 3 \times 10^{10} cm^{-1}$, and its intensity is non-linearly dependent on the exciting (laser) intensity. We show that these observations can be well explained by a phenomenological model based on the existence of long-lived photoexcited electrons. These, in turn, are due to the fact that some photoexcited holes are localized and have a small radiative recombination rate with the 2DEG.

In the case of mixed type I–type II SL's, the PIA is due either to an excitonic or electronic (e1–e2) intersubband transition, depending on the energy of the exciting laser light. This provides us with a direct way to compare between the oscillator strength of these transitions, observed in the very same quantum structure.

Acknowledgments   The work at the Technion was supported by the US-Israel Binational Science Foundation (BSF), Jerusalem, Israel, and was carried out in the Center for Advanced Opto-electronics. The work at UCSB was supported by the National Science Foundation Science and Technology Center, QUEST.

References

1. S. Schmitt-Rink, D.S. Chemla and D.A.B. Miller, Adv. Phys. **38**, 89 (1989).

2. G.D. Sanders and Y-C. Chang, Phys. Rev. **B 35**, 1300 (1987).

3. G. Livescu, D.A.B. Miller, D.S. Chemla, M. Ramaswamy, T.Y. Chang, N. Saver, A.C. Gossard and J.H. English, IEEE Journal of Quantum Electronics, **24**, 1677 (1988).

4. D. Huang, H.Y. Chu, Y.C. Chang, R. Roudre and H. Morkoc, Phys. Rev. **B 38**, 1246 (1988).

5. H.W. Liu, C. Delalande, G. Bastard, M. Voos, G. Peter, R. Fischer, E.O. Gobel, J.A. Brum, G. Weimann and W. Schlapp, Phys. Rev. **B 45**, 8464 (1992).

6. L.C. West, S.J. Eglash, Appl. Phys. Lett. **46**, 1156 (1985).

7. M. Olszakier, E. Ehrenfreund, E. Cohen, J. Bajaj and G.J. Sullivan, Phys. Rev. Lett. **62**, 2997 (1989).

8. G. Bastared, "Wave Mechanics Applied to Semiconductor Heterostructures", (Les editions de physique, Les Ulis, 1988), Chap. 4.

9. R. Cingolani, K. Ploog, G. Scamarcio and L. Tapfer, Opt. and Quan. Elect. **22**, S201 (1990).

10. L. Galbraith, P. Dawson and C.T. Foxon, Phys. Rev. **B 45**, 13499 (1992).

11. C.D. Delalande, G. Bastard, J. Orgonasi, J.A. Brum, H.W. Liu, M. Voos, G. Weimann and W. Schlapp, Phys. Rev. Lett. **59**, 2690 (1987).

12. G.A. Baraff and D. Gershoni, Phys. Rev. **B 43**, 4011 (1991).

13. M. Gurioli, A. Vinattieri, M. Colocci, C. Deparis, J. Massies, G. Neu, A. Bosac chi and S. Franchi, Phys. Rev. **B 44**, 3115 (1991).

PHOTOINDUCED INTERSUBBAND TRANSITIONS IN GaAs/AlGaAs ASYMMETRIC COUPLED QUANTUM WELLS.

F. H. JULIEN, P. VAGOS, P. BOUCAUD, L. WU
Institut d'Electronique Fondamentale, URA CNRS 22
Bât. 220, Université Paris XI
91405 Orsay, France.

and R. PLANEL
L2M-CNRS, 196 Av. H. Ravera
92220 Bagneux, France.

ABSTRACT. We have investigated undoped asymmetric coupled quantum well structures. We demonstrate that the photoinduced intersubband absorption results in a finite non-thermal occupation of the E_2 subband. The intersubband absorption of infrared photons is shown to induce a near-visible $E_2 \to HH_1$ photoluminescence which closely follows the polarization selection rules and the spectral dependence of the intersubband absorption. The dependence of the induced luminescence on the pump intensity is then used to estimate the intersubband relaxation time. The intersubband relaxation is found to be dominated by electron-phonon interactions involving interface LO-phonon modes. Application of this novel up-conversion process for infrared photodetection is discussed.

1. Introduction

Observation of intersubband absorption in quantum wells requires the ground subband to be populated. For applications necessiting strong intersubband absorptions, epitaxial doping of the quantum well layers is widely used since large carrier densities can be achieved in III-V semiconductors. In addition, the choice of the doping impurity, either n-type or p-type, allows intersubband transitions in the conduction or the valence band to be studied [1,2]. Significant occupation of the conduction subbands has also been reported in resonant tunneling structures [3-5]. An alternate technique relies on optical pumping of the interband transitions of the quantum wells to generate photo-carriers in the conduction and the valence band. Because the intersubband and intrasubband relaxation times are usually much shorter than the band-to-band recombination times, the photocarriers accumulate in the ground conduction and valence subbands of the quantum wells. Intersubband absorption can then take place under proper infrared excitation. This photoinduced intersubband absorption (PIA) process was first demonstrated in undoped GaAs/AlGaAs quantum wells by Olszakier et al [6] with special emphasis on quantum well spectroscopy. Indeed, the PIA signal which is proportional to

345

H. C. Liu et al. (eds.), Quantum Well Intersubband Transition Physics and Devices, 345–359.
© 1994 *Kluwer Academic Publishers. Printed in the Netherlands.*

the photo-carrier density in the ground subband, reflects the specific properties of the quantum well structure under study, such as the resonance energies, the joint density of states, the interband and intersubband optical matrix elements or the carrier relaxation times. The transmission measurement of an infrared probe then gives valuable informations on the intersubband and the interband transitions of the quantum wells. The associated absorption spectra are obtained by either varying the probe or the pump photon energy. This spectroscopic pump-probe technique has been successfully applied to undoped and doped GaAs/AlGaAs [6-13], InGaAs/InP [14,15] and InGaP/GaAs [16] quantum wells .

The PIA process can also be applied to the relaxation spectroscopy of quantum wells. Indeed, the photo-carrier density is strongly dependent on the band-to-band relaxation mecanisms. It was shown that when the interband recombination is dominated by mono-molecular processes, such as relaxations proceeding through cristal defects or exciton radiative recombinations, the PIA varies linearly with the pump intensity while the frequency response of the photo-modulated absorption exhibits no variation. In contrast, for dominating bi-molecular recombinations, both the amplitude and the frequency bandwidth of the PIA follow a square-root like dependence on pump intensity [17]. In the experiments, a modulated AlGaAs laser diode is used to photo-pump the quantum wells and the PIA and its frequency response are recorded as a function of the laser diode power. This simple technique is attractive since both the mono- and bi-molecular recombination times can be determined without requiring time-resolved measurements.

The possibility of light-controlled-by-light devices offered by the PIA process has also been explored. Efficient all-optical modulation of the infrared transmission of GaAs/AlGaAs quantum wells at 10 μm was demonstrated at room temperature in a waveguide structure optically pumped by an AlGaAs laser diode [18,19]. Extinction ratios as high as 150:1 are obtained for a laser diode power of only 40 mW. Under on-off modulation of the pump beam, the device operates like an optical shutter for input infrared powers as high as 0.5 W without cooling requirements. The frequency bandwidth of the modulator, of the order of 30 MHz, is limitated by the interband recombination times in GaAs and can be further extended to 200 MHz by means of low-dose proton implantation of the quantum wells [20]. In another realization, the all-optical infrared modulation was investigated in biased double quantum wells [21]. The PIA was found to be strongly enhanced under application of a dc electric field resulting in significative infrared modulation at Brewster's angle incidence. Indeed, the charge separation due to opposite tunneling of electrons and holes between the two biased wells results in an increased electron density in the ground conduction subband because of the increased electron-hole recombination time. Another interesting application of the PIA process in biased double quantum wells is the realization of an optically controlled infrared detector [22].

In this paper the PIA technique is used to investigate asymmetric coupled quantum well (ACQW) structures. Under band-to-band optical excitation, the resonant excitation of the $E_1 \rightarrow E_2$ intersubband transition of the quantum wells is shown to result in a finite non-thermal occupation of the E_2 subband which gives rise to an $E_2 \rightarrow HH_1$ photoluminescence. The induced luminescence closely follows the polarization selection rules and the spectral dependence of the intersubband absorption. By analyzing the dependence of the induced luminescence on the pump intensity, insight can be gained into the intersubband scattering processes [23]. Unlike for single quantum well structures, the intersubband relaxation in the coupled quantum well structure is found to be dominated by electron-phonon interactions involving interface LO-phonon modes [24]. Since the absorption of infrared photons translates into the emission of near-visible photons, the induced luminescence may be seen as an up-conversion process. The applications of this new mechanism for unconventional infrared photodetectors are discussed.

2. The asymmetric coupled quantum well structure

The principle of the photoinduced absorption experiments in the asymmetric coupled quantum well structure is illustrated in figure 1. Near-infrared optical pumping of the interband transition HH_1E_1 is used to generate electron-hole pairs in the ground conduction and valence subbands of the quantum wells. Under resonant excitation of the intersubband transition E_1E_2, electrons are excited in the E_2 subband. Since the heavy-hole subband HH_1 is populated, one expects the occupation of the E_2 subband to result in some induced luminescence E_2HH_1. In turn, the induced luminescence may be used to probe the occupation of the excited subband. The induced emission should manifest as a peak blue-shifted by the intersubband energy from the main luminescence line E_1HH_1.

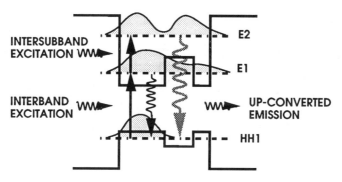

Figure 1: Photoluminescence up-conversion in asymmetric coupled quantum wells.

Since the oscillator strength of the E_2HH_1 transition is proportionnal to the square of the overlap integral between the envelope wavefunctions of subbands E_2 and HH_1, the E_2HH_1 transition is forbidden in symmetric quantum wells for parity reasons. The induced luminescence can then only be observed if some asymmetry is introduced in the structure. Different asymmetric structures including biased quantum wells, step quantum wells and coupled quantum wells have been systematically investigated within the envelope function formalism, based on the following requirements: an intersubband resonance energy compatible with excitation by a CO_2 laser, a large intersubband dipole moment and a large envelope wavefunction overlap between subbands E_2 and HH_1. The two latter conditions are somewhat contradictory. A satisfactory compromise could only be found in asymmetric coupled quantum wells of the type illustrated in figure 1, mainly

348

because the small barrier between the wells allows to separately tailor the heavy-hole and electron wavefunction extensions.

An optimized structure was grown by molecular beam epitaxy on a semi-insulating GaAs substrate. The non-intentionally-doped structure comprises 100 asymmetric coupled quantum wells separated by 10.0 nm thick $Al_{0.3}Ga_{0.7}As$ barriers. Each coupled quantum well structure consist of a 3.7 nm thick GaAs well followed by a 2.4 nm thick $Al_{0.14}Ga_{0.86}As$ barrier and by a 0.85 nm thick GaAs well. Calculations give an intersubband resonance at $E_{21} = 131$ meV and $\mu_{12} = 16.8$ e.Å for the associated dipole moment. The overlap integral is predicted to be $<HH_1|E_2> = 0.147$.

3. Experiments

3.1. EXPERIMENTAL CONFIGURATION

Figure 2 shows the experimental set-up used both for the photoinduced luminescence experiments and for the PIA measurements.

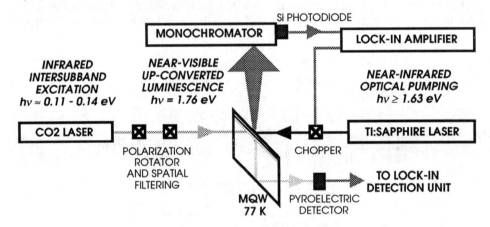

Figure 2: Experimental configuration. The focused beam waists are respectively 30 μm and 15 μm for the infrared and near-infrared beams.

The 1 mm long sample is mounted on the cold-finger assembly of a 77 K cryostat. Its facets have been polished at a 45° angle to allow total internal reflections of the infrared beam which is injected at normal incidence onto the input facet using a f/1 objective with 2.54 cm focal length. The intersubband excitation is provided by a linearly-polarized tunable cw CO_2 laser. A three-mirror assembly may be used to rotate the polarization of the injected infrared beam. The band-to-band excitation is delivered by a tunable cw Ti:Sapphire laser optically pumped by an Ar^{++} laser. The near-infrared pump beam is focused at a 45° angle onto the sample surface. The photoluminescence of the sample is collected by a f/4 lens and focused on the entrance slit of a 0.64 m monochromator. The photoluminescence is detected by a low-noise Si photodiode using a lock-in detection scheme synchronous with the pump modulation. Note that this photoluminescence detection scheme is far to be optimized since the detection sensitivity could be improved by at least six orders of magnitude by means of a photon-counting detection system, not available at the time of experiments. For the PIA measurements, the transmitted infrared beam is detected by a pyroelectric detector and a lock-in amplifier is used to extract the photoinduced signal synchronous with the pump modulation.

3.2. QUANTUM WELL SPECTROSCOPY

Preliminary spectroscopic measurements were carried out to assess the interband and intersubband energies. The 77K photoluminescence of the sample under weak excitation by an Ar^{++} laser reveals an E_1HH_1 luminescence at 1.63 eV along with a luminescence originating from the GaAs substrate at 1.51 eV. The interband transmission spectrum of the quantum well layers, shown in figure 3, was conveniently recorded by performing an excitation spectroscopy of the substrate luminescence. The assignment of the identified interband transitions is indicated in figure 3.

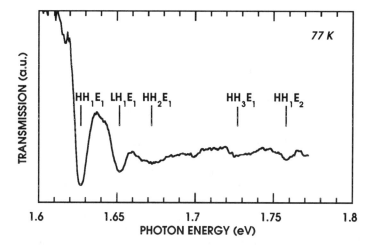

Figure 3: Excitation spectrum of the GaAs substrate luminescence which may be seen as a transmission spectrum of the quantum well structure.

As seen, the E_2HH_1 transition is found at 1.758 eV, in close agreement with calculations. The energy difference between the HH_1E_1 and HH_1E_2 transition energies gives a first estimate of the intersubband energy $E_1E_2 \approx 130$ meV. PIA spectroscopy was also performed with the pump laser set at 1.675 eV [7]. The resulting intersubband absorbance spectrum is shown in figure 4. As seen, the absorbance spectrum under p-polarized infrared excitation (full circles) exhibits a peak at 133 meV which can be attributed to the E_1E_2 intersubband resonance since it vanishes when the infrared polarization is rotated in the plane of the layers (open circles) [1]. The weak absorbance for s-polarized excitation is due to some photo-induced free-carrier absorption in the GaAs substrate although one cannot exclude some contribution due to a weakly allowed

E_1E_2 transition. The full width at half maximum of the intersubband resonance is only 7 meV which is suggestive of a good quality sample.

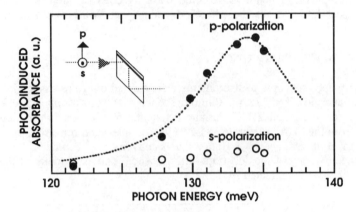

Figure 4: Photo-induced intersubband absorbance spectrum of the asymmetric coupled quantum wells at 77 K for two probe polarizations. The dotted curve is a Lorentzian fit.

3.3.   UP-CONVERSION MEASUREMENTS

The up-conversion measurements were performed with a p-polarized intersubband excitation set at 133 meV. The near-infrared excitation was set at 1.675 eV in order to excite electrons well below the second conduction subband.

Figure 5 shows a semilogarithmic plot of the sample photoluminescence (left) and a close view around the E_2HH_1 energy in a linear scale (right). The lower spectra are obtained without intersubband excitation (a) while the curves labelled b) to d) correspond to increasing excitation intensities. The pump intensity is 29 W/cm^2. The major result of figure 5 is the apparition of a well resolved emission around the E_2HH_1 energy under infrared excitation. Three main features which are not present on the spectrum without intersubband excitation can be noticed. By comparison with the transmission spectrum of figure 3, the strongest peak at 1.758 eV is clearly due to the E_2HH_1 luminescence while the smaller and somewhat broader peaks at 1.725 eV and at 1.785 eV are attributed to the E_1HH_3 and to the E_2HH_2 luminescence, respectively. Their amplitudes grow with increasing intersubband excitation intensities. Not shown in the figure, both peaks vanish when the intersubband excitation is polarized in the plane of the layers. The dependence of the induced luminescences on the photon energy of the intersubband excitation was investigated in separate experiments. Both the integrated E_2HH_1 and E_1HH_3 luminescence signals were found to closely follow the spectral evolution of the PIA signal shown in figure 4.

The occupation of the E_2 and HH_3 subbands is thus clearly related to the presence of the intersubband excitation, although one cannot exclude at this step a thermal population of both subbands. On the other hand, the occupation of the HH_2 subband can be explained by the excitation of the interband HH_2E_1 absorption at the 1.765 eV pump photon energy since the induced E_2HH_2 luminescence was found to vanish when the pump photon energy was set at the HH_1E_1 energy.

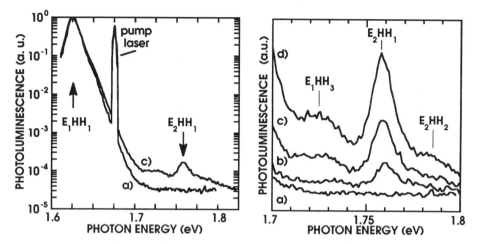

Figure 5: Photoluminescence spectrum in semilogarithmic scales (left) and a close view around the E_2HH_1 energy in linear scales (right). a) is without intersubband excitation. For curves b), c) and d), the intersubband excitation intensity is 3.6 kW/cm^2, 7 kW/cm^2 and 15 kW/cm^2, respectively. The residual signal observed above 1.7 eV in spectrum a) reflects the noise level of our detection system.

As seen in the left spectra of figure 5, an exponential decrease is visible on the high energy side of the E_1HH_1 line. Such evolution is characteristic of band-filling effects arising from non-degenerate carrier distributions [25]. Conversely, the carrier temperature which is a weighted function of the electron and hole temperatures, can be deduced from the high energy slope in a semilogarithmic plot. The carrier temperature was found to increase from 79 K without intersubband excitation to 97 K for the 15 kW/cm^2 excitation. Correspondingly, the lattice temperature in the quantum well layer, which was deduced from the energy position of the E_1HH_1 luminescence peak, was found to slightly increase from 79 K to 80 K under the same conditions. The intense intersubband excitation thus induces a significant increase of the carrier temperature respective to the lattice temperature, which is suggestive of a non-equilibrium carrier distribution in the conduction subbands and could eventually be responsible for a thermal occupation of the E_2 subband. However, note that the carrier temperature under the most intense excitation corresponds to an electron temperature of at most 101 K assuming the holes are in equilibrium with the lattice. This temperature leads to an occupation factor of the E_2 subband respective to the E_1 subband of the order of 2.2×10^{-7} which is far too small to account for the observed ratio $\approx 10^{-3}$ between the intensities of the E_2HH_1 and E_1HH_1 photoluminescences at 15 kW/cm^2 excitation. This result gives a first demonstration that the induced luminescence is due to a non-thermal occupation of the E_2 subband.

Further evidence of a non-thermal occupation of the excited subbands was obtained by measuring the dependence of the integrated induced luminescence signals on the intersubband excitation intensity. The results are shown in figure 6. As seen, the evolution of the integrated intensity of the induced luminescences normalized to the E_1HH_1 integrated intensity is linear for both the E_2HH_1 and the HH_3E_1 transitions. A thermal occupation of the E_2 and HH_3 subbands would have translated into a fast exponential growth of the photoluminescence ratio with intersubband excitation intensity.

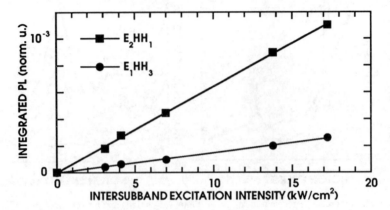

Figure 6: Integrated luminescence signal normalized to the integrated E_1HH_1 luminescence versus intersubband excitation intensity for the E_2HH_1 (squares) and E_1HH_3 (circles) induced photoluminescences.

The observed up-converted E_2HH_1 emission is thus clearly related to a linear population transfer from the first to the second conduction subband induced by the intersubband excitation. Eventual contributions from two-photon non-linear processes can be excluded since in that case the induced PL intensity would be strongly dependent on the band-to-band excitation photon energy, which is not observed. The origin of the weaker induced emission at the HH_3E_1 energy is less clear. A similar process involving a photo-induced absorption between the valence subbands HH_1 and HH_3 absorption is unlikely because of the low oscillator strength of such transition along with the large detuning involved. However, the experiments can be well interpreted if the induced occupation of the HH_3 subband originates from the reabsorption of the E_2HH_1 emission.

4.   Intersubband relaxation

We now show that the up-conversion technique can be used to estimate the intersubband scattering time. The ratio of the photoluminescence integrated intensities, I_2/I_1, corresponding to the E_2HH_1 and E_1HH_1 transitions, respectively, can be expressed as :

$$\frac{I_2}{I_1} = \frac{n_2}{n_1}\frac{f_2}{f_1} = \frac{n_2}{n_1}\frac{h\nu_2}{h\nu_1}\left|\frac{<HH_1|E_2>}{<HH_1|E_1>}\right|^2 \tag{1}$$

where n_i represents the occupation factor of the E_i conduction subband, $h\nu_i$ is the energy of the E_iHH_1 transition and f_i is the associated oscillator strength, which is proportionnal to the square of the envelope wavefunction overlap, $<HH_1|E_i>$ [5,23]. In the non-saturating regime, the occupation ratio is proportionnal to the intersubband excitation intensity, I_{IR}, and to the intersubband relaxation time, τ_{21}:

$$\frac{n_2}{n_1} \approx \frac{\sigma_{12}\,\tau_{21}}{h\nu_{IR}}\,I_{IR} \qquad (2)$$

where $h\nu_{IR}$ is the intersubband excitation energy and σ_{12} is the absorption cross-section associated with the E_1E_2 transition which was estimated from the PIA measurements to be $\sigma_{12} \approx 1.2 \times 10^{-14}$ cm^2 in our 45° angle incidence configuration. As seen from Eqs. (1) and (2), this simple model which assumes a negligible thermal occupation of the excited subband, predicts a linear increase of I_2/I_1 with the intersubband excitation intensity as observed in the experiments. In turn, the experimental value of I_2/I_1 can be used to estimate the intersubband scattering time although some corrections are expected due to both the pump and infrared attenuation and the luminescence reabsorption in the quantum well layer [23]. By accounting for these propagation effects, the intersubband relaxation time is found to be $\tau_{12} = 2.5 \pm 0.7$ ps. The τ_{12} value in our asymmetric coupled quantum well structure is significantly longer than previously reported scattering times for symmetric quantum well structures [5, 26, 27] although longer relaxation times have been observed in heavily modulation-doped quantum wells due to scattering processes proceeding through the barrier states [28]. Note that τ_{12} values of the order of 2 ps were also measured in coupled quantum wells by means of subpicosecond luminescence techniques [29].

Theoretical investigation of the electron relaxation through emission of confined slab and interface polar optical-phonon modes in our structure was carried out using the Frölich Hamiltonian formalism for dielectric slabs [24]. Results are summarized in table 1 which reveals that the dominant relaxation mecanisms involves both LO-phonon slab modes in the GaAs thick well and interface modes at the interfaces of the barrier separating the wells.

Region		τ_{slab} (ps)	$\tau_{interface}$ (ps)	τ_{both} (ps)
1	$Al_{0.3}Ga_{0.7}As$ barrier	80	105	45
2	0.85 nm GaAs well	540	70	62
3	$Al_{0.14}Ga_{0.86}As$ barrier	13	6.1	4.2
4	3.7 nm GaAs well	6.8	9.3	3.9
5	$Al_{0.3}Ga_{0.7}As$ barrier	79	105	45

Table 1: Contribution of the different LO-phonon modes to the intersubband relaxation time for each quantum well layer. Phonon emission is assumed to proceed through slab modes (second column), interface modes (third column) and both confined modes (fourth column).

The contributions from all regions can be summed using Mathiews rules to calculate the electron relaxation time which is found to increase with the initial in-plane kinetic energy of electrons within the ground subband. Calculations show that the intersubband relaxation time via slab (interface) phonon modes only, ranges from 3.7 ps (2.8 ps) to 6 ps (4.9 ps) from initially cold to very hot electrons, respectively. Correspondingly, when both interface and slab modes are considered, τ_{12} ranges from 1.6 to 2.7 ps which is close to the experimental value. These results demonstrate that the interface modes play a major role in the intersubband relaxation in the asymmetric coupled quantum well structure. Note that the opposite result, i.e. intersubband relaxation dominated by LO-phonon slab mode emission, is found in square quantum wells [24].

The up-conversion technique thus appears to be a sound method for estimating the intersubband relaxation times without requiring bleaching experiments or sub-picosecond-time-resolution measurements [27-29]. Note that this technique can also be applied here to valence band scattering by analyzing the dependence of the induced E_2HH_2 photoluminescence on infrared excitation intensity. By doing this, the intersubband relaxation time for the HH_2HH_1 transition was estimated to be of the order of 6 ps although this result is entached of a large uncertainty due to the noise level of our photoluminescence detection system.

5. Application to infrared photodetection

The quantum efficiency of the up-conversion process which is defined as the generation rate of emitted near-visible photons divided by the generation rate of absorbed infrared photons, can be expressed as:

$$\eta = \frac{\tau_{21}}{\tau^r_2} = \frac{\tau_{21}}{\tau^r_1} \frac{h\nu_2}{h\nu_1} \left| \frac{<HH_1|E_2>}{<HH_1|E_1>} \right|^2 \tag{3}$$

where τ^r_i is the radiative recombination time associated with the E_iHH_1 transition which is known to be dependent on carrier concentration and temperature [30,31]. Note that the right hand identities in Eqs 1 and 3 are only valid if both the E_1HH_1 and E_2HH_1 luminescences have the same origin, i.e. either excitonic or free-carrier recombination. By using an empirical fit of the radiative relaxation time measurements in Ref [31] for a 70 Å thick quantum well, we deduced the evolution of the quantum efficiency with temperature and with the near-infrared pump intensity, I_p, shown in Figure 7.

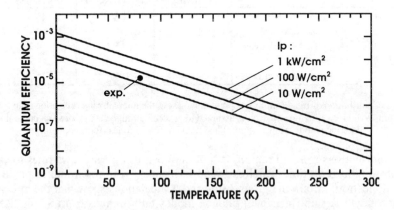

Figure 7: Calculated quantum efficiency of the up-conversion process versus temperature for three pump intensities. The dot indicates the experimental value.

As seen, η exhibits an exponential decrease with temperature along with a $I_p^{1/2}$ dependence on pump intensity [32]. For the 29 W/cm^2 pump intensity used in the experiments, η is estimated to be 1.6x10^{-5}. Quantum efficiencies of the order of 10^{-3} could be achieved by increasing the pump intensity to 1 kW/cm^2 while lowering the temperature to 10 K. Under such conditions, the absorption of 1000 infrared photons would lead to the emission of one near-visible photon. Note that a significant improvement of the quantum efficiency could also be obtained if the intersubband transition energy was set lower than the LO-phonon energy because of a slower intersubband relaxation [23].

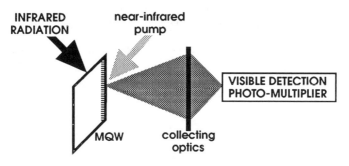

Figure 8: Infrared detection based on visible up-conversion.

We now focus on the application of the up-conversion process to infrared detection as illustrated in figure 8. Practically, since only part of the luminescence radiation can be collected on the near-visible detector, care must be taken in designing the collecting and filtering optics. The performances of such unconventional infrared detector are directly related to the near-visible detector characteristics. The visible detection can be advantageously performed by a near-visible photomultiplier unit which can offers gains exceeding $\approx 10^6$ along with large quantum efficiencies. The responsivity of the up-conversion detector can be expressed as:

$$R = \frac{h\nu_2}{h\nu_{IR}} \eta \, \eta_{abs} \, \eta_{col} \, R_{vis} \qquad (4)$$

where η_{col} is the collecting efficiency, η_{abs} is the quantum efficiency of the infrared absorption process in the quantum well layer and R_{vis} is the responsivity of the photomultiplier unit. Using realistic values of the parameters, $\eta_{col} \approx 0.05$, $\eta_{abs} \approx 0.5$, $\eta \approx 10^{-3}$, $R_{vis} \approx 10^5$ A/W, the responsivity of the up-conversion detector can be quite large, of the order of 30 A/W. Note that $g = h\nu_2 \, \eta \, \eta_{col} \, R_{vis}/e$ may be seen as the gain of the up-conversion detector by analogy with conventional intersubband photoconductors. With the parameters listed above, the gain is of the order of 8.8.

At not too low temperatures, the main source of dark current is due to the detection of some hot luminescence induced by thermal occupation of the E$_2$ subband, since the dark current of the photomultiplier unit is usually negligible:

$$I^{th}_D \approx \eta_{col} \, h\nu_2 \, R_{vis} \, (n_1/\tau^r_2) \, S \, \exp(-E_{21}/kT) \tag{5}$$

where S is the surface of the near-infrared pump spot, n_1 is the electron density in the ground subband, E_{21} is the intersubband transition energy and kT is the thermal energy. For low enough temperatures, the dark current originates from up-conversion of the incoming background photon flux and is simply given by:

$$I^b_D \approx R \, P_b \tag{6}$$

where P_b is the incident background power [33]. It follows from the dependence of both n_1 and τ^r_2 on the near-infrared pump intensity, I_p, that both I^{th}_D, I^b_D and R will increases as I_p for moderate pump intensities.

Since the detecting element operates in the visible range, the main source of noise arises from the generation-recombination noise in the quantum well sample. The noise current can be expressed as $i_s = \sqrt{4 \, e \, I_D \, g \, \Delta f}$. Note that in the BLIP condition, the NEP associated with the up-conversion detector simply reduces to:

$$(NEP)_B = 2 \sqrt{P_b \, h\nu_{IR} \, \Delta f} \, / \, \eta_{abs} \tag{7}$$

which is identical to the expression found for conventional intersubband photoconductors [33]. In fact, since we have neglected any contribution to the noise arising from the visible detector, the performances of the up-conversion detector in terms of NEP, specific peak detectivity and BLIP temperature are expected to be similar to those of intersubband photoconductors with similar carrier concentrations. At temperatures higher than T_{BLIP}, the specific peak detectivity can be expressed as:

$$D^* = \frac{\sigma_{12} \, \exp(E_{21}/2kT) \, \sqrt{\tau_{21} \, n_1}}{2 \, h\nu_{IR}} \tag{8}$$

assuming one asymmetric quantum well structure and $\eta_{abs} \approx \sigma_{12} \, n_1$. Since n_1 varies like $I_p^{1/2}$, D^* is predicted to moderately increase with pump intensity as $I_p^{1/4}$. By taking the experimental values for σ_{12}, τ_{21} and E_{21}, D^* for this non-optimized single-quantum-well detector would be 3.3×10^9 W^{-1} cm $\sqrt{Hz}$ at 77 K and 2.1×10^{10} W^{-1} cm $\sqrt{Hz}$ at 65 K assuming $n_1 \approx 10^{11}$ cm^{-2}. Note that a major improvement of D^* could be achieved by means of epitaxial n-doping of the quantum well layers.

The speed of the up-conversion detector is intrinsically limitated by the intersubband relaxation time. Practically, the time response will be governed by the response of the photomultiplier unit which can be very fast.

Other applications of the photoinduced up-conversion process could be envisaged such as remote infrared detection by means of an optical fiber link between the sample and both the visible detector and the pump laser diode, or thermal imaging making use of visible light intensifiers.

6.   Conclusion

We have investigated above the photoinduced intersubband absorption process in undoped asymmetric coupled quantum well structures. We have demonstrated that the photoinduced intersubband absorption results in a finite non-thermal occupation of the E_2 subband which manifests as a near-visible $E_2 \to HH_1$ photoluminescence whose features closely follow those of the intersubband absorption. We have shown that this up-conversion process may be used to estimate the intersubband relaxation time which is found to be of the order of 2.5 ps. The intersubband relaxation in our asymmetric coupled quantum well structure is dominated by electron-phonon interactions involving interface LO-phonon modes. Application of this novel up-conversion process for infrared photodetection have also been discussed. We have predicted large responsivities and gains for such unconventional detectors if the up-converted emission is detected by a performant photomultiplier unit. The performances in terms of detectivity and BLIP temperature should be comparable to those of conventional intersubband photoconductors.

We acknowledge J.L. Educato and J.-P. Leburton for carrying out the electron-phonon calculations and for many fruitful discussions. This work was supported in part by CNRS GDR III-V.

References

1.   L. C. West and S. J. Eglash, First observation of an extremely large dipole infrared transition within the conduction band of a GaAs quantum well, Appl. Phys. Lett. 46:1156 (1985).
2.   B. F. Levine, S. D. Gunapala, J. M. Kuo, S. S. Pei, and S. Hui, Normal incidence hole intersubband long wavelength GaAs/AlGaAs quantum well infrared photodetectors, Appl. Phys. Lett. 59:1864 (1991).
3.   M. Helm, P. England, E. Colas, F. DeRosa, ans S. J. Allen, Jr., Intersubband emission from semiconductor superlattices excited by sequential resonant tunneling, Phys. Rev. Lett. 63:74 (1989).
4.   H. C. Liu, M. Buchanan, G. C. Aers, and Z. R. Wasilewski, Single quantum well intersubband detector using asymmetrical double-barrier structures, Semicond. Sci. Technol. 6:C124 (1991).
5.   H. T. Grahn, H. Schneider, W. W. Rühle, K. von Klitzing, and K. Ploog, Nonthermal occupation of higher subbands in semiconductor superlattices via sequential resonant tunneling, Phys. Rev. Lett. 64:2426 (1990).
6.   M. Olszakier, E. Ehrenfreund, E. Cohen, J. Bajaj, and J. Sullivan, Photoinduced intersubband absorption in undoped multi-quantum-well structures, Phys. Rev. Lett. 62:2997 (1989).
7.   D. D. Yang, F. H. Julien, J.-M. Lourtioz, P. Boucaud, and R. Planel, First demonstration of room-temperature intersubband-interband double-resonance spectroscopy of GaAs/AlGaAs quantum wells, IEEE Photon. Technol. Lett. 2:398 (1990).
8.   D. Delacourt, D. Papillon, J. P. Pocholle, J. P. Schnell, and M. Papuchon, Room temperature observation of mid-infrared inter-subband absorption actived by

near-infrared illumination in multiple quantum wells, Electron. Lett. 26:279 (1990).

9. M. Olszakier, E. Ehrenfreund, E. Cohen, J. Bajaj, and J. Sullivan, Intersubband absorption by photoinduced exciton in undoped multi-quantum-wells, Surf. Science 228:123 (1990).

10. M. Olszakier, E. Ehrenfreund, E. Cohen, and L. Pfeiffer, Photoinduced absorption by excitons in multiple quantum wells, J. Lumin. 45:186 (1990).

11. Y. Garini, M. Olszakier, E. Cohen, E. Ehrenfreund, A. Ron, K-K. Law, J. L. Merz, and A. C. Gossard, Photoinduced intersubband absorption in barrier doped multiple quantum wells, Superlattices Microstruc. 7:287 (1990).

12. M. Olszakier, I. Brener, E. Ehrenfreund, and E. Cohen, Intervalence-band photoinduced absorption in undoped GaAs/AlGaAs multiple-quantum-wells, Superlattices Microstruc. 7:291 (1990).

13. Y. Garini, E. Cohen, E. Ehrenfreund, A. Ron, K-K. Law, J. L. Merz, and A. C. Gossard, Photoinduced intersubband absorption in n-type well- and barrier-doped quantum wells, Surf. Science 263:561 (1992).

14. J. Oiknine-Schlesinger, E. Ehrenfreund, D. Gershoni, D. Ritter, M. B. Panish, and R. A. Hamm, Photoinduced intersubband absorption in lattice-matched InGaAs/InP multiquantum well, Appl. Phys. Lett. 59:970 (1991).

15. E. Ehrenfreund, J. Oiknine-Schlesinger, D. Gershoni, D. Ritter, M. B. Panish, and R. A. Hamm, Intersubband transitions in InGaAs/InP quantum wells studied by photomodulation spectroscopy, Surf. Science 267:461 (1992).

16. C. Francis, M. Bradley, P. Boucaud, F.H. Julien, and M. Razeghi, Intermixing of GaInP/GaAs multiple quantum wells, Appl. Phys. Lett. 62:178 (1993).

17. D. D. Yang, P. Boucaud, F.H. Julien, L. Chusseau, J.-M. Lourtioz, and R. Planel, Laser diode modulation of 10.6 μm radiation in GaAs/AlGaAs quantum wells, Electron. Lett. 26:1532 (1990).

18. F.H. Julien, P. Vagos, J-M. Lourtioz, D.D. Yang, and R. Planel, Novel all-optical 10 μm waveguide modulator based on intersubband absorption in GaAs/AlGaAs quantum wells, Appl. Phys. Lett. 59:2645 (1991).

19 F.H. Julien, Room-temperature photo-induced intersubband absorption in GaAs/AlGaAs quantum wells, Intersubband Transitions in Quantum Wells, Edited by Rosencher E., Vinter B., et Levine B., Plenum Press, NewYork (1992).

20 P. Boucaud, P. Vagos, F.H. Julien, and J-M. Lourtioz, Modulation bandwidth enhancement of all-optical modulators based on photo-induced intersubband absorption in GaAs/AlGaAs quantum wells by proton implantation, Electron. Lett. 28:1373 (1992).

21 V. Berger, N. Vodjani, B. Vinter, E. Costard, and E. Böckenhoff, Optically induced intersubband absorption in biased double quantum wells, Appl. Phys. Lett. 60:1869 (1992).

22 V. Berger, E. Rosencher, N. Vodjani, and E. Costard, Quantum well infrared photodetection induced by interband pumping, Appl. Phys. Lett. 62:378 (1993).

23 P. Vagos, P. Boucaud, F. H. Julien, J.-M. Lourtioz, and R. Planel, Photoluminescence up-conversion induced by intersubband absorption in asymmetric coupled quantum wells, Phys. Rev. Lett. 70:1018 (1993).

24   J. L. Educato, J.-P. Leburton, P. Boucaud, P. Vagos, and F. H. Julien, Influence of interface phonons on intersubband scattering in asymmetric coupled quantum wells, Phys. Rev. B 47:12949 (1993).

25   G. Bastard, "Wave Mechanics Applied to Semiconductor Heterostructures", Les Editions de Physique, Les Ulis, (1988).

26   M. C. Tatham, J. F. Ryan, C. T. Foxon, Time-resolved Raman measurements of intersubband relaxation in GaAs quantum wells, Phys. Rev. Lett. 63:1637 (1989).

27   F. H. Julien, J.-M. Lourtioz, N. Herschkorn, D. Delacourt, J. P. Pocholle, M. Papuchon, R. Planel, G. Le Roux, Optical saturation of intersubband absorption in GaAs-Al$_x$Ga$_{1-x}$As quantum wells, Appl. Phys. Lett. 53:116 (1988) and Appl. Phys. Lett. 62:2289 (1993)

28   U. Plödereder, T. Dahinten, A. Seilmeier, and G. Weimann, Intersubband relaxation in modulation doped quantum well structures, Intersubband Transitions in Quantum Wells, Edited by Rosencher E., Vinter B., and Levine B., Plenum Press, NewYork (1992).

29   B. Deveaud, A. Chomette, F. Clérot, P. Auvray, A. Regreny, R. Ferreira, G. Bastard, Subpicosecond luminescence study of tunneling and relaxation in coupled quantum wells, Phys. Rev. B 42:7021 (1990).

30   B. K. Ridley, Kinetics of radiative recombination in quantum wells, Phys. Rev. B 41:12190 (1990).

31   M. Gurioli, A. Vinattieri, M. Colocci, C. Deparis, J. Massies, G. Neu, A. Bosacchi, and S. Franchi, Temperature dependence of the radiative and non-radiative recombination time in GaAs/Al$_x$Ga$_{1-x}$As quantum-well structures, Phys. Rev. B 44:3115 (1991).

32   We implicitly assumed that the free-carrier radiative recombination is the dominant process at the temperature of our experiments. This assumption is supported by separate PIA measurements at 77 K which shows a $\sqrt{I_p}$ dependence of the PIA signal.

33   I. Gravé, and A. Yariv, Fundamental limits in quantum well intersubband detection, Intersubband Transitions in Quantum Wells, Edited by Rosencher E., Vinter B., et Levine B., Plenum Press, NewYork (1992).

FAR INFRARED SPECTROSCOPY OF INTERSUBBAND TRANSITIONS IN MULTIPLE QUANTUM WELL STRUCTURES.

W. J. LI and B. D. McCOMBE
Department of Physics
SUNY at Buffalo
Buffalo, NY 14260
USA

ABSTRACT. We review a series of spectroscopic studies of intersubband transitions in lightly doped GaAs/AlGsAs quantum well structures. The oscillator strength was determined from measurements of intersubband absorption coefficient, photogenerated electron density, and a calculation of grating coupling efficiency. We also report recent linear and nonlinear saturation absorption experiments on coupled double-quantum-well structures, and present the results of the subband structures and hot electron relaxation in a coupled subband-Landau-level system.

I. INTRODUCTION

Intersubband transitions in multiple-quantum-well (MQW) structures and superlattices have attracted considerable attention due to the typical transition energies (infrared), large transition dipole, and narrow bandwidth. These unique properties lead to extensive potential applications as infrared (IR) devices. In recent years there has been tremendous progress in the development of intersubband photodetectors, emitters, modulators and non-linear devices.[1] To optimize device performance, it is crucial to understand in detail the single-particle subband structure, and transitions among these subbands, the influence of many electrons on the transition energies, and the dynamic relaxation processes in the excited subbands.

We have carried out systematic studies of intersubband transitions on a series of single (SQW) and coupled-double (CDQW) quantum wells in GaAs/$Al_{0.3}Ga_{0.7}As$ MQW structures with very low doping concentration (10^9 cm^{-2} per well) to 10^{11} cm^{-2} per well, and with well widths from 168Å to 420Å. Intersubband absorption was either induced by metallic grating couplers fabricated on the surface of the sample, or via subband-Landau level (LL) coupling. Detailed analysis of results from linear and nonlinear saturation spectroscopy was made in conjunction with model calculations that include the anisotropic plasma model for free electrons in QWs, the grating coupling efficiency, the bimolecular-type relaxation process for optically generated free carriers, and the eigenstates and population decay processes of electrons in coupled subband-LL systems. These results have let us to a new and deeper understanding of the nature of the quantum confinement effect, the role of electron-electron interaction, the role of compensation, and the energy loss mechanism of hot carriers in quasi-2D structures.

This paper is divided into 4 parts. The sample parameters and experimental techniques are

361

H. C. Liu et al. (eds.), Quantum Well Intersubband Transition Physics and Devices, 361–370.
© 1994 *Kluwer Academic Publishers. Printed in the Netherlands.*

described in section II. Experimental results and theoretical analysis are given for single-quantum-well structures in section III, and for coupled double-quantum-well structures in section IV. Absorption saturation spectroscopy for the coupled subband-Landau level system is presented in section V. A brief summary of conclusions in given in section VI.

II. EXPERIMENTAL DETAILS

Samples were grown by MBE on semi-insulating GaAs substrates. The MQW and superlattice structures were sandwiched between two thick (1500 ~ 2000Å) $Al_{0.3}Ga_{0.7}As$ cladding layers and capped by a ~ 100Å GaAs layer. The important sample parameters are given in Table I.

Sample	Well width (Å)	Barrier width (Å)	No. of Periods	Doping density	Doping position (width)
1	420	600	20	3×10^{10} cm^{-2}	Barrier center (δ-doping)
2	320	107	12	1×10^{16} cm^{-3}	Well top edge* (107Å)
3	240	240	30	4×10^{16} cm^{-3}	Barrier center (85Å)
4	210	150	30	1×10^{16} cm^{-3}	Well center (70Å)
5	168	48	30	1×10^{16} cm^{-3}	Well center (60Å)
6	240	600	20	3×10^{10} cm^{-2}	Barrier center (δ-doping)
7	240	600	15	6×10^{10} cm^{-2}	Barrier center (δ-doping)
8	240	600	10	1.2×10^{11} cm^{-2}	Barrier center (δ-doping)

Table I. Summary of sample parameters. Samples 6-8 are coupled double quantum wells with a two-monolayer AlAs barrier grown at the center of each GaAs well. The parameters listed here are either confirmed or obtained by fitting photoluminescence and reflectivity data. * Top edge refers to the last 1/3 of the GaAs layer grown before the next AlGaAs barrier.

To couple the transverse electric-field component of the normally incident infrared radiation into the longitudinal polarization (the direction of the intersubband transition dipole), aluminum gratings of strip width 4 μm and period 8 μm were fabricated directly on the surface of the sample. A sensitive pumping and probing technique has been employed to detect weak intersubband transition signals, in which an *in situ* red light-emitting diode (LED) or a He:Ne laser (or Ar laser) was square-wave modulated at audio frequencies and was directed onto the sample to generate excess free carriers in the wells. The IR transmitted signal coherent with the modulation was detected by a Ga:Ge photoconductive detector. The signal was lock-in amplified and Fourier transformed into the frequency spectrum. This spectrum was then normalized to a background taken with chopped IR light. The resulting spectrum is a difference spectrum between light-on and light-off. It has been found that the primary effect of cross-band-gap optical pumping is to neutralize donors the wells which are ionized by compensating acceptors. At temperature > 25K these donors produce excess free electrons in the wells.[2]

III. SINGLE QUANTUM WELLS

A typical grating-coupler-induced optically-pumped differential intersubband absorption spectrum is shown in Fig. 1 as the solid curve. A spectrum taken under similar experimental conditions from an adjacent section of the same wafer without grating coupler is also shown for comparison (dashed line). The $1 \rightarrow 2$ intersubband transition peaked at 235 cm^{-1} for this 210Å QW is clearly identified. Figure 2 displays a plot of the measured intersubband transition energy as a function of well width. Theoretical curves from a simple square-well calculation with boundary condition 1: F and $\partial F/\partial z$ (dashed line), and boundary condition 2: F and $(1/m^*)\partial F/\partial z$ (solid line) are also shown for comparison. Here F is the quantum well envelope function, and m^* is the electron effective mass. The well widths used in the calculation were determined from photoluminescence (PL) and reflectivity measurements, and the barrier height was taken as 225 meV. As expected, condition 2, which ensures the conservation of probability current, yields better agreement with the data. The lower intersubband energies for condition 2 is due to a more extended envelope function in the well as a result of the discontinuity of the slope $\partial F/\partial z$ at the boundary. For the above samples (doping densities $\leq 3 \times 10^{10}$ cm^{-1}), the shift of resonant energy from the single particle intersubband energy due to the combined depolarization and excitonic effect (final state interaction) is ≤ 3cm^{-1}.[3] The effect of exchange interaction among electrons on the ground subband is also small (≤ 3 cm^{-1}).[4] The influence of non-parabolicity is important for narrow wells. For 168Å QWs (sample 5), the non-parabolic shift is ~ 8 cm^{-1} to lower energy.[5]

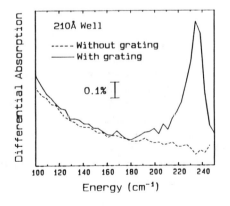

Fig. 1. Optically-pumped E_2-E_1 intersubband transition with peak resonance at 235 cm^{-1}. A spectrum from the sample without grating coupler is also shown for comparison. The monotonic decreasing background is due to free electron absorption (in the presence of impurity scattering).

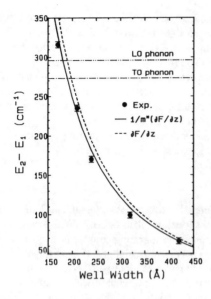

Fig. 2. Well width dependence of intersubband energy. Square-well calculations are in shown for two boundary conditions (see text). TO and LO phonon energies for GaAs are also indicated.

The grating-coupler-induced intersubband absorption coefficient is given by

$$\alpha = \frac{4\pi}{c} \epsilon^{3/2} \frac{Re(\sigma_z)}{|\epsilon_z|^2} \eta ,$$ (1)

where ϵ_z (σ_z) is the QW dielectric (conductivity) function from the anisotropic plasma model,[6] ϵ is the background dielectric constant of GaAs, and η is the grating coupling efficiency, defined as the ratio of the total electromagnetic energy density of the z-polarized wave, $<E_z^2>$, to the energy density of the incident wave, $<E_0^2>$, which is polarized in the plane perpendicular to the z-axis. The bracket $<\ >$ indicates spatial average over the x-y plane of the QW structure. The oscillator strength is defined as

$$f \equiv \frac{2m^*\omega_{21}}{\hbar} |<F_2|z|F_1>|^2 ,$$ (2)

where $\omega_{21} = (E_2-E_1)/\hbar$. In the infinite-barrier approximation, f is independent of the quantum well parameters, and is equal to 0.96. $Re(\sigma_z)$ contains the oscillator strength, the intersubband resonance frequency and absorption linewidth, and the plasma frequency $\omega_p^2 = 4\pi Ne^2/m^*L$ (N: carrier density, L: well width). Therefore the oscillator strength can be obtained from Eq.(1) from the measured intersubband absorption coefficient and the photoexcited electron density (from cyclotron resonance (CR) experiments) combined with a numerical calculation of the coupling efficiency. Detailed calculation of η is given in Ref. 7. Here we outline the key results. The calculation is for normal incidence of EM waves on a grating consisting of infinitely thin and perfectly conducting strips on the surface of a semiconductor, and is only suitable for lightly doped QW structures in which the absorption is sufficiently weak that it does not perturb substantially the grating-induced electric field pattern. This is applicable to our lightly-doped samples.

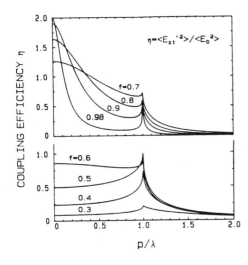

Fig. 3. Coupling efficiency vs p/λ for various metal filling factors, f. For semiconductor material with in index of refraction n, λ should be replaced by λ/n.

Figure 3 presents the plot of the coupling efficiency as a function of the ratio of the grating period to the intersubband resonance wavelength, p/λ. f is the filling factor defined as f=t/p, where t is the width of the metal strips. For small values of filling factor the metal strips are very narrow; thus the scattering of EM waves is weak, leading to a lower efficiency. With increasing filling factor, the efficiency is enhanced dramatically. For f > 0.6, the efficiency drops back again since more energy is reflected by wider metal strips. At p/λ < 0.4, energy density greater than that of the incident wave can be built up in the evanescent modes (similar to a resonant cavity). For f$\leq$0.5, the efficiency drops slowly on the p/λ< 1 side but much more rapidly on the p/λ> 1 side, in agreement with the experimental observation.[8] This phenomenon is attributed to the fact that when p/λ> 1, the wave becomes propagating, so that the energy is no longer stored in the vicinity of the grating.

In practical cases, an average of the energy density over the active quantum well region along the z-axis has to be taken in order to account for the actual effective coupling. A summary of results for the measured density, decay length of the evanescent wave, the coupling efficiency and the oscillator strength is given in Table II. Within the experimental error, f is in good

Sample	N (cm^{-2} per well)	p/λ	Decay Length δ (μm)	Coupling Efficiency η	f
2	3.4×10^9	0.28	1.33	26.5 %	1.1$\pm$0.2
3	2.4×10^9	0.48	1.46	15.5 %	1.0$\pm$0.2
5	2.7×10^9	0.89	2.79	63.4 %	1.0$\pm$0.2

Table II. Summary of results. N is the photogenerated carrier concentration from CR measurements, δ is the decay length of the grating-induced evanescent wave, η is the grating coupling efficiency, and f is the oscillator strength of the intersubband transition.

agreement with the theoretical value of 0.96. There are three reasons for a higher coupling efficiency found in sample 5: (1) p/λ is closer to 1, (2) a longer decay length, and (3) narrower wells and barriers so that the entire active medium is closer to the grating surface.

IV. COUPLED DOUBLE QUANTUM WELLS

Tailoring of intersubband separation by a central spike barrier has been proposed by two groups,[9,10] The idea is to perturb the electron wave functions by placing a thin barrier layer in the middle of the well. The ground subband wave function is greatly affected because it has maximum amplitude at the well center, whereas the second subband wave function, being zero at the center, is essentially unaffected. As a result, the ground state is shifted up dramatically compared to the first excited state. This method of controlling the intersubband separation has several advantages for practical device applications. For example, a resonant intersubband tunneling detector requires the second subband to be very close to the top of the barrier or even out of the well into the continuum.[11] For simple quantum wells both well width and height have to be varied simultaneously in order to meet this requirement, whereas for the perturbed quantum wells only the width or composition of the central barrier need to be changed. For a long wavelength detector, a very wide well is needed for the unperturbed quantum wells. This may not be possible for certain strained layer structures in which the thickness of the well is constrained by the critical thickness. The perturbed quantum wells do not have this problem.

Three CDQW samples (samples 6-8) have been studied. The nominal well width is 240Å. The central spike barrier is an AlAs layer of 5.66Å thickness. Intersubband absorption induced by a grating coupler (t=15.3 μm, p=25.5 μm) is shown in Fig. 4. This spectrum was obtained by taking the ratio of the spectrum under cross-band-gap optical excitation to that without optical excitation. The intersubband absorption occurs at 57 cm^{-1}. Compared with a 170 cm^{-1} intersubband energy measured from sample 3, which has an identical nominal well width of 240Å but no perturbing barriers, E_2-E_1 is shifted by 113 cm^{-1} in the presence of a central AlAs layer.

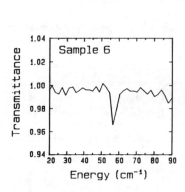

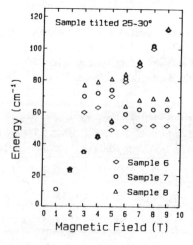

Fig. 4. Grating-coupler-induced intersubband transition for CDQW structures.

Fig. 5. Subband-LL anti-level crossing for samples 6-8. These samples have the same QW parameters but different doping concentrations.

Further evidence of this effect has been provided by reflectance and PL measurements (see Ref. 12 and 13 for details). E_2-E_1 can also be measured through subband-Landau level interaction by tilting the sample away from the direction of magnetic field. Plotted in Fig. 5 are free electron transition energies vs total magnetic field. The intersubband resonance energy, obtained by extrapolating the upper branch to zero field, are 57 ± 3 cm^{-1}, 65 ± 3 cm^{-1}, and 75 ± 3 cm^{-1}, for sample 6-8, respectively. Considering 30% uncertainty in doping concentration, the increased resonance energy with higher electron density can be accounted for by the depolarization shift, final state interaction, and exchange interaction.

V. SATURATION SPECTROSCOPY OF COUPLED DOUBLE-QUANTUM-WELLS

Saturation spectroscopy is a useful tool to probe relaxation processes of hot carriers. In principle, by measuring the intensity dependence of the resonant absorption, the relaxation rate (and thus the lifetime) of carriers in the excited states can be extracted. Our goal is to study the energy loss mechanism for electrons in the excited subband-LL coupled state. Figure 6 shows the energy levels calculated for our CDQW structure in the presence of a magnetic field at a 25° tilt angle.[14] Level repulsion takes place between states with adjacent subband quantum numbers and LL quantum numbers. Strong resonant pumping of transition I and II will saturate these transitions as levels 2 and 3 are substantially populated. At the same time transition III and IV will set in. The resonance of these transitions may occur at different magnetic fields.

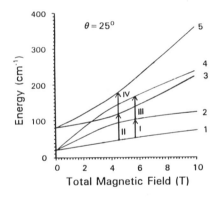

Fig. 6. Numerical solution of the energy levels of free electrons in CDQW structures in the presence of a magnetic field at a 25° tilt angle.

Saturation measurements were made with the Free Electron Laser (FEL) at the University of California at Santa Barbara in conjunction with a 7T superconducting magnet system. The laser pulse duration was 2.75 microseconds, and the repetition rate was 1 Hz. The laser beam was divided into two by a beamsplitter. One beam was focused onto a pyroelectric detector as the reference signal, and the other was focused onto the sample, which was immersed in liquid helium. The transmitted signal was detected by a sensitive Ge bolometer, the output of which was amplified and stored in a digital oscilloscope. The spectra were recorded as the ratio of the signal to the reference.

Absorption saturation spectra were taken at 4 laser frequencies: two (42 cm^{-1} and 51 cm^{-1}) are below the intersubband separation, and the other two (61 cm^{-1} and 71 cm^{-1}) are above the intersubband separation. The sample was tilted at 18°. Transmittance for sample 1 at various laser intensities is shown in Fig. 7. For frequency $\nu=51$ cm^{-1}, a broad absorption profile (transition I) peaked at 4.9 T is seen at low laser intensities. At 18 W/cm^2, the absorption becomes much sharper, and is resonant at 4.55 T (transition III). For $\nu=61$ cm^{-1}, the asymmetric absorption feature (transition II) at low laser intensities is replaced by a sharper, symmetric feature at 32 W/cm^2 (transition IV). The resonant position is shifted from 4.5 T to 5.3 T. The saturation intensity, at which the absorption strength is reduced to half its value at low laser intensity, occurs at ~ 2.6 W/cm^2 for the upper branch; for the lower branch, the coarse spacing in laser intensity, from 0.075 W/cm^2 (still in the linear region) to 18 W/cm^2 (already strongly saturated), is too large, there is only a line shift observable, indicating transition I being replaced by transition III. The intensity dependence of the relative transmittance is plotted in Fig. 8.

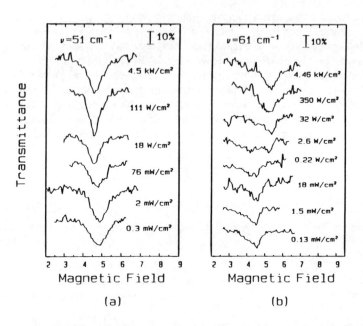

Fig. 7. An evolution of FEL spectra at different laser intensities for laser frequency at 51 cm^{-1} ($<E_2$-E_1) and 61 cm^{-1} ($>E_2$-E_1). Peak resonance positions are shifted under intense laser excitation.

To extract the population decay time, a two-level rate equation was solved with the electron density and absorption linewidth determined from the experiments. For multi-layer structures, layer-by-layer calculation of the absorption coefficient is necessary.[15] By fitting the absorption intensity vs laser intensity to the measurements with input values of the decay time τ (Fig. 8), we obtain $\tau=0.7\pm0.4$ nsec for laser photon energy $> E_2$-E_1, and $\tau=2.0\pm0.4$ nsec for laser photon energy $< E_2$-E_1. For these transitions, the initial state, $|i>$, is predominantly a pure Landau state; the final states, $|f>$, are linear combinations of the 1st excited LL of the ground subband, $|N=1, i=1>$, and the ground LL of the 1st excited subband, $|N=0, i=2>$, i.e. $|f> =a|N=1,i=1> +b|N=0,i=2>$. The dipole transition rate ($\sim |<i|\mathbf{A}\cdot\mathbf{p}|f>|^2 \sim 1/\tau$) is comprised of the intersubband transition rate and the inter-LL transition rate. Thus the lifetime

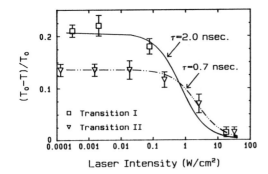

Laser Intensity (W/cm²)

Fig. 8. Relative peak transmittance for transition I (low branch) and II (upper branch) as a function of laser intensity. Theoretical curves with fitting parameter τ are also shown.

extracted from the data is a mixture of the lifetimes of the two interacting states. However, since a and b are comparable in the strong coupling region, the measurement is dominated by the shorter decay time of the two. The present result is consistent with direct saturation measurement of intersubband absorption on a wide QW (400Å) in which a decay time of 0.6 nsec is inferred;[16] it is also consistent with results from time-resolved Raman scattering on a 215Å MQWs in which a intersubband relaxation time of 0.57 nsec was measured. In all these cases the intersubband separations are less than the LO-phonon energy of GaAs, 297 cm⁻¹, thus the energy loss of hot electrons is primarily through emission of acoustic phonons, which is a much slower process than through emission of optical phonons, as has been observed in narrow QW structures.[17, 18]

VI. CONCLUSION

Optically pumped subband spectroscopy in single- and coupled double-quantum-well structures have been presented. Analysis of experimental results combined with theoretical calculations provides a direct test of the validity of the envelope function approximation and the boundary conditions. The oscillator strength of the intersubband transitions extracted from the measurements is in good agreement with the theoretical value. The modification of the intersubband spacing by a thin barrier at the well center is dramatic. This effect may have important applications in intersubband IR detectors. Saturation spectroscopy is also presented. Results indicate that the relaxation time of electrons from the excited subband to the ground subband is relatively long when the intersubband separation is less than the LO-phonon energy.

ACKNOWLEDGEMENT

It is a pleasure to acknowledge our collaborators, Drs. J. P. Kaminski and S. J. Allen, at UC Santa Barbara where saturation experiments were conducted; and Drs. M. I. Stockman, L. N. Pandey, T. F. George and Mr. L. S. Muratov, at Washington State University for theoretical calculation of CDQW structures. We also thank Dr. W. J. Schaff for high quality samples, and Dr. A. Petrou and his students for PL and reflectivity measurements. This work was supported by ONR/SDIO under the MFEL program.

REFERENCES

1. See for example, Intersubband Transitions in Quantum Wells. Proceedings of a NATO Advanced Workshop, edited by E. Rosencher, B. Vinter and B. Levine, (Plenum Publishing Corp. New York, 1991)
2. W. J. Li, B. D. McCombe, F. A. Chambers, G. P.Devane, J. Ralston, and G. Wicks, Phys. Rev. B, **42**, 11953, (1991).
3. A. Pinczuk, S. Schmitt-Rink, G. Danan, J. P. Valladares, L. N. Pfeiffer, and K. W. West, Phys. Rev. Lett. **63**, 1633 (1989).
4. M. O. Manasreh, F. Szmulowicz, T. Vaughan, K. R. Evans, C. E. Stutz, and D. W. Fisher, Phys. Rev. B **43**, 9996 (1991).
5. D. F. Nelson, R. C. Miller, and D. A. Kleinman, Phys. Rev. B, **35**, 7770 (1987).
6. W. P. Chen, Y. J. Chen, and E. Burstein, Surf. Sci., **58**, 263 (1979).
7. W. J. Li, and B. D. McCombe, J. Appl. Phys. **71**, 1038, (1992).
8. M. Helm, E. Colas, P. England, F. Derosa, and S. J. Allen, Appl. Phys. Lett. **53**, 1714 (1988).
9. W. Trzeciakowski and B. D. McCombe, Appl. Phys. Lett. **55**, 891 (1989).
10. F. M. Peeters and P. Vasilopoulos, Appl. Phys. Lett. **55**, 1106 (1989).
11. K. W. Goossen and S. A. Lyon, J. Appl. Phys. **63**, 5149 (1988).
12. W. J. Li, S. Homes, W. Trzeciakowski, B. D. McCombe, X. Liu, A. Petrou, M. Dutta, P. G. Newman, and W. Schaff, Superlattices and Microstructures **8**, 163 (1990).
13. T. Schmiedel, B. D. McCombe, A. Petrou, M. Dutta, and P. G. Newman, J. Appl. Phys. **72**, 4753 (1992).
14. W. J. Li, B. D. McCombe, J. P. Kaminski, S. J. Allen, M. I. Stockman, L. S. Muratov, L. N. Pandey, T. F. George, and W. J. Schaff, 8th International Conference on Hot Carriers in Semiconductors, August, 1993, Oxford, UK.
15 R. Ranganathan, J. Kaminski, W. J. Li, J.-P. Cheng, and B. D. McCombe, Proceedings of the SPIE, **1678**, 120 (1992).
16. K. Craig, C. Felix, J, Heyman, A. Markelz, M. Sherwin, K. Campman, A. Gossard, and P. Hopkins, 8th International Conference on Hot Carriers in Semiconductors, August, 1993, Oxford, UK.
17. A. Seilmeier, H.-J. Hübner, G. Abstreiter, G. Weimann, and W. Schlapp, Phys. Rev. Lett. **59**, 1345 (1987).
18. F. H. Julien, J.-M. Lourtioz, N. Herschkorn, D. Delacourt, J. P. Pocholle, M. Papuchon, R. Planel, and G. LeRoux, Appl. Phys. Lett. **53**, 116 (1988).

OBSERVATION OF INTERSUBBAND TRANSITIONS IN ASYMMETRIC δ-DOPED GaAs, InSb AND InAs STRUCTURES

C. C. PHILLIPS, H. L. VAGHJIANI, E. A. JOHNSON, P J P TANG and R. A.
STRADLING. Solid State Group, Physics Dept., Imperial College, London SW7 2AZ, UK.

J. J. HARRIS. Interdisciplinary Research Centre for Semiconductor Materials, Imperial
College, London, SW7 2AZ, UK.

M. J. KANE.,Defence Research Agency, St. Andrews Rd., Great Malvern, Worcestershire,
WR14 3PS, UK.

ABSTRACT. Optical measurements in the mid infrared band on asymmetric δ-doped GaAs and
asymmetrically slab-doped InSb and InAs doping superlattices (a-nipi's) at 10K showed strong
intersubband absorption features. We also report, for the first time, the observation of narrow (FWHM
~85cm^{-1}) intersubband absorption in 10nm wide δ-doped GaAs/AlGaAs quantum wells at wavelengths
near 10μm. The observed subband spectra are compared with a multi-layer optical matrix model in
which the bare subband splitting energies are calculated self-consistently and the well is taken to be an
anisotropic oscillator with an absorption resonance blue shifted by depolarisation effects. Good
agreement is found between the calculated and the measured transition energies for the δ-doped
GaAs/AlGaAs QW. In the δ-doped GaAs a-nipi transition energies are only weakly broadened by
impurity disorder and FWHM line widths of <300cm^{-1} are found. The long recombination time
produced by the electrostatic potentials in the a-nipi samples allows us to modulate the subband electron
concentrations by optical excitation. Modulation of the intersubband absorption spectra was seen for
optical pump densities as low as 20mW cm^{-2} for the GaAs superlattices and the induced subband spectra
yield new information about the subband structure.

1. Introduction

There has been a long term interest in manufacturing low noise, large area infrared detectors
which operate at acceptable temperatures and recently attention has focused on detectors
exploiting intersubband absorption as a route to improved detector performance without the
homogeneity, fabrication and metallurgical problems associated with the Cadmium Mercury
Telluride (CMT) materials system.

The GaAs δ-doped QW system offers a convenient way of tailoring the subband energy
splitting and number of populated subbands. Recent magnetotransport studies by Harris et al.[1]
have shown that, due to the difference in the electron probability density at the position of the
dopant plane for this system, there is a considerable difference in the subband electron mobilities

H. C. Liu et al. (eds.), Quantum Well Intersubband Transition Physics and Devices, 371–378.
© 1994 Kluwer Academic Publishers. Printed in the Netherlands.

of the QW. This allows for the possibility of detection of intersubband absorption by in-plane conductivity variations.

Semiconductor doping superlattices ("nipi" crystals) are also of interest for subband detectors primarily because materials choice is not limited by the lattice matching restrictions which apply to heterostructure systems. Here we study asymmetric δ-doped doping superlattices (a-nipis) which were designed so that the n-type sheet dopant density exceeded the p-type sheet dopant density by an amount such that the Fermi energy (computed self-consistently and with non-parabolicity effects taken into account[4]) lay between the first and second subbands of the electron well and the subband transition energies corresponded to the mid IR spectral region.

Narrow gap binaries like InAs and InSb are particularly appealing for this application because large and controllable doping concentrations ($\approx$ 4 x 10^{19} cm^{-3}) are readily achieved by MBE. This allows very large electrostatic modulations of the band-edge energies to be achieved over short length scales giving large electron confinement energies. New intersubband detector schemes can be envisaged which take advantage of the low effective masses and high mobilities in these materials and there is the exciting prospect of a monolithically integrated dual wavelength device structure operating simultaneously in the 3 - 5μm (interband absorption) and 8 - 14μm (intersubband absorption) bands. Also, because the "nipi" confining potential is electrostatic, it can in principle be varied either by external bias or by interband optical excitation to tune the subband absorption energy over a wide range.

Narrow gap doping superlattices are also interesting from a theoretical point of view; the calculation of the bare subband energies requires a self-consistent approach; non-parabolicity effects are large on account of the subband confinement energies being a substantial fraction of the bandgap, and depolarisation effects are important due to the high plasma frequencies resulting from the small effective masses.

Here we summarise recent experimental and theoretical results from a number of MBE grown δ-doped and slab-doped samples and show that a sufficiently accurate methodology now exists to design asymmetrically doped "nipi" structures in GaAs, InAs and InSb which have subband absorption features in the 8 - 14μm band.

2. Experimental Details

2.1 MBE GROWTH DETAILS OF SAMPLES STUDIED.

2.1.1 *GaAs asymmetric nipi (RMB 459)*. The superlattices were grown by MBE deliberately without wafer rotation (to study the effect of varying the n- and p- dopant sheet concentrations) and consisted of:- semi-insulating GaAs substrate; 1μm GaAs contact layer (n-type, 1.2x10^{18} cm^{-3}); 10nm undoped GaAs; Be δ-doped plane at 5.3 x 10^{12} cm^{-2};50 periods of a-nipi superlattice each consisting of { 14.5nm of undoped GaAs, Si δ-doped plane at 6.7 x 10^{12} cm^{-2}, 14.5 nm of undoped GaAs, Be δ-doped plane at 5.3 x 10^{12} cm^{-2}}; 10 nm undoped GaAs; 0.3μm GaAs capping layer (n-type, 1.2 x 10^{18} cm^{-3}).

2.1.2 *δ-doped GaAs/AlGaAs quantum wells (G957 and G958)*. The 10nm wide GaAs/Al$_{0.33}$Ga$_{0.67}$As δ-doped QWs[1] were doped nominally at 2.5 x 10^{12} cm^{-2} (G957) and 5.0 x 10^{12} cm^{-2} (G958), but magnetotransport measurements indicated a single occupied subband in both samples with a populations of 2.1×10^{12} cm^{-2} and 3.6×10^{12} cm^{-2} in G957 and G958 respectively. The Si δ-plane was grown at the centre of a single QW on a semi-insulating GaAs substrate.

2.1.3 *InAs asymmetric nipi (IC200).* This was grown by MBE deliberately without wafer rotation and consisted of:- semi-insulating GaAs substrate; 0.5μm undoped GaAs buffer layer; 0.5μm undoped InAs; 2 nm InAs (Be-doped at p ≈ 3×10^{19} cm^{-3}); 200 periods of a-nipi superlattice each consisting of {4 nm InAs (n ≈ 4.6×10^{19} cm^{-3}), 5.5nm of undoped InAs, 4 nm InAs (p ≈ 3×10^{19} cm^{-3}), 5.5nm of undoped InAs}; 2 nm InAs (p ≈ 3×10^{19} cm^{-3}), 30nm undoped InAs; 2.5nm InAs capping layer (p ≈ 3×10^{19} cm^{-3}).

2.1.4 *InSb asymmetric nipi (IC94).* Grown by MBE deliberately without wafer rotation and consisting of:- p-type InSb substrate; 0.5μm undoped InSb buffer layer; 40 periods of a-nipi superlattice each consisting of {10 nm InSb (n ≈ 1×10^{18} cm^{-3}), 40nm of undoped InSb, 10nm InSb (p≈7×10^{17} cm^{-3}), 40nm of undoped InSb}; 5nm InSb capping layer.

2.2 SPECTROSCOPIC MEASUREMENT DETAILS.

2.2.1 *Intersubband absorption measurements.* IR Transmission spectra were measured using an evacuated Bomem DA3 FTIR with the samples held in a closed-cycle variable temperature helium cryostat. Intersubband measurements were generally performed in a multi-pass waveguide geometry[2]. Instrumental response and substrate absorption features were removed by ratioing the s and p polarised transmitted spectra (inset figure 1b). Some of the samples were gold coated on the epilayer side to obtain a favourable electro-magnetic mode structure within the sample for the IR electric field to couple to the subbands of the well more effectively[2].

2.2.2 *Optically induced subband transmission spectra.* These were obtained under identical experimental conditions as above, but with the electron concentration in the a-nipi samples being modulated at ≈6.6KHz by interband optical excitation with an 810nm laser diode laser through an optical fibre. At low temperatures the type-II nipi potential gives long (10's μsec) interband recombination times and substantial fractional changes in the electron concentration are possible. Using double modulation techniques, at a pump density of 70mW cm^{-2}, an induced absorption spectrum was obtained using the FTIR spectrometer for each of the IR polarisations .

2.2.3 *Interband Photoluminescence.* The samples were mounted in the same closed-cycle cryostat used for the intersubband absorption measurements and were excited with low power density 810nm radiation modulated at a few kHz. Luminescence was collected with large aperture reflecting optics, dispersed with a 0.25m grating monochromator and detected with a 77K CMT photodiode.

3. Results and discussion.
3.1 GaAs ASYMMETRIC NIPI'S.

Figure 1a shows the normalised transmittance of the GaAs δ-doped a-nipi. The maximum small signal absorption occurs at 814 cm⁻¹ / 100meV and has a FWHM of 343 cm⁻¹ / 42 meV, largely determined by inhomogenous broadening effects caused by the impurity disorder statistics.

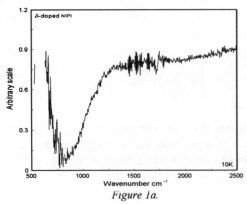

Figure 1a.

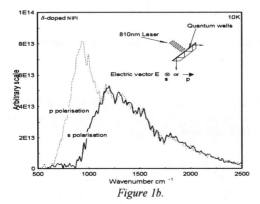

Figure 1b.

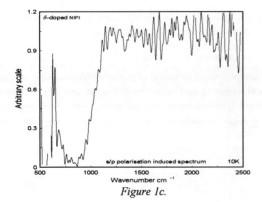

Figure 1c.

To model this data we first calculate the "bare" transition energy, E_{01}, between the i=0 and i=1 subband levels. The electrostatic part of the problem is handled with an iterative self-consistent computer program, and the effects of non-parabolicity are included by computing the mean energy of the electron in a given subband above the conduction band minimum and using a standard k.P expansion to calculate the appropriate electron effective mass.

The observed optical transition frequency, ω, is blue shifted by the the depolarisation (screening) effect which is calculated here (using the formalism, of ref. 7) as $\tilde{\omega}_{10}^2 = \omega_{10}^2 + \Omega_{p10}^2$ where the screening energy is given by $\Omega_{p10}^2 = 2N_s e^2 \omega_{10} S_{11} / \varepsilon_0 \varepsilon_r \hbar$. The length tensor S_{11} is computed from the appropriate integral (along the growth direction z) of the calculated subband wavefunctions and from the intersubband dipole matrix element $\langle i|z|f \rangle$. Values of these parameters derived for the various samples studied, together with the calculated optical resonance energies ($E'_{10} = \hbar \omega$) are listed in table 1.

Sample	E_{10} "bare" transition energy (cm^{-1}/meV)	$\langle 0\|z\|1\rangle$, dipole matrix element (nm)	S_{11}, tensor (see ref.7) (nm)	Length	d_{eff} (nm)	$\tilde{E}'_{10}$, calculated optical transition energy (cm^{-1}/meV)	Depolarisation energy, $\hbar\,\Omega_{p10}$ (cm^{-1}/meV)
RMB 459	1020/127	2.10	0.532		8.31	1115/138	451/56
G957	840/104	2.28	0.666		7.83	1009/125	560/69
G958	893/111	2.36	0.676		8.24	1173/146	760/94
IC 94	374/46	7.75	2.01		29.9	424/53	200/25

Table 1. Fitting parameters derived from measured optical absorption features in the samples studied.

Assuming the nominal dopant density figures, the measured and calculated absorption resonance energies differ by some 30%. but experimentation with the model revealed that departures from the nominal doping values of only 5% are required to bring the caclulated and measured values into agreement

The induced absorption spectrum of the same sample (figure 1b, p polarisation) is composed of two features arising from the two equal components of the 45° electric field vector normal and parallel to the sample growth direction. The normal component couples to the subband transitions and produces the sharp feature peaking around $\lambda\approx12\mu m$ and the parallel component gives rise to the broader non-resonant feature at shorter wavelengths. The spectrum for s polarised light shows just the non-resonant feature produced by the parallel electric field vector with a strength and spectral shape the same as that seen for the in-plane component of the p polarised radiation.

When the two induced absorption spectra are ratioed (figure 1c) a clear polarisation dependent feature is observed whose peak and FWHM correspond closely to that seen in the transmittance data. This implies that the effect of the extra photoinduced electron population on the *shape* of the electrostatic confining potential in the a-nipi is negligible; the majority of the induced change in subband absorption occurs from extra electrons with similar z-dependent wave function components to the doped-in electrons.

3.2. δ-DOPED GaAs/AlGaAs QUANTUM WELLS.

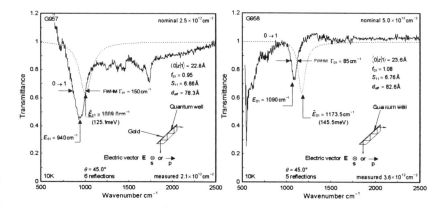

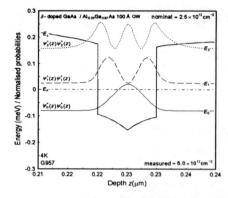

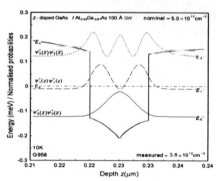

Figure 2.Experimental absorption spectra (solid lines) of the δ-doped GaAs/AlGaAs quantum wells G957 and G958 taken in the multipass waveguide geometry. Dotted lines, calculated absorption spectra as described in the text.

The modelling approach adopted was the same as that described in section 3.1, and the appropriate parameters (calculated using the sheet electron concentrations measured in the Shubnikov de Haas experiments, section 2.1.2) are listed in table 1.

Also listed is the effective thickness $d_{eff} = \langle i|z|f \rangle^2 / S_{11}$ of the 2D electron layer[8], from which it can be seen that the extra confining potential associated with the δ-doping has resulted in a more tightly bound electron layer than would result from the 10nm wide QW layer alone.

Figure 3 calculated subband wavefunctions of a) G957 and b) G958

The absolute transmittance curves (dashed lines, figure 1) were calculated by modelling the structure as an anisotropic multi-layer dielectric stack and using a standard anisotropic layered matrix model to calculate its optical properties and to take the EM boundary conditions imposed by the gold overlayer (where present) into account.

3.3 InAs ASYMMETRIC NIPI (IC200).

Figure 3a shows the transmittance data for IC200, the InAs a-nipi[5]. Unfortunately the spectrum in this sample is dominated by pronounced Fabry-Perot oscillations arising from the dielectric mismatch at the InAs and GaAs substrate and is rather more difficult to model quantitatively. Nevertheless the calculated absorption feature (dotted line), modelled using the same self-consistent approach described above for the GaAs a-nipis provides an acceptable fit to the data. Non-parabolicity effects are strong and the depolarisation term is more important than in the case of GaAs due to the small electron mass.

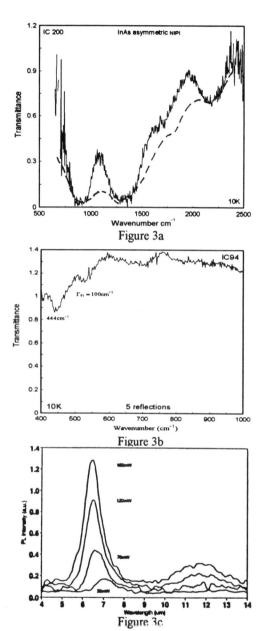

Figure 3a

Figure 3b

Figure 3c

Reasonable agreement was found between calculated and observed absorption features using doping and layer thickness values within the sample growth inaccuracies, although the calculations are complicated by an unintentionally high electron concentration ($\approx 6.4 \times 10^{12}$ from Shubnikov de Haas measurements) and the fact that the second subband was partially occupied at these densities.

3.4 InSb ASYMMETRIC NIPI (IC94).

The ratioed absorbance spectrum of the homoepitaxial InSb a-nipi[6] (fig.3b) shows a pronounced and sharp (100 cm^{-1} / 12 meV FWHM) subband absorption feature at 444 cm^{-1}/55 meV. The bare subband splitting energy (E_{01}=374cm^{-1} / 46 meV) was calculated self-consistently (in the same way as described for the GaAs a-nipi) assuming that the nominal values for the width and density of the n-doped slab were correct, and taking the net electron concentration (3×10^{11} cm^{-2}) from magnetotransport measurements on the same sample. Only the lowest subband was occupied here.

The calculated E_{10} value was rather insensitive to the effects of possible dopant diffusion; changing the dopant slab width from 10nm to 20nm in the calculation whilst keeping the sheet electron density constant (at $N_s = 3 \times 10^{11}$ cm^{-2}) resulted in a reduction of only 5meV in the calculated E_{01} value. It was also rather insensitive to the absolute value of the n- and -p-sheet dopant concentrations, provided that the difference between them (i.e. n_s as measured in the magnetotransport experiment) was held constant; changing the n-type sheet dopant concentration from 0.9 to 1.1 x 10^{12} cm^{-2} subject to this constraint only changed the calculated E_{01} value from 339 cm^{-1} / 42 meV to 395 cm^{-1} / 49 meV. In this material the final state excitonic effects are negligible on account of the small electron mass and large dielectric constant

378

and the depolarisation (screening) effect was calculated (table 1) to give a proportionately large depolarisation energy of $\hbar\Omega_{p10} = 200$ cm^{-1}/ 25meV. As can also be seen in table 1, the light electron mass results in a rather large ($\approx$30 nm) effective width for the 2DEG in this material

In this sample it proved possible to make a direct measurement of the bare subband splitting energy. A value of some 560 cm^{-1} / 69 meV was measured from the i=0 and i=1 peak separation in the low temperature interband PL spectrum (figure 3c) . The PL peaks correspond to electrons in the i = 0 and i = 1 subband states with close to zero in-plane wavevector, $k_{\parallel}$ whereas the intersubband transitions represent an average over $k_{\parallel}$ states which span approximately a third of the 2D Brillouin zone (at these electron densities). The difference between the two independently derived values of the bare subband splitting energy is thus due to the pronounced non-parabolic nature of the i = 1 subband. This gives it a heavier in-plane effective mass than the i = 0 subband and reduces the mean (when averaged over $k_{\parallel}$ space) energy separation between them.

Acknowledgements

We would like to thank Dr R Grey of Sheffield University UK for the kind provision of sample RMB 459 and the Science and Engineering Research Council, UK for financial support for this work.

References

1.   J. J. Harris, R. Murray and C. T. Foxon, *Semicon. Sci. Technol.*, **8**, 31, (1993).
2.   M. J. Kane, M. T. Emeny, N. Apsley, C. R. Whitehouse and D. Lee, *Superlattices and Microstructures*, **5**, 587 (1989).
3.   W. P. Chen, Y. J. Chen and E. Burstein, *Surf. Sci.*, **58**, 263, (1976).
4.   E. A. Johnson, and A Mackinnon, *Semicond. Sci. Technol.*, **5**, S189, (1990).
5.   C. C. Phillips, E. A. Johnson, R. H. Thomas, H. L. Vaghjiani, I. T. Ferguson and A. G. Norman. *Semicond. Sci. Technol* **8**, S373-S379, (1993).
6.   C. C. Hodge, C. C. Phillips, R. H. Thomas, S. D. Parker, R. L. Williams and R. Droopad, *Semicond. Sci. Technol.*, **5**, S319, (1990).
7.   S. J. Allen Jr., D. C. Tsui and B. Vinter, *Solid State Commun.*, **20**, 425-428, (1976).
8   T. Ando, A. B. Fowler and F. Stern, Rev. Mod. Phys., **54**, 437, (1982).

EFFECTS OF COUPLING ON INTER-SUBBAND TRANSITIONS

J.M. Xu
Department of Electrical and Computer Engineering
University of Toronto
10 King's College Road
Toronto, Canada M5S 1A4

ABSTRACT.   This paper outlines an attempt to probe into the issue of inter-subband transitions from a different angle - coupling effects. It describes some of our latest efforts in exploring the potential of electronic and photonic couplings for selective control of the inter-subband transitions and in understanding the related coupling mechanisms and effects. Specifically, three types of coupling mechanism are discussed: the spatially direct band-to-band coupling, the spatially indirect band-to-band coupling, and the long-range photonic coupling. The aim of this work is to achieve a good qualitative understanding of the primary effects and their physical origins. Detailed and quantitative modeling and calculations - tasks more suitable for separate publications - are left out of this condensed article. While emphasizing the physical origins of the coupling effects, the discussions are extended to the configurations of new quantum well structures and device consequences, where appropriate.

1. Band-to-Band Coupling Effects on Inter-subband Transitions

1.1 Basic Considerations

As it is well known, inter-subband transitions can take place in a conduction band (CB) quantum well (QW) or a valence band (VB) quantum well.

For transitions in a VB QW, it is now reasonably well known that the mixing of the valence bands, degenerated at Γ in all cubic semiconductors, can have significant impact on the intra-valence band transitions, although simplified calculations ignoring the mixing also exist in the literature. Since the subject of intravalence band transitions has been examined by many, it will not be considered in the present work. Instead, we will focus on the seemingly less complex inter-subband transitions in *conduction band quantum wells*.

For a CB QW (or wells), the inter-subband transition processes are often viewed and treated within the conduction band. The single-band approach is attractive due to its simplicity and has been widely used in the literature. Its justification relies on the belief that as long as the band gap is wide enough, a process taking place close to the conduction band edge is dominated by the conduction band characteristics. While this point of view may usually be valid and the single-band approach is often successful, they could be grossly inadequate or even obscure in many cases. This is because, as this work attempts

H. C. Liu et al. (eds.), Quantum Well Intersubband Transition Physics and Devices, 379–388.

to show, the band-to-band couplings in some cases are significant enough to make critical differences for inter-subband transition processes in CB QWs.

One case of particular interest is the inter-subband absorption or emission of light polarized in the plane of a CB QW. In a common square-well configuration, it is a popular belief that, "For light polarized in the planes, inter-subband transitions become second-class and are allowed only in the valence band, due to valence band mixing" [1a]. Hence, the normal-incident far infrared light is thought to be undetectable by conduction band square quantum wells in the usual planar detector configurations [1b]. However, the use of CB QWs is more attractive than the valence band well alternative, because of the higher mobility, lower effective mass, deeper well, and low dark current. Consequently, much effort is devoted to engineering these detectors to facilitate the Brewster angle or the 45 degree bevel angle incidences. This technological complication is, of course, very undesirable. This work is one of the attempts to seek a better solution. It finds that in this case a solution may lie within an improved understanding of the properties of the conduction subbands. We show below that by taking into account the contributions from the valence bands, the in-plane inter-subband transition rates could be non-zero and they follow a new selection rule.

Before we proceed to the theoretical analysis, it is noted that although the importance of band couplings in narrow band gap materials like InAs or InSb is relatively well-known [2], the fact that the interband coupling effect could be sizeable in common QWs made of wide band gap materials has not always been appreciated. In a GaAs QW, for example, there are two main factors contributing to the coupling effects: (1) the increased mixing of the Bloch functions of all the bands at large k has a sizeable effect on subbands located > 0.1 eV above the conduction band edge; and (2) the wave penetration into the barrier layer brings in an additional and spatially-indirect coupling to the valence band of the surrounding material whose strength depends on the band alignment.

1.2 A Two-Band Approach

To assess the impact of band-to-band couplings on inter-subband transitions in a CB QW, we consider a generic QW structure shown below:

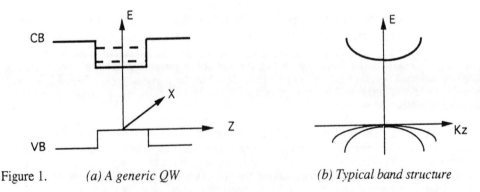

Figure 1.   (a) A generic QW   (b) Typical band structure

For the bulk material of each layer, one has the well-known band structure depicted in Fig.1b, typical of zincblend and diamond semiconductors at Γ. The expression of an exact wavefunction in *a conduction band quantum well* can be given by:

$$\Psi(r) = \Sigma_n \Sigma_k \ \phi_n(\mathbf{k}) \exp(-i\mathbf{k}.\mathbf{r}) u_{nk}(\mathbf{r}) \qquad (1)$$

It sums over all conduction and valence bands (n's) and all $\mathbf{k}$. This expression is exact but not useful, and an approximation is usually made. In standard k.p perturbation theory [2], one approximates the Bloch functions, $u_{nk}(\mathbf{r})$, of the bulk material as follows:

$$u_{nk}(\mathbf{r}) = u_{n0}(\mathbf{r}) + \Sigma_{m \neq n} \frac{\mathbf{k}.\mathbf{p}_{mn}}{m_0 \ (E_n(0)-E_m(0))} u_{m0}(\mathbf{r}) \qquad (2)$$

On the basis of this approximation, hence its underlying assumptions, Eq.(2) makes it apparent that for $k \neq 0$ there is a finite coupling of the n-th band to the other bands even in bulk. However, it will become clear below that even if one decides to neglect this effect of $\mathbf{k}.\mathbf{p}$ mixing of the Bloch functions the coupling of the bands could still have significant effect on the properties of the total wavefunction.

For the total wavefunction in the QW and within the envelope function approximation, we have [3]:

$$\Psi(r) = \Sigma_n \{F_n(\mathbf{r}) u_{n0}(\mathbf{r}) + \Sigma_{m \neq n} \frac{-i\nabla F_n(\mathbf{r}).\mathbf{p}_{mn}}{m_0 \ (E_n(0)-E_m(0))} u_{m0}(\mathbf{r}) \} \qquad (n=1,2....N) \qquad (3)$$

This N-component total wavefunction Ψ is quantized into subband states Ψ_l in a QW, $\Psi_l(\mathbf{r})$. Therefore, the transitions between subbands in a QW inherently involve all the "bulk" band, via both the envelope functions $F_{n,l}$ and the Bloch functions u_{n0}. This is still true without the second term in Eq.(3), which contributes to the non-parabolicity at $k \neq 0$. Another point worth stressing is that no assumption of the existence of a local crystal or strain deformation potential or the interface roughness is made here, in contrast to the case studied in [4].

Since the total eigen functions in a CB well also contain valence band components, a multi-band Hamiltonian becomes necessary for treating the system. In principle, the 8-band Luttinger-Kohn Hamiltonian, together with the boundary conditions [3]. can be used to determine the eigen energies E_l and eigen states Ψ_l of the subbands and the inter-subband transitions. This is a rather cumbersome process of mathematical manipulations and has been done in our recent work [5].

To make the effects of band-to-band coupling more apparent - a primary aim of this article - simplifications are necessary. Below, it will be shown that a qualitative understanding of the band-to-band coupling effects on intersubband transitions can be established whitin a two-band representation. In particular, it will also become clear that some transition processes that are prohibited in the single-band theory are in fact allowed when the band-to-band coupling is considered. The two bands included in this work are the conduction and the valence bands; the heavy hole band is neglected as it does not interact at all with the other bands at $k_{\|} = 0$ and interacts very little for small $k_{\|}$ [3]. The quantized total wavefunctions and the approximate two-band Hamiltonian, the minimum for the modelling of interband coupling effects, for a simple QW, are given by Equations (4) and (5) below.

$$\Psi_l(\mathbf{r}) = F_{c,l}(z) u_{c0}(\mathbf{r}) + F_{v,l}(z) u_{v0}(\mathbf{r}) + \frac{a_c}{E_g} u_{c0}(\mathbf{r}) + \frac{a_v}{E_g} u_{v0}(\mathbf{r}) \qquad (4)$$

$$\begin{bmatrix} E_c + \dfrac{k_z^2}{2m^*} & -i\sqrt{(\tfrac{2}{3})}\,pk_z \\[4mm] i\sqrt{(\tfrac{2}{3})}\,pk_z & E_v - \dfrac{1}{2}(\gamma_1 + 2\gamma_2)k_z^2 \end{bmatrix} \qquad (5)$$

Since our emphasis is on a qualitative understanding of the coupling effect and a number of approximations have been made, the last two second-order terms in Eq.(5) will be omitted hereafter without affecting the main conclusion. Hence, we have

$$\Psi_1(\mathbf{r}) \approx F_{c,1}(z)u_{c0}(\mathbf{r}) + F_{v,1}(z)u_{v0}(\mathbf{r}) \qquad (6)$$

At this point, we draw attention to the fact that the conduction band Bloch function u_c transforms like the $|s\rangle$ atomic orbital while the light hole u_v ($|3/2, 1/2\rangle$) is a superposition of the $|x\rangle$, $|y\rangle$ and $|z\rangle$ atomic orbitals; and $\langle u_{c0}|u_{v0}\rangle = 0$, $\langle u_{c0}|p_x|u_{c0}\rangle = \langle u_{v0}|p_x|u_{v0}\rangle = 0$, but $\langle s|p_x|x\rangle \neq 0$ [6].

Now, let us look into the specific issue of the in-plane transitions, and see if the matrix element $\langle \Psi_1(\mathbf{r})|p_x|\Psi_{1'}(\mathbf{r})\rangle$ between two subbands 1 and 1' can be non-zero. Keeping in mind that in-plane field operator p_x now operates on both the envelope functions F's and the Bloch basis functions u's and that $\sum_i \langle F_{i,1} | F_{i,1'}\rangle = \delta_{11'}$, , one has the following general conclusion:

$$\langle \Psi_1(\mathbf{r})|p_x|\Psi_{1'}(\mathbf{r})\rangle \approx \langle F_{c,1}|F_{v,1'}\rangle\langle u_{c0}| p_x |u_{v0}\rangle + \langle F_{v,1}|F_{c,1'}\rangle\langle u_{v0}| p_x |u_{c0}\rangle \qquad (7)$$

$$\neq 0 \qquad (\text{ if } \langle F_{c,1}|F_{v,1'}\rangle \neq 0,\ \langle F_{v,1}|F_{c,1'}\rangle \neq 0,\ \text{and } k_x \neq 0)$$

Consistent with the common knowledge, the part of the momentum matrix element associated with the envelope functions remains zero in this two-band model. But, the other part of the matrix element originated from the Bloch functions is finite, as long as the envelope functions have the same parity.

Here, two points are worth stressing. First, in the foregoing analysis no assumption of the existence of a local deformation and interface roughness or high-k nonparabolicity was made. Of course it does assume $k_x \neq 0$. Second, there is a general selection rule of $\Delta l = (l - l') = 2, 4, 6...$ for the allowed in-plane transitions. This finding is in agreement with that of the more rigorous 8-band treatment described in our recent article [5]. This is not surprising because the two-band model retains the essential physics of band-to-band coupling and the primary coupling terms. However, there would a difference in the calculated oscillator strength as expected; and naturally the more accurate (but less clear to the understanding) 8-band model should have been utilized if quantitative results were the focus of interest.

We note that the $\Delta n = 1$ transitions, the focus of most experiments, are still prohibited in the 2-band model. This may explain why the conventional wisdom of no in-plane transitions remained unchallenged in the past. However, some recent experiments have provided evidence of the occurrence of $\Delta n = 1$ in-plane transitions [4,7,8], which are attributed to a local crystal deformation or interface roughness [4,8] that break the inversion symmetry and facilitate in-plane transitions.

Although the 8-band model appears to be the preferred general approach for quantitative calculations of transition rates, the standard Luttiger-Kohn formulation in the literature is quite complex mathematically. It becomes even more complex in the cases of quantum wires (or dots) in which we anticipate a growing interest in the inter-subband transitions. For this reason, I shall draw attention to a general coordinate-free representation [9,10] of the Kane model before we leave this subject. This coordinate-free method takes into account the band-to-band coupling, employs SU(2) group operators and retains the symmetry (invariance) of the system without reference to any particular coordinate system. The advantages of this coordinate-free model are that for a specific low-dimensional system, most of the mathematical manipulations can be performed algebraically without reference to any particular coordinate system. It thus avoids most of the great mathematical complexity otherwise associated with solving the Luttinger-Kohn Hamiltonian for those low-dimensional structures. Projection to a most convenient coordinate system is done at the last step to obtain the quantitative results. This general model can be readily applied to the problem of inter-subband transitions. Examples of band structure calculations of quantum wires and dots can be found in our early work [9,10].

2. Spatially indirect band-to-band couplings:

As we become more aware of the potential benefits of band-to-band coupling, one may wish to have a stronger band-to-band coupling effect in a material than the material itself can provide. So, the question is: is this just wishful thinking ?

The answer lies in the barrier region. In a usual QW structure, such as AlGaAs/GaAs/AlGaAs, one would expect little enhancement of the band-to-band coupling effects from the wide band gap barriers. However, in a *leaky quantum well* depicted in Figure 2, an enhancement can be achieved with a spatially indirect coupling between the conduction band of the well layer and the valence band of the outer barrier layer.

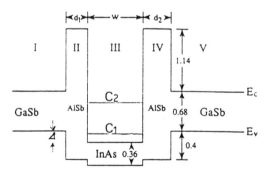

Figure 2   *A typical leaky QW with a leaky window $\Delta=0.15eV$ at the bottom of the well.*

Clearly, the strength of this indirect band-to-band coupling depends on the thickness of the inner barrier layer. It could be strong enough to make the ground state from bound to quasi-bound and even change its character from a conduction band type to one that is mostly valence band type [11]. Through the leaky window Δ at the bottom of the conduction band well, the wave function of the ground state C_1 couples to the valence band continuum in the outer barrier layer, hence possessing a broader linewidth and a shorter lifetime τ_1 than that of the upper state C_2. This lifetime inversion has direct consequences in inter-subband transitions. One example proposed recently [12] is the possibility of achieving population inversion between subbands C_2 and C_1 and facilitating inter-subband lasing via such spatially direct band-to-band couplings.

The fact that this leaky QW is constructed in Type-II heterostructure system may be discouraging to some who are skeptical about the practicality of using "exotic" materials in real applications. Fortunately, leaky QWs can also be created in the common Type-I heterostructures. One example of a leaky QW in Type-I system is illustrated below:

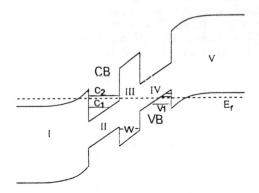

Figure 3.   *A leaky quantum well created in a Type-I heterostructure - Resonant Interband Tunneling Diode [13].*

This would be a simple PN junction without the two narrow band gap layers inserted on each side of the junction. The CB QW on the N-side is coupled through the junction barrier to the VB QW on the P-side. Due to this spatially indirect band-to-band coupling, the conduction subbands C_1 and C_2, etc., have significant contributions from both bands, i.e., $\Psi_{Ci} = (F_c(z), F_v(z))_{Ci}$. Similarly, the valence subbands V_1 and V_2, etc., are $\Psi_{Vi} = (F_c(z), F_v(z))_{Vi}$. By applying external bias, one can move the P-side up or down with respect to the N-side, thereby changing the alignment between the two wells and changing a conduction subband from in-resonance to off-resonance with a valence subband. In this way, the strength of the band-to-band coupling becomes externally controllable which is rather attractive for device applications. Specifically, the matrix element, $M_{VC} = \langle \Psi_C |$ $H_p(V_a) | \Psi_V \rangle$, is dependent on the applied voltage, V_a, where Ψ_C and Ψ_V are the two-component subband wave functions at zero bias.

Now, let us consider the DC-biased structure in a superimposed photon field, $F_z = 2 F_0 \cos(\omega t)$, where $h\omega = E_{C2} - E_{V1}$. One possible inter-subband transition is: an electron absorbs a photon and makes a transition to the conduction subband C_2 from the valence subband V_1 which is in resonance or (near resonance) with the conduction subband C_1. This transition could be direct from V_1 to C_2 or indirect from V_1 to C_1 then to C_2. The time-dependent Schrodinger equation in the form of a transfer Hamiltonian is then given, as in [14], by:

$$ i\hbar \frac{d}{dt} \begin{bmatrix} \Psi_{V1}(t) \\ \\ \Psi_{C2}(t) \end{bmatrix} = \begin{bmatrix} E_{V1} & M_{V1C2} \\ \\ M_{V1C2} & E_{C2}+2eF_0Wcos(\omega t) \end{bmatrix} \begin{bmatrix} \Psi_{V1}(t) \\ \\ \Psi_{C2}(t) \end{bmatrix} \tag{8} $$

and the formal solution to this equation is:

$$ \Psi_{C2}(t) = \Psi_{C2}(0) \exp \{ -i\frac{E_{C2}t}{\hbar} - i\frac{2eWF_0}{\hbar\omega}\sin(\omega t) \} $$

$$ \approx \Psi_{C2}(0) \exp(-i\frac{E_{C2}t}{\hbar}) \{ J_0(\frac{2eWF_0}{\hbar\omega}) + 2J_2(\frac{2eWF_0}{\hbar\omega})\cos(2\omega t) +.... $$

$$ -2iJ_1(\frac{2eWF_0}{\hbar\omega})\sin(\omega t) -2iJ_3(\frac{2eWF_0}{\hbar\omega})\sin(3\omega t) -..... \tag{9} $$

The only term in the above expansion series of Bessel functions that contributes to the net transition rate is the term with time dependence of $\sin(\omega t)$. The other terms oscillate and have no net contribution to the real transition. The effective dipole moment is then $M_{V1C2}J_1(\frac{2eWF_0}{\hbar\omega})$ and the photocurrent I is thus proportional to $| M_{V1C2}J_1(\frac{2eWF_0}{\hbar\omega}) |^2$.

Therefore, in this spatially indirect coupled system the spatially indirect photon-assisted tunneling process, the photocurrent is tunable by V_a through M_{V1C2}, and is nearly linear for small normalized optical power $(\frac{2eWF_0}{\hbar\omega})^2$ and becomes highly nonlinear at larger power, as shown in the following plot.

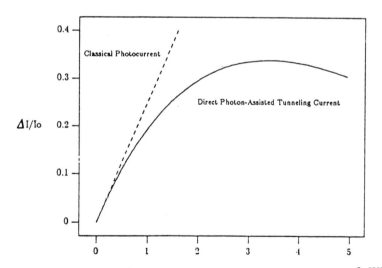

Figure 4   *Normalized photocurrent $\Delta I/I$ versus normalized optical power $(\frac{2eWF_0}{\hbar\omega})^2$*

due to a photon-assisted transition from subband V_1 to subband C_2 of the spatially-indirect coupled quantum wells in Figure 3.

We note that the reverse transition - emission of photons can be treated similarly.

In closing this discussion on the effects of spatially indirect band-to-band coupling, I would like to mention that leaky QWs can also be created without the use of any type of heterojunction. For example, a δ-doped PNPN junction, first proposed as one of three basic configurations of resonant interband tunneling diodes in [13], is such a leaky QW where the spatially indirect band-to-band coupling has significant effect on the inter-subband transitions within the same well or between wells.

3. Long-Range Photonic Coupling:

In addition to the short-range electronic couplings, one may consider long-range photonic coupling for the purpose of gaining control over certain inter-subband transition processes. In the remaining part of this work, I briefly discuss how this might be achieved in a new device structure proposed here, as an example.

Consider a common layered structure with a multi-quantum-well (MQW) region sandwiched in between two doped layers, as depicted in Figure 5.

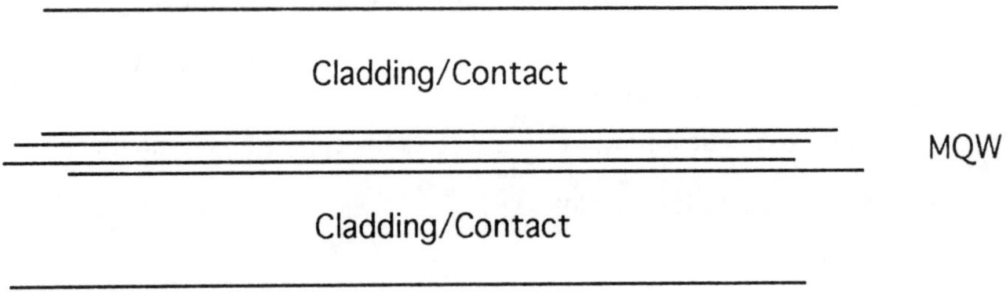

Cladding/Contact

MQW

Cladding/Contact

Figure 5. *A common layered MQW structure for inter-subband emission/absorption.*

The MQW region is where the inter-subband emission or absorption of photons is to take place. The doped layers act as both optical cladding layers to guide light emission/absorption along the MQW in the z-direction and as electrical contact layers to supply carriers into the MQW region. The specific features of an optimized device design are similar to an edge-emitting laser of long wavelength ($\sim 10\mu m$), but are not the focus of the present work. The focus and the new feature here is the incorporation of a long-range photonic coupling mechanism into this otherwise standard device structure. This coupling is to be electrically tunable and can be used to switch on or off an inter-subband transition process. To this end, the layered structure needs to be processed into a periodic mesa structure, like in a gain-coupled laser, with a period L corresponding to the wavelength of the particular transition. This periodic mesa structure is schematically illustrated in Figure 6. This structure itself can be, like for an ordinary laser, embedded in either a ridge waveguide or buried heterostructure to form an edge emitting laser. Also sketched in this figure is the fact that since the period L is chosen to be the photon wavelength of the particular inter-subband transition, a standing wave would be built up and be in-phase with the mesa structure. The in-phase condition implies a maximum overlap between the photon field and active media, therefore a maximum transition rate (emission or absorption).

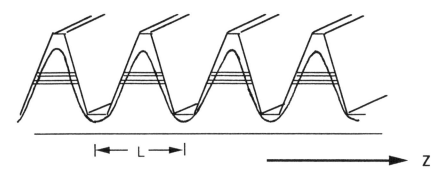

Figure 6 *Periodic mesa structure in which the MQW sections are coupled by light wave.*

There is also a complimentary situation where the photon field is out-phase with the mesa structure, namely, it peaks in the center of each passive section and has a minimum gain. In between these two extreme cases is a so-called photonic gap region where no emission or absorption is possible as light is repelled from the structure. This is perhaps best illustrated by the photon dispersion diagram illustrated in Figure 7. The normal photon dispersion diagram is a con represented by the dashed line. In a sufficiently long periodic structure like the one shown in Figure 6, the con is intersected at $k_z = \pm\pi/L, \pm 2\pi/L, ..etc$ by planes parallel to the ω-$k_\parallel$ plane. At the intersecting points, a photonic gap is created. Below the gap is the "red" branch whose top end is the exact in-phase condition and has a zero group velocity. Above the gap is the "blue" branch whose low end is the exact out-phase condition and also has a zero group velocity.

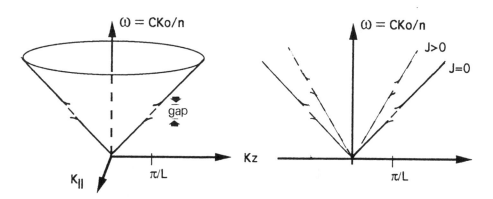

Figure 7 *The photon dispersion diagrams for the periodic photon coupled structure in figure 6, with and without current injection.*

Since $\omega = \dfrac{k_0 C}{n}$, we can tune the modal index n of the waveguide by current injection that will result in a shift in the photon dispersion curve. For current induced plasma effect, the dispersion curve will shift toward the vertical axis ω, as n decreases. The photonic gap is therefore moved upwards too, as shown on the right side of Figure 7. If MQW is designed in such a way that the transition energy resides near the photonic gap and, say,

388

in the red branch, then it is easy to see that we can switch this transition process on and off by varying the current injection level. This is the basic operation principle of a switchable inter-subband emitter or detector that is made possible by a long-range photonic coupling. Of cause, the actual implementation would require us to find solutions to a series of engineering problems and issues that are subjects of future work this article attempts to stimulate.

Acknowledgement: This work summarizes our recent investigations in this field, of which much has been done with Dr. M. Sweeny and later with Dr. R.Q. Yang. The investigations have been supported by OCMR, OLLRC and BNR.

References:
[1a] Andreani, L.C., (1993) Lecture Notes on "Linear Optical Properties in Bulk and Low-Dimensional Semiconductor Structures", NATO Advanced Study Institute on Confined Electrons and Photons, Erice, Italy, July 13-26.
[1b] Weisbuch C. and Vinter B., (1991) "Quantum Semiconductor Structures", Academic, Boston.
[2] Kane, E.O. , (1966) in Semiconductor and Semimetals: Physics of III-V Compounds 1, Ed. R.K. Willardson and A.C. Beer, Academic Press, New York.
[3] Altarelli, M., (1986) "Band Structure, Impurities and Excitons in Superlattice", in "Heterojunctions and Semiconductor Superlattices" Ed. by G. Allan et. al., Springer-Verlag, Berlin.
[4] Peng, L.H., Smet, J.H., Broekaert, and Fonstad, C.G.(1992), "Transverse electric and transverse magnetic polarization active intersubband transitions in narrow InGaAs quantum wells", Appl. Phys. Lett., 61(17), 2078-2080.
[5] Yang, R.Q., Xu, J.M., and Sweeny M., "Selection rule of intersubband transitions in conduction band quantum wells", submitted to Phys. Rev. B.
[6] For example, Feymann, R. (1965) "Lecture Notes in Physics", Addison-Wesley, Reading.
[7] Liu, H.C., Buchanan, M. and Wasilewski Z.R., (1991) Phys. Rev. B 44, 1411.
[8] Peng, L.H., Smet, J.H., Broekaert, and Fonstad, C.G., (1993), "Strain effects in the intersubband transitions of narrow InGaAs quantum wells", Appl. Phys. Lett., 62(19), 2413-2415.
[9] Sweeny, M. and Xu, J.M. (1989) "Hole Energy Levels in Zero-Dimensional Quantum Balls", Solid State Comm. 72(3), 301-304.
[10] Sweeny, M., Xu, J.M. and Shur, M., (1988) "Hole Subbands in 1- Dimensional Quantum Well Wires", J. of Superlattice and Microstructures, 4(4/5), 623-626.
[11] Yang, R.Q. and Xu, J.M. (1992)"Bound and Quasi-Bound States in 'Leaky' Quantum Well", Phys. Rev. B., 46(11), 6969-6974.
[12] Yang, R.Q. and Xu, J.M. (1991) "Population Inversion through Resonant Interband Tunneling", Appl. Phys. Lett., 59(2), 181.
[13] Sweeny, M. and Xu, J.M. (1989), "Resonant Interband Tunnel Diodes", Appl. Physics Lett., 54(6), 546-548.
[14] Sweeny, M. and Xu, J.M. (1989), "On Photon-Assisted Tunneling in Quantum Well Structures", IEEE Quantum Electronics, 25(5), 885-888.

ON SOME PECULIARITIES OF INTERSUBBAND ABSORPTION IN SEMICONDUCTOR QUANTUM WELLS

Z. Ikonić, V. Milanović

Faculty of Electrical Engineering, University of Belgrade,
Bulevar Revolucije 73, 11000 Belgrade, Yugoslavia

ABSTRACT. Some interesting effects appearing in intersubband optical transitions in quantum wells are discussed. In the first part, the influence of optically induced charge redistribution on intersubband separation is explored and criteria for wells sensitive or insensitive to this effect given. In the second part the influence of the effective mass position dependence on intersubband absorption is analyzed. It was found, *inter alia*, that transitions induced by in-plane light polarization become allowed as well.

1. Introduction

There has recently been strong interest in intersubband transitions related optical effects in semiconductor quantum wells (QW) [1–3]. Both linear and non-linear phenomena (absorption, second harmonic generation, optical bistability, etc) have been widely studied theoretically, as well as experimentally. Here we discuss some unusual features related to intersubband absorption in QWs: the influence of optically induced charge redistribution over bound levels on interlevel separation (Sec. 2), and the influence of the position dependent effective mass on absorption (Sec. 3).

2. Excitation Induced Levels Shifting in Quantum Wells

In order to obtain technically significant effects based on intersubband optical transitions, one usually needs QW structures with larger values of electron sheet density N_s. In such cases the self-consistency effects (charge screening) tend to play an important role in determining the electron bound states energies. When the intensities of optical fields interacting resonantly with electron system are high, as they usually are, especially in nonlinear interactions, the electron distribution over the relevant levels will significantly deviate from the equilibrium one, the one it starts to evolve from when optical fields are turned on. This electron redistribution will self-consistently induce the electron levels to change, which

389

H. C. Liu et al. (eds.), Quantum Well Intersubband Transition Physics and Devices, 389–397.
© 1994 *Kluwer Academic Publishers. Printed in the Netherlands.*

affects the interaction resonance conditions. It is therefore of interest to explore the behavior of these levels as the electron distribution departs from equilibrium.

While these effects exists in both symmetric and asymmetric structures, they will clearly be much more pronounced in the latter, because of permanent dipole moments present (in nonlinear optics they are more interesting anyway). A simple model employed to describe the effect—"charge separation dipole model"—states that, as the electron sheet density in the upper state, N_2, varies, the interlevel separation changes by a factor $e^2 N_2 (z_{22} - z_{11})/\epsilon \epsilon_0$, where z_{ii} are the permanent dipole moments ($i = 1, 2$). It was used e.g. in [3] in studying optical bistability in asymmetric coupled QWs. However, a more reliable result may be obtained by solving the Schrödinger equation self-consistently.

2.1. SELF-CONSISTENT CALCULATIONS

For a number of asymmetric QW structures known in various contexts in intersubband optical transitions physics, we made the self-consistent calculations of levels separation, as it depends on the degree of excitation. Basically, the envelope function Schrödinger equation

$$\frac{\mathrm{d}}{\mathrm{d}z} \frac{1}{m(z)} \frac{\mathrm{d}}{\mathrm{d}z} \psi + U_0(z)\psi + U_c(z)\psi + U_{exc}(z)\psi = E\psi \qquad (1)$$

where U_0, U_c and U_{exc} are the built-in, Coulomb and the exchange-correlation potentials, respectively, is solved in the iterative self-consistent way, as described in a number of papers (e.g. [2]), the only difference being that the electron distribution over the available states was preset, instead of allowing it to take the thermal equilibrium form. Some of the results are displayed in Fig. 1. The structures included in it are: (a) asymmetric step QW $Al_{0.4}Ga_{0.6}As$-$GaAs(60Å)$-$Al_{0.18}Ga_{0.82}As$ $(90Å)$-$Al_{0.4}Ga_{0.6}As$; (b) symmetric rectangular QW $Al_{0.4}Ga_{0.6}As$-$GaAs(90Å)$-$Al_{0.4}Ga_{0.6}As$, in electric field $K = 10^5$ V/cm; (c) asymmetric coupled step QW $Al_{0.44}Ga_{0.56}As$-$GaAs(16Å)$-$Al_{0.44}Ga_{0.56}As(16Å)$-$Al_{0.25}Ga_{0.75}As(48Å)$-$GaAs(48Å)$ -$Al_{0.44}Ga_{0.56}As$; (d) asymmetric coupled QW $Al_{0.4}Ga_{0.6}As$-$GaAs(20Å)$-$Al_{0.3}Ga_{0.7}$ $As(30Å)$-$GaAs(60Å)$-$Al_{0.4}Ga_{0.6}As$. The electron (i.e. doping) sheet densities N_s are (1) 10^{12} cm^{-2}, (2) $3.4 \cdot 10^{11}$ cm^{-2}, (3) 10^{11} cm^{-2}, (4) $1.44 \cdot 10^{12}$ cm^{-2}, where only the outer barriers are taken to be doped in a symmetric manner. All these structures have their E_{21} approximately matched to CO_2 laser radiation.

A remarkable point to note is that E_{21} vs. $D = \rho_{11} - \rho_{22}$ dependence is non-monotonous, and critically depends on N_s ($\rho_{11,22}$ are the diagonal elements of the density matrix). At low N_s values E_{21} decreases as the upper state population goes up, while for high N_s E_{21} increases, but only after going through local minimum at $D \approx 0.9$. QWs with N_s in the intermediate range, say $(3 - 5) \cdot 10^{11}$ cm^{-2} have their E_{21} best stabilized against the variations in levels population. The origin

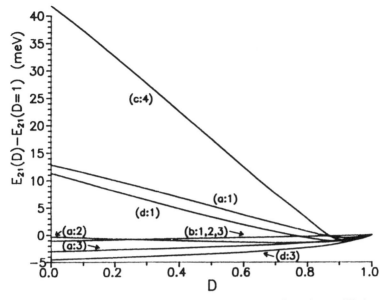

Fig. 1 Deviation of the lowest two levels separation E_{21} from its equilibrium value, plotted against the population difference $D = \rho_{11} - \rho_{22}$, for a number of QW structures and electron sheet densities (see text for notation), at $T = 77$ K.

of this nonmonotonous behavior is the inclusion of the exchange potential U_{exc} in the self-consistent Schrödinger equation. Without it, E_{21} would only increase with decreasing D. Clearly, the influence of U_{exc} is comparatively lowest at high electron densities, but even at $N_s = 10^{12}$ cm^{-2} it is quite significant: the slopes of $E_{21}(D)$ in the linear region are still $1.5 - 2$ times larger than would be the case if $U_{exc} = 0$. Furthermore, by comparing the values in Fig. 1 with those obtained by the charge separation dipole model, we find that the latter tends to overestimate the slope of $E_{21}(D)$ by a factor of $2 - 3$.

In some applications, like the resonant second harmonic generation, it is desirable for the interlevel separation to be insensitive to the excitation level to the largest possible extent, so that the resonance conditions are not perturbed by varying pump power. In this respect we may conclude that intermediate N_s values, typically $(3 - 5) \cdot 10^{11}$ cm^{-2} are best suited for such applications. That this range of N_s is optimal for second harmonic generation was also concluded in [1] from quite different requirements. If, however, different values of N_s have to be employed for some reasons, and levels walk-off has to be taken into account, this should be done by solving the Schrödinger equation self-consistently, the charge separation dipole model can hardly be considered as accurate enough. A QW structure should be designed so that specified resonance conditions are met at the electron distribution-to-be, not at equilibrium.

2.2. SOME NOTES ON OPTICAL BISTABILITY

It is well known that the optical bistability phenomenon based on intersubband transitions is possible in (especially asymmetric) QWs, relying on the excitation induced levels shifting—increased absorption positive feedback mechanism [e.g. 3]. For this purpose a strong $E_{21}(D)$ dependence is essential, and the asymmetric coupled step QW (c) in Fig. 1 seems most suitable to achieve bistability (this type of structure was proposed in [4] to get the largest Stark shift presently available, and we modified its parameters to obtain the largest $E_{21}(D)$ sensitivity, displayed in Fig. 1). At large N_s values, where $E_{21}(D)$ dependence is strong, it is also linear in D to a very good approximation, except in a small range of D values close to equilibrium. However, in the interesting range of D, where switching will occur, E_{21} can be taken as linearly dependent on D, though not with the slope given by the charge separation dipole model.

Based on this linearity Khurgin [3] has derived an approximate equation (eqn.(8) in Ref.[3]) that straightforwardly gives the minimum detuning at which bistability can appear at all, optimum detuning of laser vs. transition frequency, and minimum slope of E_{21} at this detuning. We want to point out that some of these results are in fact exact, not approximate (within the model). Indeed, introducing the dimensionless initial detuning $a = [E_{21}(D_0) - \hbar\omega]/\Delta E$, where D_0, $\hbar\omega$ and ΔE are the initial (equilibrium) population difference, laser photon energy and absorption line half width, respectively, and the dimensionless slope b, as $E_{21}(D) = E_{21}(D_0) + b(1 - D/D_0) \cdot \Delta E$, the steady-state equation relating the dimensionless light intensity $i = I/I_{sat}$ and the variable $x = 1 - D/D_0$, introduced for convenience, reads

$$i = \frac{(a^2 + 1)x + 2abx^2 + b^2 x^3}{1 - x} \tag{2}$$

Depending on a and b, the $x(i)$ dependence may be either S-shaped, giving bistability, or is a single valued function. Since eqn.(2) is not a polynomial, but a polynomial fraction, the simple cubic equation analysis cannot be used to distinguish between the two cases. To do it, $\mathrm{d}i/\mathrm{d}x$ and $\mathrm{d}^2i/\mathrm{d}x^2$, both being polynomial fractions with third order polynomial in their numerators, are required to have a common zero. Using the Sylvester determinant criterion to find conditions when this is possible, we find that a system is on the verge of bistability if $b = 8a^3/(27 - 9a^2)$, at $x = (3 - 9/a^2)/4$. From $x \geq 0$ it follows that bistability is impossible for $|a| < \sqrt{3}$, and from $\mathrm{d}b/\mathrm{d}a = 0$ that $a_{opt} = \pm 3$ (hence $b_{min} = \mp 4$). The last two results are the same as those obtained by Khurgin [3] (note that b, as defined here, is half of his K) and the first one is new.

Specifically, the asymmetric coupled step QW of Fig. 1 should be able to display bistability if the linewidth $2\Delta E \lesssim 20\,\mathrm{meV}$ could be obtained, which seems quite realistic.

3. Effective Mass Difference Induced Normal Incidence Absorption

3.1. THEORETICAL CONSIDERATIONS

In this part we explore in some more detail the influence of the effective mass difference in the well and barrier materials of a QW structure on optical absorption. Consider a well of width $2d$ embedded in barrier of width L on either side of it ($L \rightarrow +\infty$). The absorption coefficient α, due to transitions between two bound states is given by [5]

$$\alpha = \frac{1}{4\pi L n \epsilon_0 c \omega A^2} \int\!\!\!\int\limits_{-\infty}^{+\infty} |P_{if}|^2 \left(f_{FD}(E_i) - f_{FD}(E_f)\right) \delta(E_f - E_i - \hbar\omega)\, \mathrm{d}k_x \mathrm{d}k_y \quad (3)$$

where n is the refraction index (assumed to be constant), and $E_{i,f}(k_t^2 = k_x^2 + k_y^2)$ the initial and final quantized states energies. The transition matrix element P_{if} is

$$P_{if} = \int\limits_V F_i^* \hat{H}' F_f \, \mathrm{d}V \equiv \mathcal{M}_{if} \cdot \epsilon A \quad (4)$$

where ϵ is the unit polarization vector of the magnetic vector potential $\mathbf{A}$, $F_{i,f}$ the envelope wavefunctions of the two states, $\hat{H}'$ is the interaction Hamiltonian, equal to the difference of full Hamiltonians with $\mathbf{A}$ present and absent, $\hat{H}_A - \hat{H}_0$, ie.

$$\hat{H}_0 = \hat{\mathbf{p}} \frac{1}{2m} \hat{\mathbf{p}} + U_{eff}(z), \qquad \hat{H}_A = (\hat{\mathbf{p}} + e\mathbf{A}) \frac{1}{2m} (\hat{\mathbf{p}} + e\mathbf{A}) + U_{eff}(z) \quad (5)$$

where $U_{eff}(z)$ includes the real binding potential $U(z)$ and the transverse kinetic energy $\hbar^2 k_t^2 / 2m(z)$.

Due to z dependence of $m(z)$, the potentials in both $\hat{H}_0$ and $\hat{H}_A$ are actually nonlocal [6], their nonlocal parts being $(\hat{\mathbf{p}} \frac{1}{m}) \hat{\mathbf{p}}/2$ and $(\hat{\mathbf{p}} \frac{1}{m}) \hat{\mathbf{p}}/2 + e\mathbf{A}(\frac{1}{m})\hat{\mathbf{p}}$, respectively. With the velocity operator $\hat{\mathbf{v}} = i[\hat{H}, \hat{\mathbf{r}}]/\hbar$ the interaction Hamiltonian may be written as

$$\hat{H}' = e\hat{\mathbf{v}}_0 \mathbf{A} - \frac{e^2 A^2}{2m} \approx e\hat{\mathbf{v}}_0 \mathbf{A} \quad (6)$$

where $\hat{\mathbf{v}}_0$ is the electron velocity operator calculated at $\mathbf{A} = 0$, ie.

$$\hat{\mathbf{v}}_0 - \frac{1}{m}\hat{\mathbf{p}} + \frac{1}{2}\left(\hat{\mathbf{p}}\frac{1}{m}\right) \quad (7)$$

If $m(z)$ was constant, the potential in $\hat{H}_0$ would be local, and $\hat{\mathbf{v}}_0 = \hat{\mathbf{p}}/m$, ie. $\hat{H}' = (e/m)\hat{\mathbf{p}}$, therefore $\mathcal{M}_{if} \sim \int F_i^* \hat{\mathbf{p}} F_f \, dV$. However, with $m(z)$ really z-dependent, the transition matrix element takes the form

$$\mathcal{M}_{if} = e \int F_i^* \hat{\mathbf{v}}_0 F_f \, dV = -\frac{i\hbar e}{m_0} \left[\int \frac{F_i^*}{m^*} \nabla F_f \, dV + \frac{1}{2} \int F_i^* \nabla(\frac{1}{m^*}) F_f \, dV \right]$$

$$\equiv -\frac{i\hbar e}{m_0} \mathbf{M}_{if} \tag{8}$$

where m^* denotes the electron effective mass in free electron mass units.

Eqn.(8) may be derived in another way, as well. Using the expression for the complete wavefunctions $\psi_{i,f}$, calculated with the local, microscopic Hamiltonian [7]

$$\psi(\mathbf{r}) = F(\mathbf{r})u_{n0}(\mathbf{r}) + \sum_{m \neq n} \frac{-i\nabla F(\mathbf{r}) \cdot \mathbf{p}_{mn}}{m_0(E_{n0} - E_{m0})} u_{n0}(\mathbf{r}) \tag{9}$$

where $u_{n0}(\mathbf{r})$ are the cell-periodic parts of Bloch wavefunctions, and where $\mathbf{p}_{mn} = \int_{\text{cell}} u_{n0}^* \hat{\mathbf{p}} u_{m0} \, dV$ the Kane matrix element, which is material (i.e. z) dependent, and related to the (also z-dependent) effective mass via the sum rule. Now, since the Hamiltonian is local, the expression for the transition matrix element

$$\mathcal{M}_{if} = \frac{e}{m_0} \int_{(V)} \psi_i^* \hat{\mathbf{p}} \psi_f \, dV \tag{10}$$

takes, after some manipulations, the form given by eqn.(8).

Since the envelope functions in planar QW structures take the form $F(\mathbf{r}) = f(z)\exp[i(k_x x + k_y y)]$, eqn.(8) may be recast as

$$\mathbf{M}_{if} = M_z \mathbf{i}_z + i(k_x \mathbf{i}_x + k_y \mathbf{i}_y) M_{xy} \tag{11}$$

where

$$M_z = \int_{(z)} \frac{f_i^*}{m^*} \frac{df_f}{dz} \, dz + \frac{1}{2} \int_{(z)} f_i^* \frac{d}{dz}\left(\frac{1}{m^*}\right) f_f \, dz \tag{12}$$

and

$$M_{xy} = \int_{(z)} \frac{f_i^* f_f}{m^*} \, dz \tag{13}$$

Using eqn.(7), M_z may also be written in the classical dipole form

$$M_z = -\frac{m_0}{\hbar^2}(E_f - E_i) \int_{(z)} f_i z f_f \, dz \tag{14}$$

Certainly, the wavefunctions $f_{i,f}$ may be chosen as real, and then M_z and M_{xy} are also real. Writing the unit polarization vector ϵ as $\{\cos\phi\sin\theta, \sin\phi\sin\theta, \cos\theta\}$, where θ is the angle between $\mathbf{A}$ and the z-axis, $|P_{if}|^2$ takes the form

$$|P_{if}|^2 = [M_z^2\cos^2\theta + M_{xy}^2(k_x^2\cos^2\phi + k_y^2\sin^2\phi + k_x k_y\sin 2\phi)\sin^2\theta]\frac{e^2\hbar^2 A^2}{m_0^2} \quad (15)$$

Finally, since $E_{i,f}$ and $f_{i,f}$ depend on $k_t^2 = k_x^2 + k_y^2$ only, using eqn.(15) in (3), and performing the integration gives the expression for the QW fractional absorption $a = 2L\alpha$

$$a = \frac{e^2\hbar^2}{2n\epsilon_0 cwm_0^2}[f_{FD}(E_i) - f_{FD}(E_f)]\frac{M_z^2\cos^2\theta + \frac{1}{2}M_{xy}^2 k_t^2\sin^2\theta}{|\mathrm{d}[E_f(k_t^2) - E_i(k_t^2)]/\mathrm{d}(k_t^2)|}\Bigg|_{k_{t0}^2} \quad (16)$$

where k_{t0}^2 is the solution of $E_f(k_t^2) - E_i(k_t^2) - \hbar\omega = 0$. In eqn.(16) it is obviously assumed that the only source of line broadening is the z-dependent effective mass itself (i.e. variation of interlevel separation as k_t^2 varies), but other sources could be included as well, by substituting the denominator with an appropriate Lorentzian. In further analysis, however, we will use eqn.(16).

The first "part" of absorption, due to M_z^2, exists if there is a finite light electric field component along the z-axis, and is well known. We may only note that the matrix element M_z evaluation proceeds as usual if the dipole interaction form is used, eqn.(14), while the corresponding "$\mathbf{A}\hat{p}$ form" is somewhat unusual, eqn.(12), due to the coordinate dependence of the effective mass.

The second term in absorption, proportional to M_{xy}^2 will exist if the light electric field has an in-plane component, and will clearly be the only one present in case of normal light incidence. The finite value of the matrix element M_{xy} is a consequence of the effective mass z-dependence. If m^* were constant, M_{xy} would be zero because of wavefunctions orthogonality. However, when multiplied by an (essentially arbitrary) function $1/m^*(z)$, their product will not integrate out to zero, although the value of M_{xy} may not be very large. This type of absorption, to our knowledge, has not previously been discussed in the literature. Transitions leading to it are not dipole in origin, and the matrix element M_{xy} cannot be recast into a form analogous to eqn.(14).

On the quantitative side, the evaluation of M_{xy} may require some precautions. In most QWs having simple structure, large variations of the effective mass are usually associated with regions where the wavefunctions are already small, i.e. with the barriers regions. In such cases only the wavefunctions "tails" multiplied by the "exces" $1/m^*(z)$ factor contribute to M_{xy}, and its value will be rather small. To find its value accurately, then, it is essential that the wavefunctions, if calculated numerically, should be orthogonalized (e.g. by the Gram-Schmidt procedure) prior

396

to evaluation of eqn.(13). Otherwise, using wavefunctions that are not sufficiently orthogonal would introduce an error that might completely mask the real effect.

3.2. NUMERICAL CALCULATIONS

Numerical calculations of normal incidence absorption were performed for a rectangular GaAs QW in $Al_{0.4}Ga_{0.6}As$ barrier. With $m^*(z)$ being an even function, same as potential, it is obvious from eqn.(13) that only the same parity transitions will here be allowed. The well width is chosen to be 200Å, so that five bound states are supported. The barrier is taken to be uniformly doped to 10^{15} cm^{-3} and thus fixes the Fermi level, and electron distribution over the QW bound states. The corresponding electron sheet density at $T = 77$K is $N_s = 1.3 \cdot 10^{11}$ cm^{-2}. With this comparatively low doping the self-consistency effects were neglected, and the whole calculation was performed analytically, so no errors of the type discussed above could occur.

Numerical results are displayed in Fig. 2. Absorption arising from transitions $1 \rightarrow 3$, $2 \rightarrow 4$ and $3 \rightarrow 5$ is displayed, while the one corresponding to $1 \rightarrow 5$ turned out to be negligible. The linewidths (and lineshapes) in Fig. 2 (solid lines) correspond to the exact evaluation of eqn.(16), and broken lines to approximate calculation, where the denominator of eqn.(16) was replaced by its three terms

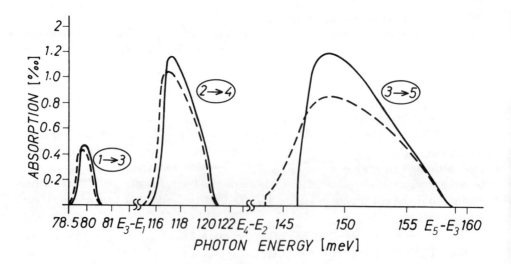

Fig. 2 The normal incidence fractional absorption vs. photon energy dependence in a 200Å wide GaAs QW embedded in $Al_{0.4}Ga_{0.6}As$ bulk, doped to 10^{15} cm^{-3}, at $T = 77$ K. Solid lines correspond to the exact calculation, eqn.(16), and broken lines to approximating the lineshape function (denominator of eqn.(16)) by its Taylor expansion at $k_t^2 = 0$.

Taylor expansion at $k_t^2 = 0$. There is some difference between the two, but it would become irrelevant when these lineshapes are convoluted with additional broadening mechanism.

The calculated values of absorption at normal incidence are clearly way below those occurring at parallel incidence, normal polarization transitions. Though maybe not of technical significance in simple QW structure, this new type of absorption should be experimentally measurable—e.g. an MQW structure with some tens of single QWs would provide a fractional absorption in the few percent range. Additionally, more complicated QW structures may be designed to enhance this effect, but we did not attempt this here. If this absorption could be enhanced by an order of magnitude, it would become technically interesting—the $2 \rightarrow 4$ transition, for example, is matched to CO_2 laser radiation.

4. Conclusion

Some interesting effects occurring in intersubband optical transitions in QWs were analyzed. In the first part of the paper the influence of optically induced charge redistribution on QW bound states energies was analyzed, and parameters of QWs designed to be very sensitive or very insensitive to this effects are pointed out. In the second part, the influence of the effective mass position dependence on intersubband transitions was discussed. Specifically, it was found that nonconstant effective mass enables the in-plane polarized (e.g. normal incidence) light absorption.

5. References

[1] J. Khurgin, *J. Opt. Soc. Am. B*, **6**, 1673 (1989).

[2] Z. Ikonić, V. Milanović, D. Tjapkin, *IEEE J. Quantum El.*, **25**, 54 (1989).

[3] J. Khurgin, *Appl. Phys. Lett.*, **54**, 2589 (1989).

[4] B. S. M. Lin, S. J. Lin, J. Hwang, *J. Appl. Phys.*, **72**, 5329 (1992).

[5] U. Bockelmann, G. Bastard, *Phys. Rev. B*, **45**, 1688 (1992).

[6] R. Girlanda, A. Quattropani, P. Schwendimann, *Phys. Rev. B*, **24**, 2009 (1981).

[7] M. Altarelli, *in Semiconductor Superlattices and Heterojunctions*, edited by G. Allan, G. Bastard, N. Boccara, M. Lannoo and M. Voos (Springer-Verlag, Berlin, 1986).

THE RELATIVE STRENGTHS OF INTERBAND AND INTERSUBBAND OPTICAL TRANSITIONS : BREAKDOWN OF THE ATOMIC DIPOLE APPROXIMATION FOR INTERBAND TRANSITIONS

M G BURT
BT Laboratories
Martlesham Heath
Ipswich IP5 7RE
United Kingdom

ABSTRACT. In assessing the relative strengths of interband and intersubband optical transitions in a quantum well it is natural to assume that the dipole matrix element for an allowed interband transition is of the order of atomic dimensions and therefore much smaller than that for an allowed intersubband transition.It is pointed out that this assumption is at variance with experiment for low bandgap materials by some orders of magnitude.It is shown theoretically how this discrepancy can be resolved.

1. Introduction

An attractive feature of optical intersubband transitions is the large dipole matrix element of the n=0 to n=1 transition[1].In the design of devices that use both interband and intersubband transitions, for example three level structures[2,3] for nonlinear optics applications,one needs to know the relative strength of the two types of transitions and the dipole matrix element is an important factor.It is natural to assume that the dipole matrix element for allowed interband transition n=0 (valence band) to n=0(conduction band) is of atomic dimensions ~ 1Å because the envelope functions change little in the transition yet the periodic atomic related part of the wavefunction changes from p symmetry to s symmetry in direct gap zincblende semiconductors.While this argument is not bad for the wide band gap materials it fails spectacularly for small band gaps as an estimate in the next section will show;dipole matrix elements an order of magnitude larger than atomic dimensions are needed to get agreement with experiment for small bandgaps.In section 3 the mathematical origin of these large interband dipole matrix elements is identified . they arise from small components of the wave function that are usually neglected,but turn out to be essentail for determining the dipole matrix element.It is also pointed out how a combination of errors produces correct answers,and consequent agreement with experiment,but helps sustain the erroneous impression that the interband dipole matrix elements are of atomic size.To avoid unnecessary complications the arguments in section 3 are given for the case of a one dimensional crystal system.Extension of the

399

H. C. Liu et al. (eds.), Quantum Well Intersubband Transition Physics and Devices, 399–402.
© 1994 *Kluwer Academic Publishers. Printed in the Netherlands.*

results to quantum wells in three dimensional crystals is straight forward.The results are summarised in section 4.

2. Estimate of the Interband Dipole Matrix Element

Take a type I quantum well composed of direct gap zincblende semiconductors and let the axis perpendicular to the interfaces be the z axis.The dipole matrix element for an optical transition with TM polarised light between a conduction band state $\Psi^{(c)}$ and a valence band state $\Psi^{(v)}$ is

$$\int \Psi^{(c)} * z\Psi^{(v)}d^3\mathbf{r} \tag{1}$$

If the states are bound in the z direction,then standard textbook manipulations allow one to express the matrix element of the z component of momentum in terms of the matrix element of z, that is

$$\int \Psi^{(c)} * p_z\Psi^{(v)}d^3\mathbf{r} = im\omega_{cv}\int \Psi^{(c)} * z\Psi^{(v)}d^3\mathbf{r} \tag{2}$$

where $\hbar\omega_{cv} = E_c - E_v$ is the energy separation of the two states.Now suppose we ignore spin orbit splitting and consider the n=0 conduction band and valence band ground states for a wide well.We further restrict our attention to the valence band state which has z symmetry for its periodic part ie the light hole n=0 state.Because the integrand of the integral on the RHS of (2) is dominated by the differentiation of the rapidly varying Bloch functions (see e.g. ref 4) the matrix element of momentum,in atomic units $\hbar = m = 1$,is approximately equal in magnitude to the Kane[5] matrix element parameter P.So in atomic units,the magnitude of the interband dipole matrix elements is approximatley P/E_g.It is now very informative to estimate this quantity for InSb.Kane[5] needed $P^2 = 0.44$ atomic units to fit the bulk absorption spectrum.Now E_g=0.23 eV for InSb and the unit of energy in atomic units is 27.2 eV (the double Rydberg) giving a value of 0.0085 for E_g in atomic units.The dipole matrix element (1) then has magnitude of 78 atomic units.Since the atomic unit of length is the Bohr radius = 0.529 Å,then our estimate of the interband dipole matrix element is about 41 Å.This is the interband dipole matrix element one would expect for the n=0 light hole to n=0 conduction band in a wide InSb quantum well for TM polarisation in the absence of spin orbit splitting.The introduction of spin orbit splitting only reduces the matrix element by $\sqrt{2/3}$ (this being the coefficient of the z like valence band periodic function in the light hole Bloch state[5]) ie to 34 Å.This is enormous by atomic standards.One can repeat this procedure for other materials using the data gathered by Lawaetz[6].Small band gap materials always lead to large interband dipole matrix elements.This is because P is roughly the same for all the direct gap zincblende materials,yet the interband dipole matrix element,as we have deduced from (2),varies as P/E_g.

3. Evaluation of the Interband Dipole Matrix Element using Envelope Functions

In this section we identify the origin of the large value of the dipole matrix element for allowed interband transitions.To avoid inessential detail we will consider the one dimensional case and the interband dipole matrix element has the simple form

$$\int_{-\infty}^{+\infty}\Psi^{(c)} * z\Psi^{(v)}dz \tag{3}$$

For a quantised state near the band edge,the wavefunction can be reasonably represented by a simple product FU where F is a slowly varying envelope function and U is the band edge Bloch function (see eg ref 7).For the conduction and valence band states we write

$$\Psi^{(c)} = F_c^{(c)}U_c \qquad (4a) \qquad \text{and} \qquad \Psi^{(v)} = F_v^{(v)}U_v \qquad (4b)$$

As an example,for the n=0 states in a wide deep well the envelope functions will have the simple approximate form

$$F_c^{(c)} = F_v^{(v)} = \sqrt{\frac{2}{L}}\cos\frac{\pi z}{L} \qquad (5)$$

inside the well ($-L/2 < z < +L/2$) and zero outside.To evaluate (3) using the approximation shown in equation (4a) and (4b) one needs to evaluate

$$\int_{-\infty}^{+\infty}\left(F_c^{(c)}U_c\right)^* zF_v^{(v)}U_v dz \qquad (6)$$

In reference 4 it is shown that this integral is zero in the limit of slowly varying envelope functions in contrast to the answer usually given namely

$$\frac{1}{a}\int U_c^* z U_v dz \qquad (7)$$

where the region of integration is over a unit cell,length a.As shown in reference 4,this answer,(7), cannot be right because it depends on how the unit cell is chosen whereas (6) has a unique value.

So for slowly varying envelope functions our first approximation (6) to the interband dipole matrix element yields the answer zero even for allowed transitions.This cannot be correct.The approximations (4a) and (4b) are clearly too crude.The author has shown[7,8] that it is possible to define an envelope function expansion that exactly represents the wavefunction so it must be possible to obtain the correct answer by taking sufficient number of terms in the envelope function expansion.As shown in reference 4 it is sufficient to write

$$\Psi^{(c)} = F_c^{(c)}U_c + F_v^{(c)}U_v \qquad (8a)$$

$$\Psi^{(v)} = F_v^{(v)}U_v + F_c^{(v)}U_c \qquad (8b)$$

The envelope functions $F_v^{(c)}$ and $F_c^{(v)}$ introduced in (8a) and (8b),can be determined[4] using envelope function theory.For the example of n=o states in a wide deep well,$F_v^{(c)}$ and $F_c^{(v)}$ (corresponding to $F_c^{(c)}$ and $F_v^{(v)}$ given by (5)) are given approximately by[4]

$$F_v^{(c)} = F_c^{(v)*} = (i\hbar p_{vc} / mE_g)(\pi / L)\sqrt{2 / L}\sin(\pi z / L) \qquad (9)$$

where E_g is the bandgap and p_{vc} is the interband momentum matrix element

$$p_{vc} = \frac{1}{a}\int U_v^* p_z U_c dz \qquad (10)$$

the integral being taken over a unit cell ie any range of length a.(Plots of all the envelope functions mentioned above for a model one dimensional quantum well problem can be found in reference 7).One sees that the envelope functions in (9) become negligible compared to the dominant ones in (5) in the limit of wide wells because of the presence of the extra factor of L in the denominator of the RHS of (9).However,the presence of the extra terms in (8a) and (8b) radically alters the result we get for the dipole matrix element (3).As shown elsewhere[4] for a wide deep well one gets

402

$$\int_{-\infty}^{+\infty} \Psi^{(c)} * z\Psi^{(v)} dz = \frac{P_{cv}}{im\omega_g} \qquad (11)$$

where ω_g is the bandgap frequency. The important terms in the integrand of the dipole matrix

element integral are $\left(F_v^{(c)} U_v \right)^* z F_v^{(v)} U_v$ and $\left(F_c^{(c)} U_c \right)^* z F_c^{(v)} U_c$ .ie products containing a large term in

one wavefunction with the small term in the other wavefuction. Because both $U_c^* U_c$ and $U_v^* U_v$ have mean value unity over a unit cell, they do not contribute significantly to the integral for slowly varying envelope functions and these factors may be dropped from the integral. The envelope functions in these products have opposite parity and hence lead to a nonzero contribution to the dipole matrix element integral.

In the above, we have pointed out the error in equating the dipole matrix element (3) with the integral (7), an error that reinforces the erroneous impression that the matrix element (3) is of atomic dimensions. Another error seen in the literature is to assume that (2) (or rather its one dimensional equivalent) also applies to the periodic functions U_c and U_v ; this is wrong since in deriving (2) boundary terms arising from integration by parts, which can be neglected for bound state functions, are nonzero for periodic functions. However, if these two errors are combined, then one arrives at the correct result (11), and hence agreement with experiment, but with the erroneous impression that the dipole matrix element is of atomic dimensions. The fact that the RHS of (11) is hardly ever evaluated explicitly, because it is just one factor in a complicated formula for, say the absorption coefficient, helps maintain this state of affairs.

4. Summary

We have shown that the dipole matrix element for allowed interband transitions in a quantum well can be at least an order of magnitude larger than atomic dimensions for low bandgap zincblende materials. Using a well known textbook theorem it has been shown that the interband dipole matrix element is proportional to P/E_g where P is the corresponding momentum matrix element and E_g is the band gap. P is roughly independent with bandgap for direct gap zincblende materials and hence the dipole matrix element is roughly inversely proportional to the bandgap. This in turn leads to large values for the interband dipole matrix element for lowbandgap materials. We have shown how to get the correct result using envelope function/effective mass theory; small admixtures of contributions from other bands to the wavefunction are crucial. It has also been pointed out how a combination of errors can lead to correct answers, and agreement with experiment, but still sustain the erroneous impression that the interband dipole matrix element is of atomic dimensions.

5. References

1. L C West and S J Eglash   Appl Phys Lett **46** 1158 (1985)
2. M Walmsley R A Abram and M G Burt   Semicond Sci Technol **8** 268 (1993)
3. I Morrison M Jaros and M W Beavis   Appl Phys Lett **55** 1609 (1989)
4. M G Burt   J Phys : Condens Matter **5** 4091 (1993)
5. E O Kane   J Phys Chem Solids **1** 249 (1957)
6. P Lawaetz   Phys Rev **B4** 3460 (1971)
7. M G Burt   J Phys : Condens Matter **4** 6651 (1992)
8. M G Burt   Semicond Sci Technol **3** 739 (1988)

Optical Transitions and Energy Level Ordering for Quantum Confined Impurities

S. R. Parihar and S. A. Lyon
Department of Electrical Engineering
Princeton University
Princeton New Jersey
USA 08544

ABSTRACT. Optical transitions on impurities confined in a quantum well are considered as an alternative to intersubband transitions. Because of the discrete energy level structure, if the impurity energy level spacing is pushed beyond the optical phonon energy, the nonradiative phonon relaxation should be effectively suppressed. Radiative processes then have a better chance to produce more efficient light emission and detection than is currently obtained through intersubband transitions. Two examples are presented to illustrate the concept.

1. Introduction

The intersubband transition in a quantum well (QW) is very flexible in that it can be varied over a wide energy range in the infrared (IR) by changing the well width and barrier heights. This versatility is evident in the variety of QW IR detectors that have been demonstrated to date. Unfortunately the complementary process, light emission via subband transitions, has proven to be far more difficult to achieve. So far, only the case of far IR emission has been demonstrated. (Helm, 1988,1989).

For light emitting schemes the ultimate goal is to make an IR laser. In Figure 1a) we sketch out a model four level laser system. The main point is that between levels 2 and 3 (the lasing levels) one wants the nonradiative relaxation to be relatively slow compared to the radiative relaxation in order to obtain reasonable quantum efficiency. The other transitions between 2 and 1 and 4 and 3 should be fast to aid in population inversion. All four levels can be readily produced in a QW as shown in Figure 1b). The problem is that all the QW transitions are dominated by very fast nonradiative phonon relaxation, including the lasing transition. The reason for this is shown in Figure 2. The in plane dispersion of carriers in a quantum

403

H. C. Liu et al. (eds.), Quantum Well Intersubband Transition Physics and Devices, 403–410.
© 1994 *Kluwer Academic Publishers. Printed in the Netherlands.*

404

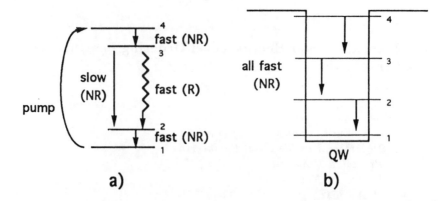

Fig. 1. a) Schematic four level laser system and b) realization in a QW. Radiative and nonradiative transitions are labelled by R and NR respectively.

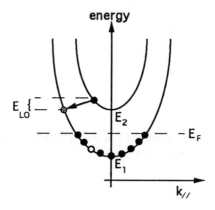

Fig. 2. Illustration of the nonradiative intersubband transition via LO phonon emission from subband 2 to subband 1 in a QW.

well gives nested subbands and does not possess a real energy gap in the mid or far IR. As a result phonon transitions (especially longitudinal optical (LO) phonon emission in III-V materials) easily occur leading to about 1ps lifetimes for subband separations greater than the LO phonon energy. Such short lifetimes necessitate high injection current levels to achieve inversion (Borenstain and Katz, 1989). Recent schemes have employed tunneling in attempts to obtain fast carrier transfer to the upper lasing level and removal from the lower one in order to compensate for the very quick relaxation rate. However, inversion may still be difficult to obtain (Helm and Allen, 1990). We wish to approach the problem at a more basic level - namely how to avoid the phonon relaxation in the first place. If we increase the confinement of

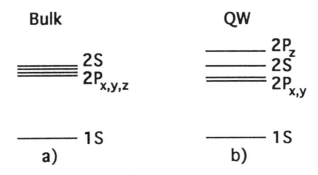

Fig. 3. Impurity energy level ordering in a) bulk and b) a QW.

the carriers so that they are fully confined in all three dimensions the energy levels become discrete as in atomic systems. Bockelmann and Bastard (1990) have shown in this case the calculated phonon scattering between levels is reduced. Furthermore, if we take the zone center optical phonon energy to be approximately the highest single phonon energy possible, then by having the level spacing greater than this ensures no single phonon transitions are allowed. Radiative transitions would then compete against the much lower probability multiphonon events.

The current work on quantum dots indicates that the fabrication of structures small enough ($\sim$ 10nm) to provide confinement energies in the mid IR is still difficult. Therefore, we must look elsewhere for alternatives. The hydrogenic states of impurities are a natural choice and there has been much work in recent years on their behaviour when confined in QWs (Reeder, 1988). In particular, we are interested in the modification of the bulk energy level ordering and below we show how this might profitably be used for IR light emission and detection purposes.

2. Impurity Energy Levels - Donor in Si QW

We begin by showing in Figure 3a) a typical impurity energy arrangement in the bulk (for isotropic bands). The lowest state is S-like followed by P-like states and excited S states. The gap between S and P states is on the order of the binding energy. For donors in III-V materials this is only a few meV - much smaller than the LO phonon energy while for acceptors it is $\sim$ 25 meV as a result of the heavier mass. In Figure 3b) the effect of a QW confining potential is included. This lifts the degeneracy of the bulk 2P levels leaving the $2P_z$ level higher than the $2P_{x,y}$. We are interested in optical transitions

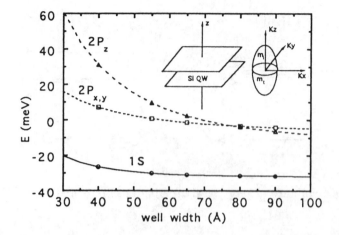

Fig. 4. Low lying S and P levels for a donor in a Si QW.

from the ground 1S state to the 2P$_z$ state for detector applications because this state has the correct momentum to tunnel out of the well and produce a photocurrent (the wave function has a node along the growth axis). However, because the state is not the lowest one above the ground level, relaxation to the 2P$_{x,y}$ level limits its lifetime. There are two ways to invert the ordering of the 2P levels to make the 2P$_z$ level lowest. The first is to have an anisotropic effective mass while the second is to insert the impurity in a potential spike. For the donor case we will consider the mass anisotropy and leave the demonstration of the potential spike to the acceptor case.

For a donor in an X valley conduction band semiconductor like Si the effective masses are fairly large so that we can expect binding energies similar to those of acceptors. The individual conduction band ellipsoids have a simple parabolic form but the interaction between the six valleys causes the effective mass ground state to split into several levels. Therefore we simplify the problem by considering a Si QW grown on a relaxed SiGe buffer with 30 % Ge. The QW ends up under tensile stress which lifts the four in plane conduction valleys up ($\sim$ 200 meV) leaving the two valleys along the growth direction lowest in energy. Next, we consider the donor placed in the middle of the QW. Figure 4 shows the lowest S and P levels as a function of well width. We note that of the P states, the 2P$_z$ now lies below the 2P$_{x,y}$ in energy at large well widths due to the heavy mass normal to the QW plane. It then rises rapidly because of the confinement from the QW at smaller well widths crossing the 2P$_{x,y}$ at 80 Å well width. Thus we have inverted the 2P$_{x,y}$ and 2P$_z$ ordering for well widths larger than 80 Å. This is in contrast to the situation with an isotropic effective mass where the 2P$_z$ remains above the 2P$_{x,y}$ at all well widths.

If it was not for the two degenerate conduction band valleys this system would be quite attractive since there are many donors in Si with strong central cell corrections that could move the ground state far down in energy. However, the ground S state will be split because of the interaction between the two valleys into a symmetric and antisymmetric combination. The symmetric combination lies lowest in energy at a value determined by the central cell potential for the donor while the antisymmetric state remains at approximately the calculated effective mass energy. Thus while the central cell potential of the impurity can be used to pull down the ground state, there is no low-lying excited state to act as the lower level of the lasing transition. With appropriate engineering a donor in a Si QW may be useful as a detector but it appears to be an unlikely candidate for an IR laser.

3. Acceptor Case: Mn in GaAs QW

In order to avoid the problems of the multivalley impurity we now consider the situation for acceptors where only the zone center valence bands are involved. We wish to see if the energy levels of the acceptor can be put in the form of the four level system discussed earlier on. To get increased binding energy we must employ acceptors with substantial central cell corrections such as Mn which has a bulk binding energy around 95 meV in GaAs. More importantly, Mn has the same excited level structure of shallow effective mass acceptors so that we can use effective mass theory to calculated its states in a QW. We have carried this out using a variational calculation along the lines of Fraizzoli (1991) and accounted for the deep ground state of Mn with a short range central cell potential. The result is given in Figure 5 for a 400 meV deep QW and shows the various energy levels plotted versus well width. (These are electron energies - the pictures can simply be inverted to see hole energies.) The states are labelled according to parity (S or P) and the axial component of angular momentum in units of $\hbar$ The energy level structure is considerably more complicated than in the previous example because of the greater degeneracy of the valence bands. The ground S states are now separated from the excited levels by an energy gap greater than the LO phonon energy. The ground states are also split into a heavy hole (S3/2) and light hole (S1/2) because of the symmetry of the QW. We note that due to the much smaller Bohr radius the ground states shift much less in energy than the more extended excited states as the well is narrowed.

The acceptor in a QW naturally gives the general level structure appropriate for a four level laser - the ground state with a low-lying excited state and a group of high-lying excited states separated from the low-lying one by more than a phonon energy. However, if one looks more carefully, the simple case of an acceptor in a QW is not a good candidate for a laser. First, the splitting between the ground S1/2 and S3/2 states is very small making population inversion difficult except at extremely low temperatures

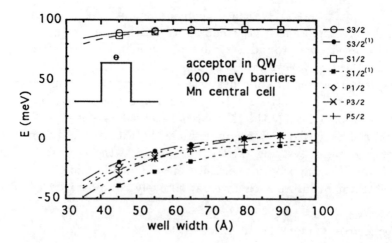

Fig. 5. Energy levels of Mn at center of GaAs QW. The zero of energy corresponds to the top of the bulk valence bands. The S3/2 state is the heavy hole ground state while the S1/2 is the light hole ground state. Similar assignments can be made to the P states depending on whether they tend to track the heavy or the light hole QW subband. Superscripts indicate excited levels of the same symmetry.

(to reduce the S1/2 population). Second, the lowest of the high-lying states (the excited S3/2 and the P1/2) have small optical matrix elements to the S1/2.

A solution which addresses both these problems is to insert a spike in the center of the QW and put the Mn impurities in it. The wave function of the P3/2 state has a prominent node in the growth axis direction so that it is not pushed up by the spike as much as the other excited states (Trzeciakowski, 1989). The spike also inverts the ordering of the heavy and light hole ground states and increases their splitting - the light hole is now the lowest level. The new level ordering is shown in Figure 6 for a 100 meV high 10 Å wide spike in the center of the well. The final energy level configuration now resembles the form we showed in Figure 1a) with the S3/2 and P3/2 forming the radiative pair 2 and 3. The ground state is the light hole S1/2 and the remaining group of excited levels form our uppermost level 4. Thus we have shown it is possible to get the impurity energy level ordering correct for a four level scheme. However, because the spike tends to counteract the central cell potential, the separation between the S3/2 and P3/2 levels is not as large as in the bulk acceptor, though it can still exceed the phonon energy. In order to regain the large energy separation in bulk we can move the Mn impurity off center and outside the spike. The disadvantage is that this gives low oscillator strength between the S and P states due to the reduced overlap. For a 70 Å well the variational wave functions we use to

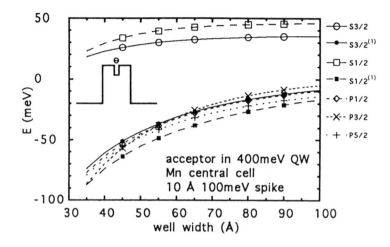

Fig. 6. Mn energies with spike in the center of the well.

calculate the energy levels give us an estimate of $f_{osc} = 0.01$ for the 2→3 transition with the Mn on center and dropping to below 0.005 when the impurity moves outside the spike.

4. Conclusions

Quantum well intersubband transitions suffer from fast nonradiative phonon relaxation especially when the subband spacing is greater than the optical phonon energy. This means a radiative quantum efficiency of $\sim 10^{-6}$ and a correspondingly low probability of achieving lasing action in the mid IR region. Going to impurity based optical transitions eliminates single phonon relaxation when the energy level separation exceeds the optical phonon energy. The tradeoff, though, is a low oscillator strength because of the small ground state. However, many solid state lasers are based on transitions on impurity atoms with oscillator strengths much lower than our calculated values (eg. Nd:YAG has $f_{osc} \sim 10^{-5}$). They work because the accompanying nonradiative relaxation is also weak. We do not have quantitative estimates of the relaxation rates between impurity levels for two phonon and higher transitions, but we expect the rate to be small enough to more than compensate for the low oscillator strength. Thus impurity transitions, when confined in a QW, offer the possibility of a laser system in the mid IR. Furthermore, the calculations presented also demonstate the ability to alter the regular energy level ordering of the impurity by selective placement in the heterostructure. This could prove useful in tailoring the impurity optical response for other applications such as IR detection.

410

Acknowledgements

This work was supported by NASA through the Jet Propulsion Laboratory under subcontract #958997 and by the Advanced Technology Center for Photonics and Optoelectronic Materials established by the State of New Jersey and Princeton University.

References

Helm, M., Colas, E., England, P., DeRosa, F., and Allen, S.J. (1988) 'Observation of grating-induced intersubband emission from GaAs/AlGaAs superlattices', *Appl. Phys. Lett.*,**53**,1714-1716.

Helm, M., Colas, E., England, P., DeRosa, F., and Allen, S.J. (1989) 'Intersubband Emission from Semiconductor Superlattices Excited by Sequential Resonant Tunneling', *Phys. Rev. Lett.*,**63**,74-77.

Helm, M. and Allen, S.J. (1990) 'Can barriers with inverted tunneling rates lead to subband population inversion ?', *Appl. Phys. Lett.*,**56**, 1368-1370.

Borenstain, S.I. and Katz, J. (1989) 'Evaluation of the feasibility of a far-infrared laser based on intersubband transitions in GaAs quantum wells', *Appl. Phys. Lett.*,**55**,654-656.

Bockelmann, U. and Bastard, G. (1990) 'Phonon scattering and energy relaxation in two-,one-,and zero-dimensional electron gases', *Phys. Rev. B*,**42**, 8947-8951.

Reeder, A.A., Mercy, J.M., and McCombe, B.D. (1988) 'Effects of Confinement on Shallow Donors and Acceptors in GaAs/AlGaAs Quantum Wells', *IEEE J. Quantum Electron.*,**24**,1690-1697.

Fraizzoli, S. and Pasquarello, A. (1991) 'Infrared transitions between shallow acceptor states in GaAs-AlGaAs quantum wells', *Phys. Rev. B*, **44**,1118-1127.

Trzeciakowski, W. and McCombe, B.D. (1989) 'Tailoring the intersubband absorption in quantum wells', *Appl. Phys. Lett.*,**55**,891-893.

SUBBAND STRUCTURES OF SUPERLATTICES UNDER STRONG IN-PLANE MAGNETIC FIELDS

Weichao Tan, J. C. Inkson and G. P. Srivastava
Department of Physics, University of Exeter,
Exeter, EX4 4QL, UK

ABSTRACT. We present the first full band structure calculation of the electronic states of a GaAs/AlGaAs supperlattice under strong magnetic fields using a new approach based on a pseudopotential method. Both the Landau level dispersion relations and the wavefunctions have been calculated. Strong nonparobolic effects and anticrossings are observed in Landau level spectra. The wavefunctions of the Landau levels show strong mixing between Γ and X derived states, especially near the anticrossing. The band structure effects on magneto-optical properties of the supperlattice are also discussed

1. Introduction

The electronic structures of superlattices under in-plane magnetic fields have been a subject of many experimental[1-5] and theoretical studies[6-12]. When an in-plane magnetic field is applied to a superlattice, the motion of the electron in the direction parallel to the field is still free particle-like. However, the motion in the plane normal to the field is now influnced by both the magnetic field and the quantum well potential of the superlattice. Since the magnetic field destroys the crystal transitional symmetry, Bloch's theorem does not hold even within a given layer of the system, and standard techniques no longer apply for calculation of electronic states. For this reason, previous works on electronic structure of superlattices in magnetic field are invariably performed within simple envelope function models, such as effective mass [2,6,7, 9,11] or **k.p** theory[8,12].

As long as experiments are performed with weak fields in large period superlattices and only the states around the edges of the conduction band are concerned, the applications of such simple models are justified. In the last decade, however, there have been a numbers of developments which require more information than these techniques can provide. Firstly, experiments in short period superlattices[2-4,13,14] have produced electron and hole states well into the conduction and valence bands and utilised magnetic field up to 100 tesla[15]. In such systems not only is the structure itself complicated but non parabolicity, multiband effects[13,17] and magnetic break down become important and the need for a technique to calculate magnetic effects starting from a pseudopotential or other microscopic basis is acute. Secondly the development of ab-initio and quasiparticle techniques have reached the stage where they now

411

H. C. Liu et al. (eds.), Quantum Well Intersubband Transition Physics and Devices, 411–420.
© 1994 *Kluwer Academic Publishers. Printed in the Netherlands.*

require that experiments such as Shubnikov - De Haas and De Haas Van Alphen effects, which measure effective masses, gyromagnetic ratios, Fermi surface parameters etc., be interpreted as directly as possible in terms of the microscopic potentials so that comparison with theory can be made.

Recently, we have developed a microscopic technique based upon a layered pseudopotential method to study the Landau level states in bulk semiconductor, heterostructures or even, to a good approximation, lateral superlattice and quantum wires[18]. The method for the first time allows us to take into account the full band structures of component materials and the magnetic field on equal footing. In this paper, we have applied the method to study the electronic structures of a short period GaAs/AlGaAs superlattice under strong in-plane magnetic fields (up to 40 tesla). Both the Landau level spectra and wavefunctions are calculated. The full band structure effects on the magneto-optical properties of the superlattice are also discussed.

2. Theory

Consider a GaAs/AlGaAs superlattice grown along z-direction. An in-plane magnetic field **B** is applied along the x-axis. The Schrodinger equation then can be written as

$$\left(\frac{1}{2m}(-i\hbar\nabla + eA)^2 + U(x,y,z)\right)\psi(x,y,z) = E\psi(x,y,z) \tag{1}$$

with

$$U(x,y,z) = \begin{cases} V_{GaAs}(x,y,z), & \text{for } z \in GaAs \text{ layer,} \\ V_{AlGaAs}(x,y,z), & \text{for } z \in AlGaAs \text{ layer,} \end{cases} \tag{2}$$

where m and e(>0) are the mass and charge of an electron respectively. **A** =(0,-Bz,0) is the vector potential in the Landau gauge. $V_{GaAs}(x,y,z)$ is the microscopic crystal potential of bulk GaAs, and $V_{AlGaAs}(x,y,z)$ is the microscopic crystal potential of bulk AlGaAs within the virtual crystal approximation in which the effects of alloy scattering in AlGaAs are ignored.

Although the magnetic field breaks the translational symmetry in z direction, there is crystal symmetry in x-y plane. The wave function of the quantum well system may be written as

$$\psi_{k_\parallel}(\mathbf{r}) = \sum_g \phi_{k_\parallel,g}(z)e^{i(g+k_\parallel)\cdot\rho} \tag{3}$$

with the summation over the set of 2D reciprocal lattice vectors $\{g=g_x\mathbf{i}+g_y\mathbf{j}\}$. Here $\rho=x\mathbf{i}+y\mathbf{j}$ and $k_\parallel = k_x\mathbf{i}+k_y\mathbf{j}$ are the 2D position vector and wave vector in x-y plane respectively. The Schroginger equation can then be decoupled into

$$-\frac{d^2\phi_{k_\parallel,g}(z)}{dz^2} + \frac{\hbar^2}{2m}\left(k_\parallel + g + \frac{eA(z)}{\hbar}\right)^2 \phi_{k_\parallel,g}(z) + \sum_{g'}V_{g-g'}(z)\phi_{k_\parallel,g'}(z) = E\phi_{k_\parallel,g}(z), \text{for all } g \tag{4}$$

where $V_g(z)$ is the 2d Fourier transformation of the total crystal potential U(r).

To solve equation (4), we divide each section of the supperlattice into thin artificial layers parallel to the x-y plane and digitise the vector potential A(z) to { $A_i=(0,-z_iB,0)$ }, where z_i is the centre position of the ith layer, as shown in figure. 1. Within the ith artificial layer, equation (4) becomes

$$-\frac{d^2\phi_{\mathbf{k}_\parallel,\mathbf{g}}(z)}{dz^2} + \frac{\hbar^2}{2m}\left(\mathbf{k}_\parallel + \mathbf{g} + \frac{e\mathbf{A}_i}{\hbar}\right)^2 \phi_{\mathbf{k}_\parallel,\mathbf{g}}(z) + \sum_{\mathbf{g}'} V_{\mathbf{g}-\mathbf{g}'}(z)\phi_{\mathbf{k}_\parallel,\mathbf{g}'}(z) = E\phi_{\mathbf{k}_\parallel,\mathbf{g}}(z), \quad \text{for all } \mathbf{g}. \quad (5)$$

Within each layer the Hamiltonian is now, within a constant vector potential $\mathbf{A}_i$, the same as for the zero field case, which has the full crystal symmetry. We can therefore calculate the complete set of states in each layer for a given energy (E) and in-plane wavevector ($\mathbf{k}_\parallel$) using standard layer method[16,13,19]. Finally these basis states can be used to build up the complete wavefunction in the system by matching across the interface between layers

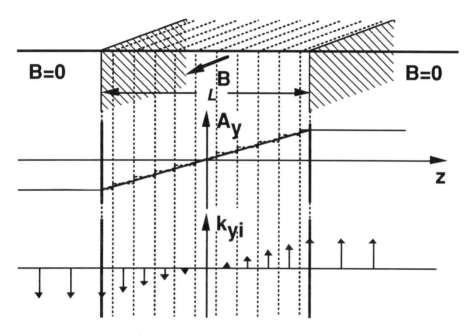

Fig. 1 A schematic illustration of the method.

By considering the gauge transformation required at the interface, it is easy to see that the vector potential A_i change from layer to layer is equivalent to a change of in plane wavevector $k_\parallel$ such that $k_{\parallel i} \rightarrow k_\parallel + \frac{e\mathbf{A}_i}{\hbar}$. This means that the matching of states in adjacent ith and (i+l)th layers has to be performed by matching the wave functions calculated with different wave vectors $k_{\parallel i} \rightarrow k_\parallel + \frac{e\mathbf{A}_i}{\hbar}$ and $k_{\parallel i+1} \rightarrow k_\parallel + \frac{e\mathbf{A}_{i+1}}{\hbar}$ but the same energy.

This process has a very simple physical interpretation: the digitisation of the vector potential concentrates the magnetic field into a series of magnetic sheets such that the net momentum impulse given to the electrons as they cross the sheet is the same as the integrated change in momentum produced by the magnetic field through the original layers. This is reflected in the change of $k_{\parallel i}$. It has been shown that, provided the layers are thin compared the magnetic length $\ell=(\hbar/eB)^{1/2}$ the digitisation does not influence the results[18].

In practical calculation, the superlattice in magnetic field is terminated by zero magnetic field regions on either side to avoid unphysical infinities, as shown in Fig 1. These are set far enough apart to avoid surface effects in the calculated states, in practice this means that L is taken to be much larger than the largest cyclotron orbit radius to be considered. The electronic properties of the magnetic field region are then obtained by considering an incident electronic state and calculating the resulting transmission and reflection.

We use the scattering matrix based layer method [16,13,19] to calculate the scattering matrix of the system from the matching coefficient generated at each layer, and then obtain the transmission and reflection coefficients. The Landau level energies are given by the energies of the resonant peaks in transmission. By varying the state type (Γ, L, X etc.), incident in-plane momentum ($k_{\parallel}$) and energy for the incoming state, we can study the electronic states in the magnetic field region over the whole Brillouin zone and energy range of interest.

3. Results and discussion

In this section, we present results for a GaAs/AlGaAs superlattice grown along the [001] direction. The superlattice consists of 7 monolayers of GaAs and 5 monolayers of $Ga_{0.5}Al_{0.5}As$, similar to that used by Sasaki, Miura and Horikoshi (SMH) in a recent magneto-optical experiment [4]. They also compared their experimental results with an effective mass calculation for the conduction band and an **k.p** calculation for the valence band.

3.1 DEVELOPMENT OF SUBBAND STRUCTURE IN THE ABSENCE OF MAGNETIC FIELD

We first discuss the subband structure of the superlattice in the absence of a magnetic field, since it forms the basis for understanding the cyclotron states in the superlattice.

Fig 2 shows the dispersion relations for the first two lowest sub-band of the superlattice calculated using a scattering matrix layer pseudopotential method.[13] (All energies measured from the conduction band edge of GaAs).The development of these two subbands may be understood as follows:

The superlattice consists a series of quantum wells in GaAs for Γ electron and in AlGaAs for X electron. In each GaAs section, there is a quasi-bound Γ state at 215meV. The lowest quasi-bound X state in each AlGaAs section lies at 292meV. If we first ignore the effect of intervalley scattering, the degenerate quasi-bound states in GaAs wells couple together to form a broaden Γ derived subband ranged from 140 to 346.5 meV. The degenerate quasi- bound X states also form a narrow subband around E=292 meV.

The intervalley couplings in superlattice arises from the interface scattering. The X-Γ coupling creates a small gap at the crossing between the X-derived subband and the Γ derived

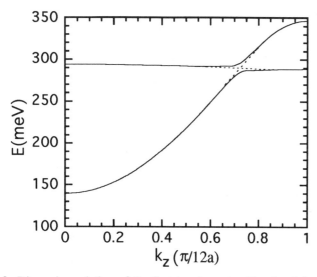

Fig. 2 Dispersion relation of the first two lowest subbands of the superlattice in the absence magnetic field. The anticrossing between the X-derived subband and Γ derived subband is schematically show by the dotted line. k_z is in unit of $(\pi/12a)$, where a is the width of a GaAs monolayer.

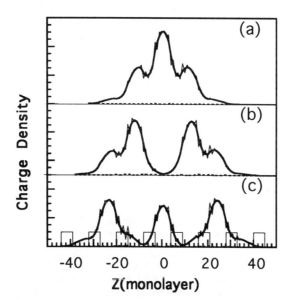

Fig. 3 Charge densities of the full wave functions (thin line) and their Γ (thick line) and X (dotted line) components of the three lowest Landau levels at energies 163 (a), 212 (b) and 254 meV (c) in the conduction band of the superlattice under a 40 tesla in-plane magnetic field. The rectangles schematically show the superlattice potential profile.

one and results in the subband structure shown in Fig. 2. The width of the splitting near the crossing, which is about 5meV in Fig. 2, represents the strength of X-Γ coupling in the system.

3.2 ELECTRONIC STATES IN STRONG IN-PLANE MAGNETIC FILED

We now consider electronic structure of the GaAs/AlGaAs superlattice with a 40 tesla magnetic field applied along the [100] direction. We have calculated the transmission coefficients as a function of energy for an input Γ state with $k_x=0$ and different k_y values. For energy up to 290meV, we find three peaks in transmission curves at energies 163, 212 and 254 meV, which are nearly independent on incident k_y (or position of cyclotron orbital centre). These are the three lowest Landau level states within the first subband.

All these three levels are below the X valley bottom of $Al_{0.5}Ga_{0.5}As$ and can be well described by a single Γ state approximation as in the effective mass calculation of SMH[4], in which a single parabolic band is used. The differences we find in energy from their results are due to a much stronger non parabolic the effect (especially the strong non parabolicity and anisotropy of the evanescence states in $Al_{0.5}Ga_{0.5}As$). Fig. 3 shows the charge densities associated with the full wavefunctions and their X and Γ components of these three Landau levels centred at the middle of GaAs well. It does show that these three levels are Γ dominated. Even for these states, however, we can still clearly see the effect of intervalley mixing between the X component, which contributes less than 3% of the total charge density, and the dominate Γ component resulting in a rapid fluctuation of the total charge density.

For energies higher than 290meV, a rather complicated energy level structure is revealed due to the presence of intermixed X state Landau levels. A part of the calculated transmission coefficient of an input Γ states as a function of energy for $k_x=0$ and different k_y values is shown in Fig. 4(a). The peak positions vary with incident $k_∥$ in a non simple way and there are in addition a number of anti-resonances. The latter are generally the hallmark of intervalley coupling in this case (Γ-X). The resulting Landau level energy as a function of the cyclotron orbit centre $Z_c = \hbar k_y / eB$ obtained from Fig. 4(a) is shown in Fig. 4(b).

The structure arises as follows; There are two further Γ originated Landau levels within the Γ subband, these are near the subband top and have a small dispersion. The lowest X Landau state appears at E=292meV. The X states are essentially localised in the $Al_{0.5}Ga_{0.5}As$ barriers as discussed in sect 3.1. Ignoring the coupling with other states, we expect that each of these states in magnetic field forms a Landau level with a nearly parabolic dependence on orbit centre[6,7]. Hence Landau levels originating from the X states can be easily distinguished by their highly dispersive feature. The intra- and inter-valley coupling removes the level crossings and results in the energy spectra seen in Fig. 4(b).

In Fig.5(a)-(h)we show the charge densities associated with the full wavefunctions as well as the X and Γ components for the states at energy momentum positions A, B, C, D, E, F and H as marked in Fig.4(b). The cyclotron orbit centres of all these states are at the middle of the GaAs well.

Let us first consider states A, B and C shown in Fig 5(a), (b) and (c). Both A and C are strongly mixed states. Their charge densities exhibit rapid fluctuations. The contributions from X and Γ are about 75% and 25% respectively in state A while they are 25% and 75% in state C. All the Γ components of the charge density in the three states have the similar spatial distributions which, as expected, have the overall structure for the fourth Landau level of the Γ subband. The X components in these three states also have the similar structures which mainly originate from the quasi-bound X states in the two AlGaAs section nearest to the cyclotron orbit centre. Hence

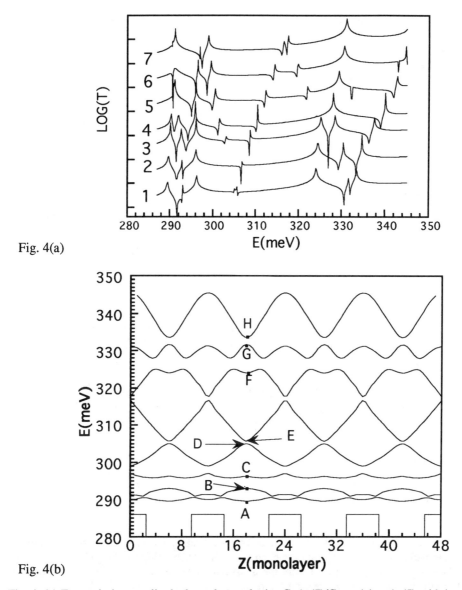

Fig. 4(a)

Fig. 4(b)

Fig. 4 (a):Transmission amplitude through superlattice GaAs(7)/$Ga_{0.5}Al_{0.5}As$(5) with in-plane field B=40T along x-direction for an input Γ state vs. energy. Curves (1) to (7) correspond to different k_y giving a range of cyclotron orbit centres Zc varying from the middle of the well to the middle of the barrier; (b) Landau level energy vs. cyclotron centre Zc (or equivalently k_y) in superlattice GaAs(7)/$Ga_{0.5}Al_{0.5}As$(5) with in-plane field B=40T. The superlattice profile is shown schematically by the rectangles.

the first three lowest Landau levels shown in Fig. 4(b) arise from the mixing of the fourth Γ derived Landau level and the two quasi-bound X states nearest to the cyclotron orbit centre. It is interesting to note that between two strongly mixed states A and C, state B is nearly a pure X state. Using a three states model, we can easily show that this is because the interaction between the Γ derived state and each of these two bound X states is much stronger than that between the two X states.

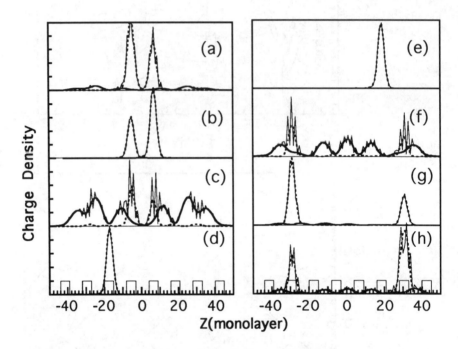

Fig. 5 The charge densities associated with the full wavefunctions(thin lines) as well as the X(dotted lines) and Γ(thick lines) components. (a), (b), (c), (d), (e), (f), (g) and (h) correspond to states at energy momentum positions A, B, C, D, E, F and H marked in Fig.4(b) respectively. (The wavefunctions have not been normalized).

From Fig. 4b, we see that states D and E are well separated from either the fourth or the fifth Γ derived levels, therefore they are nearly pure X like. Fig 5 (d) and (e) clearly show that states D and E are simply derived from the quasi-bound X states in the two AlGaAs barriers which are the second nearest to the orbit centre.

Similar to states A, B and C, states F, G and H show strong Γ-X mixing. From the spatial distributions of the charge densities shown in Fig 5(f), (g) and (h), we see that these three states result from the mixing of the fifth Landau level of the Γ derived subband and the two quasi-bound X states in the two AlGaAs sections which are the third nearest to the orbit centre.

The results discussed above explicitly show that the inter-valley scattering plays an important role in the formation of Landau levels in superlattice: It results in anticrossings in Landau spectra and rapid fluctuations in charge densities of the cyclotron states. Compared to the bulk system at the same field and at an energy where the X band is available [18], the mixing in the superlattice is very much stronger. What we see here, therefore, is not primarily magnetic breakdown but rather a result of the interface scattering. However, magnetic field squeezes the electronic states and increases the overlapping between states derived from different valleys, therefore interface scattering can be significantly enhanced in strong magnetic field. Performing the calculation for different magnetic field strengths (not shown hear), we indeed find that the effects of X-Γ become stronger with increasing the magnetic field.

3.3 BAND STRUCTURE EFFECTS ON MAGNETO-OPTICAL TRANSITION

Having obtained both the energey spectra and the wavefunctions of the Landau levels, we can calculate the matrix elements for optical transition, but here we will restrict to a qualitative discussion of the band structure effects on magneto-optical transitions.

One of the most important band structure effects is due to the non parabolic dispersion relation of the component materials. For the AlGaAs/GaAs superlattice considered here, the non parabolic effects can be clearly seen from the non equal separation of the three lowest Landau levels in the first subband, which have also been confirmed by MSH in their magneto-optical experiment [4].

The inter-valley scattering also has important consequence on the magneto-optical transition in the superlattice. Firstly, for energy higher than 290 meV, the Γ derived Landau levels are strongly mixed with neighbouring X derived states, therefore the optical transition from other Γ derived states to these states have much more complicated structures than that predicted with singe band effective mass theory. Secondly, due to the mixing nature of the Landau levels, optical transitions between the Γ-dominated Levels below 290 meV and the X-dominated levels becomes allowable, though they may be very weak.

It should be pointed out that the splittings resulting from the Γ-X coupling, which are typically less than 10 meV, are difficult to be resolved from inter-band absorption spectra. A more efficient way to study the detail structures due to Γ-X mixing may be intra-band magneto absorption (or cyclotron-resonance). In such experiment, the contribution of the complicated valence band Landau levels are excluded, and the assignation of the transition spectra becomes much easier. Superlattices with thinner GaAs wells or type II superlattices, in which structures resuting from Γ-X mixing will be more prominent and closer to the lowest subband edge, would be the most suitable to be used in intra-band magnato absorption measurements

4. Conclusion

In this paper we have presented a microscopic calculation of electronic states in an short period GaAs/AlGaAs superlattice under a strong in-plane magnetic field. In our calculation, two very different length scales associated with the magnetic and atomic potentials are treated within one computational scheme, therefore the full band structures of the component materials are included in a natural way. In the first two conduction subband, quite flat Γ-dominated Landau levels and very dispersive X-dominated Landau levels are found. It is show that Γ-X mixing results in strong anticrossing in Landau level spectra and rapid fluctuation in charge densities of the cyclotron states. Compared with the result of simple envelope function theory, the Landau level spectra obtained from microscopic calculation are much more structured. It will be interesting to observe these detailed structures in future magneto-optical experiment.

420

REFERENCES

1. Chang L. L., Mendez E. E., Kawai N. J., Esaki L.(1982) 'Shubnikov-de Haas oscillations under tilted magnetic-fields in InAs/GaSb superlattices', Surf. Sci. 113, 306
2. Belle G, Maan J. C., and Weimann G.(1986) 'Observation of magnetic levels in a superlattice with a magnetic-field parallel to the layers', Surface Sci. 170, 611
3. Sasaki S. and Miura N.(1992) 'Magneto-optical investigation of energy band structure in GaAs/AlGaAs short period supperlattices under pulsed high magnetic fields', Surf. Sci., 263, 600
4. Sasaki S., Miura N. and Horikoshi Y.(1990) 'High-field magneto-optical study of GaAs/AlGaAs short period supplattices', J. Phys. Soc. Jpn. 59, 3374;
5. Miura, N, Yamada. K, Kamata N., Osada T. and Eaves L. (1991) 'Magneto-tunneling spectroscopy of GaAs-AlGaAs double barrier tunneling device in pulsed high magnetic field up to 40T', Supperlattice and Micostructures 9, 527
6. Ando T. (1981) 'Electronic properties of a semiconductor super-lattice:.3, Energy-levels and transport in magnetic-fields', J. Phys. Soc. Jpn. 50, 2978(1981)
7. Maan J. C.(1987) 'Magneto-optical properties of heterojunctions, quantum wells and superlattice', NATO ASI Series B 170, 347
8. Altarelli M., Platero G. (1988) 'Magnetic hole levels in quantum wells in a parallel field', Surf. Sci. 196. 540
9. Xia J. B., and Fan W. (1989) 'Electronic structures of superlattices under in-plane magnetic field', Phys. Rev. B 40, 8508
10. Zaslavsky A., Li Y. P., Tsui D. C., Santos M., Shayegan M. (1990) 'Transport in transverse magnetic-fields in resonant tunneling structures', Phys. Rev. B 42, 1374
11. Brey L., Platero G., Tejedor C. (1988) 'Efffect of a high transverse magnetic-field on the tunneling through barriers between semiconductors and superlattices', Phys. Rev. B 38, 9649
12. Wu G. Y, Hung K. M. and Chen C. J. (1992) 'Effective-mass theory of p-type heterostrucyures under transverse magnetic fields', Phys. Rev. B 46,1521
13. Ikonic Z., Inkson J. C, Srivastava G. P. (1992) 'Ordering of conduction-band states in (GaAs)n (AlAs)n [001] and [110] superlattices', Semicond. Sci. and Tech.7 , 648
14. Moore.K. J., Duggan G., Dawson P., and Foxon C.T.(1988) 'Short-period GaAs-AlAs superlattices: Optical properties and electronic structure', Phys. Rev. B, 38, 5535
15. Miura N.(1992) 'Recent progress of semiconductor physics at the Meggauss Laboratory of University of Tokyo', in "High Magnetic Fields in Semiconductor Physics III", ed. G. Landwehr, Springer-Verlag, Berlin, 675
16. Ko D. Y. K., and Inkson J. C. (1988) 'Matrix-method for tunneling in heterostructures - resonant tunneling in multilayer systems', Phys. Rev. B 38, 9954
17. Morrison I., Brown D. L. and Jaros M. (1990) 'Valley-mixing effects in $(GaAs)_l(AlAs)_m$ superlattices with microscopically imperfect interfaces', Phys. Rev. B 42, 11818
18. Inkson J. C., Tan W. C., an Edwards G.(1993) 'The calculation of band structure effects on electronic states in a magnetic field', to be published
19. Ko D. Y. K., and Inkson J. C. (1988) 'Microscopic calculation of electric-field effects in GaAs/AlGaA/GaAs tunnel structures', Phys. Rev. B38, 12416; D. Y. K. Ko, J. C. Inkson and G. Edwards (1990) in "Resonant Tunneling: Physics and Application", ed. L. L. Chang, E. E. Mendez and C. Tejedor, New York

TEMPERATURE DEPENDENT INTERSUBBAND DYNAMICS IN N - MODU-LATION DOPED QUANTUM WELL STRUCTURES

A. SEILMEIER, U. PLÖDEREDER, J. BAIER, and G. WEIMANN*

*Institute of Physics, University of Bayreuth,
D-95440 Bayreuth, Germany
*Walter-Schottky-Institute, TU München,
D-85747 Garching, Germany*

ABSTRACT. Intersubband scattering in n-modulation doped GaAs/Al$_x$Ga$_{1-x}$As quantum well structures is systematically investigated by an infrared bleaching technique. The temperature dependence of the time constants is found to be sensitive to the self - consistent potential determined by the doping concentration and the well width. In samples with a shallow potential minimum in the barrier located above the excited subband of the well the direct intersubband transition from the excited to the lowest subband of the well is observed at T = 10K ($\tau \sim$ 2 ps). The measured time constants increase with rising temperature. In samples, where the energy of the excited subband is higher than that of the lowest barrier subbands (high doping concentration, narrow well width), ground state recovery times of several picoseconds are found at T = 300K. At low temperatures additional nonexponential components may appear extending over ~ 100 ps. The results are discussed taking into account transfer processes to the barrier.

1. Introduction

So far experimental data on intersubband scattering of free carriers have been taken predominantly on individual samples. Several methods have been used to obtain information on intersubband relaxation times and on the population dynamics of subbands: a Raman technique [1,2], an infrared bleaching technique [3] and techniques which use the spectrally selective observation of interband absorption [4,5] or photoluminescence [6-8] changes. These investigations give valuable information on intersubband scattering, however, systematic studies in which the relaxation is investigated as a function of various sample parameters are still required to get detailed information on the relevant mechanisms.

Experiments on electrons in undoped GaAs/Al$_x$Ga$_{1-x}$As samples performed at low temperatures give relaxation times in the order of 1 ps [2,7,8] whereas in modulation doped samples at room temperature considerably longer decay times up to 14 ps have been observed [9]. The question now

421

H. C. Liu et al. (eds.), Quantum Well Intersubband Transition Physics and Devices, 421–431.
© 1994 *Kluwer Academic Publishers. Printed in the Netherlands.*

arises whether the longer time constants are due to modulation doping or due to higher temperatures.

A comparison with theoretical calculations [10-12] shows that the short time constants observed in undoped samples are consistent with the values expected for direct intersubband relaxation via electron polar-LO phonon scattering. The relaxation in modulation doped structures appears to be more complex. The excited subbands may be influenced by the potential minima in the barriers created by the ionized donors. There exist wave functions which exhibit also a high amplitude within the barrier making possible an efficient transfer of electrons to the barrier states and vice versa[9,11-13]. This real space transfer has been discussed as the delaying process in intersubband ground state recovery in modulation doped quantum well structures [9,11,12]. So far the influence of the sample temperature on the delayed relaxation has not been studied. In this paper experimental results taken on modulation doped structures are presented which demonstrate the crucial importance of the sample temperature in the discussion of intersubband relaxation of free carriers.

2. Experimental

The intersubband scattering is investigated by an infrared bleaching technique [3]. An intense infrared picosecond pulse bleaches the intersubband absorption band. The absorption recovery is monitored by a second weaker pulse of the same frequency. Infrared pulses at the corresponding wavelengths between 5 μm and 18 μm are generated by difference frequency mixing [14]. A Nd:glass laser pumps a traveling wave infrared dye laser. The difference frequency between the fundamental of the glass laser and the traveling wave dye laser is produced in a GaSe crystal. The energy of the nearly bandwidth limited pulses are in the order of 1 μJ, the pulse duration is 1 to 2 ps.

Table 1. Parameters of the samples investigated and ground state recovery times τ for T = 10K and room temperature.

sample	well width	barrier width	doping density	periods	τ[10K]	τ[300K]
W 1757	5.8 nm	36 nm	3.2×10^{11} cm^{-2}	50	2.3 ps	4 ps
W 1756	5.9 nm	36 nm	7.5×10^{11} cm^{-2}	50	2.3 ps	6 ps
W 1755	5.9 nm	36 nm	1.4×10^{12} cm^{-2}	25	nonexp.	8 ps
W 1685	5.2 nm	40 nm	6.1×10^{11} cm^{-2}	50	11 ps	11 ps

The investigations are performed in n-type modulation doped GaAs/Al$_{0.35}$Ga$_{0.65}$As samples. They are grown by molecular beam epitaxy on (100) semi-insulating GaAs substrates of 350 μm thickness and consist of 25 or 50 thin, undoped GaAs layers of 5.9 nm or 5.2 nm thickness embedded in 36 nm to 40 nm thick Al$_{0.35}$Ga$_{0.65}$As layers in which the central 8 nm are doped by Si. The doping concentration is varied between 3.2×10^{11} cm^{-2} and 1.4×10^{12} cm^{-2}. The total multi-quantum well structures are clad between 1 μm thick Al$_{0.35}$Ga$_{0.65}$As layers to avoid surface depletion and substrate effects. The relevant sample parameters are summarized in Table 1.

For the time resolved measurements we use a special waveguide geometry with a single reflection at the quantum well layers and an exit angle close to the Brewster angle to avoid multiple excitation [3]. The samples are mounted in a closed-cycle He cryostat which allows investigations between 10 K and 300 K.

3. Results

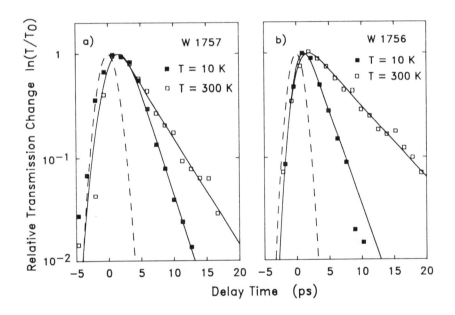

Figure 1. Normalized transmission change as a function of the delay time. a) Sample W 1757 (n = 3.2×10^{11} cm^{-2}): The absorption recovery time is τ = 2.3 ps for T = 10K and τ = 4 ps for T = 300 K. b) Sample W 1756 (n = 7.5×10^{11} cm^{-2}): An absorption recovery time of τ = 2.3 ps and τ = 6.5 ps is found for T = 10K and T = 300K, respectively. The broken lines represent autocorrelation curves of the infrared pulses.

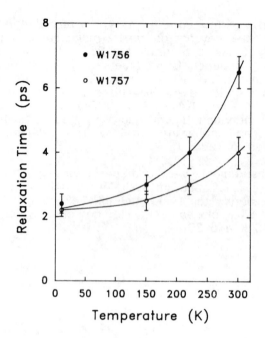

Figure 2. Ground state recovery times of samples W 1756 and W 1757 as a function of the temperature. The time constants rise more than linear with temperature.

We compare data on samples of different doping densities and different well widths at various temperatures. In all experiments the pump intensities are limited to values, which do not allow to excite more than ~ 30 % of the carriers, to reduce the influence of dynamical shifts of the absorption frequency [9,16]. The bleaching experiments are performed exactly at the frequency positions of the absorption maxima located between 1200 cm^{-1} and 1270 cm^{-1} depending on the sample and the temperature. Absorption spectra of the samples at T = 300K are presented elsewhere [3,15].

In Fig. 1 and 3 the transient transmission change

$$\ln(T/T_0) = \sigma \, N \, (\, n_2 - n_1 + n_0 \,)$$

is plotted as a function of time (T_0 is the transmission without excitation). It is a function of the instantaneous carrier density/well in the upper state n_2 and on the depletion of the lower state ($n_0 - n_1$). (n_0 and n_1 are the total carrier density/well and the instantaneous carrier density/well in the lower state, respectively; σ is the absorption cross section, N is the number of quantum wells.) The signal is a direct measure of the rapid depletion of the lower state and its reoccupation which proceeds on a longer time scale.

We start with experimental data taken on three samples of identical well width of 5.9 nm but different doping density. In Fig. 1 the signal is shown a) for sample W 1757 (lowest doping concentration n = 3.2 x 10^{11} cm^{-2}) and b) for sample W 1756 (intermediate doping concentration n = 7.5 × 10^{11} cm^{-2}) for T = 10 K and T = 300 K. At T = 10K a - within the experimental error - exponential decay is found with τ = 2.3 ± 0.5 ps in both samples; at T = 300K the signal decreases with τ = 4 ± 1 ps and τ = 6.5 ± 1 ps in the sample of lowest and intermediate doping concentration, respectively.

The temperature dependence of the intersubband relaxation in the samples W 1757 and W 1756 is depicted in Fig. 2 into more details. At low temperatures both samples exhibit the same time constant of τ = 2.3 ps. The rise of the time constants with temperature is more pronounced in the sample of higher doping concentration.

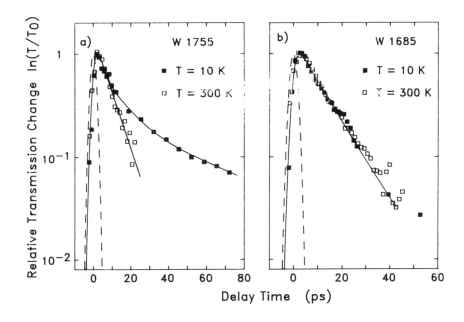

Figure 3. Normalized bleaching signal as a function of the delay time. a) Sample W 1755 (well width 5.9 nm; n = 1.4×10^{12} cm^{-2}):A decay time of 8 ps is observed at T = 300K and for short delay times at T = 10K. At long delay times and at low temperatures a nonexponential tail is measured. b) Sample W 1685 (well width 5.2 nm; n = 6.1×10^{11} cm^{-2}): The solid line represents a decay time of 11 ps.

Fig. 3a shows the basically different result for sample W 1755 (highest doping concentration n = 1.4 x 10^{12} cm^{-2}). At T = 300 K a time constant of approximately 8 ps is observed which is followed at T = 10 K by a nonexponential decay with a long tail extending over 70 ps.

In Fig. 3b data are presented on sample W 1685 with a smaller well width of 5.2 nm and an intermediate doping concentration (n = 6.1 × 10^{11} cm^{-2}). The signal decays with a time constant of approximately 11 ps with a weak nonexponential tail at long delay times. The decay curve does not depend on the temperature.

The experimentally observed time constants for T = 10K and T = 300K are summarized in Table 1.

4. Discussion

It is generally accepted that carrier relaxation in thin GaAs/Al$_x$Ga$_{1-x}$As heterostructures (subband separation > LO phonon energy) is governed by electron polar-LO phonon interaction. For the quantum wells of 5.9 nm thickness investigated here the time constant for intersubband transitions from the first excited to the lowest subband has been calculated to be ~ 2 ps [10,12]. This value nicely compares with the time constant of 2.3 ps observed in our experiments for the two not so highly doped samples W 1757 and W 1756 at a temperature of T = 10 K (Fig. 2). At higher temperatures and in highly doped samples (e.g. W 1755), however, longer time constants are measured which are believed to be connected with a transfer of excited electrons to the potential minima of the barriers. The transfer back to the wells is assumed to be the delaying process in the ground state recovery.

The temperature behavior of the decay time depends on the relative location of the excited subband of the well E_{W2} with respect to the lowest state originating from the potential minimum of the barrier E_{B1}. In our discussion the sign of the energy separation $E_{W2} - E_{B1}$, which is determined by the well width and the electron concentration, is used to characterize the temperature dependence of the reoccupation of the lowest subband. In highly doped samples of narrow well width, e.g. for n ≳ 7×10^{11} cm^{-2} at a well width of 5.9 nm [13], the optically occupied well subband is above the lowest barrier subband, i. e. $E_{W2} - E_{B1} > 0$. In the following it is assumed that all of the donors are ionized, i. e. the density of free carriers is equal to the doping density, which is a good approximation for room temperature and low densities. In the low temperature experiments the density of free carriers may be substantially lower, in particular, if illumination by visible light is avoided as in our experiments [13].

4.1. $E_{W2} - E_{B1} < 0$. In the samples W 1757 and W 1756 the excited GaAs well subband is energetically located below the bound states of the

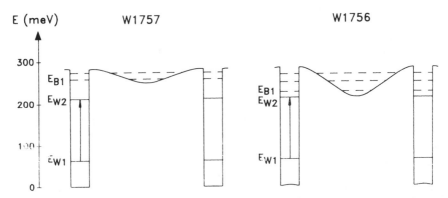

Figure 4. Structure of the conduction band of the samples W 1757 and W 1756 determined by selfconsistent calculations (T = 300K). Solid lines: subbands arising from the quantum wells; broken lines: subbands arising from the barrier potential minima. The intersubband excitation by the infrared pulse is indicated by an arrow. In both samples the excited well subband E_{W2} is located below the subband E_{B1} with a wavefunction localized almost entirely in the barrier.

potential minima of the barrier. In Fig. 4 the results of self-consistent band structure calculations together with the lowest eigenstates are presented for the two samples at room temperature using the parameters given in Table 1. The ionized donors give rise to the potential minima in the barriers. The depth of the minima is more pronounced with increasing doping concentration. Solid lines indicate subbands in which the carriers are fairly well localized in the well, whereas the dashed lines represent subbands originating from the subbands of the barrier potential minima. The arrows indicate the transitions investigated in the bleaching experiments. The relevant parameter E_{W2} - E_{B1} amounts to approximately -15 meV and -45 meV for the samples W 1756 and W 1757, respectively.

Fig.4 clearly shows that an absorption of LO phonons is required to excite carriers from the initially populated excited subband E_{W2} to higher lying states like E_{B1}. At T = 10 K the thermal LO-phonon population is not sufficient for a substantial transfer of carriers to subbands dominantly localized in the barrier. Consequently, the time constant of 2.3 µs represents the direct intersubband relaxation from the excited subband E_{W2} to the lowest subband of the GaAs well.

With increasing temperature the decay times rise. This effect is more pronounced in the sample with the larger doping concentration, W 1756, indicating an influence of the depth of the barrier potential minimum. The nearly exponential character of the rise of the time constant with temperature indicates thermally assisted transfer processes to higher lying subbands localized mainly in the barrier. They may give rise to a bottleneck in the relaxation process. A considerable amount of energy (> kT in sample W 1757) is required to reach the next higher subband. It may be supplied by thermally excited carriers and by an absorption of thermally excited phonons. A simplified model, in which only electron - (bulk) LO phonon scattering between the relevant states is taken into account, does not give the observed decay times of several picoseconds in Fig. 1. The calculations, which neglect dynamic changes of the potential, predict a biexponential decay with the intersubband time constant of ~ 2 ps at short delay times and a longer time constant of several picoseconds due to electrons transferred to the barrier and vice versa (see also [11]). The short time constant is not observed experimentally. Obviously, the model has to be improved by taking into account additional effects: i) Electron - electron scattering and excited state absorption may enhance the carrier transfer to the barrier. ii) Dynamic changes of the absorption frequency may cover the rapid population changes at short delay times [16]. iii) Phonon confinement in the 2D - systems may influence the scattering rates [11,12,17-19].

4.2. $E_{W2} - E_{B1} > 0$. A situation, where the excited subband with the strong absorption cross section E_{W2} is located above the lowest barrier state, is realized in sample W 1755. Fig. 5a shows the band structure of the sample with the highest doping concentration exhibiting a deep potential minimum in the barrier. The solid and broken lines represent the subbands originating from the wells and the barriers, respectively. The energy of the excited well subband E_{W2} is separated from the E_{B1} subband by approximately 60 meV (T = 300K). Excitation of the E_{W2}

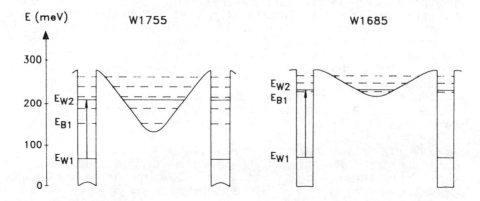

Figure 5. Result of selfconsistent calculations of the conduction band of the samples W 1755 and W 1685 (T = 300K). Solid lines: subbands arising from the quantum wells; broken lines: subbands arising from the barrier potential minima. The intersubband transition induced by the infrared pulse is marked by an arrow. In sample W 1755 the energy of the excited well subband E_{W2} is higher than that of the lowest subband originating from the barrier E_{B1}. In sample W 1685 the subbands E_{W2} and E_{B1} are close to resonance.

subband gives rise to an efficient carrier transfer to the barrier where they relax rapidly into the lowest subband E_{B1}. Ground state recovery of these carriers with time constants in the picosecond regime is only possible via higher lying states requiring thermal excitation. The observed decay time of 8 ps at room temperature (Fig. 3a) is explained fairly well by model calculations taking into account the delocalization of the excited subbands, scattering processes between the subbands, and dynamic changes of the self-consistent potential[11,12].

At lower temperatures (T $\leq$ 180 K) the same time constant is found for delay times shorter than 10 ps. It follows a non exponential decay extending over 70 ps. The curve for T = 10K (Fig. 3a) may be explained by dynamic changes of the potential structure which influence the efficiency of thermally assisted relaxation channels. The transfer of electrons to the barrier reduces the depth of the potential well in the barrier shifting the dashed subbands in Fig. 5a to higher energies. Simultaneously, scattering of excited carriers to the strongly localized barrier subband E_{B1} via emission of an LO-phonon results in an increase of the carrier temperature in the order of at least hundred K due to the residual excess energy of ~ 25 meV. In this way the transfer back to the well is not affected by the low thermal energy of T = 10K and a similar time constant is found as for T = 300K. At delay times longer than 10 ps the carrier temperature decreases and the depth of the potential minimum in the barrier is growing again. Now the low sample temperature becomes important reducing the efficiency of the thermally assisted relaxation processes and the long tail shown in Fig. 3a for T = 10 K results.

4.3. E_{W2} - E_{B1} ~ 0. Finally, we want to discuss the results taken on the sample W 1685 (see Fig. 3b) where the subbands E_{W2} and E_{B1} are close to resonance. The bandstructure calculated selfconsistently is presented in Fig. 5b. The upper subband of the well which is occupied by the infrared pump pulse is close to the lowest subband of the barrier. Strong mixing between the two states takes place producing an efficient transfer channel between the well and the potential minimum of the barrier. Thermal excitation is not required for the transfer process to the barriers and vice versa. Consequently, we expect an intersubband relaxation which does not depend on the temperature. Indeed, the experimental result in Fig. 3b is identical for T = 300K and T = 10K within the experimental error. The signal decays with a time constant of 11 ± 1 ps and shows a slightly nonexponential character at long delay times. These findings are consistent with the result of Monte - Carlo calculations performed for the same sample at T = 300K [12].

5. Conclusion

The experimental data presented here show that the sample temperature critically influences the intersubband relaxation in n-modulation doped multiple quantum well structures. The following situations depending on the well width and the doping concentration have to be distinguished:

i) $E_{W2} - E_{B1} > 0$. The situation is realized in samples of high doping concentration ($n \gtrsim 7 \times 10^{11}$ cm^{-2} at a well width of $\sim$ 6 nm). A relatively long time constant in the order of 10 ps is observed at T = 300 K. At low temperatures (T < 180 K) a nonexponential relaxation is found extending over $\sim$ 100 ps.

ii) $E_{W2} - E_{B1} < 0$. At low temperatures a time constant is found which is consistent with the value calculated for direct intersubband relaxation via electron LO phonon scattering. The observed time constants increase with temperature which is not completely described by the present simulations.

iii) $E_{W2} - E_{B1} \sim 0$. The time constant is in the order of ten picoseconds independent of the sample temperature.

There is indication that carrier transfer from the wells to the barriers and vice versa delays ground state recovery. In many situations thermally assisted processes appear to be most efficient leading to the temperature dependent intersubband scattering rates. Additional experimental and theoretical work is required to elucidate the detailed transfer processes.

6. Acknowledgement

The work was supported by the Deutsche Forschungsgemeinschaft.

7. References

1. D. J. Oberli, D. R. Wake, M. V. Klein, J. Klem, T. Henderson, and H. Morkoç (1987) *Phys. Rev.Lett.* **59** 696-699
2. M. Tatham, J. F. Ryan, and C. T. Foxon (1989) *Phys. Rev. Lett.* **63** 1637-40
3. A. Seilmeier, H. J. Hübner, G. Abstreiter, G. Weimann, and W. Schlapp (1987) *Phys. Rev. Lett.* **59** 1345-8
4. R. J. Bäuerle, T. Elsässer, W. Kaiser, H. Lobentanzer, W. Stolz, and K. Ploog (1988) *Phys. Rev. B* **38** 4307-4310
5. J. A. Levenson, G. Dolique, J. L. Oudar, and I. Abram (1990) *Phys. Rev. B* **41** 3688-94
6. B. Deveaud, A. Chomette, F. Clerot, P. Auvray, A. Regreny, R. Ferreira, and G. Bastard (1990) *Phys. Rev. B* **42** 7021-32
7. A. P. Heberle, W. W. Rühle, and K. Köhler (1992) in: R. Alfano (ed.), *Ultrafast Lasers Probe Phenomena in Semiconductors and Superconductors,* SPIE **1677**;
    A. P. Heberle, X. Q. Zhou, A. Tackeuchi, W. W. Rühle, and K. Köhler (1993) in Proc. of the 8th International Conference on Hot Carriers in Semiconductors, Oxford, UK
8. P. Vagos, P. Boucaud, F. H. Julien, J.-M. Lourtioz, and R. Planel (1993) *Phys. Rev. Lett.* **70** 1018-21

9. U. Plödereder, T. Dahinten, A. Seilmeier, and G. Weimann (1992) in: E. Rosencher, B. Vinter, and B. Levine (eds.), *Intersubband Transitions in Quantum Wells*, NATO ASI Series **B288** Plenum Press, New York 309-18
10. R. Ferreira, and G. Bastard (1989) *Phys. Rev.* **B 40** 1074-86
11. J. L. Educato, J. P. Leburton, J. Wang, and D. W. Bailey (1991) *Phys. Rev.* **B 44** 8365-8
12. S. M. Goodnick, and J. E. Lary (1992) *Semicond. Sci. Technol.* **7** B109-15
13. G. Abstreiter, M. Besson, R. Heinrich, A. Köck, W. Schlapp, G. Weimann, and R. Zachai (1991) in: L. L. Chang, E. E. Mendez, and C. Tejedor (eds.), *Resonant Tunneling in Semiconductors Physics and Applications*, NATO ASI Series **B277** Plenum Press, New York 505-13
    R. Heinrich, R. Zachai, M. Besson, T. Egeler, G. Abstreiter, W. Schlapp, and G. Weimann (1990) Surface Science **228** 465-467
14. T. Dahinten, U. Plödereder, A. Seilmeier, K. L. Vodopyanov, K. R. Allakhverdiev, and Z. A. Ibragimov (1993) *IEEE J. Quant. Electr.* **QE-29** to be published
15. U. Plödereder, T. Dahinten, A. Seilmeier, and G. Weimann (1992) *Phys. Stat. Sol. (b)* **173** 373-379
16. M. Zalużny (1993) in: T. Paszkiewicz, and K. Rapcewicz (eds.), *Proceedings of XXIX Winter School of Theoretical Physics*, Plenum Press, New York to be published
17. B. K. Ridley (1989) *Phys. Rev.* **B 39** 5282-6
18. J. K. Jain, and S. Das Sarma (1989) *Phys. Rev. Lett.* **62** 2305-2308
19. H. Rücker, E. Molinari, and P. Lugli (1992) *Phys. Rev. B* **45** 6747-56

Ultrafast Dynamics of Electronic Capture and Intersubband Relaxation in GaAs Quantum Well

D. Morris[@], B. Deveaud, A. Regreny, P. Auvray
France Telecom, Centre National d'Études des Télécommunications,
22301 Lannion

ABSTRACT The dynamics of carrier capture and intersubband relaxation in GaAs quantum well have been measured, using a time-resolved photoluminescence technique. At low carrier densities ($\leq 2 \times 10^{10}$ cm^{-2}), electron capture times are shown to oscillate as a function of well width and a quantum-mechanical resonance is observed when one confined level is 36 meV below the barrier edge. Holes capture times are fast (< 10 ps), and depend weakly on the structure. No significant changes of the carrier capture times are observed for carrier densities up to 1×10^{11} cm^{-2}. For well widths < 150 Å, a fast intersubband ($2 \rightarrow 1$) relaxation time (of typically 3 ps) is measured. This relaxation time is shown to increase by an order of magnitude at longer well width (220 Å). The observed variation of both the electron capture times and the intersubband relaxation times with well width, demonstrate the importance of LO phonon quantum-mechanical scattering processes.

1- Introduction

Carrier relaxation mechanisms play an important role in the performance of quantum well- (QW) based devices such as lasers and intersubband photodetectors. For QW lasers, a higher gain and thus a lower threshold current is obtained owing to efficient relaxation of active electrons and holes initially injected by external contacts into the barrier layers[1]. For photodetectors, a slow carrier relaxation favors the photocarrier collection and thus improves the responsivity of the devices[2]. For both practical and fondamental interest it is important to get information on the times involved in carrier capture and intersubbband relaxation processes.

The relaxation mechanisms in quantum wells, have been recently considered within the framework of a quantum-mechanical description, where the fastest process is the scattering by LO phonons. The corresponding scattering rate depends on two main terms : the Fröhlich matrix element which gives a $1/q^2$ dependence (where q is the wave vector of the emitted phonon), and the overlap of the initial and final wave functions squared. In the case of capture, published quantum-mechanical calculations[3-6] predict strong resonances (variation of two orders of magnitude) of the carrier capture time as a function of well width. A first kind of resonance is expected when the energy difference between the barrier ground state and one of the QW levels equals the LO-phonon energy, because the phonon wave vector q is small. The second kind of resonance originates from the increased wave-function overlap when a continuum state is approaching the barrier band-edge energy. In the case of intersubband transitions, the relaxation rate is expected to vary by one order of magnitude when the thickness of the well goes from 200 Å to 60 Å. This behavior is mainly due to the variation of the q vector of the emitted phonon. Furthermore, very large difference should be observed as soon as it is not any more possible to emit one LO phonon (two orders of magnitude change).

433

H. C. Liu et al. (eds.), Quantum Well Intersubband Transition Physics and Devices, 433–442.
© 1994 *Kluwer Academic Publishers. Printed in the Netherlands.*

Various time-resolved optical experiments have been used for the study of carrier relaxation in quantum wells[7-10]. Among these techniques, time-resolved photoluminescence (PL) allows quite easily to monitor the capture process as the barrier luminescence basically disappears as fast as the carriers are captured into the well. In the same manner, intersubband scattering can be studied by observing the dynamic of the n=2 PL transition. Althougth theory predicts strong oscillations of the relaxation times with well width, the first time-resolved PL experiments reported in the literature[11-13], do not give any evidence for resonant capture mechanisms. On the other hand, it has been proposed that for thick barriers or for high carriers densities, the coherence of the electron is not preserved long enough for a simple quantum-mechanical description to be valid, and therefore simple diffusion mechanisms apply[14-16]. For intersubband relaxation, Raman scattering studies have succesfully demonstrate the importance of LO-phonon scattering[17-18]. However, few experiments specifically address the relaxation time behavior as a function of well width, and the influence of carrier densities. The possible validity of the quantum-mechanical description in the case of small period MQW structure and the influence of the excitation conditions on the measured relaxation times are questions of interest which need further experimental studies.

In this paper, we present studies of the dynamics of carrier capture and intersubband relaxation in different configurations of GaAs/Al$_x$Ga$_{1-x}$As multiple-quantum-well structures.

2. Experiments

Our samples grown by molecular beam epitaxy (MBE) consist of 20 repetitions of the structure shown in Fig. 1a) or 1b). For capture, we have studied two series of MQW samples. In the first series (Fig. 1a), the barriers are 200-Å-thick with $x_{Al} \approx 0.25$. The limited value of 200 Å for the barrier thickness is important to ensure the coherence of the electrons over at least one period of the MQW. For the second series of samples, edge spikes on either side of the wells (Fig. 1b) are added. The role of these layers will be explain later in the text. These thin layers (≈ 23 Å) consist of an Al$_{0.5}$Ga$_{0.5}$As alloy obtained by the deposition of 4 periods GaAs/AlAs monolayer superlattice. For intersubband relaxation, we have studied a series of four MQW samples (Fig. 1a) having 70-Å-thick barriers, with $x_{Al} \approx 0.30$. Most of our structures are characterized by x-ray diffraction. The structural parameters (alloy composition and thicknesses of the different layers) are deduced from a fit to the x-ray diffraction patterns and to the observed energies of the PL transitions. These parameters are given in Table 1 for the edge spike's samples, and in Table 2 for the samples used for the study of intersubband relaxation.

Photoluminescence spectra are obtained at low temperature (100 K) by an up-conversion technique with a time resolution up to 120 fs [19]. We obtain 600 fs pedestal free pulses by the use of

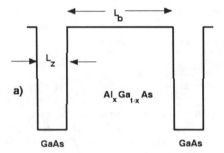

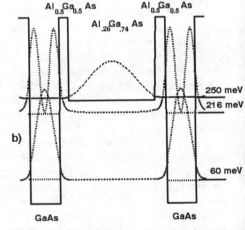

Fig. 1 : Schematic electron band configuration of our samples. In (b), the energies and the wave functions squared of the first three levels are plotted for the 73-Å sample (nearest-resonant condition).

a hybrid dye laser, synchronously pumped by the doublet output of a mode locked YLF:Nd laser. By compression of the pulses with a fibre/prism arrangement, we generate, when needed, pulses down to 100 fs. The sensitivity of the experiment allows the observation of the barrier and well luminescence without any band-filling effects; dynamical behavior of the luminescence signals can be observed at excitation density as low as $2x10^{10}$ cm^{-2} per well. For most of the spectra, the photoexcitation wavelength was 600 nm (2.05 eV). Some capture times have been obtained (see Barros et al.[20]) using a pump and probe experiments under resonant excitation conditions.

3. Electronic capture

In a first series of measurements, we have investigated the quantum-mechanical capture mechanisms by studying a complete series of multiple quantum well (MQW) structures with 200-Å-thick barriers (Fig. 1a) and well widths (L_z) covering the whole range between 30Å and 90Å. Results of this study have been published elsewhere[20,21] and are summarized in Fig. 2. The squared symbols correspond to the measured decay-times of the barrier luminescence (BL). These times, all ranging between 1 and 2 ps, correspond to the capture times of the fastest species, electron or hole. The quantum well luminescence (QWL) rise-times are found to be smaller than 4 ps in all cases [21]. This value of 4 ps is an upper limit for the capture times of both electrons and holes and differs considerably from the earlier theoretical results[3-6]. In Fig. 2, we also reproduce the results of the measured capture times at 300 K, obtained using a pump probe experiment under resonant excitation conditions[20]. These results are explained below.

The absence of long capture times is explained theoretically by considering the competing influence of LO-phonon scattering and impurity scattering, and the effect of carrier spreading in energy over a Maxwell-Boltzmann distribution. The result of our calculation for the electron capture times is plotted in Fig. 2. Details of the calculation are published elsewhere[21]. The scattering rate is computed using the Fermi golden rule. For LO-phonon scattering, we have considered only phonon emission and we neglect the effect of confined phonon modes. For impurity scattering, a uniform concentration level of $2x10^{15}$ imp./cm^3, determined by independent measurement, is assumed. The temperature of the carriers (300 K) is taken into account by a proper average over the initial states. No extra fitting parameter has been used in the calculations. Considering the approximation of the calculation, the agreement between experimental results and the theoretical curve is excellent. The very short times (≈ 100 fs)

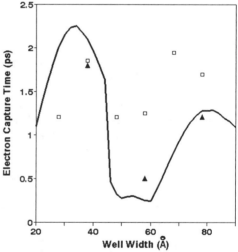

Fig. 2 : Measured electron capture times (triangles: pump-probe; squares: luminescence). The solid curve correspond to the result of the LO-phonon capture calculations including thermal averaging over the initial state distribution.

predicted when the electrons are scattered by q=0 phonons (for $L_z \approx 60$ Å) cannot be observed in luminescence because the electrons, created at 2.05 eV, first scatter to the X valley and only return slowly to the Γ valley (≈ 2 ps). However, pumping at the barrier-edge energy allows us to measure a time as short as 650 fs, for the 60-Å sample.

We have extended this first study of carrier capture by looking at a different series of MQW samples, using the same time-resolved PL technique. Compared to usual MQWs, the band configuration of the structures has been modified in order to increase the electron capture times above the limiting value of 2 ps. This has been done by adjoining edge spikes on either side of the wells (see Fig. 1b)). This scheme was first proposed by Fujiwara and al. [22]. The wave functions overlap is reduced by adjusting the height (the Al content) and the width of these edge spikes. A 23-Å-thick and a $x_{Al} = 0.50$ $Al_xGa_{1-x}As$ spike layers have been chosen to obtain an increase of the electron capture times by roughly a factor of 15. Strong oscillations of electron capture times as a function of well width and a 1.5 ps value at resonance are predicted for such structure. The stronger signal of the BL resulting from the enhanced confinement of the carriers in the barriers, allows us to study the capture mechanisms at different excitation densities, from the very low value of 2×10^{10} cm^{-2} up to 1×10^{12} cm^{-2}.

Figures 3a) and 3b) show the temporal behaviour of the QWL and BL signals. These figures compare the spectra of three samples with well widths close to the value where a resonance in the electron capture rate is expected. The intensity of the BL peak decays exponentially with time over two decades. The QWL rises upon cooling of the carriers photoexcited in the wells, and upon capture of carriers coming from the barriers. Under low excitation conditions and at 100 K, the carrier cooling process does not limit the observation of the short capture times in that particular structure. A faster decay of the BL signal and a faster rise of the QWL signal are clearly observed in the 70-Å sample. Electron and hole capture times are deduced from a simultaneous fit of the QWL and BL temporal behaviour, using the same fitting parameters. Our model considers the evolution equations for the populations of the different levels. Details of the fitting procedure are given elsewhere[23]. The deduced capture times are reported in Table 1, with fitting curves (solid lines) plotted in Figure 3 for three samples. All together the uncertainty given in Table 1 is about ±20%, whereas the electron capture times vary by almost 2 orders of magnitude.

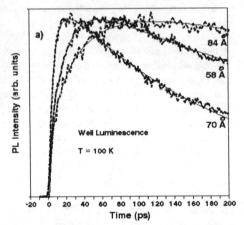

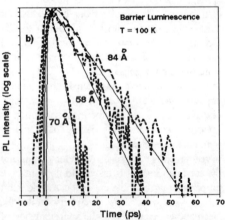

Fig. 3 : Time-delay behavior at the energy of the well ground states (a) and at the energy of the barrier peak (b) for three samples having a well width of 58Å, 70Å and 84Å (dashed lines). The solid lines correspond to fitting curves.

The electron capture time are calculated using the same calculation as the one discuss above. Again, the temperature of the carriers (100 K, in that case) is taken into account by a proper average over the initial states. The results of this calculation are plotted in Fig. 4 together with the experimental points. For a given Al composition in the barriers, two kinds of resonance are observed. The first resonance (around 63Å for $x_{Al}=0.27$) originates from increased barrier and well wave functions overlap when $E_{barr} \approx E_{well}$ (n=2). The second resonance (around 71Å for $x_{Al}=0.27$) corresponds to the situation where the wave vector of the emitted phonon is approaching zero ($E_{barr} \approx E_{well}$ (n=2) + 36meV). The position of the resonances is very sensitive to the structural parameters; a difference in the alloy composition from 0.27 to 0.31 changes the position of the resonances by almost

10Å. At elevated temperatures the effect of hot carriers tends to smooth out these resonances (see theoretical curve of Fig. 2).

It is more difficult to estimate the hole capture times. Quantum-mechanical calculations give capture times longer than for electrons and numerous very narrow resonances. However, our calculations suffer from rendering correctly the complex hole band mixing. The band non-parabolicity combined with the well width fluctuations tend to smooth out the resonances and give capture times typically >100 ps. Such quantum-mechanical calculations are questionable because the coherence length for heavy hole is smaller than the period of the structure. In our modified MQWs, the edge spikes tend to localize the heavy holes in the barriers. A semi-classical model gives a better estimation of the hole capture time. We calculate the tunneling time across the edge spike between continuum states without conservation of momentum. Once in the wells, the holes are very easily scattered into lower energy well states by emission of LO-phonon ensuring efficient hole trapping by the wells. The hole capture time estimated in this way is about 3 ps and nearly independent of well width.

In order to interpret the results, we have to consider the capture of both electrons and holes. The observed decay times of the barrier luminescence are weakly dependent of the structure. Fast decay times (< 3ps) are only observed in the two samples where LO-phonon electron capture resonance is expected (Nos. 6 and 7). In all other samples, decay times of typically 10ps are observed. This result is explained by a semiclassical model for hole capture process which gives short hole capture times nearly independent of the structure, and by a quantum-mechanical model for LO-phonon assisted electron capture mechanism which gives strong capture time resonances with well width. So in all cases, except for samples 6 and 7, the holes are captured more efficiently than the electrons.

Table 1 compares the experimental data with the theoretical electron capture times.

TABLE 1 List of the edge spike's samples (Fig. 1b) for the capture study, together with the experimental and theoretical data points

Sample No.	x_{Al} ± 0.005	Well Width (A)	$(\delta E_{el})^a$ (meV) ± 5	exp. data τ_{hole} (ps)	exp. data τ_{el} (ps)	theory τ_{el} (ps)
1	0.305	29	103	8 ± 1	50 ± 10	35
2	0.310	53	202^b	10 ± 2	10 ± 2	24
3	0.290	77	77	8 ± 1	62 ± 15	28
4	0.255	50	143	9 ± 2	118 ± 30	145
5	0.275	58	180^b	12 ± 2	21 ± 7	38
6	0.270	70	31	10 ± 2	3 ± 1	5
7	0.260	73	34	9 ± 2	2 ± 1	3
8	0.270	84	81	13 ± 2	55 ± 8	39
9	0.275	96	117	15 ± 2	41 ± 8	57

a) energy difference between the barrier ground state and the highest confined level

b) a barrier wavefunction buildup in the well is expected for those samples

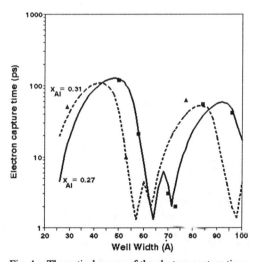

Fig. 4 : Theoretical curves of the electron capture time as a function of well width for the barrier configuration shown in Fig. 1 b) for x_{Al}=0.27 (solid line) and x_{Al}=0.31 (dotted line). Electron capture times deduced from the fitting procedure marked with rectangles (triangles) have to be compared with the theoretical solid (dotted) curve.

Also given in Table 1 is the energy difference between the electron barrier ground state and the highest confined level (δE_{el}). The fitting procedure gives hole capture times smaller than 15 ps and electron capture times oscillating between 2 ps and 120 ps. The agreement between experimental and theoretical electron capture times are remarkably good. Results clearly evidence the q=0 LO-phonon resonance (for well width around 71Å) corresponding to $\delta E_{el} \approx 36$ meV, and confirm our first series of measurements.

There are still some confusion in the scientific community regarding the interpretation of the optical measurements. This confusion originates mainly from the fact that relaxation times depend on the carrier distribution which in turns depends on the experimental conditions (excitation density, lattice temperature and doping level). For time-scale of the order of 1 ps or longer, thermal distribution of carriers within each subband have to be considered[24]. With increasing excitation density both the carrier temperatures and the occupation of the different levels increase. At high densities, band-filling effect has to be considered. Intersubband carrier-carrier scattering rate also depends on the carrier distribution functions and may affect the carrier capture time[25]. We have found in the preceding analysis that the thermal distribution of carriers in the barriers plays a significant role as it tends to smooth out the oscillation of the capture times as a function of the well width. Also, at this conference Seilmeier et al.[26] present results of the effect of doping and the influence of the lattice temperature on the intersubband relaxation times. In the following, we have investigated the effect of excitation density on the capture times.

Fig. 5 shows the time delay behavior of the QWL and the BL signals as a function of the excitation density. These densities correspond to 2×10^{10}, 8×10^{10}, and 4×10^{11} carriers/cm^2 per well. The decay-time of the BL, which is mainly dictated by the fast hole capture process for the edge spike's structures, is typically 10 ps and does not change with excitation density. The same behavior is observed for all samples in this series. Up to 1×10^{11} carriers/cm^2, no significant changes of the QWL rise-times are observed; these values correspond approximately to the electron capture

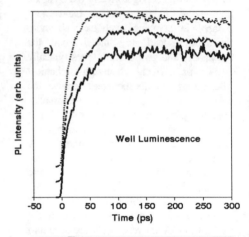

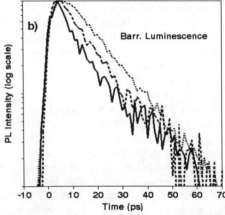

Fig. 5 : Time-delay behavior at the energy of the n=1 well peak (a) and at the energy of the barrier peak (b) for the 50-Å sample, obtained for three excitation conditions: 2×10^{10} / cm^2 (solid lines), 8×10^{10} / cm^2 (dashed lines), and 4×10^{11} / cm^2 (dotted lines).

times given in Table 1 (118 ps, for the 50-Å sample). At low densities, our results indicate that intersubband carrier-carrier scattering plays a negligeable role in the capture process. This statement is also supported by the fact that the agreement between experiments and the calculations which include only the LO-phonon scattering, is remarkably good. At high density (4×10^{11}/cm^2), we observe faster QWL rise-times (about 35 ps, for the 50-Å sample). Band-filling effects may well explain this result.

The nearly constant PL intensity seen at short times (Fig. 5a) is a strong indication that the low-energy states become fully occupied[27]. On the other hand, we may also argue that the higher concentration of holes in the wells at short times accelerates the electron capture process. At very high excitation densities one would expect such ambipolar capture process[25]. The carrier capture mechanisms are not fully understood under high excitation conditions.

4. Intersubband relaxation

We have studied a series a four samples with well widths ranging from 100 Å to 220 Å (see Table 2). Fig. 6 shows the time-resolved spectra obtained for the 100-Å and the 220-Å samples, at different time delays (1, 2, 5, 10 ps). Owing to the excellent sensitivity of the experiment, the n=2 PL peak can be seen under moderate excitation conditions ($\approx 8 \times 10^{10}/cm^2$) even for large energy separation between the subbands ($\delta E_{el} = 89$ meV, for the 100-Å sample). Fig. 7 shows the time-delay behaviors at the energy of the n=2 PL peak. The experiment has been performed at 100 K in order to minimise the carrier cooling time (≈ 3 ps) and the influence of holes. A fast hole relaxation time is assumed and we consider a nearly constant hole population of the second QW subband given by the thermal population of this level. The electron intersubband relaxation time is then deduced from the decay time of the n=2 PL intensity. This luminescence peak disappears with a characteristic time of 3 ps for the 100-Å sample. The decay times are found to increase with well width from 3 ps up to 40 ps. The results are given in Table 2.

We have carried out the calculations of the (2→1) intersubband relaxation times considering the LO-phonon scattering and the effect of hot carriers. We have assumed only phonon emission and bulk phonon modes. The thermal distribution of the carriers is taken into account. When the energy separation between the subbands is smaller than the LO-phonon energy (220-Å sample), relaxation occurs via emission of acoustical

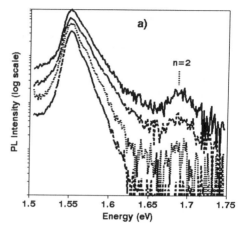

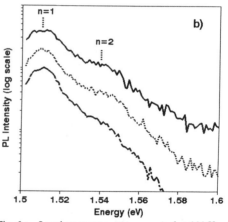

Fig. 6 : Luminescence spectra measured at 100 K, at different time delays for samples having a well width of 100 Å (a), and 220 Å (b). The time delay increases from the top to the bottom spectrum (1, 2, 5, 10 ps).

phonons. In that case, the relaxation time is approximately 300 ps at 0 K (see Bastard et al.[28]) This time is reduced roughly by a factor of 2 if we consider the fast LO-phonon relaxation of those carriers sitting at higher energy in the thermal distribution. Results of our calculations are presented in Table 2.

Our experimental results are in fairly good agreement with the measurements of Tatham et al. [18], and Seilmeier et al.[26]. However, they differ significantly from the theoretical values. We do not have a full explanation for these discrepancies. For well widths ≤ 180 Å, the contribution of the temperature of the carriers prevents the observation of resonant process (expected when $\delta E_{2 \rightarrow 1} \approx 36$ meV) but it is not enough to explain the long relaxation times observed experimentally. The gradual increase of the intersubband relaxation times with well width is quite surprising. Band-filling effects might explain this behavior but we do not see any evidence of band-filling in our PL spectra. On the other hand, we have neglected in the calculations the impurity-scattering processes and the contribution of holes. These extra mechanisms reduce, in all cases, the relaxation times. This may explain the short time observed for the 220-Å sample but cannot explain the results observed for the thinner QW. Therefore, we are left with the conclusion that hot population of LO and acoustical phonons must play an important role in the relaxation. The effect of phonon reabsorption becomes increasingly important when the energy separation between the n=2 and the n=1 levels decreases, in good agreement with our experimental results. Nevertheless, further experiments are needed to confirm this conclusion.

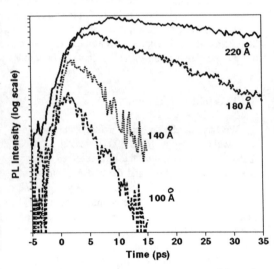

Fig. 7 : Time-delay behavior at the energy of the n=2 well peak for different samples.

5. Conclusions

In summary, we have used time resolved-luminescence spectroscopy to measure the electron and hole relaxation times as a function of well width in different configurations of multiple quantum well structures. In the first series of measurements, results show that both electrons and holes are captured efficiently in all cases ($\leq$ 4 ps) . Precise computation of different scattering processes explains these short times. By using a pump-probe experiment, we clearly evidence a resonant electron capture mechanism for a MQW sample having a 60-Å well width. In a second series of measurements, results for the edge spike's samples show that hole capture times are fast (< 10 ps), and depend weakly on the structure. For that particular structure, hole capture process is well described by a semiclassical model. On the other hand, at low carrier density (2×10^{10} cm^{-2}), electron capture times are shown to vary between 2 and 118 ps

TABLE 2 List of the samples for the intersubband relaxation study, together with the experimental and theoretical data points

Sample No.	x_{Al} ± 0.01	Well Width (Å)	$\delta E_{2 \to 1}$ (meV) ± 5	exp. data $\tau_{2 \to 1}$ (ps)	theory $\tau_{2 \to 1}$ (ps)
10	0.30	100	89	3 ± 1	0.8
11	0.30	140	56	5 ± 1	0.6
12	0.30	180	37	12 ± 2	0.4
13	0.30	220	26	40 ± 8	180[a]

a) in that case, the relaxation time is mainly determined by the acoustical-phonon scattering rate.

as a function of well width. Again, a quantum-mechanical resonance is observed when one confined level is 36 meV below the barrier edge. No significant changes of the carrier capture times are observed for carrier densities up to 1×10^{11} cm^{-2}. At high excitation densities, the carrier capture mechanisms are not fully understood mainly because the band-filling effects complicate the interpretation of the temporal spectra.

For the 100-Å sample, a fast intersubband $(2\rightarrow 1)$ relaxation time of 3 ps is measured. This relaxation time increases by an order of magnitude for larger well widths. The transition from a dominant LO-phonon scattering process (for $\delta\,E_{2\rightarrow 1} \geq 36$ meV) to a dominant acoustical-phonon scattering process (for $\delta\,E_{2\rightarrow 1} \leq 36$ meV) is clearly seen. The unexpected short value of 40 ps measured for the 220-Å sample may be due to the contribution of holes and the impurity-scattering mechanisms. For thinner wells (≤ 180 Å), the discrepancies between experimental and theoretical relaxation times may be explain by the effect of phonon reabsorption which slows down the effective LO-phonon scattering mechanism.

The observed variation of both the electron capture times and the intersubband relaxation times with well width demonstrates the importance of LO-phonon quantum-mechanical scattering processes.

Acknowledgments :

The authors wish to express their thanks to M. Barros, P. Becker, and F. Beisser for having performed the pump and probe measurements. We also thank A. Chomette, F. Clérot, C. Guillemot, and B. Lambert for useful discussions. Thanks are also due to M. Baudet for her work in the x-ray diffraction pattern fits. One of us (D.M.) thanks Fonds FCAR for financial support.

@ now at University of Sherbrooke, Dept. of Physics, Sherbrooke, Québec, Canada J1K 2R1

References :

[1] P.W.M. Blom, J.E.M. Haverkort, J.H. Wolter, Appl. Phys. Lett. **58**, 2767 (1991)
[2] A. Fraenkel, A. Brandel, G. Bahir, E. Finkmann, G. Livescu and M.T. Asom, Appl. Phys. Lett. **61**, 1341 (1992)
[3] J.A. Brum, G. Bastard, Phys. Rev. **B33**, 1420 (1986)
[4] M. Babiker, B.K. Ridley, Superlatt. and Microstruct. **2**, 287 (1986)
[5] J.A. Brum, T. Weil, J. Nagle, B. Vinter, Phys. Rev. **B34**, 2381 (1986)
[6] S.V. Kozyrev, A. Ya Shik, Sov. Phys. Semicond. **19**, 1024 (1985)
[7] J. Shah, T.C. Damen, B. Deveaud, D. Block, Appl. Phys. Lett. **50**, 1307 (1987)
[8] A. Seilmeier, H.J. Hübner, G. Abstreiter, G. Weimann, and W. Schlapp, Phys. Rev. Lett. **59**, 1354 (1987)
[9] P.C. Becker, D. Lee, A.M. Johnson, A.G. Prosser, R.D. Feldman, R.F. Austin, and R.E. Behringer, Phys. Lett. **68**, 1876 (1992)
[10] D.S. Kim, J. Shah, T.C. Damen, W. Schäfer, F. Jahnke, S. Schmitt-Rink, K. Köhler, Phys. Rev. Lett. **69**, 2725 (1992)
[11] J. Feldmann, G. Peter, E.O. Göbel, K. Leo, H.J. Polland, K. Ploog, K. Fujiwara, T. Nakayama, Appl. Phys. Lett. **51**, 226 (1987)
[12] B. Deveaud, J. Shah, T.C. Damen, W.T. Tsang, Appl. Phys. Lett. **52**, 1886 (1988)
[13] D.Y. Oberli, J. Shah, J.L. Jewel, T.C. Damen, N.D. Chand, Appl. Phys. Lett. **54**, 1028 (1989)
[14] A. Weller, P. Thomas, J. Feldman, G. Peters, and E. Gobel, Appl. Phys. **A48**, 509 (1989)
[15] S. Morin, B. Deveaud, F. Clérot, K. Fujiwara, and K. Mitsunaga, IEEE J. Quantum Electron. QE-27, 1669 (1991)

442

[16] S. Weiss, J.M. Wiesenfeld, D.S. Chemla, G. Sucha, M. Wegener, G. Eisenstein, C.A. Burrus, A.G. Dentai, U. Koren, B.I. Miller, H. Temkin, R.A. Logan and T. Tanbun-Ek, Appl. Phys. Lett. 60, 9 (1992)

[17] D.Y. Oberli, D.R. Wake, M.V. Klein, J. Klem, T. Henderson, H. Morkoc, Phys. Rev. Lett. 59, 696 (1987)

[18] M.C. Tatham, J.F. Ryan, C.T. Foxon, Phys. Rev. Lett. 63, 1637 (1990)

[19] T.C. Damen and J. Shah, Appl. Phys. Lett. 52, 1291 (1993)

[20] M.R.X. Barros, P. C. Becker, D. Morris, B. Deveaud, A. Regreny, and F. Beisser, Phys. Rev. B47, 10951 (1993)

[21] B. Deveaud, A. Chomette , D. Morris, and A. Regreny, Solid State Comm. 85, 367 (1993)

[22] A. Fujiwara, S. Fukatsu, Y. Shiraki, R. Ito, Surf. Science 263, 642 (1992)

[23] D. Morris, B. Deveaud, A. Regreny, Phys. Rev. B47, 6819 (1993)

[24] a thermalized distribution of carriers is observed within the first 100 fs at excitation densities as low as 10^{17} cm^{-3}, see : T. Elsaesser, J. Shah, L. Rota, and P. Lugli, Phys. Rev. B66, 1757 (1991)

[25] P.W.M. Blom, J.E.M. Haverkort, P.J. van Hall, and J.H. Wolter, Appl. Phys. Lett. 62, 1490 (1993)

[26] these results have been presented at this conference by A. Seilmeier, NATO Advanced Research Workshop on Quantum Well Intersubband Transition Physics and Devices (7-10 Sept. 1993, Whistler Canada)

[27] B. Deveaud, A. Chomette, F. Clérot, P. Auvray, A. Regreny, R. Ferreira, and G. Bastard, Phys. Rev. B42, 7021 (1990)

STRUCTURE-DEPENDENT ELECTRON-PHONON INTERACTIONS

B.K. RIDLEY
Department of Physics
University of Essex
Colchester, Essex CO4 3SQ
U.K.

ABSTRACT. The recent development of a continuum model of the optical vibrations in quantum wells and superlattices which successfully reproduces the results of microscopic theory allows more reliable estimates of intra- and inter-subband scattering rates to be made. In this model the electron interacts in polar material with triple hybrids consisting of mixed LO, TO and IP (interface polariton) modes whose vibrational patterns are described by simple analytic expressions. The effect of hybridization depends upon the degree of polarity of the material and on conditions at the interfaces. For non-polar material there are only double hybrids consisting of LO and TO modes. For highly polar material (e.g. AlAs) and for polar slabs with free surfaces the interface modes are virtually decoupled from the confined LO/TO components. In polar superlattices with strongly hybridized modes there is strong anisotropy. All of this makes the scattering rates dependent on detailed structure and this opens up the possibility of phonon engineering. Some examples of scattering rates will be presented.

1. Introduction

Quantum-well devices typically involve hot electrons, whose energy relaxation takes place principally by the net emission of optical phonons. The rate at which this occurs affects the speed of the device, and it is extremely useful to have available a reliable theory of the electron-phonon interaction which does not involve excessive computational time in order to model the performance of the device. While electron states have been reasonably well described by envelope-function formalism, phonon states have required extensive numerical analysis of the lattice dynamics of the heterostructure in order to describe the effects of confinement. Recently, a number of analytical models of phonon confinement have been advanced [1-3] which to varying degree reflect the mixing which occurs between longitudinally-polarized optical (LO) and interface polariton (IP) modes as revealed by microscopic calculations [eg. 4-6]. These models ignored the rôle of transversely-polarized optical (TO) modes, which were shown to be essential ingredients in satisfying mechanical boundary conditions in non-polar material [7] and in satisfying both mechanical and electrical boundary conditions in polar material [8,9,10]. A continuum model of confined phonons now exists which describes confinement in terms of triple hybrids (LO/TO/IP) in the case of polar material, and double hybrids (LO/TO) in the case of non-polar material, and it allows a straightforward analytic formulation of the interaction with electrons to be made.

H. C. Liu et al. (eds.), Quantum Well Intersubband Transition Physics and Devices, 443–456.
© 1994 *Kluwer Academic Publishers. Printed in the Netherlands.*

A brief description of this model is given in Section 2 and some of the features of the unscreened interaction of hybrid phonons with electrons are outlined in Section 3.

2.   Optical-Mode Hybrids

The equation of motion which describes long-wavelength optical vibrations in bulk, isotropic material is

$$\rho \, \frac{\partial^2 u}{\partial t^2} = -\rho \, \omega_{TO}^2 \, u + e_i \, E - c_{11} \, \nabla \, (\nabla . u) + c_{44} \, \nabla \times \nabla \times u \qquad (1)$$

where u is the relative displacement of the ions, ρ is the reduced density, ω_{TO} is the natural frequency, e_i is the effective ionic charge density, E is the macroscopic electric field, and c_{11}, c_{44} are optical-mode elastic constants which quantify the macroscopic stress in the medium and determine the dispersion. Elastic anisotropy can easily be taken into account, but here we will ignore it from the outset since, ultimately, averages over direction would have to be taken to describe scattering rates. It is assumed that a continuum picture suitable for bulk material is applicable to the quantum-well, superlattice and slab cases. Coupled with the equation of motion are Maxwell's equations.

The displacements can be factorized into longitudinal ($\nabla \times u_L = 0$) and transverse ($\nabla . \, u_T = 0$) parts, and three distinct travelling-wave solutions emerge, one longitudinal (LO) and two transverse (TO and IP). With reference to an interface these solutions further divide into two types: those with polarizations lying in the plane of incidence (p-modes) and those with polarization at right-angles (s-modes). The latter can only be a TO mode, and it has no difficulty in satisfying boundary conditions without mixing. The same is not true for p-modes. The requirement that both mechanical and electromagnetic boundary conditions be satisfied forces these modes to form a linear combination, leading to a triple hybrid. The principal properties of these modes are given in Table 1.

TABLE 1. Principal properties of optical modes

Mode	Dispersion	Electric Field	Permittivity
LO	$\omega^2 = \omega_{LO}^2 - v_L^2 k^2$	$E_L = -\dfrac{e_i}{\varepsilon_\infty} u_L$	$\varepsilon(\omega) = 0$
TO	$\omega^2 = \omega_{TO}^2 - v_T^2 k^2$	$E_T = 0$	$\varepsilon(\omega) \to \infty$
IP	$\omega^2 = k^2/(\mu_0 \, \varepsilon(\omega))$	$E_p = s \, E_L$	$\varepsilon(\omega) = \varepsilon_\infty \left(\dfrac{\omega_{LO}^2 - \omega^2}{\omega_{TO}^2 - \omega^2}\right)$

$\omega_{LO}^2 = (\varepsilon_s/\varepsilon_\infty) \, \omega_{TO}^2 \, , \qquad e_i^2 = \varepsilon_\infty \rho \, (\omega_{LO}^2 - \omega_{TO}^2) \, , \qquad s = (\omega^2 - \omega_{TO}^2)/(\omega_{LO}^2 - \omega_{TO}^2) \, ,$

$v_L^2 = c_{11}/\rho \, , \quad v_T^2 = c_{44}/\rho \, , \quad \varepsilon_s , \; \varepsilon_\infty = $ low and high frequency permittivities,

$\mu_0 = $ permeability.

The situation is analogous to that for acoustic modes where it is well known that similar mixing is responsible for Rayleigh waves and for mode conversion. Indeed, the optical analogue of a Rayleigh wave has been described [7]. There is, however, an important difference between optical and acoustic hybrids regarding frequency. In both cases each element of the hybrid must have the same frequency and while this is always possible for long wavelength acoustic modes it is only possible for optical modes by invoking complex branches of the vibrational spectrum. Thus for a TO component to have the same frequency as an LO component its wavevector must be complex and often large. The necessity for large wavevectors clearly stretches any continuum approach to its limits, but fortunately a large imaginary wavevector implies that any effect is confined to a small region near the interface, and it therefore does not seriously affect the major features of the hybrid, apart from ensuring boundary conditions are satisfied.

For wavevectors large compared with those for light at the same frequency, wave velocities are such that the velocity of light, by comparison, can be taken to be infinite, and we work in the so-called unretarded limit. The electromagnetic boundary conditions reduce to the continuity of E_x and $\varepsilon(\omega)E_z$, where the x-axis lies in the plane of the interface and the z-axis lies perpendicular to it. The elastic boundary conditions are the continuity of $\mathbf{u}$ and of the z-component of the stress tensor, T_{zi}. Two simple cases occur: rigid boundaries, in which case the elastic condition is simply $\mathbf{u} = 0$; and stress-free surfaces, in which case $T_{zi} = 0$. Here we will focus on the rigid boundary case as this is approximately the situation in AlAs/GaAs and similarly mechanically mismatched systems.

In the case of a quantum well, the optical vibrations inside the well take the form [8-10] (see Appendix):

$$u_x = ik_x A e^{i(k_x x - \omega t)} \left[\sin k_L z - s_T \sinh k_T z - s_p \sinh k_x z \right]$$

$$u_z = k_L A e^{i(k_x x - \omega t)} \left[\cos k_L z - (k_x^2/k_L k_T)s_T \cosh k_T z - (k_x/k_L)s_p \cosh k_x z \right] \tag{2}$$

and a second form exists with cosine and sine etc. interchanged. Satisfying the boundary conditions leads to explicit expressions for the amplitudes s_T and s_p and to a dispersion relation for k_L. In the barriers there are no ionic displacements at the frequency of the hybrid (by hypothesis), but there are fields from the LO and IP components. In the same way there are no ionic displacements in the well at the characteristic frequencies of the barrier, but there are fields at the barrier frequency. The amplitude A is obtained in terms of "hybridon" annihilation and creation operators via a quantization procedure.

The dispersion diagram (Fig.1) shows LO-like confined modes with $k_L L \approx n\pi$ near $k_x L = 0$, where L is the well-width, with crossings and anticrossings at larger values of $k_x L$ associated with the classic dispersion of the antisymmetric interface mode [11]. Similar effects occur with the dispersion of the symmetric interface mode at frequencies near ω_{TO}. This interplay of LO and IP dispersion results in a division of hybrids into LO-like and IP-like: LO-like modes are those distant from an anticrossing and IP-like modes are those at or near an anticrossing. At frequencies near ω_{LO} the TO component is localised near the interfaces and to an excellent approximation the hybrid can be regarded as a double (LO/IP) hybrid which does not have to satisfy $u_x = 0$, which is equivalent to ignoring the elastic boundary condition involving shear, but at frequencies near ω_{TO} the role of the TO mode cannot be ignored. Fig.2 shows LO-like modes near $k_x L = 0$ and how the LO1 mode

446

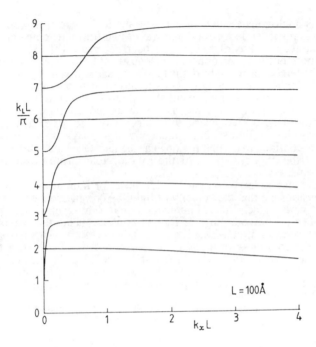

Fig.1   Dispersion of hybrid optical modes in a 100Å GaAs quantum well with rigid barriers [10].

transforms to an LO3-like mode after passing through an anticrossing whereas the LO2 mode is unchanged after passing through a crossing.

Application of hybrid theory to a superlattice yields identical forms for the ionic displacement (eq.2) for the cases $k_z(a+b) = 0$ and π, where k_z is the Bloch vector and a, b are the well and barrier widths, but, in general, symmetric and antisymmetric patterns become mixed. The dispersion becomes anisotropic being dependent on the angle $\theta = \tan^{-1} (k_x/k_z)$. Comparison of this angular dispersion with the results of microscopic theory shows excellent agreement (Fig.3) [12]. This anisotropic dispersion is accessible in experiment via the techniques of micro-Raman scattering [13], and good agreement with hybrid theory (in the double-hybrid approximation) has been demonstrated (Fig.4) [14]. The anisotropy is evident in the differing structures of the electric potential in the wells for well-modes travelling in the plane of the superlattice $k_z = 0$ and waves travelling more nearly along the growth axis $k_x (a+b) = \pi$. (Fig.5) [15]. For $k_z = 0$ the waves are IP-like in the lower orders, but LO-like for $k_z(a+b) = \pi$. This anisotropy is a consequence of the interaction between the periodic elements of the superlattice via the evanescent fields, and it makes an estimate of the electron-phonon interaction more difficult to make.

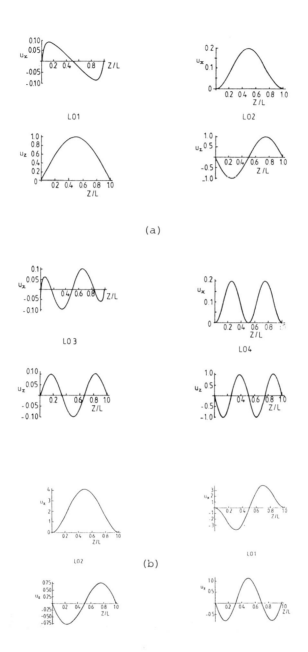

Fig.2   Relative ionic displacements for lower order LO-like modes [10].
(a) $k_xL \approx 0$.   (b) $k_xL = 3$.

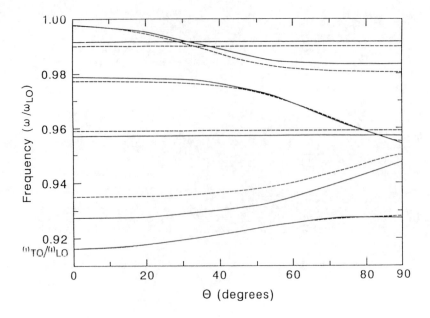

Fig.3   Angular dispersion in a 22.4Å GaAs/16.7Å AlAs [100] superlattice (dashed lines) compared
with the results of a microscopic model for a GaAs$_7$ AlAs$_7$ superlattice [12].

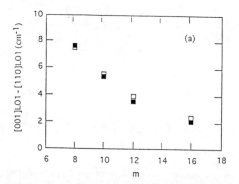

Fig.4   The predicted frequency shift of the LO1 mode from on-axis to in-plane propagation in a
GaAs$_m$ AlAs$_m$superlattice (open squares) compared with micro-Raman data (closed squares)
[14].

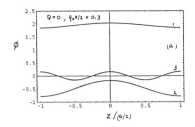

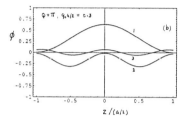

Fig.5   Potential profiles of the lower order GaAs modes in a GaAs/AlAs superlattice for propagation
(a) in-plane ($Q = k_z(a+b) = 0$) and (b) $Q = \pi$. ($k_x = q_0 = 2.5 \times 10^6$ cm^{-1}, $q_0a/2 = 0.3$, a = well width, b = barrier width = 2 monolayers) [15].

In the case of the AlAs/GaAs system microscopic calculations and evidence from Raman scattering show that the displacement of GaAs modes disappears at the first AlAl atom at the interface rather than at the interface itself. This can easily be accommodated by a shift in the effective width of the well. There is also an uncertainty in the magnitude of the LO velocity (v_L) which determines the dispersion. Values of 3.8×10^5 cm s^{-1} [14] and 2.8×10^5 cm s^{-1} [12] have been used.

3.   The Interaction with Electrons

One of the interesting consequences of hybridization is the complexity of interaction with electrons, each element contributing its characteristic coupling. In general the LO component will contribute a deformation potential and a Fröhlich potential, the TO component a deformation potential, and the interface mode a deformation potential (not normally considered) and an electromagnetic **A.p** interaction (where **A** is the vector potential and **p** is the momentum). In most of the III-V semiconductors the deformation potential can be ruled out for electrons, but not for holes. In the non-retarded limit the electromagnetic interaction of the interface mode can be shown to be replaceable by an electrostatic potential, at least as far as first-order perturbation theory is concerned [16], and so the potentials contributed by the LO and IP components are additive. For example, the scalar potential, ϕ, for the quantum well example in eq.(2) is

$$\phi = \frac{e_i}{e_\infty} A e^{i(k_x x - \omega t)} [\sin k_L z - s\, s_p \sinh k_x z] \qquad (3)$$

and a symmetrical potential exists with $\sin k_L\, z$, $\sinh k_x\, z$, s_p, replaced by $\cos k_L\, z$, $\cosh k_x\, z$ and $\dot{c}_p$.

Scattering rates can be determined using first-order perturbation theory. An electron is scattered by potentials deriving from well-modes and barrier modes, and the scattering is intrasubband or intersubband. We quote a few examples calculated with the simplifying assumption that the electron is totally confined in the well, and that the TO component contributes negligibly to the energy of the hybrids principally involved.

(a)

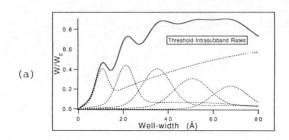

(b)

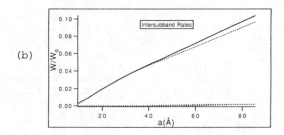

Fig.6   Electron scattering rates for the interaction with GaAs hybrids in a quantum well as a function of well-width:

(a) intrasubband rates (electron energy = $\hbar\omega_{LO}$)

(b) intersubband rates (electron energy = bottom of subband 2).

The dashed lines show the contribution from individual modes.

$W_0 = (e^2/4\pi\hbar)\,(2m^*\omega_{LO}/\hbar)^{\frac{1}{2}}\,(\varepsilon_\infty^{-1} - \varepsilon_s^{-1})$, m^* = electron effective mass.

The intrasubband rate in GaAs produced by the emission of well-modes is shown in Fig.6a as a function of well-width for an electron energy equal to $\hbar\omega_L$. The rate diminishes towards zero well-width as the increasing confinement enhances the magnitude of the wavevector of the LO component (k_L) thereby reducing the potential. This behaviour was predicted long ago for pure LO confined modes, and it is seen here to occur also for hybrids. But superimposed on this general trend are oscillations associated with the enhanced interaction with IP-like hybrids, whose effect is to keep the rate elevated over that expected for LO-like hybrids[10]. Similar oscillations occur in the intersubband rate (Fig.6b) but they are much smaller and indiscernable on the scale shown, by far the major contribution coming from the lowest order antisymmetric mode (the "converted" LO1 – nominally LO3, but actually $k_L a \simeq 2.5\pi$ over much of the range).

In GaAs the dispersion of the LO mode is large enough so that the band of LO frequencies spans the frequency gap ω_{LO}–ω_{TO} at the zone centre, and this allows the hybridization of LO and IP modes without having recourse to the complex band-structure at the zone edge. This is not the case of AlAs nor for highly polar semiconductors in general. When this is true the complex LO/LA component becomes localised at the interfaces, like the TO component, and the interface mode becomes approximately a classical interface polariton over the frequency range in between the zone-edge LO frequency ω_{LZ}, and ω_{TO}, and is strongly hybridized only in the frequency range ω_{LO} to ω_{LZ}. Another way of putting this is that for a given in-plane wavevector k_x, the higher frequency IP mode is "classical" for well-widths such that

$$L \geq \frac{2}{k_x} \tanh^{-1} \left[r_0 \frac{\omega_{LO}^2 - \omega_{LZ}^2}{\omega_{LZ}^2 - \omega_{TO}^2} \right] \qquad (4)$$

where $r_0 = \varepsilon_{\infty B}/\varepsilon_W$, $\varepsilon_{\infty B}$, $\varepsilon_{\infty W}$ are the high-frequency permittivities of the barrier and well. Thus, hybridization of this mode occurs over the whole range whatever the well-width provided that

$$\omega_{LZ}^2 \leq \omega_{TO}^2 \frac{\varepsilon_{\infty W} + \varepsilon_{sB}}{\varepsilon_{\infty W} + \varepsilon_{\infty B}} . \qquad (5)$$

For the lower-frequency IP mode the condition is simply $\omega_{LZ}^2 \leq \omega_{TO}^2$. In the case of AlAs $\hbar\omega_{LZ} = 48$ meV in the $\cdot\langle 100 \rangle$ direction [17,18], and the lower-frequency mode is essentially classical over the whole range (barring effects near ω_{TO} due to the TO component), while the higher-frequency mode is classical for $L \geq 38$Å(with $k_x = 2.51$ x 10^6 cm^{-1}, which is the in-plane vector at emission threshold in GaAs). It will therefore be a reasonable approximation to ignore hybridization altogether in this case and exploit the "classical" results [19,20,21].

In a double heterostructure the higher frequency barrier (AlAs) IP mode contributes a symmetric potential in the well and is therefore responsible for intrasubband transitions, the rate for which near threshold is shown in Fig.7a. The contribution is comparable with that of the well modes. The contribution to the intersubband rate from the lower frequency AlAs IP mode, which provides an asymmetric potential in the well, is shown in Fig.7b, which is small at the well-widths shown (but becomes substantial at well-widths near where the difference in energy between the subbands is $\hbar\omega_{LO}$ [20,21]).

In the case of a superlattice the marked dispersive anisotropy influences the interaction with electrons, which itself becomes markedly anisotropic. An example is shown in Fig.8, where the barrier width is taken to be a constant magnitude

452

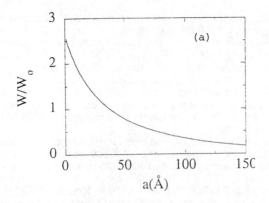

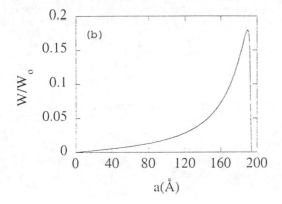

Fig.7   Contribution of the AlAs modes to the scattering rate for electrons in the GaAs quantum well

   (a) intrasubband (energy = 1.4 $\hbar\omega_{LO}$ (GaAs)
   (b) intersubband (energy = bottom of subband 2)

2 monolayers) less than the well-width. Well-hybrids travelling with $k_z = 0$ are IP-like and interact strongly, whereas hybrids with $k_z (a+b) = \pi$ are LO-like and interact moderately. The pattern of anisotropy is very much a function of the superlattice structure, specifically of the ratio of barrier-width to well-width (b/a) following the dispersive characteristics of superlattice polaritons [22].

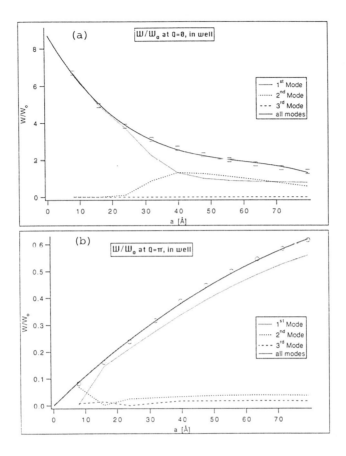

Fig.8   Anisotropy of scattering rates in a GaAs(aÅ)/AlAs (2 monolayer) superlattice for the interaction with GaAs modes for (a) Q = 0 (b) Q ≈ π.

4 .   Conclusion

The theory of hybrid optical phonons in quantum wells and superlattices based on an isotropic continuum model has been shown to give results in reasonably good agreement with those derived from numerical simulation of lattice dynamics. Its application to the interaction with electrons has served to illuminate the role that structure and dispersion has to play, and it has opened up the possibility of engineering desirable phonon properties via the choice of material.

The continuum theory described above deals solely with bare lattice vibrations and is applicable only to situations in which the carrier density is small. In order to deal with arbitrary carrier densities the theory must be extended to include coupling with plasma vibrations. This programme is currently under way. A general drawback to any continuum theory is its inapplicability to structures consisting of a few unit cells only, unless its elements are slowly varying. Usually, the main slowly varying element is the potential associated with an IP mode, and where this is dominant the theory's validity is extended accordingly.

Appendix

The process of hybridization begins by taking a linear combination of LO, TO and IP p-polarised modes, all at the same frequency and all with the same in-plane wavevector:

$$u_x =$$

$$\left[k_x \left(Ae^{ik_Lz} + Be^{-ik_Lz} \right) + k_T \left(Ce^{k_Tz} + De^{-k_Tz} \right) + k_x \left(Fe^{k_xz} + Ge^{-k_xz} \right) \right] e^{i(k_xx - \omega t)}$$

$$(A1)$$

$$u_z =$$

$$\left[k_L \left(Ae^{ik_Lz} - Be^{-ik_Lz} \right) - ik_x \left(Ce^{k_Tz} - De^{-k_Tz} \right) - ik_x \left(Fe^{k_xz} - Ge^{-k_xz} \right) \right] e^{i(k_xx - \omega t)}$$

In the case of infinitely rigid barriers there are only electric fields for $|z| > L/2$:

$$E_x = k_x He^{k_xz} e^{i(k_xx - \omega t)} \qquad = k_x Je^{-k_xz} e^{i(k_xx - \omega T)}$$

$$E_z = -ik_x He^{k_xz} e^{i(k_xx - \omega t)} \qquad = ik_x Je^{-k_xz} e^{i(k_xx - \omega T)}$$

$$z \le -L/2 \qquad\qquad z \ge L/2 \quad (A2)$$

The boundary conditions are then:

$$\mathbf{u} = 0, \ E_x \text{ and } \varepsilon E_z \text{ continuous}, \qquad (A3)$$

leading to eight homogensous equations for eight unknowns. The determinant of the 8 x 8 matrix must be zero, and this gives the dispersion relation. All amplitudes can be expressed in terms of A, and one obtains the anti-symmetric form given by eq.(2) plus a symmetric form.

The Hamiltonian consists of mechanical and electrical energy. It turns out that the mechanical energy is dominant and thus:

$$H = \frac{1}{2} M\omega^2 \int U^* (r,t) \cdot U(r,t) \, dr/v_o \tag{A4}$$

where M is the reduced mass and

$$U(r,t) = \int (u \, a_k + u^* \, a_k^\dagger) \, dk \ , \tag{A5}$$

Here a_k, $a_k^\dagger$ are annihilation and creator operators and u is given by cq.(2). From this

$$A_i = \left(\frac{2}{NK_i^2} \right)^{1/2} \tag{A6}$$

where N is the number of unit cells and K_i^2 is a complicated function of k_L, k_T and k_x. In the limit $k_T \to \infty$,

$$k_a^2 = k_L^2 \left(1 + 3 \frac{\sin k_L L}{k_L L} + 4 \frac{\sin^2 k_L L/2 \cdot \coth k_x L/2}{k_x L} \right) + k_x^2 \left(1 + \frac{\sin k_L L}{k_L L} \right)$$

$$\tag{A7}$$

$$k_s^2 = k_L^2 \left(1 - 3 \frac{\sin k_L L}{k_L L} + 4 \frac{\cos^2 k_L L/2 \cdot \tanh k_x L/2}{k_x L} \right) + k_x^2 \left(1 - \frac{\sin k_L L}{k_L L} \right)$$

where the subscripts a, s refer to asymmetric and symmetric solutions.

Acknowledgement

I would like to thank Dr. M. Babiker and N.C. Constantinou for many useful discussions. Thanks are also due to the U.S. Office of Naval Research for supporting my work in this area.

456

References

1. Huang, K. and Zhu, B. (1988) Phys. Rev. **B38** 13,377.
2. Zianni, X. Butcher, P.N. and Dharssi, I. (1992) J. Phys. Condens.Matt. **4** L77.
3. Nash, K.J. (1992) Phys. Rev. **B46** 7723.
4. Richter, E. and Strauch, D. (1987) Solid St. Commun. **64** 867.
5. Ren, S`.F. Chu, H.Y. and Chang, Y-C (1989) Phys. Rev. **B40** 3060.
6. Molinari, E. Baroni, S. Giannozzi , P.and de Gironcoli, S. (1992) Phys. Rev. **B45** 4280; and 6747.
7. Ridley, B.K. (1991) Phys. Rev. **B44** 9002.
8. Ridley, B.K. (1992) Proc. SPIE 'Compound Semiconductor Physics and Devices - High Speed Electronics and Optoelectronics' Somerset N.J., U.S.A. **1675** 492. .
9. Trallero-Giner, C. Garcia-Moliner, F. Velasco, V.R. and Cardona, M. (1992) Phys. Rev. **B45** 11944.
10. Ridley, B.K. (1993) Phys. Rev. **B47** 4592.
11. Fuchs, R. and Kliewer, K.L. (1965) Phys. Rev. **140A** 2076.
12. Chamberlain, M.P. Cardona, M. and Ridley, B.K. Phys. Rev. **B** (at press).
13. Haines, M. and Scamarcio, G. (1992) 'Phonons in Nanostructures' (NATO ARW St. Felio de Guixols, Spain).
14. Constantinou, N.C. Al-Dossary, O. and Ridley, B.K. (1993) Solid St. Commun. **86** 191.
15. Letran, T. Schaff, W. Eastman, L. and Ridley, B.K. J. Appl. Phys. (submitted).
16. Babiker, M. Constantinou, N.C. and Ridley, B.K. (1993) Phys. Rev. **B48** 2236.
17. Baroni, S. Gianozzi, P. and Molinari, E. (1990) Phys. Rev. **B41** 3870.
18. Mowbray, D.J. Cardona, M. and Ploog, K. (1991) Phys. Rev. **B43** 1598.
19. Mori , N. and Ando, T. (1989) Phys. Rev. **B40** 6175.
20. Al-Dossary, O. Babiker, M. and Constantinou, N.C. (1992) Semicon. Sci. Technol. **7** B91.
21. Constantinou, N.C. Al-Dossary, O. and Babiker, M. (1993) J. Phys: Condens.Matt. **5** 5581.
22. Camley, R.E. and Mills, D.L. (1984) Phys. Rev. **B29** 1695.

Second Harmonic Generation in p-type Quantum Wells

Zhiwei Xu and Philippe M. Fauchet
Laboratory for Laser Energetics
Department of Electrical Engineering
University of Rochester
Rochester, NY 14623

Gary W. Wicks
The Institute of Optics
University of Rochester
Rochester, NY 14627

Michael J. Shaw and Milan Jaros
Department of Physics
University of Newcastle upon Tyne
Newcastle Upon Tyne, UK

Bruce Richman and Chris Rella
W. W. Hansen Laboratory of Physics
Stanford University, Stanford CA 94305

ABSTRACT. We report the first measurements of second harmonic generation (SHG) in p-type quantum wells. The measurements are performed using a free electron laser tunable from 3 to 6 μm. The samples are ultranarrow GaAs/Al$_{0.5}$Ga$_{0.5}$As stepped quantum wells separated by AlAs barriers. The enhancement of the SHG signal in the quantum wells over that in bulk GaAs is one order of magnitude. The dominant component of the $\chi^{(2)}$ tensor is $\chi^{(2)}_{xyz}$, which is to be contrasted to the results obtained in n-type quantum wells, where $\chi^{(2)}_{zzz}$ dominates. The experimental results are in qualitative agreement with the calculated spectrum of $\chi^{(2)}$.

1. Introduction

Second harmonic generation in quantum wells has been an active field of research since the first observation of a very large enhancement of $\chi^{(2)}$ in n-type GaAs quantum wells [1]. Several types of quantum well structures have been investigated, such as square wells under DC bias [1], stepped wells [2], and coupled quantum wells [3]. An enhancement up to three orders of magnitude has been obtained [4] when the structure is

457

H. C. Liu et al. (eds.), Quantum Well Intersubband Transition Physics and Devices, 457–466.
© 1994 *Kluwer Academic Publishers. Printed in the Netherlands.*

doubly resonant, i.e. the incident photon energy is very close to the energy separation between the 1st energy level and the 2nd energy level and the energy separation between the 2nd energy level and the 3rd energy level. However, until now, most experiments have been performed in n-type quantum wells, where only the $\chi^{(2)}_{zzz}$ component is enhanced when the symmetry of the crystal in the z-direction (growth direction) is broken [5]. Unfortunately, the large refractive index of GaAs makes the z-component of the electric field inside the wells very small even at Brewster's angle of incidence. As a result, complicated experimental configurations such as a waveguide geometry have to be used in order to obtain a large conversion efficiency. In contrast, due to the complexity of the valence band, absorption at normal incidence is allowed [6], so that components other than $\chi^{(2)}_{zzz}$ in the $\chi^{(2)}$ tensor may also be enhanced. We reported our first experimental results obtained in p-type stepped quantum wells using the tunable, high peak power picosecond pulses generated by a free electron laser in reference [7]. In this paper, we present experimental results and compare our results to theoretical calculations.

2. Sample description

The quantum well (QW) samples used in our experiment were grown by molecular beam epitaxy on (001) GaAs wafers. A typical sample structure is shown in Figure 1. The stepped well is made of 5 monolayers of GaAs and 1 monolayer of $Al_{0.5}Ga_{0.5}As$. The AlAs barrier has a thickness of 4 nm. The middle 2 nm of the barrier is p-type doped with a concentration of $2 \times 10^{18} cm^{-3}$. The sheet hole concentration of a single quantum well is $4 \times 10^{11} cm^{-2}$ assuming that all the holes are transferred from the barrier to the well. By doping the barriers, we avoid the formation of DX centers which trap holes. The total thickness of the epitaxial layer is 1.6 microns, consistent with the Fabry-Perot fringes seen in the FTIR spectrum. Pure AlAs is used as the barrier material to maximize the valence band discontinuity and in principle allow resonance at shorter wavelengths. The FTIR measurements performed in the single pass Brewster's angle of incidence geometry show less than 1% linear absorption near resonance.

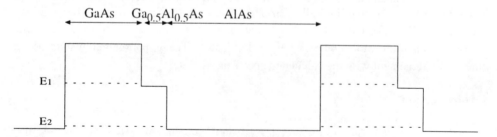

Figure 1. The valence band of GaAs/AlAs quantum wells. The AlAs barriers are approximately 4 nm wide. For 5 monolayers of GaAs and 1 monolayer of $Ga_{0.5}Al_{0.5}As$, the calculated energy separation between the 1st and 2nd heavy-hole subbands is approximately 380 meV.

A typical room temperature photoluminescence spectrum is shown in Figure 2. Two peaks are observed around 540 nm and 640 nm. The first peak corresponds to recombination involving Γ electrons in the quantum well while the second peak corresponds to recombination of the X electrons in the barrier. Half of the sample has the epitaxial layer etched away in electronic grade HF. The measurements are performed on the etched and the unetched halves of the sample, so that the contribution from the quantum wells can be obtained by subtraction. A simple calculation based on the square well approximation indicates that there are two bound heavy hole states separated by an energy of 380 meV which corresponds to a wavelength of 3.3 μm. However this simple calculation can only be treated as a guide, especially for the upper energy level where quantum confinement is weak.

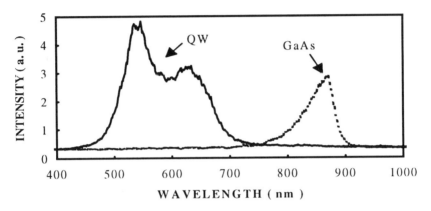

Figure 2. The room temperature photoluminescence spectrum of the sample in Figure 1. The sample is excited by the 488 nm line of an Argon laser with power of about 7 mW.

3. Theoretical discussion

<u>3.1 General theory</u>

Nonlinear optical processes can be well described by susceptibility tensors. The second-order susceptibility tensor which describes the SHG process is given by:

$$\chi_{ijk}^{(2)}(-\omega_3;\omega_1,\omega_2)=\frac{-ie^3}{\varepsilon_0 V 2\hbar m^3 \omega_3\omega_1\omega_2}\sum_{abc,k} f_0(k)\times$$

$$[\frac{p_{ab}^i p_{bc}^j p_{ca}^k}{(\Omega_{ba}-i\Gamma_{bc}-i\Gamma_{ca}-\omega_1-\omega_2)(\Omega_{ca}-i\Gamma_{ca}-\omega_2)}+\frac{p_{ab}^j p_{bc}^i p_{ca}^k}{(\Omega_{ba}+i\Gamma_{ba}+\omega_1)(\Omega_{ca}-i\Gamma_{ca}-\omega_2)}+$$

$$\frac{p_{ab}^k p_{bc}^j p_{ca}^i}{(\Omega_{ba}^{(1)}+i\Gamma_{ab}+\omega_2)(\Omega_{ca}+i\Gamma_{ab}+i\Gamma_{bc}+\omega_1+\omega_2)}+\frac{p_{ab}^i p_{bc}^k p_{ca}^j}{(\Omega_{ba}-i\Gamma_{bc}-i\Gamma_{ca}-\omega_1-\omega_2)(\Omega_{ca}-i\Gamma_{ca}-\omega_1)}+$$

$$\frac{p_{ab}^k p_{bc}^i p_{ca}^j}{(\Omega_{ba}+i\Gamma_{ba}+\omega_2)(\Omega_{ca}-i\Gamma_{ca}-\omega_1)}+\frac{p_{ab}^j p_{bc}^k p_{ca}^i}{(\Omega_{ba}+i\Gamma_{ab}+\omega_1)(\Omega_{ca}+i\Gamma_{ab}+i\Gamma_{bc}+\omega_1+\omega_2)}] \quad (1)$$

where the summation is over all possible states, p is the momentum matrix element, f_0 is the state population, and Γ_{ij} are the damping terms related to the relaxation processes. Generally speaking, if there is any kind of resonance, i.e. the energy difference between two energy levels equals the incident photon energy, $\chi^{(2)}$ will be enhanced. However, due to the complexity of the valence band, the maximum enhancement in the $\chi^{(2)}$ spectrum may not coincide with the peak absorption. This will be discussed later. The asymmetric quantum wells studied in this paper belong to the $C_{2v}(2mm)$ point group. For crystals in this point group, there are seven nonzero elements in the $\chi^{(2)}$ tensor [8]. Due to the permutation rules, we have

$$\chi^{(2)}_{ijk}(-2\omega; \omega, \omega) = \chi^{(2)}_{ikj}(-2\omega; \omega, \omega)$$

Thus, there are only five independent elements in the SHG susceptibility tensor:

$$\chi^{(2)}_{zzz}$$

$$\chi^{(2)}_{xyz} = \chi^{(2)}_{yxz} = \chi^{(2)}_{xzy} = \chi^{(2)}_{yzx}$$

$$\chi^{(2)}_{zxy} = \chi^{(2)}_{zyx}$$

$$\chi^{(2)}_{xxz} = \chi^{(2)}_{yyz} = \chi^{(2)}_{xzx} = \chi^{(2)}_{yzy}$$

$$\chi^{(2)}_{zxx} = \chi^{(2)}_{zyy}$$

In n-type quantum wells, the transition matrix element for light with polarization parallel to the xy plane is zero, and the only nonvanishing element is $\chi^{(2)}_{zzz}$. However, in the valence band, normal incidence absorption is allowed because of the mixing between the heavy hole band and the light hole band. In general, all the components allowed by the symmetry rule in the $\chi^{(2)}$ tensor will be nonzero.

3.2 Theoretical results

A semi-empirical peseudopotential calculation is employed in the modeling of the bandstructure. These exact calculations indicate that there are several bands of importance in the quantum wells: heavy-hole 1 (HH1), light-hole 1 (LH1), HH2, split-off band (SO), and the continuum states. We found that the energy separation between the HH1 and HH2 at $\mathbf{k} = 0$ is 317 meV, and the energy separation between the HH1 and SO at $\mathbf{k} = 0$ is 360 meV. The second order susceptibility is obtained from Eq. (1). The details of the calculation are discussed in reference 9. The calculated SHG susceptibility at room temperature for our sample is shown in Figure 3.

Let us examine two parameters included in the calculation. The first parameter is the position of the Fermi level. For a two-dimensional ideal hole gas the Fermi energy is given by:

$$E_F = -kT\ln\left(e^{\frac{N\pi\hbar^2}{m^*kT}} - 1 \right) \qquad (2)$$

where N is the sheet hole density, m^* is the hole effective mass for the motion perpendicular to the growth direction of the quantum wells, and T is the temperature. Using $N = 4\times10^{11}$ cm^{-2}, $m^* = 0.5\ m_0$, and $T = 300$ K, we estimate that the Fermi level lies about 65 meV above the valence band edge. The position of the Fermi level is very important in our calculation of $\chi^{(2)}$. Generally, when the Fermi level is closer to the valence band edge, the nonlinear susceptibility is larger. The second parameter included in our calculation is the damping terms Γ_{ij}. Because of the lack of experimental information for the intra-valence band transitions, we have used the value of 3 meV [9]. The value of the damping term is closely related to the amplitude of the peak value of the $\chi^{(2)}$ tensor, with a smaller Γ leading to a larger $\chi^{(2)}$.

The calculations show that $\chi^{(2)}_{xyz}$ is the dominant component for most of the wavelengths of interest. The maximum peak is at 340 meV with $\chi^{(2)}_{xyz} = 1.4\times10^{-6}$esu. This energy does not relate to any specific transition. The main reason for this is the complicated structure of the valence band. The summation in Eq. (1) includes contributions from the whole band, which explains why the spectrum of the second order susceptibility shows very complicated features, including several different peaks, such as the one at 320 meV. Nevertheless, we can identify the most important process as the SO (split-off band) - HH1 - lower lying continuum states - SO transition.

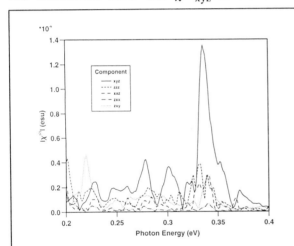

Figure 3. The spectrum of the five nonzero independent tensor components of $\chi^{(2)}(-2\omega;\omega,\omega)$ for the 5GaAs/1Ga$_{0.5}$Al$_{0.5}$As/14AlAs multiple quantum wells at room temperature. The susceptibilities are in esu and the photon energies in eV.

Similar calculations of $\chi^{(2)}$ employing the same method were also performed for bulk GaAs in order to obtain the enhancement of $\chi^{(2)}$, which is what we measure in our experiment, rather than the absolute value of $\chi^{(2)}$. The result in bulk GaAs does not show any resonance in the $\chi^{(2)}$ spectrum over the wavelength region of interest. The value of $\chi^{(2)}$ we obtain is 0.28×10^{-6}esu. The enhancement at the peak value of $\chi^{(2)}$ is thus about 6.

4. Experimental results

4.1 Experimental arrangement

The SCA free electron laser (FEL) at Stanford University is used as the laser source in our experiments. It generates macropulses at a repetition rate of 10 Hz. Each macropulse is made of a micropulse train, with a single micropulse duration of 2-3 ps and pulse-to-pulse separation of 80 ns. The duration of a macropulse is 1 ms. The wavelength tuning range of the FEL extends from 3 μm to 6 μm. The output of the FEL is linearly polarized and has an average power of 30 mW. The experimental arrangement is shown in Figure 4. The sample is placed at the focus between two paraboloidal reflectors with the QW layer facing the incident beam. The spot size at the focus is 270 microns at a wavelengths of 5 μm. The incident laser power is measured by a PbSe detector. The SHG light is measured by a PbS detector which is sensitive to wavelengths shorter than 3 μm. The output of both detectors is sent to a data acquisition system which can directly take the average of the ratio of the SHG power to the square of the incident power.

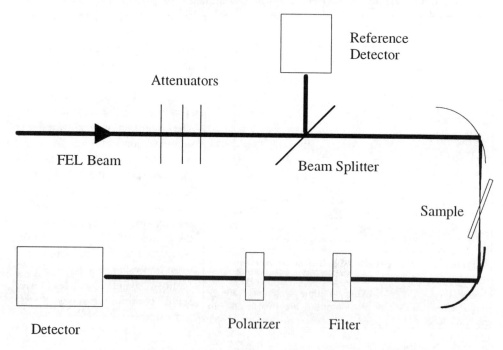

Figure 4. Layout of the SHG experiment. Two paraboloidal reflectors focus the incident beam onto the sample and recollimate the SHG signal. The reference signal could be either the incident laser power or the SHG from a reference sample.

The sample is mounted on a rotation stage such that both θ and φ angles can be adjusted. The angles θ and φ are defined in Figure 5. θ is the angle between the (001) crystallographic direction of the GaAs substrate and the incident fundamental wave propagation direction, and φ is the angle between the (110) direction of the GaAs substrate and the plane determined by (001) direction of the GaAs substrate and the wave propagation direction.

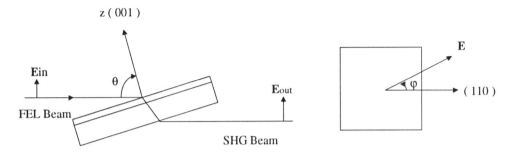

Figure 5. Geometry of our experiment. θ is the angle between the (001) crystallographic direction of the GaAs substrate and the incident fundamental wave propagation direction, and φ is the angle between the (110) direction of the GaAs substrate and the plane determined by (001) direction of the GaAs substrate and the wave propagation direction.

4.2 Modeling of the experiments

The contributions to the signal from the different components of the $\chi^{(2)}$ tensor have a different behavior. For example, in n-type GaAs quantum wells where the dominant component is $\chi^{(2)}_{zzz}$, the φ dependence of the SHG signal shows four uneven peaks as φ changes from 0 to 360 degrees [5]. In p-type quantum wells, the dominant component is $\chi^{(2)}_{xyz}$, and the φ dependence will have the same behavior as bulk GaAs, where only $\chi^{(2)}_{xyz}$, $\chi^{(2)}_{yzx}$, and $\chi^{(2)}_{zxy}$ are nonvanishing. Under these conditions and taking into account the fact that the signal from our quantum wells sample is a coherent superposition of the contributions from the bulk substrate and from the quantum well layer, we can derive the following expression for SHG with polarization parallel to that of the fundamental beam:

$$\frac{I_{2\omega}}{I_\omega^2} \propto T_{2\omega}(\theta_{in})|T_\omega(\theta_{in})|^2 \left|\eta l + \frac{e^{i\,\Delta k l}-1}{i\,\Delta k}e^{i\,\Delta k l}\right|^2 F^2(\theta_{in},\varphi_{in}) \qquad (3)$$

where $I_{2\omega}$ is the second harmonic intensity; I_ω is the fundamental intensity; $T_{2\omega}$ and T_ω are the transmission coefficients from GaAs to air at the second harmonic and from air to GaAs at the fundamental, respectively; Δk is the phase mismatch between the second harmonic and the fundamental; d is the thickness of the substrate; l is the thickness of quantum well layer; $F(\theta_{in}, \varphi_{in}) = \sin\theta_{in}\cos^2\theta_{in}\sin^2\varphi_{in}$, where θ_{in} and φ_{in} are internal angles related to the outside angles θ and φ by $n\sin\theta_{in} = \sin\theta$ and $\varphi_{in} = \varphi$ where n is the refractive index. The SHG enhancement $\eta = \chi^{(2)}_{QW}/\chi^{(2)}_{GaAs}$ is a complex number.

4.3 Experimental results and data analysis

We report the results of experiments performed at two wavelengths: $\lambda = 3.81$ μm and $\lambda = 5.08$ μm. The θ dependence and the φ dependence of the SHG signal were measured for both wavelengths. The φ dependence measured at $\lambda = 4.75$ μm is shown in Figure 6. Similar results are obtained at $\lambda = 3.81$ μm and $\lambda = 5.08$ μm. The presence of four even peaks is a clear proof of the dominant role of $\chi^{(2)}{}_{xyz}$ in contrast to the n-type quantum wells where the $\chi^{(2)}{}_{zzz}$ component dominates. We use Eq. (3) to fit our data with two adjustable parameters: d and η. Figure 7 shows the experimental data and the best fit. The enhancement at $\lambda = 3.81$ μm is 8.5, while at $\lambda = 5.08$ μm it is 5. These values are in good agreement with the theoretical calculation. The difference can be traced to several approximations made in the theoretical calculation, such as the exact position of the Fermi level, the value of the hole effective mass, the actual sheet hole density in the quantum wells, and the broadening of the energy levels. More careful calculations are necessary to fully understand our data.

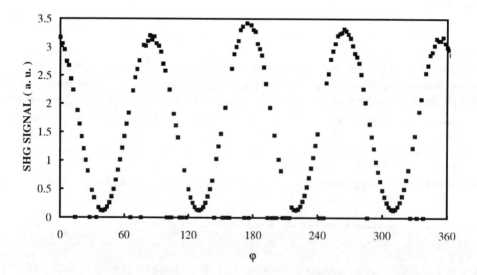

Figure 6. φ dependence of the SHG signal, defined as $I_{2\omega}/I_{\omega}{}^2$, at $\lambda = 4.75$ μm and θ = 60°. Similar reuslts are obtained at $\lambda = 3.81$ μm and $\lambda = 5.08$ μm

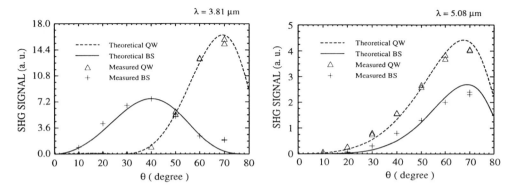

Figure 7. θ dependence of SHG signal at λ = 3.81 μm and λ = 5.08 μm, and φ=45°. BS refers to the bare substrate signal and QW to the unetched sample signal. Both fit and experimental data are plotted.

5. Summary

In summary, SHG in p-type quantum wells has been measured at two wavelengths. Both experimental results and theoretical calculation show an enhancement of one order of magnitude of the $\chi^{(2)}_{xyz}$ component. The differences between results obtained with p-type and n-type quantum wells can be traced to the complexity of the valence band and in particular to the mixing between the heavy hole band and the light hole band which allows absorption even for light with polarization parallel to the epitaxial layer.

This work is supported by the U.S. Office of Naval Research contract N00014-92-J-4063. We thank K. Marshall and J. A. Varriano for technical assistance, and Prof. H. A. Schwettman for constant encouragement.

6. Reference

1. M. M. Fejer, S. J. B. Yoo, and R. L. Byer, A. Harwit, and J. S. Harris, Jr., Phys. Rev. Lett. **62**, 1041 (1989)
2. S. J. B. Yoo, M. M. Fejer, R. L. Byer, and J. S. Harris, Jr., Appl. Phys. Lett. **58**, 1724 (1991)
3. C. Sirtori, F. Capasso, D. L. Sivco, S. N. G. Chu, and A. Y. Cho, Appl. Phys. Lett. **59**, 2302 (1991)
4. E. Rosencher, P. Bois, J. Nagle, and S. Delaitre, Electron. Lett. **25**, 1063 (1989)
5. S. J. B. Yoo, Ph. D. Thesis, Stanford University (1991)
6. B. F. Levine, S. D. Gunapala, J. M. Kuo, S. S. Pei, and S. Hui, Appl. Phys. Lett. **59**, 1864 (1991)
7. Z. Xu, L. X. Zheng, J. V. Vandyshev, G. W. Wicks, P. M. Fauchet, B. Richman, and C. Rella, SPIE Proc. **1854**, (1993)

8. Y. R. Shen, *The Principles of Nonlinear Optics* (Wiley, New York, 1984)
9. M. J. Shaw, M. Jaros, Z. Xu, and P. M. Fauchet, to be published

RESONANT HARMONIC GENERATION NEAR 100μm IN AN ASYMMETRIC DOUBLE QUANTUM WELL

J. N. HEYMAN, K. CRAIG, M. S. SHERWIN
Quantum Institute, University of California at Santa Barbara
Santa Barbara, CA 93106
USA

K. CAMPMAN, P. F. HOPKINS, S. FAFARD, A. C. GOSSARD
Dept. Materials Science Engineering, University of California at Santa Barbara

ABSTRACT. We have observed the generation of resonant second- and third-harmonic radiation near 100 μm in an AlGaAs/GaAs asymmetric coupled double-quantum well (ADQW) which approximates a two-level system. Resonant effects are studied by Stark tuning the system through the pump laser frequency. We measure much larger nonlinear susceptibilities than reported in structures optimized for nonlinear optics in the mid-infrared, and we observe saturation of the second-harmonic generation. These measurements constitute the first resonant harmonic generation experiments in a heterostructure in which many-electron effects are important. We find that the second-order susceptibility can be modeled satisfactorily by single-electron perturbation theory when the intersubband spacing is replaced with the depolarization-shifted resonance frequency.

1. Introduction

There is currently considerable interest in the study of non-linear optical phenomena in semiconductor heterostructures. This is because these structures may be designed to possess resonant electronic polarizabilities associated with intersubband transitions at mid- and far-infrared frequencies which are extremely large compared to the polarizabilities of electrons in atoms and molecules. The primary reason for these enhancements are the large electric-dipole matrix elements associated with intersubband transitions, which in turn arise from the small electron effective-mass.

The resonant enhancement of MIR optical nonlinearities using intersubband transitions was first proposed by Gurnick et. al.[1] and others[2]. Large, resonant second-order susceptibilities near $\lambda = 10$ μm have been observed in AlGaAs/GaAs[3-12] and GaInAs/AlInAs[13-15] heterostructures through the measurement of second-harmonic generation[4-7, 9, 11, 14, 15] and optical rectification[10, 11]. Resonant third-order susceptibilities near $\lambda = 10$ μm have been studied through measurements of the intensity-dependent refractive index[16] (Kerr effect), and third-harmonic generation[17]. Multiphoton ionization processes have also been observed[18] in such structures. Recently, large non-resonant susceptibilities in the FIR have been deduced from measurements[19] of second-

467

H. C. Liu et al. (eds.), Quantum Well Intersubband Transition Physics and Devices, 467–476.
© 1994 *Kluwer Academic Publishers. Printed in the Netherlands.*

468

harmonic generation at the 2-D electron gas at an AlGaAs/GaAs interface, and in a half-parabolic AlGaAs/GaAs quantum well.

Far-infrared resonant hetero-structures can be designed to have larger electric-dipole matrix elements and longer relaxation times than resonant MIR systems. Thus much larger nonlinear susceptibilities should be achievable[8]. In addition, because the electron effective plasma frequency can be comparable to the inter-subband energy in FIR systems, they should exhibit strong many-electron effects. Finally, because of the much

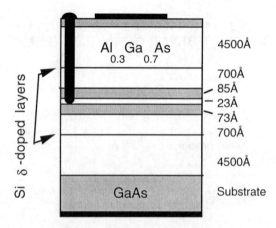

Fig.1. Asymmetric coupled double quantum well structure. The shaded regions are GaAs, while the barrier regions are $Al_{0.3}Ga_{0.7}As$. Electrical contact is made through a Schottky gate and diffused indium contacts. The gate and the backside metallization confine the FIR electric field in the sample.

stronger coupling to light, the non-perturbative regime of the light-matter interaction may be reached at modest laser intensities. Resonant FIR systems should thus be well suited to the study of strongly driven quantum systems.

2. Experimental Methods

2.1 SAMPLE GROWTH

The structure used in these measurements (Figure 1) consists of a single period of an asymmetric double quantum well (ADQW) with 85Å and 75Å GaAs wells separated by a 25Å $Al_{0.3}Ga_{0.7}As$ barrier. The structure was grown by MBE on a semi-insulating GaAs substrate. The barrier region on either side of the well consists of 5200Å $Al_{0.3}Ga_{0.7}As$ grown as a digital alloy with a 20Å period (14Å GaAs and 6Å AlAs). The digital alloy was used to improve the interface smoothness. The well is remotely doped on both sides with Si delta doped layers 700Å from the well. Figure 2 shows the results of a simulation of the conduction band potential in the well, which was used to estimate the subband energies and envelope functions. The results were obtained by self-consistently solving the Schrodinger and Poisson equations in the well for a carrier density of $2 \cdot 10^{11}$ cm^{-2}. Exchange-correlation terms were neglected in this simulation. The calculated subband spacings are $E_{12} = 8.7$ meV, $E_{13} = 106$ meV and $E_{14} = 145$ meV. For FIR excitation at hν ≈ E_{12} the system can be approximated by a two subband model.

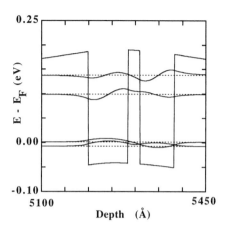

Fig. 2. Simulation of the self-consistent potential in the asymmetric double well. The simulation was performed for $2 \cdot 10^{11}$ cm^{-2} electrons in the well and 0V DC bias. The calculated subband energies and envelope wavefunctions are also shown.

Table 1: Properties of the ADQW structure at 0 V bias.

	Value	Technique
N_s	$2.0 \cdot 10^{11}$ cm^{-2}	Hall effect, C(V) (4.2K)
μ_n	10^5 cm/V·s	Hall effect (4.2K)
$\hbar\omega_{12}$	11 meV	Photoluminescence (4.2K)
$\hbar\omega_{12}*$	14.3 meV	FIR Absorption (1.4K)

2.2. CHARACTERIZATION

The low temperature (T<50K) charge density and electron in-plane mobility in the well measured by Hall effect are $2 \cdot 10^{11}$ cm^{-2} and 10^5 cm^2/Vs. Photoluminescence (PL) and Photoluminescence excitation (PLE) measurements performed at 4.2K were used to determine the intersubband spacings. As there is no parity selection rule in our asymmetric structure, an E_{12} electron subband spacing of 11 meV was extracted directly from the spacing between the E_1-HH$_1$ and E_2-HH$_1$ PL peaks.

For further measurements, an aluminum Schottky gate was fabricated on the surface of the structure, and indium-alloy ohmic contacts were made to the quantum well. This contact structure allows us to impose a DC electric field across the well by applying a reverse bias to the gate. The charge in the well is also depleted by the reverse bias. The charge density depends linearly on the electric field, and the structure is fully depleted at -1.6V bias, which corresponds to a DC electric field across the well of $1.5 \cdot 10^4$ V/cm. The back side of the substrate was metallized so that sample could support a TEM waveguide mode.

2.3. LINEAR AND NONLINEAR SPECTROSCOPY

Linear absorption measurements were made on a gated sample at 1.4K using a Bomem DA3.002 Fourier Transform Spectrometer. FIR radiation was coupled into the edge of the sample using a Winston cone concentrator. The transmitted radiation was passed through a polarizer, and the component with electric field parallel to the growth direction was detected with a Ge bolometer. The sample and bolometer were mounted in the same cryostat. Transmission spectra of the sample were measured as a function of gate bias, and absorbance spectra were computed using a spectrum of the fully depleted sample as a reference.

The FIR pump used in our harmonic generation experiments was the UCSB Free Electron Laser (FEL). The UCSB FEL is an electrostatic free electron laser which provides a coherent FIR beam tunable between hν = 0.5 - 20 meV. The FEL produces pulses of linearly polarized FIR radiation 1-10 μs long at a repetition rate of 1 Hz. The pulses are thus much longer than the time scale of the processes we are investigating (< 1ns). Peak

470

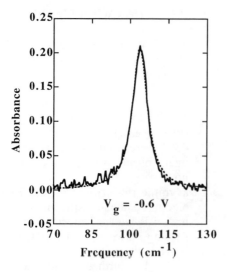

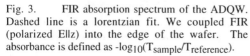

V$_g$ = -0.6 V

Fig. 3.   FIR absorption spectrum of the ADQW. Dashed line is a lorentzian fit. We coupled FIR (polarized E∥z) into the edge of the wafer. The absorbance is defined as -log$_{10}$(T$_{sample}$/T$_{reference}$).

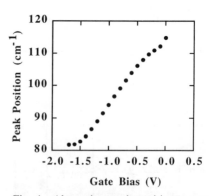

Fig. 4. Absorption peak position versus gate bias. We observe an approximately linear shift of 0.29 meVcm/kV. This large linear electro-optic effect demonstrates the large second-order susceptibility of this structure.

powers of 1-4 kW and peak intensities of order 1MW/cm^2 can be achieved. In typical measurements, the FIR electric field was parallel to the growth direction in the quantum well. The power in the FIR beam was controlled with a set of calibrated attenuators.

In our experiment (Figure 5) the FIR beam was coupled into the edge of our sample using focusing optics and a 2-D Winston cone. The sample was mounted in a variable temperature cryostat and measurements were typically performed at T = 10 - 15K. The radiation emitted by the sample was collected and its frequency was measured with a scanning Fabry-Perot spectrometer. The intensity of the harmonic radiation was measured with a 4.2K Ge bolometer. The experiment was performed in a N$_2$ purged atmosphere to reduce absorption due to water vapor.

Two FIR filters are required in our experiment. Because the FEL produces its own harmonic radiation, the beam was passed through a 100 cm^{-1} low-pass filter before reaching the sample. An efficient high-pass filter is also required to separate the weak harmonic radiation produced by the sample from the strong transmitted fundamental. A waveguide cutoff filter consisting of an array of 60μm diameter holes in a 300 μm thick brass foil, and having a transmission at the fundamental of order 10^{-8} served as a high-pass filter, and was placed in front of the bolometer.

3.   Experimental Results

3.1.   LINEAR FIR ABSORPTION SPECTROSCOPY

The FIR absorption spectrum consists of a single line with an approximately lorentzian lineshape (Figure 3). At 0V gate bias the peak position is 14.3 meV (115 cm^{-1}). We ascribe the 3.3 meV shift between the infrared absorption peak and the bare intersubband spacing measured by PL to the depolarization shift, which arises from the dynamic screening of the FIR field in the well by the charge. The magnitude of the shift is comparable to the shift (4.3 meV) we calculate from our simulation. At -0.6V bias the peak position is 12.9 meV (104 cm^{-1}) and the FWHM is 0.85 meV. Using an effective thickness for the ADQW of 200Å, we obtain an effective absorption coefficient in the well of 10^4 cm^{-1} at resonance.

We find an approximately linear Stark shift of the absorption peak of approximately 2.7 meV/V (2.9 10^{-4} meV cm/V) (Figure 4), and the resonance can be tuned from 14.3 to 10.1 meV before the well is completely depleted.

3.2   SECOND-HARMONIC GENERATION

We observe resonant second-harmonic generation (SHG) by illuminating the sample with the FIR laser at a fixed frequency and using the Stark effect to tune the intersubband energy spacing (Figure 6). The same technique has been applied previously by Sirtori et. al.[15].

A well-defined resonance is observed at $V_g = -0.67$ V when the pump frequency is 51.5 cm^{-1}. At this bias the laser frequency is one-half the frequency of the infrared absorption peak ($E_{12}^* = 2\,h\nu_{pump}$), and the energy of a second-harmonic photon equals the depolarization-shifted resonance. The experimental lineshape gives an excellent fit to a lorentzian multiplied by the square of the charge density. As discussed below, this is the lineshape predicted by perturbation theory.

We have measured the second harmonic intensity at resonance as a function of pump intensity for intensities inside the sample up to 100 kW/cm^2 (Figure 7). At intensities below 8 kW/cm^2, the SHG power is proportional to the second power of the pump intensity. The SHG resonance lineshape (Figure 8) is independent of pump intensity in this regime. At higher pump powers, the dependence of the SHG on pump intensity is sub-quadratic, and saturation is observed. In this regime the resonance lineshape also changes. The resonance broadens and the center of the line shifts to smaller gate biases. In the high power regime, the lineshape can no longer be fit by the perturbative form.

In addition, we have measured the dependence of the SHG power on the polarization of the pump beam. The SHG power is approximately proportional to the fourth power of the electric field component along the growth direction. The polarization of the radiation emitted by the quantum well is found to be parallel to the growth direction.

Table 2: Properties of the ADQW structure at -0.65V bias determined by Fourier transform spectroscopy. The charge density was determined by C(V). The absorption coefficient was calculated using an effective thickness of 200Å.

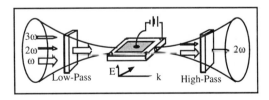

Fig. 5. Experimental geometry for harmonic generation experiments.

$h\omega_{12}^*$	12.8 meV
$h\Gamma$	0.42 meV
z_{12}	45.3 ± 5Å
z_{22}	29 ± 5Å
α	10^4 cm^{-1}
N_s	(1.2±0.4)·10^{11} cm^{-2}

These results indicate that $\chi^{(2)}_{zzz}(2\omega,\omega,\omega)$ is the only non-zero component of the tensor $\chi^{(2)}(2\omega,\omega,\omega)$.

Finally, we have observed resonant third-harmonic generation (THG) in our asymmetric double quantum well by illuminating the sample with the FEL at 35 cm^{-1} and using the Stark shift to tune the intersubband spacing (Figure 9). The THG resonance occurs when the system is tuned so that the laser frequency is one-third the frequency of the infrared absorption peak ($E_{12}^* = 3h\nu_{pump}$). The signal is sufficiently weak that it can only be observed under pump intensities ≥ 20 kW/cm^2. Thus we have not yet determined if we are measuring THG in the perturbative regime.

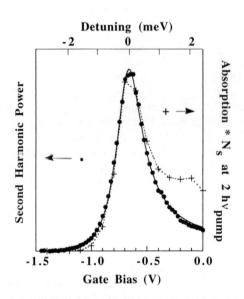

Fig. 6. (•) Second-harmonic generation from ADQW versus gate bias. The laser frequency was fixed at 51.5 cm^{-1}. The solid line is a fit to the lineshape predicted from perturbation theory. The detuning is the energy difference between the second harmonic frequency (103 cm^{-1}) and the peak absorption at that gate bias.

(+) Absorption times charge density measured at two times pump frequency.

4.   Discussion

4.1   SECOND-ORDER SUSCEPTIBILITY FOR HARMONIC GENERATION

To analyze our results we have assumed that the fundamental radiation propagates entirely in the TEM mode in our sample, while the second harmonic contributes to all modes The second harmonic power is then given[20] by:

Table 3: $X^{(2)}_{(2\omega;\omega,\omega)}$ at resonance in the ADQW (Vg = -0.65 V) as measured by second-harmonic generation and as calculated from the measured matrix elements. The ADQW 3-D susceptibility was calculated using an effective thickness of 200Å.

$X^{(2)}_{(2\omega;\omega,\omega)}$	value	source
ADQW $X^{(2)}_{2-D}$	$(4 \pm 2) \cdot 10^{-8}$ cm^2/statV	SHG data
ADQW $X^{(2)}_{2-D}$	$(1.2 \pm 0.4) \cdot 10^{-7}$ cm^2/statV	Calculation
ADQW $X^{(2)}_{3-D}$	$(2 \pm 1) \cdot 10^{-2}$ cm/statV	SHG data
GaAs $X^{(2)}_{3-D}$	$(1.4 \pm 0.9) \cdot 10^{-7}$ cm/statV	reference [a]

[a] A. Mayer and F. Keilmann, Phys. Rev. B **33**, 6954 (1985).

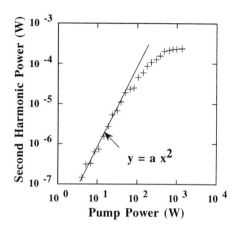

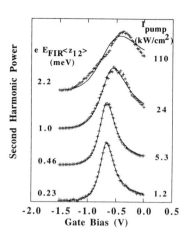

Fig 7.   Second harmonic power at $V_g = -0.67$ V versus pump power. Pump and second harmonic powers given are not corrected for losses in the cryostat. Peak conversion efficiency is $7 \cdot 10^{-7}$, but is estimated to be 10^{-5} if losses are included.

Fig 8.   Second harmonic line shape versus pump intensity. The curves are been normalized and offset for clarity. The solid lines are fits to the perturbative line shape (lorentzian x (charge density)^2). Also shown is the strength of the perturbation due to the FIR electric field for different intensities.

$$P_{2\omega} = \frac{32\pi^3 \omega^2 \, l^2}{c^3 \, \varepsilon_\omega \varepsilon_{2\omega}^{1/2} \, A} \frac{\left| \chi_s^{(2)}(2\omega;\omega,\omega) \right|^2}{a^2} \, P_\omega^2 \left[\frac{\sin^2\left(\Delta k_0 l/2\right)}{\left(\Delta k_0 l/2\right)^2} + 2\sum_{m=1}^{\infty} \frac{k_{2\omega}^{(0)}}{k_{2\omega}^{(m)}} \frac{\sin^2\left(\Delta k_m l/2\right)}{\left(\Delta k_m l/2\right)^2} \right]$$

where a and L are the thickness and length of the sample, respectively, A is the area of the FIR spot, P_ω is the power of the pump beam inside the sample, $k_{2\omega}^{(m)}$ is the propagation constant of radiation at 2ω traveling in waveguide mode m, and $\Delta k_m = k_{2\omega}^{(m)} - 2k_\omega$.. The second-order susceptibility per unit area for harmonic generation at resonance, $\chi_s^{(2)}(2\omega,\omega,\omega)$, can thus be extracted from our low-excitation intensity data (Table 3). The large experimental error in this quantity is due to uncertainties in the absolute FIR power at the sample. The equivalent volume susceptibility $\chi^{(2)}(2\omega,\omega,\omega) = 0.02$ cm^2/statV (8.5μm/V) is approximately 10^5 times the value for bulk GaAs.

4.2   ESTIMATES FROM PERTURBATION THEORY

The polarizability of each electron in our doped quantum well can be calculated using time-dependent perturbation theory for a two-state system[21]. Neglecting terms which are resonant at $\omega<0$ we have for the first-order susceptibility:

$$\xi^{(1)}(\omega) = \frac{e^2}{\hbar} z_{1,2} z_{2,1} \left\{ \frac{(\omega_{2,1} - \omega) - i\Gamma}{(\omega_{2,1} - \omega)^2 + \Gamma^2} \right\}$$

where $1/\Gamma$ is the dephasing time, and the $z_{1,2}$ is the off-diagonal matrix element $z_{1,2} = <1|z|2>$. In the dilute limit, the total susceptibility per unit area is $\chi_s^{(1)}(\omega) = N_s \xi^{(1)}(\omega)$. At finite charge densities the relationship between the microscopic and macroscopic susceptibilities must include the potential created by the oscillating charge. The self-consistent problem has been solved to first order[22, 23]. It is found that the resonance of the linear absorption is shifted to a frequency $\omega_{1,2}*$. In systems with more than two states, the effective oscillator strengths are also renormalized.

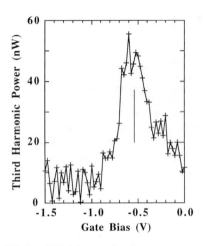

Fig 9. Third harmonic power versus gate bias. The pump frequency is 35 cm^{-1} and the intensity is ≈ 20 kW/cm^2. The marker shows the position hν = E$_{12}*/3$ where the pump frequency is one third of the of the frequency of the absorption peak position.

At present no theory exists for the macroscopic susceptibilities of a two level system which is self-consistent to second order, as required to discuss second-harmonic generation. We are presently formulating such a theory. In its absence we have chosen to analyze our results using formulas which are valid in the dilute limit, but replacing the resonance frequency ω_{12} by $\omega_{12}*$, the depolarization shifted frequency. In the dilute limit[21], the magnitude of the second-order susceptibility for second-harmonic generation in a two-state system at T=0 (neglecting terms resonant at $\omega<0$) is

$$|\chi^{(2)}(2\omega;\omega,\omega)| = N_s \frac{e^3}{h^2} z_{1,2} z_{2,2} z_{2,1}$$

$$\times \left| \frac{1}{(\omega_{1,2} - \omega - i\Gamma)(\omega_{1,2} - 2\omega - i\Gamma)} + \frac{1}{(\omega_{1,2} - \omega + i\Gamma)(\omega_{1,2} + \omega - i\Gamma)} \right|.$$

Assuming no permanent polarization of the quantum well[20], we have chosen our origin such that $z_{1,1} = 0$. It can be seen that in the limit $\Gamma/\omega_{1,2} \ll 1$, the lineshape of the second harmonic power near resonance at $\omega \approx \omega_{12}/2$ will be a lorentzian times the square of the charge density

$$P_{2\omega} \propto |\chi^{(2)}(2\omega;\omega,\omega)|^2 \propto N_s^2 \left\{ \frac{1}{(\omega_{1,2} - 2\omega)^2 + \Gamma^2} \right\} \quad ; \quad \omega \approx \frac{\omega_{1,2}}{2}.$$

As noted above, this form gives an excellent fit to our SHG data. It is interesting to note that the product of the absorption coefficient and the charge density should also have this form. Comparing the two sets of data (Figure 6), it is seen that the resonance positions and linewidths are in fact equal within experimental error, although the absorbance data deviate from the perturbative lineshape at small reverse biases.

All of the parameters required to predict $|\chi^{(2)}(2\omega;\omega,\omega)|$ from perturbation theory can be extracted from our linear absorption data. The resonance frequency $\omega_{2,1}{}^*$ and the dipole matrix element product $z_{12} \cdot z_{21}$ are determined by the position and strength of the linear absorption peak. The matrix element z_{22} can be determined from the Stark shift. This can be seen by examining the linear electro-optic coefficient, which describes the change in the susceptibility at frequency ω in response to a DC electric field. It can be shown that in a two-level system:

$$2\,\chi^{(2)}(\omega;0,\omega) = \frac{\partial}{\partial E_{DC}}\chi^{(1)}(\omega) \quad ; \quad e\,\langle z\rangle_{2,2} = \hbar\,\frac{\partial \omega_{1,2}}{\partial E_{DC}} \quad .$$

The $|\chi^{(2)}(2\omega;\omega,\omega)|$ extracted from this analysis is of the same order as our value deduced from second-harmonic generation (Table 3) when we substitute $\omega_{2,1}{}^*$ for $\omega_{2,1}$.

4.3 SATURATION

We are currently studying the saturation of the SHG at high pump powers. We have made preliminary measurements of the harmonic generation at elevated temperatures (T = 20K - 105K), and far from resonance ($h\nu_{pump}$ = 35 cm^{-1}, Vg = 0V). Our results suggest that saturation only occurs when the driving frequency is near resonance. The non-resonant SHG does not saturate at the highest pump powers we have achieved ($\approx$ 30 kW/cm^2). In addition the high temperature resonant SHG data retain the perturbative lineshape, even when a substantial temperature induced broadening and shift of the line are observed.

The observation of resonant saturation would be interesting because there are no first- or second-order processes which produce resonant absorption at $h\nu_{pump} = \omega_{12}/2$. If saturation is occurring through resonant heating leading to the equalization of the populations in the upper and lower states, then possible mechanisms include two-photon absorption and a related mechanism, the reabsorption of second-harmonic light. However, we also expect line broadening and saturation to appear in a strongly driven two-level system due to the AC Stark shift. Further measurements should clarify the saturation mechanism in our system.

5. Conclusion

We have designed and studied an asymmetric double-quantum-well structure which exhibits extremely large second- and third-order nonlinear susceptibilities in the far infrared. The physical system is simple, approximating a two-level system. We have observed resonant second- and third-harmonic generation in a single heterostructure and performed measurements in a system in which many-electron effects are important. We find that resonant second and third harmonic emission occur at the depolarization shifted frequency, and that the second-order susceptibility can be approximated from ordinary time-dependent perturbation theory for a two-state system, when the bare resonance frequency is replaced by the depolarization shifted frequency.

6. Acknowledgments

The authors gratefully acknowledge useful discussions with B. Bewley, C. Felix, A. Markelz, and S. J. Allen. This work was supported by the Army Research Office ARO-7IP29185 (JNH and MSS), the Office of Naval Research ONR N00014-92-J-1452 (JNH, KC, and MSS), the Alfred P. Sloan Foundation (MSS), the Air Force Office of Scientific Research AFOSR 91-0214 (KC, PFH, and ACG), and the NSF Science and Technology Center for Quantized Electronic Structures, Grant No. DMR 91-20007 (SF).

7. References

1. Gurnick, M.K. & DeTemple, T.A. IEEE J. Quantum Electron. **QE-19**, 791 (1983).
2. Yuh, P.F. & Wang, K.L. J. Appl. Phys. **65**, 4377 (1989).
3. Mayer, A. & Keilmann, F. Phys. Rev.*B* **33**, 6954 (1986).
4. Fejer, M.M., *et .al.* Phys. Rev. Lett. **62**, 1041 (1989).
5. Boucaud, P., *et al.* Appl. Phys. Lett. **57**, 215 (1990).
6. Boucaud, P., *et al.* Optics Letters **16**, 199 (1991).
7. Julien, F.H. in *Intersubband Transitions in Quantum Wells* (eds. Rosencher, E., Vinter, B. & Levine, B.) 173-181 (Plenum Press, New York and London, 1992).
8. Rosencher, E. & Bois, P. Phys. Rev. B **44**, 11315 (1991).
9. Yoo, S.B.J., *et .al.* Appl. Phys. Lett. **58**, 22 (1991).
10. Rosencher, E., *et .al.* Appl. Phys. Lett. **55**, 1597 (1989).
11. Rosencher, E., *et .al.* Appl. Phys. Lett. **56**, 1822-1824 (1990).
12. Rosencher, E. & Bois, P. in *Intersubband Transitions in Quantum Wells* , *op. cit.* 183-196.
13. Capasso, F., Sitori, C., Sivco, D. & Cho, A.Y. in *Intersubband Transitions in Quantum Wells* , *op. cit.* 141-149
14. Sirtori, C., *et .al.* Appl. Phys. Lett. **59**, 2302 (1991).
15. Sirtori, C., *et .al.* Appl. Phys. Lett. **60**, 151 (1992).
16. Sa'ar, A., *et .al.* in *Intersubband Transitions in Quantum Wells* , *op. cit.* 197-207
17. Sirtori, C., *et .al.* Phys. Rev. Lett. **68**, 1010 (1992).
18. Sirtori, C., *et .al.* Appl. Phys. Lett.. **60**, 2578 (1992).
19. Bewley, W.W., *et .al.* Phys. Rev. B **38**, 2376 (1993).
20. Shen, Y.R. *The Principles of Nonlinear Optics* (Wiley, New York, 1984).
21. Ward, J.F. Rev. Mod. Phys. **37**, 1 (1965).
22. Allen, S.J., Tsui, D.C. & Vinter, B. Solid State Commun. **20**, 425 (1976).
23. Ando, T., Fowler, A.B. & Stern, F. Rev. Mod. Phys. **54**, 436 (1982).

Second Harmonic Generation in GaAs-AlAs and Si-SiGe Quantum Well Structures

E. Corbin, M.J. Shaw, K.B. Wong and M.Jaros,
Department of Physics,
The University of Newcastle Upon Tyne,
Newcastle Upon Tyne,
UK.

Abstract

We present full scale relativistic pseudopotential calculations of both the first and second order susceptibility in p-type GaAs-AlAs and Si-SiGe structures. The frequency dependence of the linear and non-linear response due to transitions between valence mini-bands is calculated and the microscopic origin of the peaks determined. We show that simple particle-in-a-box models are unable to correctly describe the observed peak positions or the mechanisms involved. In particular, many important contributions come from areas of the Brillouin Zone away from the zone centre.

Introduction

In this paper we consider the possibilities of using both GaAs-AlAs and Si-SiGe p-type quantum well structures to design infra-red detectors, modulators and switching devices. The separation of the mini-bands in such structures can be tuned by altering parameters such as well width, chemical composition and the symmetry of the system. Thus the mini-band structure in these systems can be modified to cover a wide range of energies and offers an opportunity to design optical devices that would operate in the mid-to-far infrared range of wavelengths.

The polarization angle of the incident light is of importance in the design of devices. For transitions between conduction mini-bands to be useful it is necessary for the velocity vector of the incoming light to lie parallel to the growth direction. It is considerably more convenient for devices to be illuminated normal to the interface planes. We therefore consider optical effects occurring between valence mini-bands. In the valence band the mini-band structure is complicated by the near degeneracy of the mini-bands at the top of the band and by the relativistic spin orbit effect.

In Si-SiGe systems there is also the effect of strain, caused by the lattice mismatch between the two constituents, which mixes the states characterised by the familiar bulk bands and gives rise to new selection rules.

We shall begin with a study of absorption in order to understand the role of the band structure in the optical sums in the simplest possible case. Our main concern is, however, to develop guidelines for second harmonic generation. The mathematical formalism governing second order susceptibility is well known and will not be repeated here. Three states are needed for the process to occur. For example, there are two transitions between adjacent levels. The third transition then connects levels that are not adjacent. Since in symmetric wells, at least in terms of the simple particle-in-a-box model, only optical transitions between adjacent levels are allowed, some degree of asymmetry is needed in order to make the strength of the jump between non-adjacent states finite. We therefore choose asymmetric

477

H. C. Liu et al. (eds.), Quantum Well Intersubband Transition Physics and Devices, 477–491.
© 1994 *Kluwer Academic Publishers. Printed in the Netherlands.*

wells and perform a full scale calculation of the second order susceptibility for second harmonic generation. We have considered two quantum well systems, GaAs-AlAs and Si-SiGe. We predict the magnitude and the position of the main peaks in the response functions as a function of the light frequency and identify their microscopic (band-structure) origin. It is hoped that a more convenient arrangement with light incident normally to the interfaces can be achieved when the excitations occur in the valence band.

1. Absorption in SiGe quantum wells

1.1 TECHNICAL DETAILS

The initial structure studied, shown in Figure 1, was a $Si_{0.85}Ge_{0.15}/Si$ quantum well, consisting of 20 mono-layers (27 Å) alloy well with a 40 mono-layer (54 Å) Si barrier grown on a (100) Si substrate. This structure was chosen to obtain absorption in the 10-15μm range.

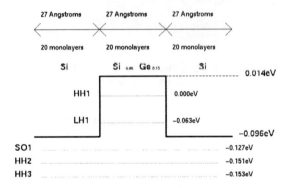

Figure 1: *Schematic representation of $Si_{0.85}Ge_{0.15}/Si$ quantum well showing the positions of the energy levels at* Γ*, the zone centre. The growth direction is along the z axis. The zero of energy is taken to be at the top of the superlattice valence band.*

The strain induced in the system due to the lattice mismatch of the alloy and the barriers was considered to be confined to the alloy well and the silicon barriers remain unstrained. A local empirical pseudo-potential method was used to calculate the band structure (Jaros 1984, 1985). The form of the pseudo-potential used was that of Freidel et al. (1988). In order to evaluate the susceptibility it is necessary to calculate optical sums which account for contributions coming from all points of the Brillouin Zone. In our calculations only the region close to the zone centre in the plane parallel to the interface contributes significantly. The sampling can be further reduced by using point group and time reversal symmetry allowing the sampling to be restricted to the irreducible segment of the zone. For the $\chi^{(1)}$ calculations only the region up to 0.2 ($\frac{2\pi}{a}$) (where a is the bulk lattice constant of the structure), in the x and y directions was sampled. The zero of energy was taken to be the top of the first heavy hole (HH1) band at Γ and the z-axis assumed to be in the growth direction. The valence band offset was taken to the 120meV, from linear interpolation of the results of Van der Walle and Martin (1986).

Both spin-orbit coupling and strain effects were considered resulting in mixing between the valence mini-bands. The exchange interaction was included as a rigid shift applied to the HH1 state only, as this is the only state with significant carrier population. The magnitude of the exchange shift was estimated from the work of Bandara et al. (1988). Other many body effects such as direct Coulomb interaction and correlation effects were not included at this stage.

All the results presented in this section were calculated at 0K. The dependence of the band structure and absorption on temperature and doping concentration has not yet been fully explored for the structures discussed in this section. However, the microscopic origin of the peaks is unlikely to change with temperature, although a small shift in energy and a broadening of the peaks may to occur.

Several different 'quasi' Fermi levels were used, (20meV, 10meV and 5meV below the top of HH1 at Γ) corresponding to various doping concentrations. Park et al. (1991) states that a doping of $1 \times 10^{19} cm^{-3}$ gives a Fermi level of -20meV. At the Fermi levels considered only the HH1 mini-band is above the Fermi level and hence only this band is capable of supplying carriers for transitions.

The linear susceptibility of the system, $\chi^{(1)}$, was calculated using the formula:

$$\chi_{\mu\sigma}^{(1)}(-\omega_\sigma;\omega_1) = -\frac{e^2 N}{V m \epsilon_o \omega_\sigma \omega_1} \delta_{\mu\alpha} + \frac{e^2}{V \epsilon_0 \hbar m^2 \omega_\sigma \omega_1} \sum_k \sum_a f_a \sum_b$$

$$\times \left[\frac{p_{ab}^\mu p_{ba}^\alpha}{\Omega_{ba} - i\Gamma_{ba} - \omega_1} + \frac{p_{ab}^\alpha p_{ba}^\mu}{\Omega_{ba} + i\Gamma_{ba} + \omega_1} \right] \tag{1}$$

where p_{ab}^μ is the optical matrix element between states a and b for incident light polarized in the μ direction, ω_1 is the frequency of the incoming beam, ω_σ is the frequency of the response, Ω_{ba} relates to the energy separation of the levels a and b, and V is the volume of the crystal. f_a is the equilibrium distribution function for electrons which is assumed to be of Fermi-Dirac form and Γ_{ba} is a de-phasing factor which accounts for relaxation processes leading to the equilibrium distribution f_a.

The absorption of the system is given by the imaginary part of the susceptibility i.e Im $[\chi^{(1)}]$. The response of the quantum well can be seen to be dependent on the polarization of the incident light. Light incident parallel to the growth direction yields $Im[\chi_{zz}^{(1)}]$. The absorption when light is incident perpendicular to the growth direction is taken to be the average of $Im[\chi_{xx}^{(1)}]$ and $Im[\chi_{yy}^{(1)}]$, denoted $Im[\chi_{xx}^{(1)}]$ in Figures. By restricting the summation over indices a and b in equation (1) the contribution of individual processes to the total absorption can be determined.

1.2 RESULTS AND DISCUSSION

The mini-band structure along the P-Γ-X symmetry lines (i.e. from the zone edge in the growth direction to the zone centre and out along the k_x axis) is shown in Figure 2.

The mini-bands are labelled based on the dominant bulk-like momentum signature at Γ. Away from the zone centre the mixing between valence band states caused by the the difference in the potentials of silicon and the alloy, strain and spin-orbit coupling effects means that the bands can no longer be considered as purely heavy hole, light hole or spin split-off bands. No attempt has been made to smooth the band structure diagram as that would mask the detailed features.

The diagram indicates distinct non-parabolicity of the mini-bands. This non-parabolicity will also contribute towards mixing between the bands. The spin split-off mini-band (SO1) and the second heavy hole mini-band (HH2) can be seen to cross along the growth direction.

The absorption spectrum obtained when light is incident parallel to the growth direction (z polarization) and the Fermi energy is set at -20meV can be seen in Figure 3. The microscopic origin of the peaks, i.e the minibands between which the transitions occur, can be seen.

Simple models suggest that the spectrum should be dominated by the HH2-HH1 transition occurring at energies around that of the zone centre separation of 151meV. The response is dominated, however, by a peak at ≈ 120meV (≈ 11 μm). It appears that much of this peak is due to a transition between SO1 and HH1. However, as indicated by arrow A on Figure 2, the area of the zone in which the energy levels are ≈ 120meV apart, the

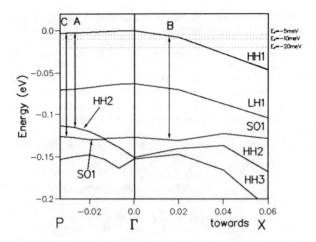

Figure 2: *Mini-band structure of the $Si_{0.85}Ge_{0.15}/Si$ quantum well structure along the P-Γ-X symmetry lines. See text for explanation of notes*

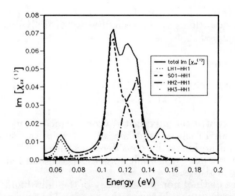

Figure 3: *Microscopic origin of absorption spectrum obtained for light incident parallel to growth direction with a Fermi energy of -20meV*

HH2 mini-band is above the SO1 mini-band. As the states are labelled on the basis of their nature at Γ transitions occurring in this area of the zone are likely to be mislabelled. As the optical matrix elements between the two heavy hole mini-bands are an order of magnitude greater than those between SO1 and HH1 in this region, it seems probable that the vast majority of the contribution to the absorption is from the HH2-HH1 transition. The HH2 state lies just outside the well for this structure thus the dominant transition is a free to bound transition. Thus the expected transition dominates but the peak response occurs at an energy different from the zone centre separation of the levels.

The variation with Fermi level of the absorption spectrum produced when light is incident normal to the growth direction is presented in Figure 4.

Varying the Fermi energy gives a good indication of the area of the zone in which the transition arises. The curvature of the HH1 mini-band is much smaller along the growth direction than in the x-y plane. The calculations of Man and Pan (1992) indicate that states with large $k_{\parallel}$ will be the most mixed, thus transitions which are only allowed in the presence of mixing between mini-bands are more likely to occur between states away from the zone centre. As the Fermi level is raised to -5meV, states with large $k_{\parallel}$ are not available and any absorption must occur in the region round Γ and in the growth direction where the mixing of states is less pronounced.

The large peak at $\approx$ 125meV is due to transitions between SO1 and HH1 around the areas indicated by arrows B and C on Fig. 2. This transition is weakly allowed at Γ. Away from Γ the mixing between the SO1 and HH2 states means that the envelope function of SO1 will acquire some of the character of a HH state and hence the transition probability will be enhanced. The height of this peak drops sharply as the Fermi level is raised to -5meV, demonstrating that much of the peak was due to transitions around B on Figure 2, i.e. those with large $k_{\parallel}$.

Figure 5 shows the variation of absorption spectrum with polarization of the incident light. It should be remembered that although the peaks occur at similar energies, their microscopic origin is different, (HH2-HH1 for parallel incidence, and SO1-HH1 for normal incidence). The height of the parallel incidence peak is about five times larger. This is because there is considerably less overlap of the envelope functions for a transitions between two different valence mini-bands. For the parallel incidence case the transition is between two mini-bands with envelope functions of the same type, hence their overlap will be much larger.

The band structure and absorption calculations were then repeated for both 15% and 30% germanium concentration in the well, with well widths of 28 mono-layers ($\approx$ 38Å), 36 mono-layers ($\approx$ 49Å) and 52 mono-layers ($\approx$ 70Å). As the well width is increased the mini-bands move closer together as would be expected from the particle-in-a-box model. In the limit of infinite well width the bulk structure should be recovered. The most notable change in the band structure as the well width is increased is that the order of the bands changes. The SO bands move more slowly and are 'overtaken' by the light and heavy hole bands. This is because of the difference in the effective masses of the holes in different bands. The results of varying well width for the two germanium concentrations are summarised in Figure 6. In addition the separation of the HH2 and HH1 levels at Γ for each structure are shown. All the peak positions refer to peaks due to transitions between HH1 and HH2 mini-bands. The movement of the peaks to lower energies is as expected from the decrease in energy separation of the mini-bands in the band structure. As the well width is increased more states are confined and the HH1-HH2 transition becomes a bound-to-bound transition, therefore has larger overlap of wavefunctions and thus the intensity of the absorption is greater. These results agree well with those of Fromheiz et al (1993) both in line width and energy of the peaks.

It can be seen that the zone centre separation follows the general trend of the peak positions but is offset slightly. This difference between the curves is most pronounced at narrower well widths. Thus a calculation performed only at Γ is unlikely to give a good representation of the structure.

Returning to the case of normal incidence, figure 7 shows the the variation with Fermi

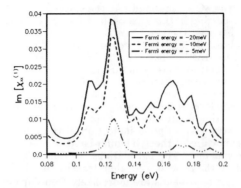

Figure 4: *Variation of absorption spectrum with Fermi level for light incident normal to the growth direction*

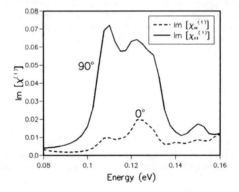

Figure 5: Variation of absorption spectrum with the polarization angle of the incident light

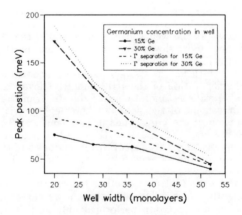

Figure 6: *Variation of the peak positions with well width for 15% and 30% Germanium concentrations The separation of the HH1 and HH2 states at Γ is also shown*

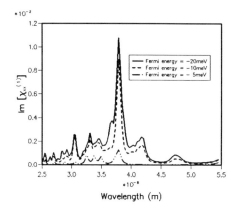

Figure 7: *Variation of the absorption with Fermi energy of a 28 mono-layer well containing 30% germanium illuminated with light incident normal to the growth direction.*

energy of the absorption spectrum of a 28 mono-layer well containing 30% germanium illuminated with light incident normal to the growth direction.

The intensity of the absorption is significantly less than that obtained with parallel incident light but this structure could still be of some relevance. The large peak at 3.75μm is due to a bound to free transition between the LH2 and HH1 bands. The peak again decreases rapidly with decreasing Fermi levels suggesting that the transition again occurs away from the zone centre.

These results demonstrate the possibility of infra-red detection using the valence band of p-type Si-SiGe quantum well structures. However, there are several areas in which the calculations need to be improved before a definitive statement on the ideal structure can be made. In particular, the temperature dependence, doping concentration and distribution of carriers need to be considered more fully, and other many body effects such as the direct coulomb interaction and correlation effects may be important, particularly at high doping concentrations.

2. Second Harmonic Generation in GaAs-AlAs Structures

Optical non-linearities were calculated for asymmetric GaAs-AlAs quantum well structures. An empirical pseudo-potential method was again used to calculate the band structure and density matrix theory (see, for example, Bloembergen (1965) or Flytzanis (1975)) was used to describe the optical susceptibilities. For second harmonic generation to occur three transitions are needed, one of which must connect non-adjacent states. In a symmetric well such a transition is either weak or strictly forbidden. Therefore an asymmetric structure was used which relaxes the selection rules governing transitions between non-adjacent states.

A schematic representation of the structure used in this section is presented in Figure 8 and the mini-band dispersion in figure 9.

The frequency and directional dependence of $\chi^{(2)}(-2\omega;\omega,\omega)$ at different temperatures was calculated and the microscopic excitation sequences which make significant contributions to the second harmonic generation processes were determined. It is shown that the band structure of the superlattice is critical in determining its $\chi^{(2)}(-2\omega;\omega,\omega)$ response.

Simple effective mass theory predicts that the zzz component will dominate, due to a large contribution from $HH3 \rightarrow HH2 \rightarrow HH1 \rightarrow HH3$ processes. The $HH1 \rightarrow HH2$ and $HH2 \rightarrow HH3$ are allowed for z-polarised incident light for both symmetric and asymmetric wells. In contrast, the $HH1 \rightarrow HH3$ transition is forbidden due to parity in symmetric well structures. The presence of the alloy layers breaks the parity of the states in the asymmetric

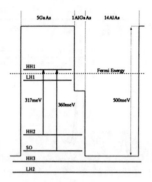

Figure 8: *A schematic representation of the structure used for* $\chi^{(2)}$ *calculations*

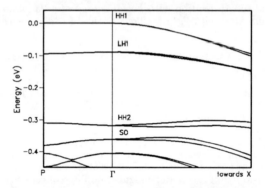

Figure 9: *Mini-band structure along principal symmetry axis for structure shown in Figure 8*

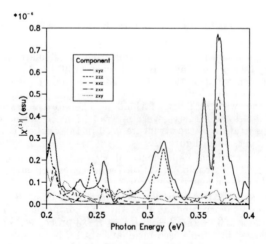

Figure 10: *The non-zero components of* $\chi^{(2)}$ *for the structure shown in figure 8*

structure and the $HH1 \rightarrow HH3$ transition becomes allowed for z-polarized light. Thus one would expect the zzz component of the susceptibility to dominate. However, the mixing between valence mini-bands results in a relaxation of the selection rules and many transitions between mini-bands become allowed. It was found that while many processes gave a fininte contribution to χ^2, the response was dominated by the contributions from just a few. Figure 10 shows the separate contributions to $\chi^{(2)}(-2\omega;\omega,\omega)$ calculated at 0K. The response is clearly dominated by the xyz component. The susceptibility of the superlattice is determined by the precise detailed form of the band structure and the overall picture depends on the exact manner in which the contributions from the various parts of the zone combine.

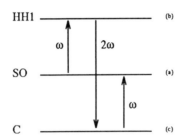

Figure 11: *Schematic representation of the processes contributing to the $\chi^{(2)}$ response. SO refers to the spin split-off band and C to low lying continuum states*

The most important processes were those of the form shown schematically in Figure 11. It can be seen that such processes will result in a doubly resonant term in the expression for the susceptibility when the energy separation of the states a and b is equal to that of states a and c. The particular processes which result in the peaks of $\chi^{(2)}_{xyz}$ at $\approx 350\text{meV}$ and $\approx 370\text{meV}$ in Figure 10 are those in which the original state (state a) is the spin split-off band, SO. The processes involve excitations from SO to the unoccupied HH1 state ('absorbing' one photon from the ω field), from the HH1 to lower lying continuum states ('emitting' a 2ω photon) and finally returning to the initial SO state ('absorbing' a second photon from the ω field) satisfying conservation of energy. It should be stressed that these are all virtual processes and there is no movement of the electrons between the mini-bands. The separate peaks at 350meV and 370meV arise from processes involving different low lying states. We note that these peaks do not occur at the zone-centre SO-HH1 transition energy of 360meV, despite the fact that both result from processes based on this transition. This is because the processes contributing to this peak are occurring away from the zone centre where, if we look back at the band structure in Figure 9, we can see that the SO-HH1 transition energy is reduced from its zone-centre value.

2.1 TEMPERATURE DEPENDENCE

We now move on to consider the effect of increasing the temperature on both the first and second order response. We first consider the absorption response in order to gain insight into the behaviour of the carriers as the temperature increases, and then move on the discuss the more complex situation of the second harmonic generation.

Figure 12 shows the variation of the linear susceptibility, $\chi^{(1)}_{zz}$, with temperature. At 0K the spectrum is dominated by a peak at $\approx 315\text{meV}$ which is due to the HH1-HH2 transition. The much smaller peak at $\approx 360\text{meV}$ can be associated with the SO1-HH1 transition.

From the band structure it can be seen that the curvature of HH1 is much smaller in the growth direction than it is across the plane. At 0K the carriers are confined within the Fermi surface. The carriers will therefore occupy states across the full width of the Brillouin

486

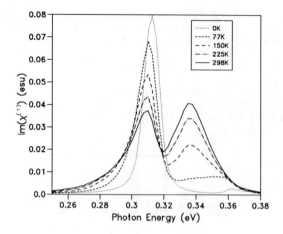

Figure 12: *The variation of the linear susceptibility with temperature*

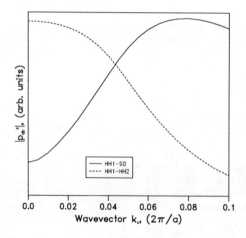

Figure 13: *Change in the relative strengths of the optical matrix elements across the zone*

zone in the growth direction, but are confined to a small region parallel to the interfaces. As temperature increases, carriers are excited and will occupy states lying further from the zone-centre according to the Fermi-Dirac distribution.

Thus at higher temperatures the susceptibility includes contributions from k-points with relatively large parallel wave vector components, and a reduced zone-centre contribution. The total number of carriers is not increased by the increase in temperature in our model, since we constrain the number of carriers to be equal to the number of dopant ions, so a change in temperature merely alters the distribution of the carriers throughout the Brillouin zone. The dispersion of the optical matrix elements for the z-polarized HH1-HH2 and HH1-SO1 transitions is shown in Figure 13. It can be seen that while at the zone centre HH1-HH2 is much stronger than HH1-SO1, the situation is reversed away from the zone-centre. Thus at 0K, where the zone centre contribution dominates, the HH1-HH2 peak is very much larger than HH1-SO1 peak. At higher temperatures when more carriers occupy states away from the Γ the HH1-SO1 peak becomes comparable and eventually larger than the HH1-HH2 peak. It can also be seen that the HH1-SO1 peak shifts to lower energy as the temperature increases. This is expected since away from the zone centre the HH1-SO mini-band separation gradually decreases.

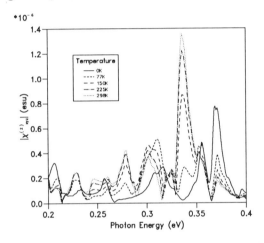

Figure 14: *The effect of varying temperature on the second order susceptibility*

In the case of the second-order susceptibility the situation is more complicated since the contributing processes each involve three transitions. The effect of the redistribution of carriers with temperature depends on the delicate balance between the changes in transition strengths and energies of the various transitions involved, throughout the Brillouin Zone. Figure 14 shows the variation of the dominant component, i.e. xyz component with temperature. Consider the behaviour of the peaks at ≈ 350meV and ≈ 370meV. Both peaks were identified as being due to processes involving SO1-HH1 transitions and intuitively would be expected to increase in magnitude as the carriers occupy states further from the zone centre. However, since so many other competing factors will also affected by the change in carrier distribution (such as the strengths of the other transitions, and the degree of the resonant enhancement), the contributions need not necessarily increase. Indeed, from figure 14, it can be seen that the 350meV peak does increase in magnitude (and shifts in energy in a manner similar to the SO1-HH1 peak in Figure 12), The magnitude of the 370meV peak, however, is rapidly reduced from its 0K value as the temperature rises. This is because the resonance conditions which hold at the zone centre for the excitation sequence responsible for this peak are no longer met away from Γ where the miniband separation alters.

The damping constant Γ_{ab} which represents the relaxation processes in the expression for the susceptibilities depends on the bands between which the excitation occurs, the position in the zone and the temperature. Obviously to include this accurately by means of a full many body calculation would be an enormous task. We therefore use a single value to represent all damping terms. The lack of experimental data for relaxation constants makes even this empirical approach difficult. We can see from figure 15 that as the damping constant is increased the peaks broaden and the height of the peaks decreases and that the smaller the damping constant the better resolved are the peaks. Further study is obviously needed on this point.

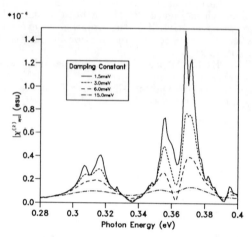

Figure 15: *Variation of the second order response with damping constant.*

3. Second harmonic generation in Si-Ge quantum wells

The second order susceptibilities were studied for a Si-SiGe quantum well consisting of 18 monolayers of $Si_{0.85}Ge_{0.15}$ well, with a 40 monolayer silicon barrier separated by 2 monolayer $Si_{0.925}Ge_{0.075}$ layer which produces a step in the valence band discontinuity. This structure is very similar to that studied for the absorption but again with the slight asymmetry introduced to enhance the non-linear response.

A schematic energy level diagram for this structure is shown in figure 16, with the principal zone centre transitions indicated and the mini-band dispersion is presented in figure 17. The light and heavy hole mixing in this case will be stronger in this superlattice than in the GaAs-AlAs structure, due to the strain induced momentum mixing which will alter the non-linear optical interactions. The calculation was performed at 0K with a Fermi level chosen to lie 26meV below the top of the uppermost valence band. The five independent non-zero components of $\chi^{(2)}(-2\omega; \omega\omega)$ for this structure are presented in Figure 18. We can see from this spectrum that the largest component is χ_{zzz}, which peaks at an energy of $\approx$ 100meV, peak B on Figure 18. The excitation sequences which dominate the contribution to this peak are not the $HH3 \rightarrow HH2 \rightarrow HH1 \rightarrow HH3$ expected from simple models but C→SO→HH1→C and SO→HH1→C→SO where C refers to low lying continuum states. The nature of these processes is the same as that described earlier for figure 10. Unlike the GaAs-AlAs system both of these processes contribute almost equally to the total susceptibility. The smaller peak, peak A, at 93meV is also due these processes. The arrows on the band structure diagram, Figure 17, indicate the areas of the zone in which these transitions are likely to occur. It can be seen that in all cases it is areas away from the zone centre which contribute most strongly to the non-linear processes.

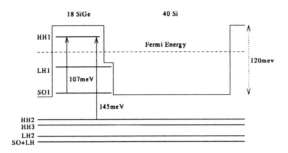

Figure 16: *Schematic representation of the Si-Ge structure used to study second order susceptibility*

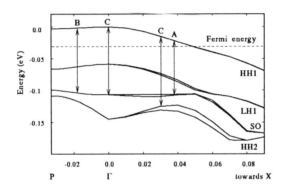

Figure 17: *The mini-band dispersion of the structure shown in figure 16. See text for explanation of notes*

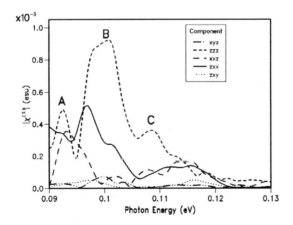

Figure 18: *The non-zero contributions to $\chi^{(2)}$ for the Si-Ge structure*

The peak at 107meV, peak C, would appear to correspond to the zone centre SO→HH1 energy separation. However, the separation of the HH1 and HH2 levels at the edge of the Brillouin Zone in the growth direction is also 107meV and there is a significant contribution to this peak from the processes HH2→HH1→C→HH2 and C→HH2→HH1→C. The magnitude of such a peak could not be predicted without including contributions from all possible transitions associated with various intersub-band processes.

The other none zero components of $\chi^{(2)}$ are also shown in Figure 18. The zxx component which peaks at $\approx$ 95meV, describes an induced polarization in the z-direction in response to fundamental fields applied in the x-direction (parallel to the interfaces). Normal incidence is the preferred direction for device applications. This peak is also due to virtual excitation sequences involving SO-HH1. The shift from the zone centre separation of 107meV is again due to the curvature of the bands and the increased amount of mixing between mini-bands further from the zone centre.

Several areas of uncertainty remain. In particular there is the problem with lack of data about the relaxation constants discussed earlier, and the inclusion of the exchange interaction as a rigid shift, which is obviously a primitive approximation of the true situation. Other many body effects such as the direct Coulomb interaction and the correlation or resonant screening effects have not yet been considered and may be important, particularly at high doping concentrations. Uncertainties in the details of curvature of the valence mini-bands and corresponding uncertainties with the density of states. Small changes in the potential can result in large changes in the shape of the response spectra. The uncertainties are not in the position of the ground state but that of the excited states in particular the spin split off bands. The relative position of the split off bands is important in determining the degree of mixing between mini-bands and hence the strength of the normal incidence response. Thus, although the precise details of the shape of the frequency response are open to debate, the comparisons between various structures remain valid.

Also we have only considered ideal structures. Imperfections at the interfaces and other defects can alter the symmetry and the degree of mixing between mini-bands. However, none of these effects should alter our main conclusions, but more work needs to be done.

Conclusion

In conclusion, we have provided a detailed description of the processes involved in first and second order susceptibility in GaAs-AlAs and Si-SiGe quantum well structures. We found that for all structures considered, the linear and non-linear response is dominated by transitions occurring away from the zone centre. We have shown that transitions between valence mini-bands in Si-SiGe structures can be used to generate strong absorption and second harmonic generation with normal incident light in both the 10-15μm and 3-5μm ranges.

Acknowledgements

We would like to thank the Defence Research Agency, Malvern, the Science and Engineering Research Council of Great Britain, the ESPRIT-basic research European program and the Office of Naval Research.

References

K.M.S.V. Bandara, D.D. Coon and O. Byungsung. (1988) *Appl. Phys. Lett.*, **53**, 1931

N. Bloembergen, 'Non-linear Optics', W.A. Benjamin, New York, (1965)

C. Flytzanis 'Quantum Electronics', edited b Rabin H., and Tang C.L. Academic Press, New York (1975).

P.Freidel, M.S. Hybertsen and M. Schlüter, (1988), *Phys. Rev. B*, **39**, 7974

T. Fromherz, E. Koppensteiner, M. Helm, G. Bauer, J.F. Nützel and G. Abstreiter, *Conference paper presented at Garmsih, August 1993*

M. Jaros, K.B. Wong, *J. Phys. C*, **17**, L765 (1984)

M. Jaros, K.B. Wong, and M.A. Gell, *Phys. Rev. B.*, **31**, 1205, (1985)

P. Man and D.S. Pan (1992), *Appl. Phys. Lett.*, **61**, 2799

J.S. Park, R.P.G. Karunasiri and K.L. Wang, *Appl. Phys. Lett*, **60**, 103, (1992)

C.G. Van de Walle and R.M.Martin, (1986), *Phys. Rev., B*, **34**, 5621

Non-Resonant Two-Photon Absorption in Quantum Well Infrared Detectors

E. Dupont, P.B. Corkum, P.W. Dooley*
National Research Council
Steacie Institute for Molecular Sciences
Ottawa
Ontario K1A 0R6
Canada

and

H.C. Liu, P.H. Wilson**, M. Lamm**, M. Buchanan, Z.R. Wasilewski
National Research Council
Institute for Microstructural Sciences
Ottawa
Ontario K1A 0R6
Canada

ABSTRACT. We demonstrate that a symmetric GaAs/GaAlAs quantum well photodetector, for which the peak responsivity is centered at $8.1\mu m$, shows a quadratic response at $10.6\mu m$. This is attributed to an intersubband two-photon transition. The wavelength and voltage dependence of the photodetector signal is qualitatively accounted for by the linear and the quadratic absorption and the related escape probabilities.

1. Introduction

The performance of quantum well (QW) photoconductors is close to that of the HgCdTe detectors. Quantum well detectors could be an advantageous alternative for mid-infrared detection, especially for the fabrication of focal plane arrays. To further improve the performance, most of the present research focuses on understanding the photoconductive gain, the escape and capture probabilities, and the dark current. These detectors are based on one-photon (1P) intersubband transitions and until now the possibilities of non-linear response have not been extensively studied. Extremely high non-linear susceptibilities in QWs have been reported (Rosencher 90) but, as far as we know, only Sirtori et al. (Sirtori 92) showed a non-linear behavior of the photocurrent. The process was based on a three-photon resonant emission in coupled AlInAs/GaInAs quantum wells. Recently, Khurgin and Li (Khurgin 93; Li 93) calculated the two-photon (2P) absorption coefficient, and the associated changes of index of refraction in asymmetric quantum wells and in superlattices. Here, we report that in a relatively "standard" symmetric quantum well detector, non-resonant two-photon

---

* present address: Dept. of Physics, University of Waterloo, Waterloo, Ontario N2L 3G1, Canada
** affiliation: the Centre for Electrophotonic Materials and Devices, McMaster University, Hamilton, Ontario L8S 4M1, Canada

H. C. Liu et al. (eds.), Quantum Well Intersubband Transition Physics and Devices, 493–500.
© 1994 *Kluwer Academic Publishers. Printed in the Netherlands.*

intersubband transitions can produce a significant quadratic photocurrent. After calculations of the quantum efficiency, the measured responsivities will be interpreted in terms of photoconductive gain.

2. Sample Description and Two-Photon Experiments

Since we were interested to observe a two-photon emission involving a "quasi-free" final state in the continuum, we used a structure having only two bound states $| 0 >$ and $| 1 >$, with an energy separation *far detuned* from the 117meV photon energy of the $10.6\mu m$ radiation. In this way, significant population in level $| 1 >$, which when combined with a second single photon step would imply a loss of coherence of the 2P process, should be avoided.

The sample was grown by molecular beam epitaxy on a (001) semi-insulating GaAs substrate. It consists of 32 periods of 5.8nm Si-δ doped ($9 \times 10^{11} cm^{-2}$) GaAs wide well and a 34.6nm $Ga_{0.7}Al_{0.3}As$ barrier. These parameters have been refined by fitting the x-ray double diffraction profiles. The whole structure is clad by 0.75 and $0.4\mu m$ 1.5×10^{18} cm^{-3} n^+ doped layers. Using standard photolithographic techniques $240\mu m$ square mesas were etched, and finally alloyed NiGeAu ohmic contacts were processed onto the doped layers.

We used a 45° edge facet configuration to couple the optical field with the intersubband dipoles. The absorption line at room temperature is centered at 152 meV. We could measure no absorption at $10.6\mu m$ with a Brewster's angle coupling configuration. The edge of the absorption is at $9.3\mu m$. Thus, with the CO_2 laser we used, it will be possible to study the 1P response of the device as well as the 2P response for longer wavelengths.

The response of the device has been tested with a grating tuned hybrid Lumonics TEA CO_2 laser. Single longitudinal mode operation was achieved with an intracavity low-pressure (4 Torr) section gain. Typically, a 100ns pulse with a peak power of 0.5MW is produced by this laser system. The repetition rate was set at 7Hz. After attenuation with calibrated CaF_2 windows, the beam was focused on the sample. A small voltage was applied across the quantum well detector: typically $-50mV$. In our two-photon experiments, the intensity $\mathcal{I}$ on the sample at $10.6\mu m$ was in the $200-500kW/cm^2$ range. This intensity could be obtained with a cw-laser but in that case, undesirable thermal effects might affect the results. The pulse energy is checked by a fast photon drag detector (reference signal). The photocurrent of our device was amplified through a 50Ω input 20-dB 500MHz amplifier. These two signals were simultaneously recorded with a digital oscilloscope or gated integrators.

Figure 1 shows the pulse shape obtained from the photon-drag detector (dashed line) and the quantum well device (solid line) when the laser was tuned to $10.6\mu m$. The pulse widths differ by a factor of 1.46: this is close to $\sqrt{2}$, which is the reduction of the pulse duration in the case of a doubled gaussian time-dependent beam. To confirm that our observations correspond to a two-photon absorption process, the QW photoresponse was compared with the square of the reference signal from the photon drag detector: the agreement is excellent.

The nature of the two-photon $10.6\mu m$ response can be tested by looking at the polarization dependence of the signal. If ϕ is the angle between the polarization and the plane defined by the normal of the 45° facet and the [001] direction, the active electric field regarding the intersubband dipoles is $E_0/\sqrt{2}\cos(\phi)$. Thus, we expect a $\cos^2(\phi)$ and a $\cos^4(\phi)$ dependence respectively for the linear and quadratic intersubband absorptions. The polarization was rotated by a CdS/CdSe $9-11\mu m$ half-wave plate. Two-photon data were averaged over 30 shots provided the input $10.6\mu m$ pulse energy fell within a 10% energy window. Figure 2 shows normalized signals for two wavelengths: 9.28μm and 10.6μm. The two solid lines representing the $\cos^2(\phi)$ and $\cos^4(\phi)$ curves fit very well the two sets of data. In this way, we confirm that the 10.6μm photoresponse is due to intersubband two-photon absorption.

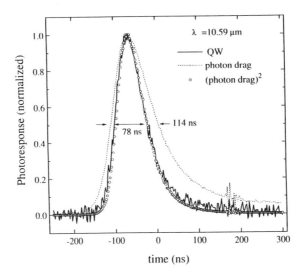

Fig. 1. pulse shape by the QW detector ($\mathcal{V} = $ -80mV) and the reference detector

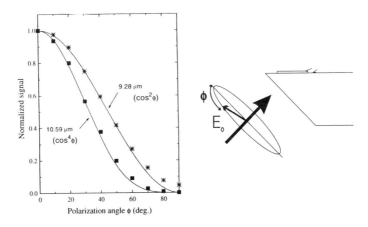

Fig. 2. 2P signal at 10.6μm and 1P signal at 9.28μm vs. the angle of polarization

Others measurements for higher voltages showed that the QW response deviated from the quadratic behaviour. Figure 3 shows the ratio between the quadratic and the linear components of the photocurrent versus the voltage for an intensity of about 250kW/cm^2 at 10.6μm. The quadratic response drops very fast with the polarization: at -0.4V (3kV/cm) the quantum well behaviour is mainly linear and finally, at -1.6V the response is perfectly

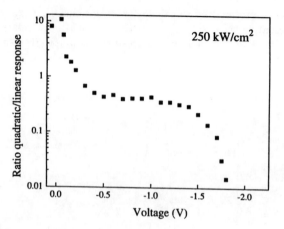

Fig. 3. Ratio between the quadratic component and the linear component of the signal vs. the voltage at 10.6μm. This ratio is proportional to the intensity.

linear. The origin of the plateau between -0.5 and -1.3V is still not clear. The intensity used at 9.28μm was about 16kW/cm^2. For such an intensity the linear absorption is the dominant process. By increasing the intensity by a factor of 3, we were able to detect a small non-linear component in the signal.

The measured responsivities are listed in Table I. The values of the responsivities at 10.6μm and 9.28μm are quite comparable, but in one case the response is quadratic and in the other it is linear. To understand such a high value for the two-photon response, we calculated the efficiency of this non-linear absorption.

TABLE I

CO_2 laser measured responsivities at $\mathcal{V}= -50$mV

10.6μm	$\mathcal{I} = 480$kW/cm^2	$\mathcal{R} = 3.8 \times 10^{-6} A/W$	(quadratic)
9.28μm		$\mathcal{R} = 1.3 \times 10^{-5} A/W$	(linear)

3. Analysis

In this section, we estimate the 2P absorption at 10.6μm using the second-order perturbation theory (Yariv 89; Delone 84). The transition probability rate $\mathcal{P}^{(2)}$ from the ground level to the continuum states is given by:

$$\mathcal{P}^{(2)}\left(\omega\right) = \beta E_z^4 = \beta \frac{\cos^4 \phi}{4} E_0^4$$

$$\beta = \frac{\pi e^4}{8m^{*4}} \frac{1}{\omega^4} \sum_f L_\gamma \left(\omega_{f0} - 2\omega\right) \left|\sum_n \frac{\delta z_{fn} \delta z_{n0}}{\omega_{n0} - \omega - i\gamma}\right|^2$$

where L_γ stands for the normalized Lorentzian function with a FWHM of 2γ to take into account the dephasing effect of the excited electrons, δz_{ji} represents the matrix element of the $\partial/\partial z$ operator between the states i and j, f and n are all the possible final and intermediate states of the structure, and m^* is the effective mass in GaAs. Taking into account the Gaussain shape of the beam, $E_0 = E \exp(-r^2/a^2)$ (a being the beam spot size), the efficiency of absorption $\eta^{(2)}$ is easily deduced :

$$\eta^{(2)}(\omega) = 2 \frac{2\hbar\omega \int_\Sigma n_s \mathcal{P}^{(2)}(\omega)\, dS}{\int_\Sigma \frac{1}{2} n \cos 45^\circ \frac{E_0^2}{Z_0}\, dS} = 2 \frac{\cos^4 \phi}{4} \frac{n_e}{\cos 45^\circ} 2\hbar\omega \frac{4Z_0^2}{n^2} \beta \frac{P}{\pi a^2}$$

where Z_0 is the vacuum impedance (377Ω), n is the index of refraction, Σ is the area of the detector, and n_s is the 2D density of electrons. The additional factor of 2 is due to the reflection of the beam on the top electrode. The factor $P/\pi a^2$ is identified as the internal intensity $\mathcal{I}$. In the following, $\eta^{(2)}$ will be normalized by this intensity.

The spot size a in our experiments is about 35μm. After fitting the absorption spectrum at 300K, we deduced a dephasing time $\gamma^{-1} = 0.073 ps$. To calculate $\eta^{(2)}$ at 80K, we assumed a increasing factor of 1.3 upon γ^{-1} and a 229 meV barrier potential ($\Delta E_c = 61\%$). The one-photon (dashed line) and two-photon (solid line) absorption efficiencies are plotted on Figure 4. The two-photon curve has two peaks. The low energy side of the low energy

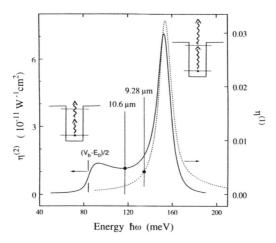

Fig. 4. Simulated 2P and 1P spectral absorption efficiencies at 82K. Here, the 2P efficiency $\eta^{(2)}$ is normalized by the incident intensity. This calculation takes into account the two paths of the beam per QW and the 45° configuration ($\psi - 0$).

peak starts when the final state of the 2P process meets the continuum, i.e. when $\hbar\omega = (V_b - E_0)/2$. Its shape is quite similar to the oscillator strength density for bound-to-continuum transitions because the summation over the intermediate states is dominated by the term $n = 1$. The second and Lorentzian peak is centered at $\hbar\omega = E_1 - E_0$ and corresponds to the situation where the intermediate state of the 2P process is resonant with the first excited level. For the range of optical power we used, the 2P efficiency is about 2-3 orders of magnitude lower than the 1P absorption at 10.6μm. We estimated the ratio between the two photoconductive gains $g_{10.6}^{(2)}$ and $g_{9.28}^{(1)}$:

$$\frac{g_{10.6}^{(2)}}{g_{9.28}^{(1)}} = \left(\frac{\mathcal{R}_{10.6}^{(2)}}{\mathcal{R}_{9.28}^{(2)}}\right)_{ex.} \left(\frac{\eta_{9.28}^{(1)}}{\eta_{10.6}^{(2)}}\right)_{theo.} \frac{1}{\mathcal{I}} \frac{2h\nu_{10.6}}{h\nu_{9.28}} \approx 380$$

Note that here g represents the global photoconductive gain i.e. the ratio between the escape probability p_e and the capture probability p_c (Liu 92; Levine 92). The electron-LO phonon scattering time being very fast (Lobentanzer 87; Kash 85), we can assume that the capture probabilities at 10.6μm (2P) and 9.28μm (1P) are similar. Thus, the important ratio $g_{10.6}^{(2)}/g_{9.28}^{(1)}$ means that a high escape probability of the 2P excited electrons might compensate the small 2P absorption efficiency. The photoconductive gain is a characteristic of the transport of the excited electrons, not of the process by which the electrons are excited. As a result, we expect the optical gain at 10.6μm with a 2P process to be the same as that at 5.3μm with a 1P process, i.e. $g_{10.6}^{(2)} = g_{5.3}^{(1)}$. Thus, single photon measurements give us an alternative perspective on this important parameter.

4. Single photon experiments

The linear response of this device is measured using a Fourier Transform Infrared spectrometer (FTIR) and absolute responsivities are deduced using a 1000K black body source and a narrow pass filter tuned at 8μm. Figure 5 shows the responsivity spectrum for the voltage we usually used in the 2P experiments. Two features are cleary observed: a narrow

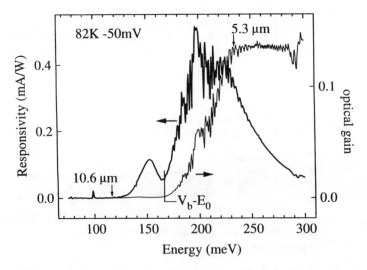

Fig. 5. Responsivity spectrum and the deduced photoconductive gain at $\mathcal{V} = -50$mV and T=82K

peak centered at the intersubband resonance and at high energies, a large and broad peak. Both are polarization sensitive. The main peak starts exactly at the barrier potential i.e. when $\hbar\omega = V_b - E_0$, which effectively confirms that the carriers are easily extracted from the attractive potential of the well when they are photoexcited above the barrier (Luc 93).

We used the 1P absorption efficiency $\eta^{(1)}$ which is displayed in Figure 4 to deduce the spectral photoconductive gain $g^{(1)}$. The result is given in Figure 5. The general shape is similar to a Heaviside function and the step coincides with the potential barrier. A more careful look on this curve indicates also an increase of g below the intersubband resonance, this was previously reported by Martinet (Martinet 92). The ratio between $g^{(1)}_{5.3}/g^{(1)}_{9.28}$ should be the same as $g^{(2)}_{10.6}/g^{(1)}_{9.28}$. Although it is difficult to be more precise we find:

$$\frac{g^{(1)}_{5.3}}{g^{(1)}_{9.28}} \approx 300.$$

This apparent discrepancy is probably due to the absorption model $\eta^{(1)}$ which differs from the actual absorption of the sample. Indeed, it was difficult to find a good fit of the room temperature absorption line up to $10.6\mu m$. The measured ratio $g^{(1)}_{5.3}/g^{(1)}_{9.28}$ confirms that the strong 2P signal is due to a significant increase of the escape probability at high energies.

We also compared the response of the device at different voltages. An example is given in Figure 6. By increasing the voltage, we noticed that the valley between the resonant peak (i.e. at 154 meV) and the broad peak vanishes. Below the barrier, the photoexcited

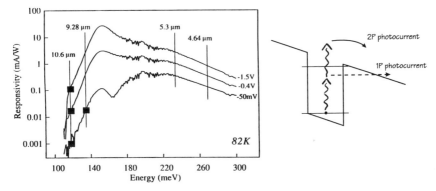

Fig. 6. Responsivities spectra for three different voltages. T=82K

electrons which give rise to the photocurrent undergo a tunnel process which is exponentially dependent on the voltage (Levine 92), while far above the barrier, the photoconductive gain is significantly less sensitive to the electric field. As a result, at $10.6\mu m$, an increase of the voltage across the device will favor the 1P component of the photocurrent over the 2P component. At $-0.4V$ (3kV/cm), the linear response can no longer be neglected. Thus, for such small voltages, a strong linear behavior of the device will dominate, unless we use longer wavelengths. This explains why at $10.6\mu m$ we observed quadratic responses only with vanishing electric fields.

5. Conclusion

In summary, we have shown that a non-resonant two-photon ionization can be observed in "standard" quantum well infrared photoconductors. In a $8\mu m$ detector, and with intensities of few hundreds of kW/cm^2 the two-photon absorption at $10.6\mu m$ is about 2-3 orders of magnitude lower than the linear absorption. Nevertheless, in such conditions, a quadratic response was easily observed and identified with intersubband transitions. In our sample, the large difference between the escape probabilties above the barrier and below 1P ionization threshold made the two-photon experiments possible. To maintain such a large difference, these experiments had to be carried out in a weak field regime. In this way, the low tunneling rate prevents the collection of the 1P photoexcited electrons below the barrier.

Acknowledgements

The authors want to thank D. Joines and P. Dietrich for technical support and advice as well as M. Ivanov for useful discussions.

References

Delone N.B., Krainov V.P.: 1984, "Atoms in Strong Light Fields", *Springer-Verlag*, Berlin, p. 32

Kash J.A., Tsang J.C., Hvam J.M.: 1985, "Subpicosecond time-resolved Raman spectroscopy of LO phonons in GaAs", *Phys. Rev. Lett.*, **54**, 2151

Khurgin J., Li S.: 1993, "Two-photon absorption and nonresonant nonlinear index of refraction in the intersubband transitions in the quantum wells", *Appl. Phys. Lett.*, **62**, 126

Levine B.F., Zussman A., Gunapala S.D., Asom M.T., Kuo J.M., Hobson W.S.: 1992, "Photoexcited escape probability, optical gain, and noise in quantum well infrared photodetectors", *J. Appl. Phys.*, **72**, 4429

Li S., Khurgin J.: 1993, "Two photon confined-to-continuum intersubband transitions in the semiconductor heterostructures", *J. Appl. Phys*, **73**, 4367

Liu H.C.: 1992, "Noise gain and operating temperature of quantum well infrared photodetectors", *Appl. Phys. Lett.*, **61**, 2703

Lobentanzer H., Rühle W.W., Stolz W., Plog K.: 1987, "Hot carrier-phonon interaction in three and two dimensional Ga$_{0.47}$In$_{0.53}$As", *Solid State Comm.*, **62**, 53

Luc F., Rosencher E., Bois Ph.: 1993, "Intersubband optical transients in multi-quantum-well structures", *Appl. Phys. Lett.*, **62**, 2542

Martinet E., Luc F., Rosencher E., Bois Ph., Costard E., Delaître S., Böckenhoff E.: 1992, "Electric effects on intersubband transitions in symmetric and asymmetric resonant ionization in GaAs/GaAlAs quantum wells", in "Intersubband transitions in quantum wells", edited by E. Rosencher, Vinter B., Levine B.F., *Plenum*, New York

Rosencher E, Bois Ph., Nagle J.: 1990, "Compositionally asymmetrical multiquantum wells: "Quasi-molecules" for giant optical nonlinearities in the infrared (9-11μm)", *Proc. SPIE*, **1273**, 138

Sirtori C., Capasso F., Sivco D.L., Cho A.Y.: 1992, "Resonant multiphoton electron emission from a quantum well", *Appl. Phys. Lett.*, **60**, 2678

Yariv A.: 1989, "Quantum Electronics", *John Wiley & Sons*, New York, p. 64

Optical Saturation of Intersubband Transitions

Lawrence C. West and Charles W. Roberts

AT&T Bell Laboratories, Room 4G518
Crawfords Corner Road
Holmdel, NJ 07733-1988

INTRODUCTION

A direct intersubband transition was discovered and observed[1, 2] in the search for better nonlinear optical materials for optical logic,[3] and this transition was termed a quantum well envelope state transition, "QWEST." Of direct importance to optical logic devices is the lifetime and saturation intensities of this type of intersubband transition. Fortunately, many direct measurements of the saturation powers[4, 5] and lifetimes[4, 6, 7] have been performed over the last few years. Nevertheless, wide variations have been found in the saturation intensities between[5] 300 kW/cm^2 and[4] 2 MW/cm^2. Furthermore relaxation times for various processes within the conduction band have been measured from 0.2 picoseconds[8] to many 10's of picoseconds[7] with many other timescales in between.[4, 6] We attempt here to explain all these results with a simple model for the intersubband relaxation based on a uniform temperature of the hot electrons throughout the conduction band which is much higher than the lattice temperature.

We note first that within a subband, the electrons are known to quickly relax to a thermal distribution,[8] with times in the low 100's of femtoseconds. In bulk Al$_x$Ga$_{1-x}$As the scattering times vary from 13 to 330 femtoseconds, with the faster times occurring at the higher kinetic energies.[9] As such, the subbands are expected to thermalize effectively immediately. The thermalization times between subbands will necessarily be longer than thermalization times within the subbands, as the momentum changes are larger. Yet recent experiments have demonstrated[6, 7, 10, 11] that these times are also very fast, under 1 picosecond.

In contrast, experiments have also observed that relaxation of hot electrons tend to maintain a thermal distribution as they relax back to the temperature of the host lattice over timescales of many picoseconds.[7, 12, 13] In light of these observations, we propose a model for intersubband optical saturation which occurs via resonant infrared heating of the electron gas to very high temperatures, uniform over all subbands, as opposed to the usual two level model.[5]

501

H. C. Liu et al. (eds.), Quantum Well Intersubband Transition Physics and Devices, 501–510.

FREE ENERGY OF AN ELECTRON GAS IN A QUANTUM WELL

The density of states $\rho(E)$ of a quantum well is given as a function of electron kinetic energy, E, in the well by

$$\rho(E)dE = \sum_{n=1}^{\infty} \frac{g_s m^*}{2\pi\hbar^2} Step(E - E_n)dE,$$

where g_s is the electron spin degeneracy, $g = 2$, m^* is the electron effective mass, and E_n is the ground energy of the n'th subband. The $Step(E)$ function is zero for negative values and unity for positive values of the argument. We assume a Fermi-Dirac occupation function, $f(E)$, given by

$$f(E) = \frac{1}{e^{(E-\mu)/kT} + 1}$$

where μ is the Fermi energy and T is the temperature. The free energy $U(T)$ of the electron gas is then given by

$$U(T) = \int_0^{\infty} E f(E)\rho(E)dE - \int_0^{E_f} E\rho(E)dE$$

where the energy has been set to be zero at zero electron temperature. The Fermi energy, μ, is determined by

$$N = \int_0^{\infty} \rho(E)f(E)dE$$

and the Fermi energy at zero temperture, E_f, is defined by

$$N = \int_0^{\infty} \rho(E)dE,$$

where N is the two dimensional electron density per well.
If we now assume $E_1 < E_f < E_2$, as is usually the case for AlGaAs quantum wells with $E_2 - E_1$ over 30 meV and doping under $2 \times 10^{12}/cm^2$, then the zero temperature integral of the energy, U_0, defined below, becomes

$$U_0 \equiv \int_0^{E_f} E\rho(E)dE = \frac{m^*}{2\pi\hbar^2} E_f^2.$$

TWO SUBBAND APPROXIMATION

We will now assume that we only have two sublevels. This is valid in the typical case where $E_3 - E_2$ is larger than $E_2 - E_1$ and we are optically saturating the n=1 to n=2 subband. The free energy of the two level quantum well becomes

$$U(T) = \frac{m^*}{\pi\hbar^2} \left\{ \int_{E_1}^{\infty} \frac{EdE}{e^{(E-\mu)/kT} + 1} + \int_{E_2}^{\infty} \frac{EdE}{e^{(E-\mu)/kT} + 1} \right\} - U_0.$$

The Fermi energy, μ, is found from

$$N = \frac{m^*}{\pi\hbar^2}\left\{\int_{E_1}^{\infty}\frac{dE}{e^{(E-\mu)/kT}+1} + \int_{E_2}^{\infty}\frac{dE}{e^{(E-\mu)/kT}+1}\right\}.$$

This integral can be solved simply in two dimensions to give

$$N = \frac{m^*}{\pi\hbar^2}kT\left\{\ln\left[1+e^{-(E_1-\mu)/kT}\right] + \ln\left[1+e^{-(E_2-\mu)/kT}\right]\right\}.$$

TWO SUBBANDS AT HIGH TEMPERATURE

We now assume that the electron distribution is nondegenerate. Again, this is a fair approximation when the electron densities are under $2\times10^{12}/\text{cm}^2$ and the electron gas is heated to temperatures as high as 1000 Kelvin. In this case, $e^{(E-\mu)/kT}$ is much greater than unity and the Fermi-Dirac distribution reduces to a simple Boltzman distribution. E_1 is referenced to zero for mathematical convenience and E_2 now becomes E_{21}. The previous equations reduce to

$$U(T) = \frac{m^*}{\pi\hbar^2}(kT)^2 e^{\mu/kT}\left\{1 + e^{-E_{21}/kT}\left(1+\frac{E_{21}}{kT}\right)\right\} - U_0$$

and

$$N = \frac{m^*}{\pi\hbar^2}(kT)e^{\mu/kT}\left\{1 + e^{-E_{21}/kT}\right\}.$$

The above two equations can be combined to give

$$U(T) = NkT\left\{\frac{1 + e^{-E_{21}/kT}(1+\frac{E_{21}}{kT})}{1 + e^{-E_{21}/kT}}\right\} - U_0.$$

Note that in the high temperature approximation, $U_0 = 0$.

ABSORPTION VS ELECTRON TEMPERATURE

The absorption per unit distance, α, of a collection of electrons in the two subbands is given by

$$\alpha = \sigma(N_2 - N_1)$$

where σ is the absorption cross section and N_1 and N_2 are the three dimensional electron densities. The cross section has an ambiguous definition because the interaction strength depends on the polarization of the infrared light and the manner in which a two-dimensional electron density is handled in three dimensions. We define the cross section with the optimum polarization so that the cross section at other polarizations are related by a simple cosine-squared multiplier proportional to the square of the electric field component in the direction of the allowed dipole. The relationship between two and three dimensional electron densities is very dependent on the optical configuration. We look at two common cases. In the first case, we have a plane wave traveling with its polarization normal to a stack of quantum wells with layer separation d. Here the average attenuation defines a planar cross section σ_p by

$$\alpha = \sigma_p(N_{s2} - N_{s1})/d$$

where N_{s2} and N_{s1} are the two dimensional surface densities of the electrons in the quantum wells for the lowest two quantum states. When defined this way the cross section will remain independent of the quantum well layer separation, d. Attenuation in a sample consisting of a number W of wells in which a beam is incident on the surface at Brewster's angle can now be defined in terms of this cross section between the m and n'th subband states by[2]

$$AF_{mn} = \sigma_p^{mn}(N_{sm} - N_{sn})W/n\sqrt{n^2 + 1}$$

where the refractive index factors, $n\sqrt{n^2 + 1}$, adjust for the polarization of the beam not being aligned with the intersubband dipole. The absorption fraction, $AF \equiv -\ln(Transmission)$, is related to the absorbance, $Abs \equiv -\log(Transmission)$, by $AF = Abs \ln(10)$. This cross section can be found in terms of fundamental parameters for a Lorentzian lineshape with dephasing time T_2 by[2]

$$\sigma_p^{mn} = \frac{e^2 T_2}{2\epsilon_0 m_e cn} f_{mn} \frac{1}{1 + (\omega - \omega_0)^2 T_2^2}$$

where e is the electron charge, m_e is the free electron mass, c is the speed of light, n is the refractive index, ϵ_0 is the free space dielectric constant, $\omega - \omega_0$ is the difference of the laser frequency from the intersubband resonance, and f_{mn} is the defined oscillator strength of the intersubband resonance.[1]

Using the cross section defined as above, the absorption of a collection of independent oscillators with two levels would saturate with increasing optical intensity, I, according to the function[14]

$$\alpha(I) = \frac{\alpha(0)}{1 + I/I_{sat}}$$

with the saturation intensity, I_{sat} depending on the upper state lifetime T_1 as given by

$$I_{sat} = \frac{E_m - E_n}{2\sigma_p^{mn} T_1}.$$

But as discussed before, the intersubband saturation is not simply a collection of independent oscillators. In the two level model above, the saturation relaxes with a single transition back to the lower state. In contrast, the conduction band electrons have been observed to scatter rapidly among themselves with electrons making many transitions back and forth between subbands before the entire electron distribution relaxes back to the lattice temperature.

Working for the moment in planar geometries with the infrared beam polarized in the direction of the intersubband dipole, we find the absorption per length from an m to n'th band transition is given by

$$\alpha = \sigma_p^{mn}/d \int_0^\infty [\rho(E_m)f(E_m)(1 - f(E_n)) - \rho(E_n)f(E_n)(1 - f(E_m))]\, dE_n$$

where $E_n = E + E_m$ because of transverse momentum conservation.[1, 2] For the lowest transition, the absorption as a function of electron temperature becomes

$$\alpha^{12} = \frac{m^*}{\pi\hbar^2}\sigma_p^{12}/d \int_0^\infty \frac{e^{(\epsilon+E_{21}-\mu)/kT} - e^{(\epsilon-\mu)/kT}}{\left(e^{(\epsilon+E_{21}-\mu)/kT} + 1\right)\left(e^{(\epsilon-\mu)/kT} + 1\right)}\, d\epsilon$$

where ϵ is a dummy variable running over the range of possible transverse kinetic energies. At high temperatures, this expression reduces to

$$\alpha^{12}(T) = \frac{m^*}{\pi\hbar^2}\sigma_p^{12}/d(kT)e^{\mu/kT}\left(1 - e^{-E_{12}/kT}\right).$$

If we use the expression for N at high temperature, we find

$$\alpha^{12}(T) = \sigma_p^{12}N/d\frac{1 - e^{-E_{12}/kT}}{1 + e^{-E_{12}/kT}}$$

SATURATION VIA ELECTRON HEATING

The rate of energy per-well per-area growth in the electron gas is given by

$$\frac{dU(T)}{dt} = I\alpha d - R_E(T)$$

where $R_E(T)$ is the energy relaxation rate as a function of the electron gas temperature. This relaxation rate, usually via phonon emission, is a complicated function[7, 11, 12, 13] of temperature, electron density, well parameters and intersubband energies. Nevertheless, over a small range of temperatures $R_E(T)$ is sufficiently smooth that we will approximate it by a rate proportional to the electron gas energy with time constant of τ_u.

$$\frac{dU(T)}{dt} = I\alpha d - \frac{U(T) - U(T_0)}{\tau_u}.$$

In steady state saturation,

$$U(T) = I\alpha d\tau_u + U(T_0).$$

If we now use the high temperature expression for α we obtain

$$U(T) - U(T_0) = I\sigma_p^{12}N\tau_u\frac{1 - e^{-E_{12}/kT}}{1 + e^{-E_{12}/kT}}.$$

But we also know the energy $U(T)$ at high temperatures. Using this expression the electron density drops out and we obtain a simple result for the electron temperature as a function of intensity,

$$I\sigma_p^{12}\tau_u = kT\left\{\frac{1 + e^{-E_{21}/kT}(1 + \frac{E_{21}}{kT})}{1 + e^{-E_{21}/kT}}\right\}.$$

506

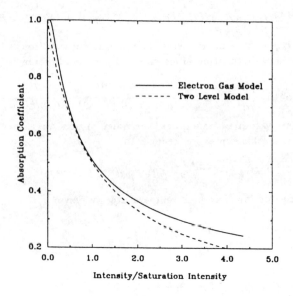

Figure 1: Comparison of the two-level saturation model with that of the hot electron model. The saturation intensity is defined to be where the absorption drops in half.

Using the expression for $\alpha(T)$, if we define I_{sat} as the intensity required to lower the absorption by a factor of two, as in the two-level model, then the temperature must be sufficiently high that $e^{-E_{12}/kT} = 1/3$, which corresponds to $kT = .91E_{12}$. For an intersubband transition resonant with a CO_2 laser at 10.6 micrometer wavelengths, this implies an electron temperature of 1235 Kelvin. Using this ratio of $kT = .91E_{12}$ at I_{sat}, we find from the above expression that

$$I_{sat} = \frac{E_{12}}{\sigma_p^{12}\tau_u} 2.23.$$

Note this is an expression very similar to that of the two-level model but with a saturation intensity predicted to be 4.46 times larger. We also note the saturation curve of absorption versus infrared intensity is sufficiently similar (see Fig. 1) that it would be difficult to discriminate experimentally between the two curves other than the large increase in saturation power observed for a given cross section and relaxation time, which are not always well characterized.

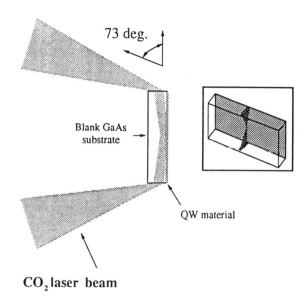

Figure 2: The cleave coupled geometry used for the saturation measurements. The infrared beam is incident at Brewster's angle and bounces once. The electric field was polarized normal to the wells and increased linearly away from the surface.

EXPERIMENTS

We conducted saturation experiments using a CW CO_2 laser tunable from 9.2 to 10.8 micrometers. The laser was chopped with an acousto-optic modulator into 10 microsecond square wave pulses with submicrosecond rise times. Our laser was automatically aligned to maintain a high quality single mode on each laser line which was also filtered with a pinhole. The laser was focussed at Brewster's angle on the cleaved edge of a GaAs substrate as shown in Fig. 2 with an aspheric lens with 2.5 inch focal length and 1.1 inch diameter. Pinhole tests determined our $1/e^2$ beam diameter was maintained in the range of 60 to 65 micrometers. After refraction on the cleaved surface, our beam diameter increases by a factor of 3.4 to 220 micrometer diameter in one dimension. The infrared beam makes one bounce on the surface of the quantum wells with its field almost perfectly aligned with the intersubband dipole. The transmitted beam refracts out of the other cleaved edge of the 0.9 millimeter wide semiconductor sliver and is collected by a power meter with absolute accuracy of 1%. The input beam is temporarily reflected 100% into an 1% accurate power meter just before the saturation measurement so as to obtain quantitative values for input power and transmission. Because of the thin sample, the infrared beam was slightly apertured at the input for a maximum throughput of only 72%, even for a test bulk GaAs substrate of similar dimensions. We were capable of putting peak CW powers of up to 18 Watts into a high quality beam into the sample. So as to prevent thermal destruction of our samples we held the duty cycle of the acousto-optic modulator under 10%.

At the surface with the air of the quantum well the infrared beam is totally internally reflected. The beam is also near total internal reflection in the AlGaAs quantum well layers and some critical phenomenon is observed at the peak negative index of refraction of the intersubband transition as the critical angle is widely varied during saturation. Spectral scans show a shift with an asymmetric increase in absorption at the longer wavelengths because of this critical phenomenon. The field in the quantum well region which is normal to the layers goes to zero at the surface because of total internal reflection and increases linearly with distance from the surface, rather than as a cosine, because of the incidence of the beam near critical angle of the AlGaAs layers.

The saturation curve presented in Figure 3 was performed on sample B331 of the original discovery[1] of the intersubband transition with a peak absorption at a wavelength of 9.67 micrometers at room temperature. This sample was heavily inhomogeneously broadened yet we obtained good saturation. Our saturation data fit a simple two-level model integrated over the lateral beam profile[5]. We found our curves fit with a saturation power of 11.2 Watts. We adjusted this power upwards by an approximate factor of 4/3 to account for the linear variation of fields away from the surface. We thus obtained a saturation power of

$$I_{sat} = \frac{4}{3} \frac{11.2W}{65 \times 225 \mu m^2}$$

$$= 103 kW/cm^2.$$

For reasons indicated below, we feel this saturation value should be multiplied by a factor of 3 or higher to a value of over $300 kW/cm^2$. We can obtain the cross section from the transmission values. The fraction of the infrared beam power, with polarization required for intersubband coupling, increases linearly away from the air surface to a peak value, near that of an antinode, of $4I_{inc}$, where constructive interference of the two counter-propagating beams occurs. The average power in the layers is therefore $4/3I_{inc}$. We can now deduce the cross section from the earlier expressions as

$$\sigma_p^{12} = \frac{3}{4} \frac{AF}{N_s W} \frac{\sin(17)}{\cos^2(17)}.$$

The angular factors compensate for the polarization of the fields and the projected area of the beam on the surface. Our absorption fraction, AF, was found to be 1.6, minus 0.35 for background. The well density is approximately $4 \times 10^{11}/cm^2$ and the number of wells, W, was 50. This gives a cross section of $1.50 \times 10^{-14} cm^2$. Using the expression in the thermal model above for I_{sat} we obtain $\tau_u = 2.23 E_{12}/\sigma_p^{12} I_{sat} < 8.6$ picoseconds.

An experiment was also performed with a similar sample grown[15] at AT&T with a 45 degree polish on both sides in the form of a parallelogram. The light was coupled at Brewster's angle on the 45 degree surface. In this geometry the saturation intensity was approximately three times larger even though the linewidth was only 8.5 meV instead of 24 meV. We are performing spectral hole burning and other experiments to aid in resolving this conflicting result.

We have also found recently that the saturation power depends on average optical power, rather than peak, indicating a thermal process assisting in the saturation. Experiments were conducted with a substrate temperature from 77 to 380 Kelvin and found no dependence of saturation behavior on semiconductor temperature. We believe the saturation is assisted by a thermally induced gradient in index as a function of distance from the QW surface, enhanced by the shallow angle reflection off the internal surfaces with the P polarization. The higher saturation power for the 45 degree polish and the dependence of saturation on optical heating but not on average temperature are evidence of this process. As such, we suspect the saturation powers of the intersubband transition are at least three times higher in the absence of the thermal gradient assistance.

CONCLUSIONS

We have developed an analytic model for the CW saturation of intersubband transitions based on the heating of the electron gas to high temperatures. We found that the saturation power is 4.46 times higher than the usual two-level model but has a similar dependence on input intensity. The short lifetimes observed[7, 10] in many measurements were not necessarily intersubband relaxation times as electrons can rescatter multiple times in a hot electron gas. The saturation intensities expected and observed were greater than 300 kW/cm^2. We are in agreement with the results on saturation intensities with the work of F. H. Julien et. al.[5].

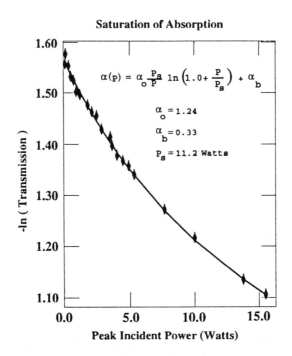

Figure 3: The absorption fraction = -ln(Transmission) as a function of peak infrared power incident on the cleaved edge of the sample at a wavelength of 9.67 micrometers and beam diameter of 65 micrometers

However, we feel the cross section cited in their paper is too low by a factor of $n\sqrt{n^2 + 1}$ or about 11.2. In their expression for cross section, it appears the external absorbance was used, which is lower than the effective internal absorbance by the reduced coupling from polarization angle in this geometry. The adjustment of this cross section along the lines discussed in this paper and the use of the higher saturation intensity of the hot electron model discussed here would put their estimates for T_1 in their paper in the range of 5 to 7 picoseconds.

The saturation intensities of approximately 2 MW/cm^2 calculated and observed by Seilemeier[4] were well above the intensities observed here. This is to be expected since the pulses used were short relative to the relaxation times of the electron gas, and much greater intensity was needed to heat the electron gas to the same temperatures and energy in the shorter time frame which effectively integrates the pulses. Seilemeier's longer lifetime processes of order 10 picoseconds is most likely from thermal relaxation, in agreement with the results of others.[12, 13] The proposed mechanisms of incomplete carrier confinement[7] for the slower times may not be necessary in light of agreement with observed thermal relaxation times.

In conclusion, we have provided a new model for intersubband relaxation based on thermalization which is in agreement with our experiments and helps to explain many other experiments as well.

References

[1] L. C. West and S. J. Eglash, Appl. Phys. Lett. 46, 1156 (1985).

[2] L. C. West, *Spectroscopy of GaAs Quantum Wells*, Stanford University Ph.D. thesis, UCRL-53681, July 1985.

[3] L. C. West, *Computer* 20, 34 (1987)

[4] A. Seilmeier, H. J. Hubner, G. Abstreiter, G. Weimann, and W. Schlapp, Phys. Rev. Lett. 59, 1345 (1987)

[5] F. H. Julien, J.-M. Lourtioz, N. Herschkorn, D. Delacourt, J. P. Pocholle, M. Papuchon, R. Planel, and G. LeRoux, Appl. Phys. Lett. 53, 116 (1988)

[6] M. C. Tatham, J. F. Ryan, and C. T. Foxon, Phys. Rev. Lett. 63, 1637 (1989)

[7] R. J. Bauerle, T. Elsaesser, W. Kaiser, H. Lobentanzer, W. Stolz, and K. Ploog, Phys. Rev. B 38, 4307 (1988)

[8] W. H. Knox, C. Hirlimann, D. A. Miller, J. Shah, D. S. Chemla, and C. V. Shank, Phys. Rev. Lett. 56, 1191 (1986)

[9] W. Z. Lin, J. G. Fujimoto, E. P. Ippen, and R. A. Logan, Appl. Phys. Lett. 51, 161 (1987)

[10] R. J. Bauerle, T. Elsaesser, H. Lobentanzer, W. Stolz, and K. Ploog, Phys. Rev. B 40, 10002 (1989)

[11] T. Elsaesser, R. J. Bauerle, W. Kaiser, H. Lobentanzer, W. Stolz, and K. Ploog, Appl. Phys. Lett. 54, 256 (1989)

[12] C. V. Shank, R. L. Fork, R. Yen, J. Shah, B. I. Greene, and A. C. Gossard, Solid State Comm. 47, 981 (1983)

[13] J. Shah and R. F. Leheny, Semiconductors Probed by Ultrafast Laser Spectroscopy, Vol. 1, p. 45, Academic Press (1984) ISBN 0-12-049901-0

[14] Amnon Yariv, *Optical Electronics*, Saunders College Publishing

[15] Moses Asom, Solid State Technology Center, Reading, Pa.

USE OF CLASSICALLY FREE QUASIBOUND STATES FOR INFRARED EMISSION

L. C. West,   C. W. Roberts,   J. Dunkel, and M. T. Asom
AT&T Bell Laboratories Rm 4G518
101 Crawfords Corner Rd
Holmdel NJ 07733
and

G. N. Henderson,   T. K. Gaylord,   E. Anemogiannis, and E. N. Glytsis
School of Electrical Engineering and Microelectronics Research Center
Georgia Institute of Technology
Atlanta GA 30332

ABSTRACT. A possible laser device is designed with the use of classically free quasibound electron states. An asymmetric semiconductor electron wave Fabry-Perot interference filter is designed with an upper electron state having much stronger confinement ($235fs$ lifetime) than the lower electron state ($76fs$ lifetime). This structure also allows for direct current pumping of the upper state and rapid depletion of the lower state under the presence of a field. Experiments demonstrate the existence of the upper quasibound state in this structure. Another structure, designed for infrared gain with current pumping, has improved parameters over the structure used in the spectroscopy measurement.

1. Introduction

1.1. BACKGROUND

The observation of large-dipole infrared transitions in semiconductor quantum wells [1] has stimulated significant interest in semiconductor heterostructure infrared detectors and emitters.[2-6] The use of semiconductor superlattices as infrared emitters was proposed as early as 1971.[7] Since this time, others have proposed configurations for intersubband infrared emitters and lasers.[3-6] Intersubband spontaneous emission has also been observed in a GaAs/AlGaAs quantum well.[5] To date, however, an intersubband infrared laser has not been developed.

1.2. INTERSUBBAND LASER CONCEPTS

A new semiconductor laser based on AlGaAs materials would be a welcome addition to the mid-infrared spectrum, where few lasers are presently available. The intersubband transition has a very strong oscillator strength,[1] ideal for a low power laser. The problem is in maintaining population inversion. The spontaneous radiative lifetimes of these states are hundreds of nanoseconds, yet the nonradiative lifetimes are in the subpicosecond range, over six orders of magnitude lower.[8-10] This fast nonradiative lifetime does not necessarily preclude highly efficient lasing. If the lower state can be depleted at a rate faster than

511

H. C. Liu et al. (eds.), Quantum Well Intersubband Transition Physics and Devices, 511–524.
© 1994 *Kluwer Academic Publishers. Printed in the Netherlands.*

this nonradiative lifetime, then the population inversion will be maintained and gain will result. The infrared intensity naturally grows within a laser to the point that the stimulated emission efficiently competes with the other relaxation processes, radiative or nonradiative. When this point is reached, the gain then saturates to a lower value which is equal to other losses in the cavity. If the relaxation process is nonradiative, the stimulated emission can nevertheless compete with this process with high efficiency. Slow spontaneous emission times are irrelevant to efficient lasing since the light will be stimulated out of the upper state at a much higher rate. This high stimulated emission rate requires a high infrared intensity. Generally, an efficient laser emits light with an intensity in the range of its saturation intensity due to the reasons mentioned above. This saturation intensity corresponds to the number of photons per cross section per lifetime. The lifetime of the intersubband transition is about 3 orders of magnitude smaller than that of an interband transition (used in typical semiconductor lasers) which are on the order of a few hundred picoseconds. On the other hand, the cross section of the intersubband transition, a few times 10^{-14}cm^2, is over two orders of magnitude larger than that of the bandgap transitions, a few times 10^{-16}cm^2. The saturation and lasing intensities of the intersubband laser then become only slightly larger than common semiconductor lasers since the shorter lifetimes are compensated by the much larger cross sections. The powers of mid-IR lasers can be potentially larger because of the larger waveguide dimensions. The waveguides for intersubband lasers at $10\,\mu m$ can be as small as $1.6\,\mu m \times 3.2\,\mu m$ with large index differences. Again, these dimensions are not much larger than conventional semiconductor laser waveguides. As will be shown in this paper, the current density requirements are also modest, only a few thousand A/cm^2, again comparable to common interband lasers in use today.

1.3. TECHNICAL PROBLEMS

Although efficient lasing is not precluded by the nonradiative relaxation, the short lifetimes create difficulties in operation. The primary difficulty is in obtaining population inversion between subband states in a quantum well.[6,11] All previously proposed schemes have relied on tunneling for injection into the upper state and for depletion out of the low-energy state.[3,4,6,11,12] As pointed out by Helm, however, it is difficult to achieve population inversion when such tunneling schemes are used.[11] This is partly because tunneling times are too long to deplete the low-energy state. The depletion of the lower state must be less than any relaxation times (associated with LO phonon emission, electron-electron scattering, etc.) from the upper state which may be as small as a few hundred femtoseconds.[8-10] A second and even greater problem is the need to create a situation where the lower-energy state has a shorter lifetime than the upper-energy states. This is especially difficult in quantum wells because the lower energy state is inherently more tightly bound. Resonant schemes for lower-state depletion do not work well because tunneling rates are always slower for lower-energy states.

1.4. SOLUTIONS WITH CLASSICALLY FREE QUASIBOUND STATES

The solution which has been investigated and is presented here is to use optical transitions between classically free quasibound states in a semiconductor heterostructure. This approach solves two main problems with confined states. First, electron spatial confinement times can be short compared to the upper state lifetimes. Second, one can use resonant electron-wave Fabry-Perot interference filters to allow stronger confinement of the upper

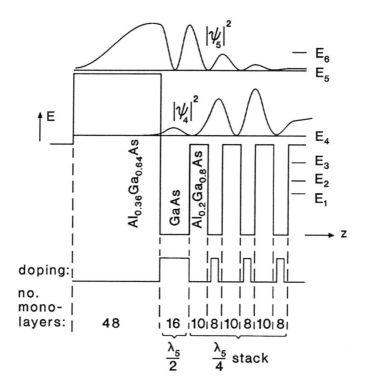

Fig. 1. Conduction band and doping profile for asymmetric electron-wave Fabry-Perot interference filter designed with two classically free quasibound states (one mono-layer = 0.282 nm). At the upper design energy (E_5) the center GaAs layer is a half-electron-wavelength resonator and the righthand wells and barriers are quarter-electron-wavelength reflectors.

quasibound state (Fig. 1). In addition, the structure described here has an asymmetry to allow upper state pumping by current injection and lower state depletion by field extraction.

2. Electron Filter Structure Design

2.1 FILTER CONCEPTS

Classically free quasibound states in semiconductor heterostructures were first proposed by Herrick and Stillinger.[13,14] More recently, by using exact analogies between electron wave propagation in semiconductors and electromagnetic wave propagation in dielectrics,[15] a quantitative procedure was developed for designing electron-wave energy filters using standard thin film optical design techniques.[16] The filter shown in Fig. 1 was designed using this procedure. This filter, the electron wave analog of an electromagnetic interference filter, was designed to be resonant at the upper state energy of 272 meV (above the GaAs conduction band edge). At this energy the 16 monolayer GaAs region is a half-wavelength "electron cavity" and the righthand layers are quarter-wavelength reflectors.[16] The barrier on the left of the structure provides selective electrical injection into and reflectivity for the upper state. An electron at $272\,meV$ is confined inside the resonant layer by constructive interference of the waves reflected by the quarter-wave stack. Recent bound-to-quasibound optical absorption measurements of a symmetric energy filter demonstrated clear confinement in the resonant state above the barriers.[17] The lower energy classically free state ($166\,meV$) in the present design is a nearly-bound state which has a low spatial confinement lifetime due to the lower reflectivity of the quarter-wave stack at this energy.

The two classically free quasibound states of this structure are analogous to leaky-modes in electromagnetic waveguides [18] or Fabry-Perot interferometers. Although the states are not orthogonal eigenstates, a strong resonance nevertheless exists. An eigenstate would have an infinite lifetime, but these states can be considered superpositions of multiple eigenstates which dephase from each other with a finite lifetime as calculated directly from the wave theory with good accuracy. The energy eigenvalues ($\mathcal{E}$) for quasibound states are complex where the lifetime (τ) of the state is $\tau = \hbar/2\Im m(\mathcal{E})$.[18,19] Using the procedure of Ref. [18], the energies and lifetimes of all the bound and quasibound states can be found under the envelope-function approximation by finding the complex energy eigenvalues.

The energies of the bound and quasibound states of a given structure are derived under the effective mass approximation using the Argument Principle Method (APM) in conjunction with the transfer matrix method as described in Ref. [18]. Band nonparabolicity effects, which can be significant for higher energy states, are included in the results as corrections to the state energies through the inclusion of an energy-dependent effective mass which is derived from the solution of the Kane model as given in Ref. [20]. Band nonparabolicity cannot be included directly in the APM root extraction scheme because the inclusion of an energy-dependent mass makes the complex dispersion equation derived from the structure transfer matrix a nonanalytic function. The APM is used to estimate the state energies using the band edge mass and these estimates are then refined using Muller's method with the energy-dependent mass included. The nonparabolicity lowers the state energies of higher quantum number states.

2.2. FILTER DESIGN

The filter structure needs to be carefully designed to optimize various requirements. First, the second classically free state must lie just above the left hand barrier energy. Second, the energy separation of the two quasibound states must be designed to be around 120 meV so as be probed with a CO_2 laser. Third, the structure must be designed to have a high lifetime ratio between upper and lower states with the use of three or less quarter-wave filter periods, since electron coherence can not be assured at room temperature beyond this distance. Finally, the structure must be designed to have a lower state lifetime under $100\,fs$ so that inversion can be assured even for the shortest upper state lifetime. [8-10]

The structure of Fig. 1 was designed to satisfy these criteria, and has three bound states at energies [$E_n = \Re e(\mathcal{E}_n)$] of $E_1 = 69\,meV$, $E_2 = 90\,meV$, and $E_3 = 122\,meV$ and three classically free quasibound states at $E_4 = 166\,meV$, $E_5 = 272\,meV$, and $E_6 = 298\,meV$ with lifetimes of $\tau_4 = 76\,fs$, $\tau_5 = 235\,fs$ and $\tau_6 = 49\,fs$. Other higher energy quasibound states exist but are extremely weak and not of practical interest. The $n = 5$ state is designed to become the upper laser state and the $n = 4$ state is designed to become the lower laser state. The spatial confinement lifetime of the upper state is greater than triple the spatial confinement lifetime of the lower state, as desired for population inversion. In addition, the spatial confinement lifetime of the lower state is much shorter than $100\,fs$

to be less than the carrier relaxation times.[8-10] The dipole matrix elements z_{ij} between the states were found by computing the complex wavefunctions ψ_i for each state, and then evaluating $z_{ij} = \int \psi_j^* z \psi_i dz$, where z is the position variable along the growth direction of the structure. The quasibound wavefunctions grow exponentially outside of the structure and thus the wavefunctions must be normalized over a fixed region. For these calculations, the wavefunctions were normalized inside the structure. This normalization extended 4.0 nm into the output region on the right so the overlap integral with the bound states would be very close to zero. The dipole matrix element for the 4-5 design transition was found to be $4.0 \pm 0.8\,nm$, which is extremely large for an intersubband transition with this energy but typical of transitions between adjacent higher levels.[1] The uncertainty in the matrix element ($\pm 0.8\,nm$) is due to the nonorthogonality of the quasibound states.[18,19] The dipole was calculated with the origin at either end of the filter structure. The choice of origin does not affect the dipole matrix element for orthogonal states, but gives the above stated variation for the quasibound states.

2.3. EPITAXIAL STRUCTURE

The filter structure shown in Fig. 1 was grown using MBE with very high purity sources.[21] A 282 nm thick $Al_{0.36}Ga_{0.64}As$ undoped buffer was grown on a (100) oriented semi-insulating GaAs substrate. Thirty periods of the filter as shown in Fig. 1 were then grown with an undoped $56\,nm$ $Al_{0.2}Ga_{0.8}As$ spacer. Doping at $1.2 \times 10^{18}cm^{-3}$ was placed only in the GaAs regions. In order to maintain a flat band, the doping was placed to balance that of the charge distribution in the lower three bound states (discussed below). The doping was placed throughout the full 16 monolayers of the half-wave region but only in the middle 1.3 nm of the remaining quarter wave stacks as shown in Fig. 1. The total doping in all GaAs wells sum to a sheet charge density of $1.0 \times 10^{12}cm^{-2}$. The structure was capped with 282 nm of undoped GaAs.

2.4. CHARGE DISTRIBUTION

The MBE grown structure described above is designed to be tested using infrared absorption measurements of the line strengths, energies, and linewidths. This intersubband absorption is between the three lower energy bound states and the two quasibound states. For sufficient absorption strength, the 30 filter structures in the stack are doped at $1.0 \times 10^{12}\,cm^{-2}$. To retain the charge within the filter structure for absorption measurements, the output region consisted of a $Al_{0.2}Ga_{0.8}As$ barrier. If a similar structure were designed for a laser, the opposite goal of depleting the charge in the lower states would suggest use of a GaAs output region instead.

The Fermi level at 300K for this filter was calculated to be $77.4\,meV$. Likewise, the relative two-dimensional populations $N_n/\sum N_n$ were found to be 57%, 30%, 10%, and 2% for levels $n = 1, 2, 3$, and 4 respectively. The 2% of the population in the $n = 4$ state allows for possible observation of the 4-5 transition due to its large dipole matrix element. The high density used here could cause a variation in the electrostatic potential of as much as $60\,meV$ or higher if the doping is not properly balanced across the structure. For instance, uniform doping across the structure would result in a net potential change of $36\,meV$, with larger variations internally. (Note that any laser structure would be doped at least two orders of magnitude lower and not have these problems.) A square-wave doping profile was chosen to compensate for the charge distribution in the three bound states ($n = 1, 2, 3$) because of the simplicity of doping within the MBE growth using a constant density. By adjusting the doping profile as shown in Fig. 1, the designed band structure was retained within 2 meV via charge balance; thus a self-consistent algorithm for calculating the wavefunctions was not needed.

3. Optical Properties

3.1. TWO DIMENSIONAL ABSORPTION

The absorption per unit distance, α, of a collection of electrons in two states of a heterostructure filter (such as the one shown in Fig. 1) is given by

$$\alpha = \sigma(N_n - N_m) \tag{1}$$

where σ is the absorption cross section and N_n and N_m are the three dimensional electron densities in states n and m, respectively. The cross section has an ambiguous definition because the interaction strength depends on the polarization of the infrared light. To solve this ambiguity, the cross section σ_z is first defined for a polarization in the growth direction z which has a maximum interaction with the intersubband transition. The cross section at other polarizations is then given as the product of σ_z and the square of the electric field component in the growth direction.

3.2. CROSS SECTION DEFINITION

For an electromagnetic plane wave traveling perpendicular to the growth direction with the electric field polarization in the growth direction, the attenuation of the field is described by a planar cross section σ_z

$$\sigma_z = \alpha d/(N_{sn} - N_{sm}) \tag{2}$$

where N_{sm} and N_{sn} are the two dimensional surface densities of the electrons in states m and n and d is the spacing between filters. With this definition, the cross section will remain independent of the separation d. This formulation is also useful in waveguides where d is small compared to the wavelength. In a given region with a uniform distribution of wells, the attenuation (or gain) can be found using Eq. 2 within that region. It should be noted that within a waveguide the polarization can vary considerably with distance across the guide, and waveguide absorption calculations should take this into account.

The cross section can be found in terms of fundamental parameters for a Lorentzian lineshape with dephasing time T_2 [1]

$$\sigma_z^{mn} \equiv \frac{e^2 T_2}{2\epsilon_0 m_e cn} f_{mn} \frac{1}{1 + (\omega - \omega_{mn})^2 T_2^2} \tag{3}$$

where e is the electron charge, m_e is the free electron mass, c is the speed of light, n is the refractive index, ϵ_0 is the free space dielectric constant, and $\omega - \omega_{mn}$ is the difference of the laser frequency from the intersubband resonance. The oscillator strength of the intersubband resonance, f_{mn}, is defined by [1]

$$f_{nm} \equiv \frac{2m_e \omega}{\hbar} (z_{mn})^2. \tag{4}$$

If all physical constants are evaluated and all material parameters are grouped, the cross section reduces to

$$\sigma_z^{mn} = 6.99 \times 10^{-14} \text{cm}^2 \frac{f_{mn}}{n \Delta E(\text{meV})} \tag{5}$$

where the full linewidth at half the maximum peak, $\Delta E(\text{meV})$, is expressed in meV. Similarly, one can use the oscillator strength to define the cross section in terms of the dipole matrix element as

$$\sigma_z^{mn} = 0.1834 \frac{\omega}{\Delta \omega} \frac{(z_{mn})^2}{n} \tag{6}$$

where the ratio of resonance to linewidth, $\frac{\omega}{\Delta \omega}$, is dimensionless.

For an infrared beam incident at an internal angle, θ_i to the growth direction, the concept of an attenuation per unit length is no longer valid. Instead the fraction of the light incident on one side of the filter which does not appear on the other side is evaluated as the absorption fraction, AF. The net attenuation is now not dependent on the spacing between filters but only on the number of filters, W. The absorption fraction, AF, is given by

$$AF_{mn} = \sigma_z^{mn}(N_{sm} - N_{sn})W \frac{\sin^2(\theta_i)}{\cos(\theta_i)} \tag{7}$$

where the $\sin^2(\theta_i)$ in the numerator adjusts for the polarization of the TM beam not being aligned with the intersubband dipole. The $\cos(\theta_i)$ factor in the denominator adjusts for the projected area of the beam on the well, or equivalently, the increased propagation distance away from normal incidence. The infrared beam with the TE polarization does not have intersubband absorption. The absorption fraction, $AF \equiv -\ln(P_{tr})$, is related to the absorbance, $\text{Abs} \equiv -\log(P_{tr})$, by $AF = \text{Abs} \ln(10)$ where P_{tr} is the power transmitted through the filter. In the case of Brewster's angle coupling on the filter surface, the internal angle θ_i is 17 degrees from normal. The angle factors then reduce, using $\tan(\theta_i) = 1/n$, to $1/n\sqrt{n^2 + 1}$. As the angle, θ_i, increases towards 90 degrees, the AF becomes infinite. In this case, Eq. 2 for the plane wave propagating along the layer should be used.

For the purposes of measuring the oscillator strength independent of the lineshape, one can integrate over the lineshape to obtain an integrated absorption fraction, IAF, given by

$$IAF_{mn} = \frac{e^2 h}{4\epsilon_{mn} m_e c} \frac{f_{mn}(N_{sm} - N_{sn})W}{n} \frac{\sin^2(\theta_i)}{\cos(\theta_i)}. \tag{8}$$

In the zigzag geometry described below, the infrared beam makes a multiple number of bounces, B, by total internal reflection inside the sample given by

$$B = \frac{L}{d \tan(\theta_i)}. \tag{9}$$

The reflected beam interferes with the incident beam so that the electric field in the growth direction goes to zero at the surface. In a uniform dielectric material, the peak of the electric field occurs at the antinode at distance equal to $\lambda_0/4n \cos(\theta_i)$. In a nonuniform dielectric material with doping, the formula is no longer simple and a multilayer calculation must be performed taking into account the heterogeneous material, free electron effects on index, and intersubband transitions on index. The electric field component in the growth direction

518

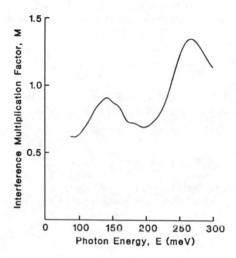

Fig. 2. The interference multiplication factor spectrum for the filter shown in Fig. 1 in the zigzag geometry shown in Fig. 3a

can be squared and averaged over the wells to give an interference multiplication factor M. M is zero for a single filter at a node, such as the surface, and equal to 2 for a single filter at an antinode where the beams constructively interfere. For a series of wells with a net thickness over $2\,\mu m$, M will average closer to unity, the typical value assumed. The value of M is wavelength dependent and is a correction factor to the spectrum. The integrated absorption fraction in the zigzag geometry, IAF^{ZZ}, can be written in terms of the single pass expression as

$$IAF_{mn}^{ZZ} = (IAF)(M(\omega_{mn}))(B) \tag{10}$$

It should also be noted that M can account for the depolarization shift. The depolarization shift is a classical phenomena which arises because the peak absorption of a resonance occurs slightly off resonance due to the refractive index changes from the resonance itself. M includes these infrared penetration effects and will cause the peak to appear shifted when multiplied by the original spectrum. Therefore, the original absorption peaks should be multiplied by M to obtain the correct spectrum. This form of the depolarization shift allows simple derivation for multiple transitions. The interference multiplication factor for the filter of Fig. 1 is shown in Fig. 2 for the zigzag geometry described below.

4. Experiments

4.1. SAMPLE GEOMETRY

The absorption was measured using an FTIR spectrometer in the zig-zag geometry as shown in Fig. 3a, where a six times beam condenser was used to focus the light onto a polished end of the sample. The beam attenuation is described by the absorption fraction [1,22] AF given in Eq. 7 In order to have a large aperture for beam coupling, the sample was polished on opposite sides at a shallow angle of $\theta_p = 17 deg$. The sample was $L = 7.9\,mm$ long and

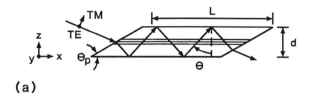

(a)

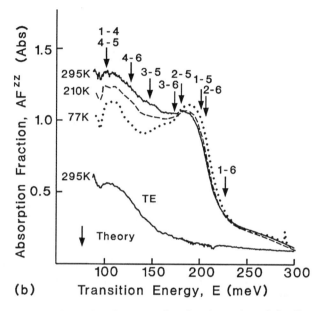

(b)

Fig. 3. (a) Experimental configuration for measuring the absorption of the filter shown in Fig. 1 using an FTIR spectrometer in the zig-zag geometry. In this geometry, only the TM polarization can excite the intersubband transitions. (b) TM and TE polarized absorption spectra. The peak at $200\,meV$ in the $77\,K$ spectrum is due to the 1-5 and 2-5 transitions. As the temperature is raised, the 3-5, 4-5 and 4-6 transitions begin to fill the valley between the 1-5 and 1-4 transitions.

$d = 503\,\mu m$ thick. The beam was incident on the polished surface at an angle of $45\,deg$ from the z direction. In this geometry, only the TM polarization can excite the intersubband transitions (because it has a component of the field in the growth direction z) [1,22].

4.2. ABSORPTION SPECTRA ANALYSIS

The absorption spectra can be calculated using the following procedure. First, the dipole matrix elements (z_{nm}) and carrier concentrations (N_n) are calculated for the states $n, m = 1, \cdots, 6$. Then the absorption fraction for each transition is calculated using Eq. 7. The linewidth (FWHM) of the Lorentzian is approximated as

$$\Delta E_{ij} = [(\hbar/\tau_i^{sc})^2 + (\hbar/\tau_j^{sc})^2 + \Delta E_I]^{1/2} \qquad (11)$$

where ΔE_I represents inhomogeneous broadening and is assumed to be $20\,meV$. The total

520

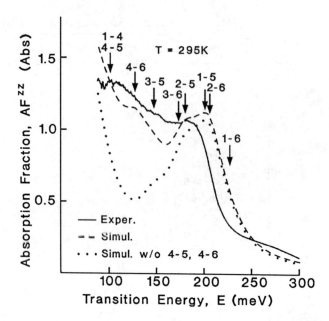

Fig. 4. Simulation versus measured spectrum for filter structure shown in Fig. 1 at 295 K. Much of the valley filling in the experiment is likely due to the 4-5 and 4-6 transitions. This point is evident in the dashed curve which shows the simulated spectrum at 295K with the $n = 4$ transitions excluded.

absorption fraction is then calculated by summing the Lorentzian absorption fractions for each transition over all n and m [1,22]. The simulated spectrum shown in Fig. 4 is multiplied by a scaling factor found using a least-squares fit to the experimental spectra. The energies of the theoretical locations of the transitions are shown with arrows in Fig. 3.

Figure 3b shows the experimental TM and TE polarized absorption spectra at multiple temperatures from 77 K to 295 K. At 77 K only the $n = 1$ and $n = 2$ states are significantly populated. As the temperature is raised, the $n = 3$ and $n = 4$ states become thermally populated. The peak at 200 meV in the 77 K spectrum is mostly due to the 1-5 and 2-5 transitions. Although the 1-6 and 2-6 transitions contribute to the peak, they have smaller dipoles than the $n = 5$ transitions. In addition, transitions to the $n = 6$ state may be weaker than calculated because this state, which is a half-wave resonance over the $Ga_{0.64}Al_{0.36}As$ barrier, requires a long electron coherence length to build up a resonance. This is evident in the simulation shown in Fig. 4, where the calculated peak occurs near 220 meV (while the experimental peak occurs near 200 meV) due to contributions from the 1-6 and 2-6 transitions. The peak near 110 meV appears abruptly in spectra taken at temperatures below 220 K in both the TE and TM polarizations. Therefore it is probably not due to intersubband transitions. As the temperature is raised, the 3-5, 4-5 and 4-6 transitions begin to fill the valley between the 1-5 and 1-4 transitions. It is not possible to resolve these individual transitions because they are very broad and closely spaced in energy. It is likely, however, that much of the valley filling is due to the 4-5 and 4-6 transitions. This point is evident in the dashed curve in Fig. 3b which shows the simulated spectrum at 295K with the $n = 4$ transitions excluded; this simulation is qualitatively similar to the 77K results. These experimental results show clear existence of the classically free resonances, even

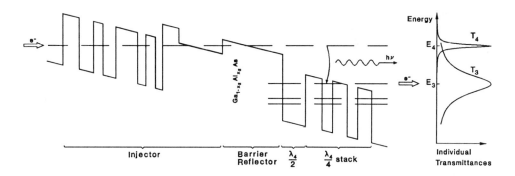

Fig. 5. Electron wave filter structures which is designed for current pumped population inversion. The injector filter structure on the left only allows one energy resonance to exist which pumps directly into the upper laser state.

at room-temperature where the electron coherence length was often considered too short for such states to exist. Recent bound-to-quasibound optical absorption measurements in a symmetric energy filter also clearly demonstrated the existence of classically free quasibound states [17].

5. Infrared Laser Design

5.1. FILTER STRUCTURE FOR POPULATION INVERSION

A structure which is to be used for observing gain is much different from the one used for the spectroscopy measurements. First, rather than having electrons confined in the bound and lower quasibound states, these states are to be depopulated. This goal is accomplished with the use of an applied field of $5 \times 10^4\ V/cm$ and a GaAs output region between periodic filter structures. The GaAs output region also allows an additional reflection interface with improved filter selectivity over the structure shown in Fig. 1 which was designed to be tested using spectroscopy. Second, the doping density is to be much lower so as to allow better matching to power levels sustainable without thermal damage to the structure. Doping to create a free electron density of $10^{16}\ cm^{-3}$ would be preferable, although difficult to assure within typical epitaxial growth capabilities. Finally, a filter must be designed to inject current directly into the upper laser state with good efficiency and narrow energy distribution. This injection filter must also prevent a buildup of charge before the injection barrier which could cause free electron absorption losses. This buildup of charge is prevented by designing the injector filter with only one miniband state which is resonant with the upper laser state. No lower energy states are allowed to exist to trap charge. A filter design for a current pumped systems is shown in Fig. 5.

This filter is designed with a lasing transition of 124 meV. The upper state has a lifetime of $208\,fs$ and the lower lasing state has a lifetime of only $33\,fs$. These times allow efficient depletion of the lower state under the shortest assumptions in inelastic upper state lifetimes.[8-10] The ratio of lifetimes of the upper to lower state are almost 9 to 1. The "bound" states are weakly confined and also leak out much faster that $100\,fs$.

5.2. GAIN CALCULATION

Using simple calculations, one can estimate the required current density for an electrically pumped laser structure using an electron wave Fabry-Perot filter similar to the one shown in Fig. 5. Using the dipole matrix element, the transition energy, and the lifetimes, one can calculate the cross-section for the lasing transition using equation 3 to be $\sigma_{34} \approx 3 \times 10^{-14} cm^{-2}$. The gain g is then found from Eq. 2 to be

$$g = \Delta N_{S34}/L_Q \sigma_{34} \qquad (12)$$

where ΔN_{S34} is the two dimensional electron inversion density, and L_Q is the spacing between filters. If one assumes that a fraction r of the electron density flowing into the upper lasing state, N_S^u, is inverted, then $\Delta N_{S34} = r N_S^u$. The current density is given by

$$J = \frac{q N_S^u}{\tau_3^{sc}} = \frac{q L_Q\, g}{r \tau_3^{sc} \sigma}, \qquad (13)$$

where q is the electron charge. If half of the electrons are inverted ($r = 0.5$), the gain in the waveguide region is taken as $g = 20\,cm^{-1}$, the spatial confinement lifetime, τ_3^{sc}, is $100\,fs$, and the spacing between the quantum well layers, L_Q, is 86 nm, then the current density is $J \approx 2 \times 10^4\,A/cm^2$ and the carrier density is $N \approx 10^{10}\,cm^{-2}$. This corresponds roughly to just under $10^{16}\,cm^{-3}$ three dimensional densities. This current density is very reasonable when compared to conventional laser diodes which have current densities as high as $10^5\,A/cm^2$. In order to ensure that the laser achieves a net gain, the free-electron losses must be less than the gain. The free electron absorption cross section is approximately $10^{-16}\,cm^2$,[20] so the carrier concentration outside the filter structure and in the electrodes must be no greater than 100 times the concentration within the filter. This will not create a problem in the gain regions. A problem does occur in the contacts which must be doped to around $10^{18}\,cm^{-3}$ to allow lateral conduction in the electrodes to the gain region and to obtain low ohmic contact. This free electron absorption loss in the electrodes can be overcome by using a waveguide structure which keeps the infrared fields away from the highly doped electrodes.

5.3. PERFORMANCE EXPECTATIONS

This laser structure has a field of $5 \times 10^4\,V/cm$ across a $4.68\,\mu m$ gain region. With other voltage drops, the overall voltage could well be as high as 35 Volts. Unlike pn diodes, these gain filters can be stacked indefinitely to create large voltages. The current could be in the low tens of Amps for initial experiments with devices structures 50 μm wide by 0.5 mm long. To prevent thermal destruction, pulsed current sources should probably be used in these large devices. The intersubband lasers proposed here should be fairly efficient so if only 30 percent of the power is converted, infrared powers as high as 100 Watts could potentially be achieved. The mesa size proposed above for initial experiments is very large compared to current laser diodes. Devices of contact area similar to that of current laser diodes could be created which would operate in the low tens of milliamps.

6. Summary

In summary, a new asymmetric Fabry-Perot electron wave filter has been designed for the use as the gain medium in an electrically pumped infrared semiconductor laser. There are a number of significant implications regarding the design and characterization of this electron wave filter: 1) The infrared wavelength was selected in advance by the design of the device. A very wide range of wavelengths in the mid- and far-infrared could have been so selected. 2) The optical transition between the two quasibound states has a very large dipole moment and thus oscillator strength. These are electron wave propagation states rather than bound states down in a quantum well. 3) This is the first structure with a high spatial confinement lifetime of the upper-state relative to the spatial confinement lifetime of the lower-state. 4) The lower-state lifetime is less than the carrier relaxation times as is needed for laser action. 5) Absorption measurements on this structure show that the electron coherence length at room temperature is larger than the device size. This represents room-temperature coherent operation in a vertical (growth direction) structure. 6) Preliminary estimates show that the current density required for laser action may be around $2 \times 10^4 \, A/cm^2$. These facts taken together clearly make this and related structures promising candidates for versatile infrared lasers.

Acknowledgements

This research was supported in part by grant no. DAAH-04-93-G-0027 from the Joint Services Electronics Program, no. ECS-9111866 from the National Science Foundation, and no. DAAL-03-90-C-0019 from ARPA.

References

[1] L. C. West and S. J. Eglash, "First observation of an extremely large-dipole transition within the conduction band of a GaAs quantum well," *Appl. Phys. Lett.*, vol. 46, pp. 1156-1158, June 15, 1985

[2] B. F. Levine, K. K. Choi, C. G. Bethea, J. Walker, and R. J. Malik, "New 10 μm infrared detector using intersubband absorption in resonant tunneling GaAlAs superlattices," *Appl. Phys. Lett.*, vol. 50, pp. 1092-1094, Apr. 20, 1987.

[3] P. Yuh and K. L. Wang, "Novel infrared band-aligned superlattice laser," *Appl. Phys. Lett.*, vol. 51, pp. 1404-1406, Nov. 2, 1987.

[4] S. I. Borenstain and J. Katz, "Evaluation of the feasability of a far-infrared laser based on intersubband transitions in quantum wells," *Appl. Phys. Lett.*, vol. 55, pp. 654-656, Aug. 14, 1992.

[5] M. Helm, E. Colas, P. England, F. DeRosa, and S. J. Allen, Jr., "Observation of grating-induced intersubband emission from GaAs/AlGaAs superlattices," *Appl. Phys. Lett.*, vol. 53, pp. 1714–1716, Oct. 31, 1988.

[6] Q. Hu and S. Feng, "Feasibility of far-infrared lasers using multiple semiconductor quantum wells," *Appl. Phys. Lett.*, vol. 59, pp. 2923–2925, Dec. 2, 1991.

[7] R. F. Kazarinov and R. A. Suris, "Possibility of the amplification of electromagnetic waves in a semiconductor with a superlattice," *Sov. Phys. Semicond.*, vol. 5, pp. 707–709, Oct. 1971.

[8] A. Seilmeier, H.-J. Hubner, G. Abstreiter, G. Weimann, and W. Schlapp, "Intersubband relaxation in GaAs-Al$_x$Ga$_{1-x}$As quantum well structures observed directly by an infrared bleaching technique," *Phys. Rev. B*, vol. 59, pp. 1345–1348, Sept. 21, 1987.

[9] W. H. Knox, C. Hirlimann, D. A. B. Miller, J. Shah, D. S. Chemla, and C. V. Shank, "Femtosecond excitation of nonthermal carrier populations in GaAs quantum wells," *Phys. Rev. Lett.*, vol. 56, pp. 1191–1193, Mar. 17, 1986.

[10] R. Ferreira and G. Bastard, "Evaluation of some scattering times for electrons in unbiased and biased single- and multiple-quantum-well structures," *Phys. Rev. B*, vol. 40, pp. 1074-1086, July 1989.

[11] M. Helm and S. J. Allen, Jr. "Can barriers with inverted tunneling rates lead to subband population inversion," *Appl. Phys. Lett.*, vol. 56, pp. 1368–1370, Apr. 2, 1990.

[12] R. Q. Yang and J. M. Xu, "Population inversion through resonant interband tunneling," *Appl. Phys. Lett.*, vol. 59, pp. 181–182, July 8, 1991.

524

[13] D. R. Herrick, "Construction of bound states in the continuum for epitaxial heterostructure superlattices," *Physica*, vol. 85B, pp. 44–50, 1977.

[14] F. H. Stillinger, "Potentials supporting positive energy eigenstates and their applications to semiconductor heterostructures," *Physica*, vol. 85B, pp. 270–276, 1977.

[15] G. N. Henderson, T. K. Gaylord, and E. N. Glytsis, "Ballistic electron transport in semiconductor heterostructures and its analogies in electromagnetic wave propagation in general dielectrics," *Proc. IEEE*, vol. 79, pp. 1643–1659, Nov. 1991.

[16] T. K. Gaylord and K. F. Brennan, "Semiconductor superlattice electron wave interference filters," *Appl. Phys. Lett.*, vol. 53, pp. 2047–2049, Nov. 21, 1988.

[17] C. Sirtori, F. Capasso, J. Faist, D. L. Sivco, S. N. G. Chu, and A. Y. Cho, "Quantum wells with localized states at energies above the barrier height: A Fabry-Perot electron filter," *Appl. Phys. Lett.*, vol. 61, pp. 898–900, Aug. 24, 1992.

[18] E. Anemogiannis, E. N. Glytsis, and T. K. Gaylord, "Bound and quasibound state and lifetime calculations for biased/unbiased semiconductor quantum heterostructures," *IEEE J. Quantum Electron.*, vol. 29, Nov. 1993. (accepted).

[19] L. D. Landau and E. M. Lifshitz, *Quantum Mechanics, Non-Relativistic Theory.* London: Pergamon Press, 1958.

[20] J. S. Blakemore, "Semiconducting and other major properties of gallium arsenide," *J. Appl. Phys.*, vol. 53, pp. R123–R181, Oct. 1982.

[21] M. T. Asom, M. Geva, R. F. Leibenguth, and S. G. N. Chu, "Interface disorder in AlAs/(Al)GaAs Bragg reflectors," *Appl. Phys. Lett.*, vol. 59, pp. 976–978, Aug. 19, 1991.

[22] J. L. Pan, L. C. West, S. J. Walker, R. J. Malik, and J. F. Walker "Inducing normally forbidden transitions within the conduction band of GaAs quantum wells," *Appl. Phys. Lett.*, vol. 57, pp. 366–368, July 23, 1990.

Evidence For LWIR Emission Using Intersubband Transitions in GaAs/AlGaAs MQW Structures

A. G. U. Perera
Department of Physics and Astronomy
Georgia State University
Atlanta GA 30303

ABSTRACT. Infrared emission using intersubband transitions in GaAs/AlGaAs multi-quantum well structures is discussed. Even for energies higher than the optical phonon values, the photon emission process which
is limited by the competition from nonradiative relaxations seems probable when the relative efficiencies are
compared. The efficiency of 10 μm emission is comparable to that of 110 μm, and could be even improved
for shorter wavelengths. Preliminary results from a two level (in the well) and a three level structure show
evidence for LWIR emission in the ranges of 8-11.5 μm and 9-14 μm respectively, which agrees with the
expected signals corresponding to 10.5μm and 11.8μm.

Infrared emission from a superlattice was first addressed by Kazarinov and Suris[1] just
after those systems were proposed by Esaki and Tsu.[2] H. C. Liu gave rough estimates for
radiative transition life times and the resonant tunneling times in proposing a novel superlattice infrared source[3] using intersubband transitions in GaAs/AlGaAs multi-quantum
well structures. More recently Helm and co-workers reported[4] low power IR emission at
wavelengths of 110, 70 and 50 μm, all of which correspond to energies lower than the optical
phonon energies. Bales et al.[5] reported optically pumped intersubband emission in a similar wavelength range. Since this IR photon emission process is limited by the competition
from non-radiative relaxations, some workers have suggested that LWIR or MWIR emission
might not be possible. However, Choe et al., have shown[6] that the <u>relative</u> efficiency of
these relaxations is the key which controls the photon emission, and both the radiative and
non-radiative relaxation rates increase at shorter wavelengths.

The efficiency of the emission process (η) can be written as[6]

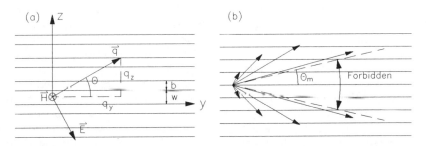

Fig. 1. (a) Wave propagation in the superlattice. b and w indicate the barrier and well thicknesses. q
is the wave vector of the photon. (b) Intensity of the radiation with direction showing the forbidden
region for emission. The arrows indicate the intensities scaled uniformly at 15° steps.

H. C. Liu et al. (eds.), Quantum Well Intersubband Transition Physics and Devices, 525–532.
© 1994 *Kluwer Academic Publishers. Printed in the Netherlands.*

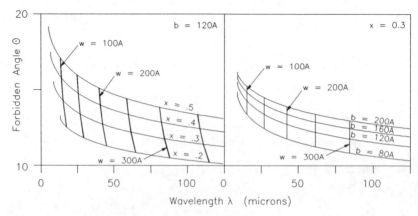

Fig. 2. Variation of the forbidden angle with the emission wavelength for different Al concentration (x) in the barrier and for different well widths (w). Also shown are the variation with the barrier thickness (b).

$$\eta = \frac{W_{\text{rad}}}{W_{\text{rad}} + W_{\text{nr}} + W_{\text{leak}}} \qquad (1)$$

where W_{nr} denotes the non-radiative (phonon) relaxation rate, W_{rad} the radiative (photon) relaxation rate and W_{leak} is the tunneling leakage rate. Liu[3; 7] and Choe et.al.,[6] both have calculated the radiative transition probability. The radiative transition rate expression obtained by Choe[6] has an angular dependence given by:

$$W_{\text{rad}} = \frac{e^2\omega^2}{12\pi\epsilon m v^3}(2 - 3\sin\theta_m + \sin^3\theta_m)f \qquad (2)$$

with

$$f = \frac{2m\omega}{\hbar}|\langle\Psi_l|z|\Psi_n\rangle|^2 \sim \frac{64}{\pi^2}\frac{l^2 n^2}{(l^2 - n^2)^3}$$

being the oscillator strength[8], ϵ the optical permittivity, ω, the angular frequency of the emitted photon, v the velocity of light in the superlattice, and m the effective mass of the electron. The angle θ_m denotes the forbidden region where no emission is possible. It is evident from the above equation, that the angle θ (see fig. 1) plays a crucial role in the device design restricting the emission angle. For example, a simple cascade emitter device designed for 10μm emission, (with device parameters $x \sim 0.3$, $a = 80$ Å, $b = 120$ Å) yields a forbidden angle $|\theta| \le 13°$, in which case $\theta_m \simeq 13°$. In addition to the emission wavelength, θ_m is sensitive to the barrier thickness as seen in fig. 2. Another, interesting feature of equation 2 is that it has a square dependence on the frequency of the emitted photon. As the wavelength becomes shorter the radiative decay rate increases, which helps in the competition with the non-radiative (phonon) processes. Using available experimental values and very conservative extrapolations we estimated the relaxation rates for non-radiative processes. WKB was used to obtain the tunneling rates for the leak. Since the nonradiative relaxation times are much faster than the other two possibilities, we can simplify eq. 1 and write

$$\eta \simeq \frac{W_{\text{rad}}}{W_{\text{nr}}}$$

λ	τ_{rad}	$\tau_{non-rad}$	Rel. Eff.	θ_m
111 μm [4]	21 μs	$\sim$ 100 ps (Acoustic)	1	8.0°
10 μm	.20 μs	$\sim$ 1 ps (Optical)	1	13.0°
5 μm	.06 μs	$\sim$ 2 ps (Optical)	6	18.4°

TABLE I

Comparison of Relative Efficiency and the forbidden angle for three different emission wavelengths. Also given are the radiative and non-radiative decay rates.

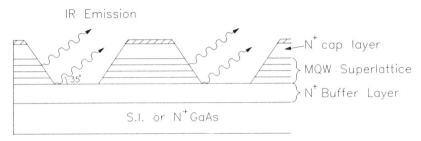

Fig. 3. Etched surface with {111} faces allowing emission near optimum angle.

The values in Table I shows the relative efficiencies for different IR wavelengths. As seen from the table, the efficiency of 10 μm emission is almost similar to that of 110 μm, and could be even improved for shorter wavelength emission.[6]

The angular restrictions mentioned above will make the edge emission very weak as shown in fig. 1(b). In order to overcome this problem, one needs to have slanted edges or have suitable optical surface gratings. In order to have efficient coupling, we have etched a surface grating structure with a period of about 4.5μm.[4] This is shown in fig. 3, and was obtained by crystallographically etched grooves delineated in a thick layer of undoped GaAs grown on top of the upper doped GaAs contact layer.

Keeping the above design concerns in mind, we have fabricated several GaAs/AlGaAs emitter structures. The design and post fabrication parameter values for the two samples which showed positive results are given in table II. The first structure which showed a positive result was a standard two level cascade structure as shown in fig. 4. The current voltage characteristics showed negative differential resistance indicating sequential resonant tunneling as seen in fig. 4 (b).

This two level structure tested positive for IR emission in the range from 4 to 14μm. The emitter sample was immersed in liquid helium, and a HgCdTe detector was mounted next to the sample with a cold shutter and a cold chopper blade in between them. The chopper was driven by a motor outside the dewar with another similar chopper blade to obtain the chopping frequency. (See fig. 5) As seen in fig. 6, the detector output increased as the emitter sample bias voltage was increased. When the emitter bias was disconnected no signal was detected. This indicates IR emission from the emitter sample in the range 4-14 μm (due to the bandpass filter in the detector path). Since the signal keeps on increasing with increasing bias the radiation could have been due to thermal heating of the sample. However, by reducing the detection range to 8 -11.5μm , a peak in the detected signal was observed with increasing bias to the emitter sample. This signal was weaker than before,

Specified Parameters				Post Fabrication Values				Well Doping
Barr. Å	Well Å	x	λ μm	Barr. Å	Well Å	x	λ μm	cm^{-3}
120	80	.35	9.8	119	80	.34	10.0	10^{17}
120	130	.5	11.7	122	130	.51	11.7	10^{17}

TABLE II

Two emitter samples (which showed signs of emission) fabricated and tested showing the design and post fabrication parameters.

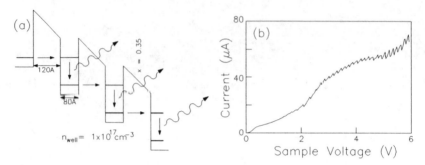

Fig. 4. (a) The quantum well structure fabricated for 11μm emission under bias showing the alignment of the energy levels. The design parameters are also given. Post fabrication measured values are compared in table II. (b) Current-voltage characteristics showing the energy level alignments for a 0.5 mm square mesa.

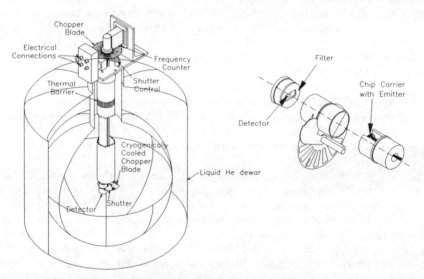

Fig. 5. The low temperature apparatus to detect IR emission from the multi-quantum well sample using the cooled chopper and the HgCdTe detector. The sample and detector area is drawn separately for clarity. The chip carrier can be used to mount the sample at different angles.

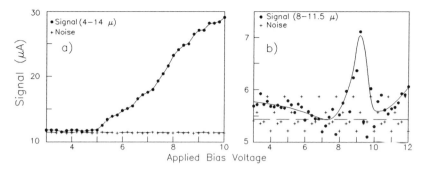

Fig. 6. a) The detected signal due to infrared emission in the range of 4-14 μm with increasing bias to the emitter sample. The dashed line is the signal when a cold shutter blocks the radiation. As the emitter bias is increased above 5 volts, the detected signal increases as expected. b) The detected signal with the narrow bandpass filter (8-11.5μm) in the path indicating the peak around 9 volts bias.

as expected, due to the reduction of the detector field of view, and possibly, removal of any thermal or other contribution outside the range. If the radiation detected is due to thermal heating it should not decrease with increasing bias. The peak indicates that with increasing bias the alignment of the energy levels starts to breakup, reducing the emission. The estimates from Wien's law gives 240 K and 360 K as temperatures which should produce the blackbody radiation peaks at 12 and 8μm wavelengths. Since the sample is immersed in liquid helium it is unlikely that the whole sample can be at a higher temperature except for a difference of few degrees in microscopic regions.

In this 2 level structure (see fig. 4) the number of ground state electrons should be less and the lifetime in the ground state should be smaller compared to the lifetime in the excited state for efficient relaxation. However at resonance, an excited state depends on the previous well ground state for the electrons for relaxation. In order to avoid this, a three level structure was fabricated as seen in fig. 7, where the emission is due to the relaxation from the second excited state to the first excited state.

A 3-level structure may be better than a 2-level structure for emission, since the required relaxation in the 3 level structure is not to the ground state but to the 1st excited state, which is less populated than the ground state. In this 3-level structure, the ground state of the nth (E_0) well will be aligned with the 2nd excited state (3rd level, E_2) of the adjoining $(n+1^{th})$ well under bias, to give resonant tunneling. (A fourth level is ignored since under bias it will be pushed to the continuum.) Then the tunneled electrons will relax via a direct or a phonon assisted transition both contributing to the photon emission. Usually the energy difference $E_2 \rightarrow E_1$ is greater than $E_1 \rightarrow E_0$, which corresponds to 106 meV (or 11.7μm) and 64 meV respectively for our structure. Therefore $E_2 \rightarrow E_1$ transition corresponds to a higher order process (slower) compared to the $E_1 \rightarrow E_0$ transition. As seen in fig. 8, the stronger peak observed in this case with a (9-14 μm) bandpass filter, is an indication of a stronger IR signal compared to the previous 2 level sample. The slope of the dashed line indicates a fall in the baseline which may be due to the liquid helium level dropping inside the dewar and associated increase in the chopper frequency. Observing a stronger signal on top of the slope is an added indication of a stronger signal which might be associated with population inversion. This is possible since the first excited state will be usually less

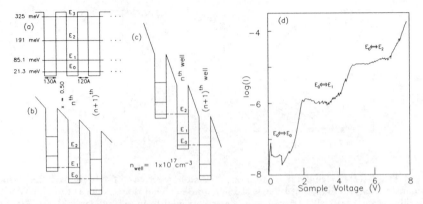

Fig. 7. (a) The three level emitter structure under zero bias. (b) The structure under bias showing the alignment of two adjacent well energy levels E_1 and E_0. (c) Alignment of the energy levels E_0 and E_2 used in emission (d) Current-voltage characteristics for the sample showing the plateaus indicating the different energy level alignments corresponding to different transitions.

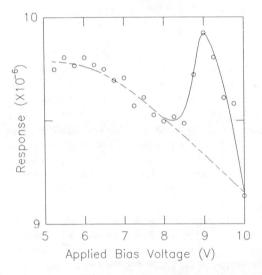

Figure 8: Detected IR (9-14 μm range) signal with increasing bias voltage showing a peak. The signal value when a low temperature shutter was used to block the emission was about 1.7×10^{-6} throughout the same bias voltage range.

populated than the second excited state due to resonant tunneling and faster relaxation. This can open the possibility for stimulated emission. At present, we are in the process of obtaining a spectral scan by designing a low temperature measurement set up with a circular variable filter instead of the bandpass filter.

The structure shown in fig. 9 is a good candidate for 10.7 μm emission. A single unit of the structure consists of 3 barriers and 2 wells. The first barrier is low and the contact is heavily doped for efficient carrier input to the excited state of the first well. The second barrier thickness is selected to reduce tunneling from the excited state but at resonance to have fast tunneling out from the ground state. The height of the second barrier is selected so that the excited (or continuum) states will not align with the excited state of the first well under bias. Following a method discussed by Choe[9], (which is simpler but faster than solving the

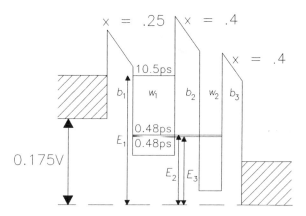

Fig. 9. The structure suggested for 10.7μm emission with two wells and 3 barriers as a unit, under a bias of 0.175V. The barrier widths are $b_1 = 52$Å, $b_2 = 50$Å, and $b_3 = 30$Å. The well widths are $w_1 = 85$Å and $w_2 = 45$Å. The resulting energy levels are $E_1 = 262$meV, $E_2 = 152$meV, and $E_3 = 149$meV. The tunneling time for the excited state in the first well is 10.5 ps, while the resonance condition leads to the fast removal of electrons from the ground state within 0.48 ps time interval. Off resonance (.1V bias) this time will increase to 0.82 ns seconds while the excited state tunneling time increases only to 21.8 ps. The doping in the well is assumed to be low ($\sim 1 \times 10^{17}$ cm^{-3}). The other device parameters are also shown.

time dependent Schrodinger equation[10]) to find the pole of the scattering matrix in the complex energy plane, for a structure (see fig. 9) with $w_1 = 85$, $w_2 = 45$ Å wells, and barriers with 0.25, 0.4, and 0.4 Al mole fraction (x) and $b_1 = 52$, $b_2 = 50$, and $b_3 = 30$Å thicknesses. We find 10.5 ps tunneling times for the first excited state electrons, while the ground state electrons will tunnel out in 0.48 ps at resonance. If the resonance is not met, for e.g. under a 0.1V bias, the ground state tunneling time will increase to 0.82 ns while the excited state tunneling time increases only to 21.8 ps. Due to this resonant tunneling, the ground state will appear almost empty for the relaxing excited state electrons which could lead to population inversion.

For a single three barrier unit structure, shown in fig. 9, increasing b_3 increases the lifetime of all the states while leaving the ratio of the lifetimes nearly unchanged. Increasing b_2 causes the condition for resonance to become much more critical. When $b_1 = 105$Å and $b_2 = 100$Å the ground states are not fully in resonance even when separated by 1 meV. The lifetime of one of the ground states is also increased off resonance. The use of multiple three barrier units in a structure could provide a means of increasing the operating voltage of the device and reducing the precision needed for the bias voltage.

Acknowledgements

This work was originated when the author was in Pittsburgh (at University of Pittsburgh and Microtronics Associates), under SBIR support from U. S. Army CECOM and NSF under grant ECS 90-06078. At present, the project is being supported by NSF grant ECS-92-96238. Author acknowledges the contributions of J.-W. Choe, Y. F. Lin, M. H. Francombe, S. Matsik and H. Yuan to this work. Earlier samples were fabricated and processed by Y. F. Lin at Westinghouse and recently by K. L. Wang and R. P. G. Karunasiri at UCLA and

532

W. Schaff at Cornell and H. C. Liu at NRC.

References

1. R. F. Kazarinov and R. A. Suris, Sov. Phys.-Semicond. **5**, 707 (1971).
2. L. Esaki and R. Tsu, IBM J. Res. Develop. **14**, 61 (1970).
3. H.C. Liu, J. Appl. Phys. **63**, 2856 (1988).
4. M. Helm, P. England, E. Colas, F. DeRosa, and S. J. Allen, Jr., Phys. Rev. Lett. **63**, 74 (1989).
5. J. W. Bales, K. A. McIntosh, T. C. L. G. Sollner, W. D. Goodhue and E. R. Brown, *Observation of Optically Pumped Intersubband Emission from Quantum Wells* in **Quantum Well and Superlattice Physics III**, SPIE, Bellingham, Wash., USA, 1990, pp. 74-81.
6. J.-W. Choe, A. G. U. Perera, M. H. Francombe, and D. D. Coon, Appl. Phys. Lett. **59**, 54 (1991).
7. H. C. Liu, J. Appl. Phys. **69**, 2749 (1991).
8. L. C. West and S. J. Eglash, Appl. Phys. Lett. **46**, 1156 (1985).
9. J.-W. Choe, et al., *Optimization of Infrared Emission in a Superlattice Cascade Process*, unpublished.
10. C. Juang, Phy. Rev. B **44**, 10706 (1991).

FAST DATA CODING USING MODULATION
OF INTERBAND OPTICAL PROPERTIES
BY INTERSUBBAND ABSORPTION IN QUANTUM WELLS

Vera B. Gorfinkel[a] and Serge Luryi[b]

[a]University of Kassel, Kassel, Germany
[b]AT&T Bell Laboratories, Murray Hill, NJ 07974

ABSTRACT. Intersubband absorption of radiation by a two-dimensional electron gas can be used to control the electron temperature and effect a significant modulation of the interband optical properties of the semiconductor in the quantum well. We discuss the implementation of a fast modulator of infrared radiation for fiber-optical communications as well as the formation of powerful and short single-mode infrared pulses. The method can also be used to modulate laser radiation by controlling simultaneously the pumping current and the optical gain in the active region. The dual modulation method allows to eliminate the relaxation oscillations and suppress the wavelength chirping in optical communication systems operating at high pulse repetition rates.

1. Introduction

It is well known that the interband (IB) absorption coefficient g in a degenerate semiconductor is a strong function of the effective carrier temperature T_e. It has been proposed[1] to use this effect for a rapid control of the semiconductor-laser gain by driving a lateral electron-heating current through the active region. Inasmuch as T_e can be independently controlled and rapidly modulated, one can envisage a number of useful applications in optical communications. We shall review some of the recent ideas for such applications, specializing to the case when the control of T_e is effected by intersubband absorption.[2] The idea is to control the optical transparency of a multiple quantum well (QW) waveguide by modulating the effective temperature of free carriers in the wells.

2. Control of Electron Temperature by Intersubband Absorption

Consider a multiple quantum well (MQW) irradiated by infrared photons $\hbar\omega$ (e.g. from a CO_2 laser, $\hbar\omega = 124$ meV) nearly resonant (within $\Delta\omega$) with the intersubband energy. The waveguide structure is illustrated in Fig. 1. The interband (IB) radiation propagates along x and the intersubband (ISB) along y directions. The ISB radiation in the waveguide is assumed TM polarized and its photon flux inside a QW will be denoted by Φ_ω. In a steady state, T_e is determined by the balance equation

H. C. Liu et al. (eds.), Quantum Well Intersubband Transition Physics and Devices, 533–545.

534

$$k \Delta T_e = w \tau_\varepsilon \hbar \omega \, , \tag{1}$$

where τ_ε is the energy relaxation time of QW electrons, $\Delta T_e \equiv T_e - T$, and w is the ISB transition rate. Evaluating w by the Golden rule, we have[3]

$$w = \frac{e^2 Q}{2\gamma} \frac{R_0}{m \, \bar{n}} \, \Phi_\omega \, \frac{\gamma^2}{(\Delta\omega)^2 + \gamma^2} \, , \tag{2}$$

where $R_0 \equiv 377 \, \Omega$ is the vacuum impedance, m the electron effective mass, and $\bar{n}$ the refractive index; $Q \approx 1$ is the oscillator strength and γ the width of the ISB resonance. A typical power flux density $\mathcal{P}_\omega = 10 \, \text{kW}/\text{cm}^2$ of a CO_2 laser corresponds to $\Phi_\omega = \mathcal{P}_\omega/\hbar\omega \approx 5 \times 10^{23} \, \text{cm}^{-2} \, \text{s}^{-1}$. We shall be considering the case of an InGaAs QW ($m = 0.041 \, m_0$). Taking $\hbar\gamma \approx 4 \, \text{meV}$ (from the data[4] in GaAs QW at 300 K) and $\tau_\varepsilon \approx 6 \, \text{ps}$ (from the energy loss rate of $7 \, \text{meV/ps}$ measured[5] in InGaAs/InP QW at $T_e = 500 \, \text{K}$ for high carrier densities), Eq. (1) gives $\Delta T_e \approx 260 \, \text{K}$ at resonance. Even longer hot electron-hole plasma cooling times (corresponding to $\tau_\varepsilon \approx 20 \, \text{ps}$) have been reported[6] in InGaAs/InAlAs MQW at $T_e \lesssim 700 \, \text{K}$ and sheet carrier concentrations of $2 \times 10^{12} \, \text{cm}^{-2}$. The relatively long energy relaxation time is usually explained by phonon bottleneck effects.[7]

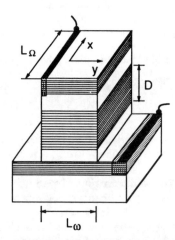

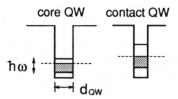

Fig. 1: Schematic diagram of the modulator (after Ref. 2). Dimensions: $L_\Omega = 500 \, \mu\text{m}$, $L_\omega = 3 \, \mu\text{m}$, $D = 1.5 \, \mu\text{m}$. The core (D) contains 50 QW's of width $d_{QW} \approx 100 \, \text{Å}$ (intersubband separation $\hbar\omega \approx 0.12 \, \text{eV}$). Narrow contact QW's do not absorb TM-polarized ISB radiation.

A desired variation of T_e can be accomplished by varying either the intensity Φ_ω or the frequency $\Delta\omega$. At a constant CO_2 power and frequency it is possible to vary the detuning $\Delta\omega$ by Stark shifting the ISB resonance. For this purpose, we need means for applying an electric field in the z direction. To minimize the loss of ISB power by free-carrier absorption in the contact layers controlling the electric field, the use of doped bulk regions should be avoided. Instead, the contact layers can themselves be multiple quantum wells, doped or modulation-doped to a high conductivity and narrow enough that the intersubband resonance in the contact QW's is far above the ISB photon energy $\hbar\omega$.

While the relatively long energy relaxation time enhances the amplitude of the T_e response under quasi-static modulation, cf. Eq. (1), it degrades the high-frequency characteristics (above $1/\tau_\varepsilon$). The deviation $\Delta T \equiv T_e - T$ of the carrier temperature T_e from the ambient temperature T satisfies an equation of the form

$$\frac{d\Delta T}{dt} = G - \frac{\Delta T}{\tau_\varepsilon} , \tag{3}$$

where $G(t) \equiv \hbar\omega w(t)$ is the absorbed power. Solution of Eq. (3), satisfying $\Delta T = 0$ at $t = 0$, is given by

$$\Delta T(t) = e^{-t/\tau_\varepsilon} \int_0^t G(t') e^{t'/\tau_\varepsilon} dt' . \tag{4}$$

For a sinusoidal $G(t) = \frac{1}{2} G_0 (1 - \cos\omega t)$, we have

$$\Delta T(t) = \frac{1}{2} G_0 \tau_\varepsilon [1 - \cos\gamma \cos(\omega t - \gamma) - e^{-t/\tau_\varepsilon} \sin^2\gamma] , \tag{5}$$

where γ is defined by $\tan\gamma \equiv \omega\tau_\varepsilon$. The quantity $T_\infty \equiv G_0 \tau_\varepsilon$ represents the temperature by which T_e would be raised by a constant G_0 equal to the signal amplitude. Figure 2 illustrates the time-dependent heating of carriers by a sinusoidal ISB irradiation, $\Phi_\omega \sin^2(\pi f t)$, where we assume $\mathcal{P}_\omega \equiv \hbar\omega \Phi_\omega = 10^4 \, \text{W}/\text{cm}^2$.

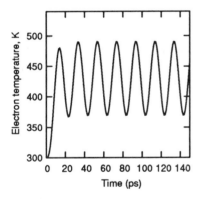

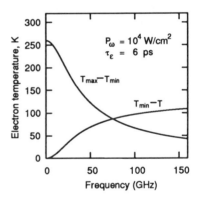

Fig. 2a: Evolution of the electron temperature T_e under CO_2 laser irradiation with $f = 50 \, \text{GHz}$. For $t < 0$ we assume $T_e = T_0 = 300 \, \text{K}$.

Fig. 2b: Frequency dependence of the T_e modulation amplitude and of the minimum T_e during a period for a sinusoidal modulation.

Next consider an input signal $G(t)$ consisting of a series of Gaussian pulses

$$G = \sum_{n=0}^{\infty} G_0(t - n\mathcal{T}) , \quad \text{where} \quad G_0(t) = (T_\infty/\tau_\varepsilon) e^{-t^2/2\Delta t^2} \tag{6}$$

and $\mathcal{T}$ is the period, $\mathcal{T} = 1/f$, with f being the pulse repetition rate. Assuming that $\Delta T(t) = 0$ for $t < 0$, we can use Eq. (4) to find:

$$\Delta T(t) = \sqrt{\frac{\pi}{2}} \; T_{\infty} \; \frac{\Delta t}{\tau_{\varepsilon}} e^{\Delta t^2/2\tau_{\varepsilon}^2} \sum_{n=0}^{\infty} e^{(nT-t)/\tau_{\varepsilon}}$$

$$\times \left[\Phi \left[\frac{nT + \Delta t^2/\tau_{\varepsilon}}{\Delta t \sqrt{2}} \right] + \Phi \left[\frac{t - (nT + \Delta t^2/\tau_{\varepsilon})}{\Delta t \sqrt{2}} \right] \right], \tag{7}$$

where Φ is the error integral $[\Phi(x) \to 1$ for $x \geq 1$ and $\Phi(-x) = -\Phi(x)]$. The sum of two Φ's in the square brackets decreases with n much faster than $\exp(-nT/\tau_{\varepsilon})$, so the overall sum converges very rapidly. Results of calculations with the formula (7) are plotted in Fig. 3.

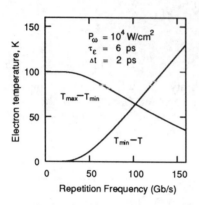

Fig. 3: Frequency dependence of the T_e modulation amplitude and the minimum T_e during a period for a sequence of Gaussian pulses of width $\Delta t = 2$ ps.

It is evident from Figs. 2b and 3 that the modulation amplitude drops at high frequencies and the mean T_e rises. We see that the coding speed of a T_e modulator is limited by ~50 Gb/s. At higher bit rates, the temperature of QW electrons does not have time to relax to the lattice temperature T_0 in between two heating pulses.

It can be shown that when the pulse is sufficiently narrow $\Delta t \ll \tau_{\varepsilon}$ then the difference $T_{max} - T_{min}$ does not depend on the repetition rate. This is easy to demonstrate as follows: for $\Delta t \ll \tau_{\varepsilon}$, we can replace the Gaussian by a δ function:

$$e^{-\frac{(t-nT)^2}{2\Delta t^2}} \approx \sqrt{2\pi} \; \Delta t \; \delta(t - nT).$$

For a single δ-function pulse, the response is of the form

$$\Delta T_n(t) = T_{\infty} \sqrt{2\pi} \; \frac{\Delta t}{\tau_{\varepsilon}} \; e^{(nT-t)/\tau_{\varepsilon}} \; \Theta(t - nT), \tag{8}$$

where $\Theta(x)$ is a unit step function. We must add responses to all pulses preceding t. For $t \gg T/\tau_{\varepsilon}$, the summation can be extended to ∞, with the result

$$\Delta T(t) = T_{\infty} \sqrt{2\pi} \; \frac{\Delta t}{\tau_{\varepsilon}} \; \frac{e^{-t_T/\tau_{\varepsilon}}}{1 - e^{-T/\tau_{\varepsilon}}}, \tag{9}$$

where t_T should be understood as the "residual time" past the last pulse, $t_T = t \, [\text{mod} \, T]$. Equation (9) can also be derived directly from (7). Any function of t_T is obviously periodic in t. The minimum occurs for $t_T = 0$ and the maximum for

$t_\tau = T - \varepsilon$, where ε is an infinitesimal. From Eq. (9) we find that $T_{min} \approx T$ for low frequency, $T \geq \tau_\varepsilon$, and $T_{min} \propto f$ for $T = 1/f \ll \tau_\varepsilon$. The T_e response amplitude equals $T_{max} - T_{min} = T_\infty \sqrt{2\pi} \, \Delta t/\tau_\varepsilon$ and does not depend on the repetition rate.

3. Control of the Gain in Steady State

Device length L_ω in the y direction should be chosen short enough that Φ_ω be approximately uniform, $\alpha L_\omega < 1$, where α is the ISB absorption coefficient,

$$\alpha = \frac{r \Gamma_\omega n_S}{d_{QW}} \frac{w}{\Phi_\omega} = \frac{\Gamma_\omega n_S}{d} \frac{e^2 Q}{2\gamma} \frac{R_0}{m \, \bar{n}} \frac{\gamma^2}{(\Delta\omega)^2 + \gamma^2} , \tag{10}$$

$d = d_{QW} + d_B$ is the MQW period, $d_{QW} \equiv r d$ the QW thickness, n_S the electron sheet concentration per period, and Γ_ω the confinement factor for the ISB radiation intensity. For an efficient operation of the modulator it is important that the steady-state density p_S of holes generated by the IB radiation be small compared to $n_S = n_0 + p_S$. The equilibrium electron density n_0 (introduced by doping) is limited by the requirement that the Fermi level be less than the ISB separation, $E_F < \hbar\omega$. At $n_0 = 2 \times 10^{12}$ cm^{-2} in the range of 300 K to 500 K one has $E_F \approx 115$ meV. Higher n_0 would result in a diminishing efficiency of carrier heating due to increasing population of the second subband.

The modulator length L_Ω in the x direction should be chosen so as to achieve a desired modulation depth of the transmitted IB beam $e^{r \Gamma_\Omega g L_\Omega}$, where Γ_Ω is the waveguide confinement factor for the IB radiation intensity. The gain function g is of the form

$$g(T_e, T_h, n_S, p_S, \hbar\Omega) = g_{max} (f_e + f_h - 1) , \tag{11}$$

where f_e and f_h are the Fermi functions of electrons and holes, respectively, at energies selected by the incident photons $\hbar\Omega$ inducing transitions between the heavy-hole and the lowest electron subbands. The value of g_{max} in an InGaAs QW is, typically,[8] $g_{max} \approx 10^3$ cm^{-1}. For transitions at the fundamental absorption edge in the QW, the Fermi factors are given by

$$f_e(n_S, T_e) = 1 - e^{-\pi\hbar^2 n_S/m \, kT_e} ; \tag{12a}$$

$$f_h(p_S, T_h) = 1 - e^{-\pi\hbar^2 p_S/m_h \, kT_h} , \tag{12b}$$

where $m_h \approx 0.5 m_0$ and T_h are the heavy-hole effective mass and temperature, respectively. Equations (12) are derived from the exact expression for the Fermi integral,

$$n_S = \frac{m \, kT_e}{\pi\hbar^2} \ln\left[1 + e^{E_F/kT_e}\right] , \tag{13}$$

in a 2D electron gas with one subband occupied. The electron quasi-Fermi level is

$$E_F(n_S, T_e) = kT_e \ln\left[e^{\pi\hbar^2 n_S/m \, kT_e} - 1\right] , \tag{14}$$

and the occupation probability at the subband bottom is given by

$$f_e (n_S, T_e) \equiv \left[1 + e^{-E_F/kT_e} \right]^{-1} = 1 - e^{-T_n/T_h} . \tag{15}$$

For electrons in InGaAs ($m_e = 0.041\, m_0$) the temperature scale in the exponent equals

$$T_n \equiv \frac{\pi \hbar^2 n_S}{m_e k} = 676.62\,K \times \frac{n_S\,[cm^{-2}]}{10^{12}} \tag{16}$$

Similarly, for holes ($m_h = 0.5\, m_0$) one has $f_h\, (p_S, T_h) = 1 - \exp(-T_p/T_h)$, where

$$T_p \equiv \frac{\pi \hbar^2 p_S}{m_h k} = 55.5\,K \times \frac{p_S\,[cm^{-2}]}{10^{12}} \tag{17}$$

Except at cryogenic temperatures, the 2D hole gas is nondegenerate even at $10^{12}\,cm^{-2}$.

The modulator can be expected to perform up to frequencies limited by the inverse energy relaxation time τ_ε, provided the slower processes associated with carrier generation by the IB radiation make negligible contribution to g. At a given value of n_0 the latter requirement puts a limit on the IB flux that is modulated.

To estimate this limit and calculate the temperature dependence of g in a steady-state, we consider the rate equations:

$$\frac{dp_S}{dt} = - \bar{c}\, g\, S - R_S\, n_S\, p_S\,; \tag{18a}$$

$$\frac{dS}{dt} = (r\Gamma_\Omega)(\bar{c}\, g) S - \tau_{ph}^{-1} (S - S_0) . \tag{18b}$$

Here S is the photon density per unit area in a single QW, $S_0 \equiv \Phi_\Omega\, d_{QW}/\bar{c}$, where Φ_Ω is the incident IB photon flux, $\bar{c} \equiv c/\bar{n}$ is the speed of light, and $\tau_{ph} = L_\Omega/\bar{c}$. The radiative recombination coefficient $R_S \equiv B/d_{QW}$, where $B \approx 10^{-10}\,cm^3/s$.

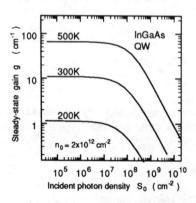

Fig. 4: Dependence of the steady-state gain on the photon density S_0 of incident interband radiation at different carrier temperatures. The assumed parameters: $g_{max} = 10^3\,cm^{-1}$, $r\Gamma_\Omega = 0.3$, and $R_S = 10^{-4}\,cm^2/s$.

Figure 4 shows the stationary gain as a function of incident power, calculated from Eqs. (18) at several carrier temperatures, assuming $n_S = n_0 + p_S$ with $n_0 = 2 \times 10^{12}\,cm^{-2}$ and $\tau_{ph} = 5\,ps$ ($L_\Omega \approx 500\,\mu m$). We take $T_h = T_e$, rather arbitrarily. Results of our calculation are not very sensitive to the choice of the hole temperature in the range $T \leq T_h \leq T_e$. It is evident from Fig. 4 that for $S_0 \leq 10^8\,cm^{-2}$ carrier generation effects can be neglected. The total modulated power $\mathcal{P}_\Omega = \Gamma_\Omega^{-1}\, \mathcal{A}\, \Phi_\Omega\, \hbar\Omega$ is related to S_0 by

$$r \Gamma_\Omega \mathcal{P}_\Omega = N L_\omega \hbar \Omega \bar{c} S_0 ,$$

where $\mathcal{A} = D \times L_\omega$ is the MQW cross-sectional area, D the core thickness, and $N = D/d$ the number of periods. Taking $N = 50$, $r \Gamma_\Omega \approx 0.3$, and $L_\omega \approx 3 \mu m$, we find that $S_0 = 10^8 \, cm^{-2}$ corresponds to a maximum modulated power of $P00 \approx 6 \, mW$. For $\mathcal{P}_\Omega \approx 1 \, mW$, the steady-state g varies from $g \approx -9 \, cm^{-1}$ at $T_e = 300 \, K$ to $g \approx -61 \, cm^{-1}$ at $T_e = 500 \, K$. For $L_\Omega = 0.5 \, mm$ and $r \Gamma_\Omega \approx 0.3$ this corresponds to a 3.5 dB modulation.

At a higher power (or lower n_0) the modulator efficiency suffers from self-induced transparency effects associated with the accumulation of electrons and holes. In the next Section, we show that these effects can be used advantageously for the formation of short high-power IB radiation pulses. For this purpose, the MQW need not be doped.

4. Formation of short pulses

Consider the situation arising at a high S_0 in the presence of ISB absorption. In what follows, we assume that the MQW is undoped, $n_0 = 0$. In the steady state there is a large number of electrons and holes $p_S (S_0, T_e)$, readily evaluated from Eqs. (18). In this state the gain has a small negative value $g (S_0, T_e)$. If the carrier heating is now abruptly terminated (by chopping Φ_ω or by shifting the ISB resonance with an external electric field) then T_e rapidly goes down to the ambient temperature and the gain function becomes temporarily positive. The excess carriers undergo stimulated recombination accompanied by a large pulse in the IB photon density. An example of such a pulse is shown in Fig. 5 The pulse shape is calculated from Eqs. (18), assuming that the initial electron heating is stopped at $t = 10 \, ps$ and the carrier temperature relaxes from $T_e = 500 \, K$ to $T_e = 300 \, K$, according to $\Delta T_e (t) = \Delta T_e e^{-t/\tau_\varepsilon}$ with $\tau_\varepsilon = 6 \, ps$. At longer times t the carrier density and the gain approach their new steady state values $p_S (S_0, T)$ and $g (S_0, T)$, respectively.

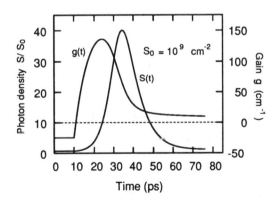

Fig. 5: Formation of short pulses by abrupt termination of carrier heating. In the presence of both IB and ISB radiations, the device is allowed to reach a steady-state with assumed carrier temperature $T_e = 500 \, K$. At $t = 10 \, ps$ the ISB absorption terminates abruptly and T_e relaxes to the ambient temperature $T = 300 \, K$. Full width at half maximum of the pulse S is 15 ps.

The pulse shown in Fig. 5 has a full width at half maximum $\Delta t_{fwhm} = 15 \, ps$ and a peak photon density of $4 \times 10^{10} \, cm^{-2}$ – corresponding to a power of 2.4 W (the pumping level is only 60 mW). It is interesting to note that the peak power varies only by 30 % over the decade variation in the pump power $10^9 \leq S_0 \leq 10^{10}$ cm^{-2}. The width Δt_{fwhm} is practically constant over the same range. Increasing S_0 in this range mainly leads to a faster device recovery in preparation for the next pulse.

Indeed, when the carrier heating is turned on, the negative gain function temporarily increases in magnitude. This results in an enhanced absorption of IB radiation and the number of carriers increases back toward the steady-state value $p_S(S_0, T_e)$. During this relatively slow process the power is stored for the next pulse. The storage time depends on the incident power and scales approximately as $1/S_0$ for $S_0 \gtrsim 10^9$ cm^{-2}. For a given value of S_0 and a given swing in the carrier temperature, the peak power is inversely proportional to its duration.[2]

5. Fast Data Coding

High-frequency characteristics of the modulator for fast data coding depend on the inertia of the response of T_e to the heating signal, discussed in Sect. 2, and on the inertia associated with carrier accumulation and depletion in the MQW, discussed in Sect. 3. For error-free information coding, it is essential that the response to an isolated pulse should be similar to that to a sequence of pulses following each other at a given bit rate. As we have seen in Fig. 3, the speed of encoding the electron temperature is limited by ~50 Gb/s − even without a consideration of the optical response. For not too high optical IB power, this turns out to be also the coding speed limitation of a T_e controlled optical modulator.

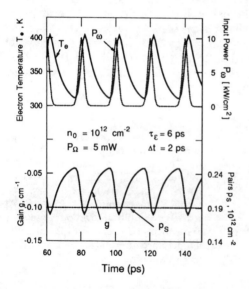

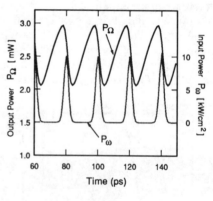

Fig. 6: Modulator response to a 50 Gb/s sequence of 2 ps pulses.

Figures 6 show the calculated room-temperature response of the modulator − the carrier temperature $T_e(t)$, the density $p_S(t)$ of photogenerated pairs, the optical gain $g(t)$, and the transmitted IB light power $P_{out}(t)$ − subject to a periodic sequence of heating pulses of the form $\mathcal{P}_\omega \exp(-t^2/\Delta^2)$, with $\mathcal{P}_\omega = 10^4$ W/cm^2 and $\Delta = 2$ ps, repeated at the rate of 50 GHz. At this repetition frequency, for not too high incident IB power (≤ 35 mW), the response is nearly identical to that for an isolated pulse of the same form. This means that the effect of accumulation of photogenerated pairs, produced during the the passage of a heating pulse, is relatively small ($p_S \ll n_S = n_0 + p_S$), where n_0 is the electron concentration in the QW's in the absence of IB radiation. The optimum value of n_0 was found to be about 10^{12} cm^{-2}. For the incident power above ~ 35 mW, the accumulation of pairs becomes important, so that the maximum coding rate becomes limited by the recombination processes and drops below 10 Gb/s.

6. High-frequency modulation and suppression of chirp in semiconductor lasers

The common method of modulating the laser radiation amplitude by varying the pumping current suffers from two drawbacks. First, it is limited to relatively low frequencies ($f \leq 10$ GHz). Second, for $f \geq 1$ GHz, it is plagued by oscillations in the wavelength of the dominant mode (chirp). Both of these problems arise from the relaxation oscillations due to an intrinsic resonance in the nonlinear laser system (the electron-photon resonance).

An alternative principle for modulating the laser output is to directly control by external means the gain coefficient g_0 of the active medium by varying the effective carrier temperature T_e in the laser active region.[1] High-frequency modulation of T_e by several tens of degrees has been demonstrated experimentally,[9] by driving an electric current through the active region. Although this method in principle allows a faster laser modulation, by itself it eliminates neither the relaxation oscillations nor the frequency chirp. The new approach, proposed in our recent work,[10] allows an enhancement of the coding frequency, suppressing the chirp at the same time. The key idea consists in a coherent combination of *two* independent means of controlling the output radiation: the pumping current I and the effective carrier temperature T_e.

Consider a semiconductor laser, subject to a time-dependent pumping current $I(t)$ and an external electron-heating power $G(t)$. The nonlinear system can be described by the standard rate equations[8] for the carrier density n and the photon density S per unit volume:

$$\frac{dn}{dt} = J - S\tilde{g} - Bn^2 \; ; \tag{19a}$$

$$\frac{dS}{dt} = (\Gamma\tilde{g} - \tau_{ph}^{-1})S + \beta B n^2 \; , \tag{19b}$$

coupled with an energy balance equation (3) for T_e. In these equations, J is the electron flux per unit volume of the active layer, $B \approx 10^{-10}$ cm^3/s is the radiative recombination coefficient, $\beta \approx 10^{-4}$ the spontaneous emission factor, and $\tilde{g} \equiv g\,\bar{c}$ the optical gain [sec^{-1}] in the active layer.

We shall use the notation, $X(t) \equiv \bar{X} + \hat{X}e^{i\omega t}$, for harmonically varying quantities $X(t)$. Let us carry out a small-signal analysis of the system (19), linearizing it about a steady state well above the lasing threshold:

$$i\omega\,\hat{n} = \hat{J} - \bar{S}\,\hat{g} - \bar{g}\,\hat{S} - 2B\,\bar{n}\,\hat{n} \;; \tag{20a}$$

$$i\omega\,\hat{S} = (\Gamma\tilde{g} - \tau_{ph}^{-1})\,\hat{S} + \Gamma\bar{S}\,\hat{g} + 2\beta B\,\bar{n}\,\hat{n} \;. \tag{20b}$$

Assume that the absorbed power $G(t)$ per carrier depends only on the ISB absorption rate w, Eq. (2), the variation of T_e is of the form

$$\hat{T}_e = \frac{G_w'\,\hat{w}\,\tau_{\varepsilon}}{1 + i\omega\tau_{\varepsilon}} \tag{21}$$

and decouples from the rest of the system. We are neglecting here possible dependences of the carrier distribution function on the photon density S due to the spectral hole burning and/or direct carrier heating by the lightwave[11, 12]. Such effects can be taken into account in a similar fashion.

$\hat{T}_e$ will be considered an independent variable in Eqs. (20a) and (20b). The small signal variation of the gain can be described by two coefficients g_n' and g_T', each of which depends on the steady-state values of concentration $\bar{n}$ and temperature $\bar{T}_e$, as well as the optical frequency Ω:

$$\hat{g} = g_n'(\bar{n},\bar{T}_e,\Omega)\,\hat{n} + g_T'(\bar{n},\bar{T}_e,\Omega)\,\hat{T}_e \;. \tag{22}$$

The steady-state coefficients in the laser generation regime $\bar{J} > J_{th}$ are related as follows:

$$\bar{g}\,\bar{S} = \bar{J} - J_{th}\;; \qquad\qquad \Gamma\bar{g}\,\tau_{ph} = 1\,, \tag{23}$$

where we have set $J_{th} = B\,\bar{n}^2$, neglecting the term $\beta B\,\bar{n}^2$.

We are interested in solutions of Eqs. (20) for which the variations $\hat{n}$ and $\hat{T}_e$ have a definite "target" relationship, $\hat{n} = \gamma\hat{T}_e$. Substituting this relationship in Eqs. (20a) and (20b) and using Eq. (23), we find

$$\hat{S} = \frac{\hat{J}}{\bar{g}}\,\frac{1}{1 - \omega^2\,\tau_{ph}\,\tau_{\gamma} + i\omega\tau_{ph}\,(1 + \tau_{\gamma}/\tau_{sp})}\,, \tag{24}$$

where

$$\tau_{\gamma} \equiv \frac{\gamma}{(g_T' + \gamma g_n')\,\bar{S}}\;; \qquad\qquad \tau_{sp} \equiv \frac{1}{2B\,\bar{n}}\,, \tag{25}$$

For $\gamma \to \infty$, Eq. (24) goes over into the "classical" dependence $\hat{S}(\hat{J})$ corresponding to a pure modulation by the pumping current. The response function in (24) contains the usual pole corresponding to the electron-photon resonance, and the laser signal power decays as $1/\omega^2$ at high enough frequencies. If we are concerned with increasing the frequency of modulation, then the target should be chosen so as to eliminate the relaxation oscillations – which corresponds to setting $\gamma = 0 = \hat{n}$. In this case, Eq. (24) reduces to

$$\hat{S} = \frac{\hat{J}}{\bar{g}}\,\frac{1}{1 + i\omega\tau_{ph}}\;; \qquad\qquad |\hat{S}| = \frac{|\hat{J}|}{\bar{g}}\,\frac{1}{[1 + (\omega\tau_{ph})^2]^{1/2}}\,, \tag{26}$$

This solution requires that the variations $\hat{j}$ and $\hat{T}_e$ be related to each other in a definite way:

$$\hat{T}_e = \frac{\hat{j}}{\bar{S}\, g_T'}\, \frac{i\omega\tau_{ph}}{1 + i\omega\tau_{ph}} \ . \tag{27}$$

When this relation is fulfilled, then there is no electron-photon resonance in the system and the modulation efficiency decays with frequency as $1/\omega$. Small-signal response of the laser output power is plotted in Fig. 7 for three types of modulation: (1) purely by the pumping current, (2) purely by the electron temperature, and (3) by their coherent combination as in Eq. (27).

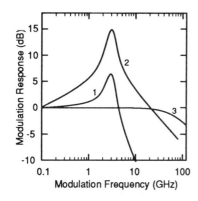

Fig. 7: Frequency dependence of the optical response $\hat{S}(f)/\hat{S}(0)$ to the variation of different parameters in a stripe MQW laser. Assumed laser parameters: five $100\,\text{Å}$ QW's, area $360\times6\,\mu\text{m}$, $\Gamma = 0.2$, $\tau_{ph} = 2.5\,\text{ps}$, $g_n' = 2.5\times10^{-16}\,\text{cm}^{-2}$, $\bar{S} = 3.5\times10^{14}\,\text{cm}^{-3}$. Curve 1: modulation by the pumping current, curve 2: modulation by the electron temperature, curve 3: dual modulation as in Eq. (27).

Suppression of relaxation oscillations of the carrier density makes possible a high repetition rate coding of information with short pulses. We have calculated[10] the laser response to a $10\,\text{Gb/s}$ series of dual current-temperature pulses, and found that the response is practically undistorted, compared to the single pulse situation, It is clear that small-signal pulses δJ of *any* shape, as well as analog signals, can be transmitted in a regime of constant n, provided the system can Fourier analyze $\delta J(t)$ and form in real time an appropriate complementary pulse $\delta T_e(t)$ from the spectrum given by Eq. (27). Large signal theory of dual modulation is discussed in a forthcoming publication.[13]

In general, the choice of a target relation for dual modulation depends on the engineering problem at hand. Instead of targeting the regime of constant carrier concentration, we may be interested in the suppression of the wavelength chirping at high modulation frequencies. Recall that this unwelcome phenomenon[8] originates from the relaxation oscillations, which lead to variations in the real part η of the refractive index $\eta_c = \eta + i\kappa$ in the active region. Variations of n affect η in two ways: by changing the free-carrier absorption and by changing the optical gain. In InGaAs lasers both effects give similar contributions, shifting the lasing mode wavelength by as much as several Å.

In our present model the gain variation $\hat{g}$ has two contributions, Eq. (22). Therefore, we should distinguish the corresponding two contributions, $\hat{\eta}_{g_n}$ and $\hat{\eta}_{g_T}$ in the refractive index variation, each given by the Kramer-Kronig relation

$$\hat{\eta}_g = -\frac{\bar{\eta}}{\pi} \, \mathcal{P} \int_0^\infty \frac{\hat{g}(\Omega') \, d\Omega'}{\Omega'^2 - \Omega^2} \, , \qquad (28)$$

where $\mathcal{P}$ denotes the integral principal value. We now see that it is possible to target a complete suppression of the total index oscillation,

$$\hat{\eta} = \hat{\eta}_{fc} + \hat{\eta}_{g_n} + \hat{\eta}_{g_T} = 0 \, , \qquad (29)$$

by judiciously choosing a relationship between $\hat{T}_e$ and $\hat{\jmath}$.

It should be emphasized that having targeted the complete elimination of chirp, we pay a penalty in the modulation frequency, since the electron-photon resonance is no longer eliminated and the modulation efficiency $\hat{S}/\hat{\jmath}$ decays as ω^{-2} at sufficiently high frequencies. On the other hand, we note that the variation of η due to variations in T_e is usually weaker than that due to variations in n. If we compare $\hat{\eta}_{g_T}$ at the point of complete elimination of relaxation oscillations with the amount of chirp $\hat{\eta}_n = \hat{\eta}_{fc} + \hat{\eta}_{g_n}$ for a purely pump current modulation with the same output signal power, we find that the latter is usually larger, because of both the additional contribution from free-carrier absorption and the essentially different frequency dependence of $g'_T(\Omega)$ and $g'_n(\Omega)$.

7. Conclusion

We have discussed the possibility of controlling the electron temperature by intersubband absorption in a semiconductor quantum well. This gives rise to a significant modulation of the interband optical properties which can be used for the implementation of a fast and chirp-free modulator of infrared radiation. The method can also be used to modulate laser radiation by controlling simultaneously the pumping current and the optical gain in the active region. This allows to eliminate the relaxation oscillations and suppress the wavelength chirping in optical communication systems operating at high pulse repetition rates. Rapid control of intersubband absorption is possible at *a constant* CO_2 flux by varying the intersubband separation with an applied electric field. For this purpose it is crucial that means for applying the electric field would themselves be transparent to CO_2 radiation, which can be accomplished with a novel scheme discussed in Sect. 2.

Obviously, the dual modulation method is not limited to the variation of T_e by intersubband absorption. In principle, any of the parameters of the system of rate equations (19) may be varied externally, though not necessarily by intersubband techniques. Besides different schemes for the variation of T_e, several alternative proposals have been discussed recently, including varying the confinement factor[13] in a stripe laser, the photon lifetime[14] in a vertical cavity laser, and even the spontaneous emission factor[15] in a microresonator. We believe that intersubband physics has a good chance to contribute to this emerging field.

References:

1. V. B. Gorfinkel, B. M. Gorbovitsky, and I. I. Filatov, "High frequency modulation of light output power in double-heterojunction laser", *Int. J. of Infrared & Millimeter Waves* **12**, pp. 649-658 (1991).

2. V. B. Gorfinkel and S. Luryi, "Rapid modulation of interband optical properties of quantum wells by intersubband absorption", *Appl. Phys. Lett.* **60**, pp. 3141-3143 (1992).

3. S. Luryi, "Photon-Drag Effect in Intersubband Absorption by a Two-Dimensional Electron Gas", *Phys. Rev. Lett.* **58**, pp. 2263-2266 (1987).

4. A. D. Wieck, H. Sigg, and K. Ploog, "Observation of resonant photon drag in a two-dimensional electron gas", *Phys. Rev. Lett.* **64**, pp. 463-466 (1990).

5. D. J. Westland, J. F. Ryan, M. D. Scott, J. I. Davies, and J. R. Riffat, "Hot carrier energy loss rates in GaIAs/InP quantum wells", *Solid-St. Electron.* **31**, pp. 431-438 (1988).

6. H. Lobentanzer, W. Stolz, K. Ploog, R. J. Bäuerle, and T. Elsaesser, "Screening of the N=2 excitonic resonance by hot carriers in an undoped GaInAs/AllnAs multiple quantum well structure", *Solid-St. Electron.* **32**, pp. 1875-1879 (1989).

7. S. Das Sarma, J. K. Jain, and R. Jalabert, "Hot-electron relaxation in GaAs quantum wells", *Phys. Rev.* **B 37**, pp. 1228-1230 (1988).

8. G. P. Agrawal and N. K. Dutta, *Semiconductor Lasers*, 2nd edition (Van Nostrand Reinhold, New York 1993).

9. S. A. Gurevich, V. B. Gorfinkel, G. E. Stengel, and I. E. Chebunina, *Proc. 13th Int. Semicond. Laser Conf.*, Sept. 21-25, 1992, Japan.

10. V. B. Gorfinkel and S. Luryi, "High-Frequency Modulation and Suppression of Chirp in Semiconductor Lasers", *Appl. Phys. Lett.* **62**, pp. 2923-2925 (1993).

11. R. F. Kazarinov and C. H. Henry, "The relation of line narrowing and chirp reduction resulting from the coupling of a semiconductor laser to a passive resonator", *IEEE J. of Quant. Electron.* **QE-23**, pp. 1401-1409 (1987).

12. K. L. Hall, G. Lenz, E. P. Ippen, U. Koren, and G. Raybon, "Carrier heating and spectral hole burning in strained layer quantum-well laser amplifiers at 1.5 µm", *Appl. Phys. Lett.* **61**, pp. 2512-2514 (1992).

13. V. B. Gorfinkel and S. Luryi, "Dual Modulation of Semiconductor Lasers", to be published.

14. E. A. Avrutin, V. B. Gorfinkel, S. Luryi, and K. A. Shore, "Control of surface-emitting laser diodes by modulating the distributed Bragg mirror reflectivity: small-signal analysis", to appear in *Appl. Phys. Lett.* (1993).

15. Y. Yamamoto, S. Machida, and G. Björk, "Microcavity semiconductor laser with enhanced spontaneous emission", *Phys. Rev.* **A 44**, pp. 657-668 (1991).

THEORY OF TERAHERTZ GENERATION DUE TO QUANTUM BEATS IN QUANTUM WELLS

S. L. Chuang and M. S. C. Luo
University of Illinois at Urbana-Champaign
Department of Electrical and Computer Engineering
1406 W. Green Street
Urbana, IL 61801

ABSTRACT. A theory for terahertz generation from bulk and quantum-well semiconductors is discussed based on a second-order nonlinear optical mechanism. For a bulk semiconductor, terahertz generation due to optical rectification is presented. For a coupled quantum-well structure, a three-level model showing contributions from both optical rectification and quantum beat is discussed. The interband and intersubband dipole moments play important roles in the terahertz generation.

1. Introduction

It has been observed that a short intense optical pulse incident on a semiconductor surface generates a transient terahertz electromagnetic signal.[1-3] It was previously believed that the signal results from the acceleration[1-2] of optically created electron-hole pairs by the depletion field in the direction perpendicular to the surface. This model is based on the electromagnetic radiation from an induced transient photocurrent J, which is proportional to the conductivity and concentration of the generated photocarriers. While such a transport picture could account for radiation on a picosecond time scale, we have recently proposed[4] a new model based on nonlinear optical rectification from the depletion field of a semiconductor surface to explain the electromagnetic pulse generation in the subpicosecond (femtosecond) regime. The depletion field is formed by Fermi-level pinning at the semiconductor surface. Optical rectification occurs because the depletion field near the surface creates an asymmetry in the semiconductor. Therefore, there is an instantaneous dipole moment since the electrons and holes separate once they are generated. The generated electromagnetic pulse is as short as the optical pulse duration. Our model is based on coherent optical rectification by the depletion field and is similar to second-harmonic generation in the second-order nonlinear susceptibility. In fact, comparison of the terahertz radiation data with the well-known second harmonic generation data[5] with crystal orientation dependencies reveals striking similarities.

Recently, terahertz generation due to charge oscillations in a coupled quantum-well structure has also been reported.[6] The generated terahertz signal contains two major components: (i) an instantaneous signal caused by the optically rectified polarization current, followed by (ii) an oscillating signal caused by the quantum beats due to charge oscillations in a coupled quantum-well structure.

H. C. Liu et al. (eds.), Quantum Well Intersubband Transition Physics and Devices, 547–552.
© 1994 *Kluwer Academic Publishers. Printed in the Netherlands.*

In this paper, we discuss the theory[4,7] for the terahertz generation in bulk semiconductors and in a coupled quantum-well structure. We show that the same density-matrix approach using the optical Bloch equations can explain terahertz generation from bulk and quantum-well semiconductors. The optically induced polarization density consists of contributions from both interband and intersubband transitions. The terahertz signal generated from a coupled-well structure can be separated into two components: one due to the net charge displacement and the population of the levels caused by the optical pump, and the other depending on the quantum beats and the intersubband transition dipole moment. We will also mention briefly the coherent control of quantum beats using two phase-locked optical pulses with a time delay equal to the charge oscillation period reported recently by Planken et al.[8,9]

2. Optical Rectification on the Surface of Bulk Semiconductors

Our model[4] for terahertz electromagnetic pulse generation from the surface of a semiconductor (Fig. 1) is based on the second-order optical rectification polarization density

$$P_i(t) = \chi^{(2)}{}_{ijk}(0; \omega, -\omega) E_j(t) E_k{}^*(t) \tag{1}$$

$$E_j(t) = E_j(t) e^{-i\omega t} + E_j^*(t) e^{i\omega t} \tag{2}$$

where the envelope function of the pump field $E(t)$ is slowly varying compared with the optical frequency ω. The depletion field-induced asymmetry is taken into account for the semiconductor in the presence of an intense optical pump beam. The induced polarization is in the z-direction because the depletion field is along the z-direction; therefore, $\chi^{(2)}{}_{ijk}$ vanishes unless $i = z$. This result explains the observation that the generated electromagnetic pulse is polarized in the plane of incidence (TM polarization). Furthermore, the induced polarization P_i can be linearly expanded in terms of the depletion field F_z,

$$P_i = \chi^{(3)}{}_{ijkz}(0; \omega, -\omega, 0) E_j E_k{}^* F_z \tag{3}$$

which explains the change in polarity of the generated electric field from n-type samples to p-type samples, since the depletion field has opposite signs. From the dependence of the induced polarization on the depletion width, we also obtain the doping concentration and bias voltage dependencies of the radiation field, which agree with the experimental observations. The observed temperature dependence in which the generated electromagnetic signal decreases with increasing temperature can also be explained using our theoretical model based on the fact that the dephasing rate increases with temperature. *The strongest evidence that the induced electromagnetic field comes from a second-order nonlinear optical rectification process is from the crystal orientation dependence.* Following a similar approach that has been used for second-harmonic generation[5] from a semiconductor surface, we find that the induced electromagnetic intensity varies as $A+B\cos 2\varphi$ for a (100) surface, where φ is measured from the (011) direction. Similarly, for a (111) surface, the azimuthal angle dependence varies as $C+D\cos 3\varphi$, where the angle φ is measured from the vertical plane containing the (100) and (111) axes. The coefficients A, B, C, and D depend on the angle of transmission.

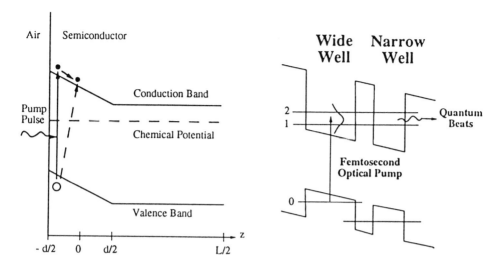

Fig. 1 Optical rectification from the surface   Fig. 2 Optical rectification and quantum
        of a semiconductor.   beat from coupled quantum wells.

Our theoretical dependencies agree very well with the experimental measurements. *The crystal orientation dependence cannot be explained by the photocarrier conduction current model[1,2] (unless the ballistic transport nature of carriers is taken into account properly), since the conductivity is isotropic for zinc-blende crystals. Our theory using the field-induced second-order nonlinear susceptibility explains all of the major features of experimental observations.* More recent work[10] using a normal incidence configuration indicates that the bulk optical rectification due to the intrinsic GaAs asymmetry produces a significant amount of terahertz generation as well, because any z-directed polarization or transport current does not radiate in the normal direction. More work is necessary to identify the relative magnitudes of the bulk and surface-field-induced optical rectifications.

To identify clearly the optical rectification mechanism for terahertz generation, quantum-well structures have been used in the femtosecond pump excitation. The pump photon energy is above the band gap in the wells and below the band gap of the barriers. It has been shown experimentally[11] that the bias electric field determines the sign and magnitude of the induced terahertz signals.

3. Optical Rectification and Quantum Beat in Coupled Quantum Wells

For a coupled quantum-well structure studied by Roskos et al.,[6] the sample consists of ten pairs of a wide-well (145 Å) and a narrow-well (100 Å) GaAs region separated by a 25 Å $Al_{0.2}Ga_{0.8}As$ barrier. The pairs are isolated by 200 Å wide buffer layers. With an applied bias voltage, the electronic states in the wide and the narrow wells can be aligned such that resonant coupling occurs between the two wells.

We use the density-matrix equations for a three-level model,[7] Fig. 2, under the rotating wave approximation since the optical pump energy is close to the band gap.   The k-dependence of the density-matrix elements for each direct transition in the k-space of the electronic band structure is taken into account from the *interband* transition energy

$$\varepsilon_{nm} = \varepsilon_n + \varepsilon_m + E_g + \frac{\hbar^2 k^2}{2m_r^*} \tag{4}$$

$$\frac{1}{m_r^*} = \frac{1}{m_c} + \frac{1}{m_v} \tag{5}$$

We then solve for the density matrix elements, $\rho_{ij}(k, t)$; the induced polarization density can be written as the sum of two terms:

$$P(t) = P_{OR}(t) + P_{QB}(t) \tag{6}$$

where

$$P_{OR}(t) = \frac{|e|}{V} \sum_k [(z_{00} - z_{11}) \rho_{11}(k, t) + (z_{00} - z_{22}) \rho_{22}(k, t)] \tag{7}$$

$$P_{QB}(t) = -\frac{|e|}{V} \sum_k 2 z_{12} \operatorname{Re}(\rho_{12}(k, t))] \tag{8}$$

Here the intersubband dipole moments

$$\mu_{nm} = -|e| <\phi_n(z) |z| \phi_m(z)> = -|e| z_{nm} \tag{9}$$

have been used, and the fast-varying components in the interband density-matrix elements such as $\exp(+i\omega t)$ and $\exp(-i\omega t)$ have been factored out explicitly such that all elements contain only the slowly varying part.

The generated terahertz electric field $E_R(t)$ can be obtained from the second derivative of the polarization density, $E_R(t) \propto \frac{\partial^2 P(t)}{\partial t^2}$. Therefore, we can write

$$E_R(t) = E_{OR}(t) + E_{QB}(t) \tag{10}$$

where $E_{OR}(t) \propto \frac{\partial^2 P_{OR}(t)}{\partial t^2}$, and $E_{QB}(t) \propto \frac{\partial^2 P_{QB}(t)}{\partial t^2}$.

In Figs. 3 (a) and 3(b), we plot the calculated terahertz signals separated into two components: $E_{OR}(t)$ in solid curve and $E_{QB}(t)$ in dashed curve, and the total radiated electric field, $E_R(t)$. We can see that the optical rectification signal $E_{OR}(t)$ only exists during the optical pump period, while the quantum beat signal $E_{QB}(t)$ continues ringing after the optical pump pulse is gone. The total signal contains the contributions of both mechanisms. From Eq. (7), we see that the polarization density due to optical rectification is due to the net displacements between the holes in the valence subband 0 and the electrons in each subband, $(z_{00} - z_{11})$ and $(z_{00} - z_{22})$, weighted by the induced population densities. On the other hand, the quantum beat signal, Eq. (8), is contributed by the intersuband dipole moment $|e| z_{12}$, which has a maximum at the resonant field. Note that the density-matrix elements depend on the interband dipole matrix elements

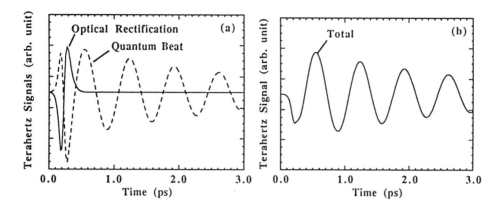

Fig. 3 (a) Terahertz radiation signals due to optical rectification (solid curve) and quantum beat (dashed curve), (b) their total signal in a coupled quantum-well structure.

as well. This theory has been used to explain the experimental data with reasonable agreement.

To produce both optical rectification and quantum-beat terahertz signals, single quantum wells using one heavy hole, one light hole and one conduction subband have also been used.[12] Valence-band mixing effects are very important to explain the nonvanishing intersubband dipole moment[13,14] between the heavy-hole and the light-hole subbands at a finite parallel wave vector (i.e., away from the band edge). This intersubband dipole moment has also been used in explaining the linear optical absorption[15] in a p-type modulation doped quantum-well structure for infrared photodetector applications.

Another interesting development is the coherent control[8,9] of charge oscillations in coupled quantum wells (or single quantum wells). It has been shown that charge oscillations caused by quantum beats in coupled wells can be enhanced or weakened using a second optical pulse with a time delay equal to an integral multiple of the charge oscillation period, depending on the relative optical phase of the two pulses. It is also possible to phase-shift the terahertz signal using a time delay equal to half of an odd multiple of the charge oscillation period and the optical pump energy at the middle of the two interband transitions.

4. Conclusions

We have briefly reviewed the theory for terahertz generation using femtosecond optical pulses from bulk and quantum-well semiconductors. In bulk semiconductors, the contribution of the optical-rectification polarization current is shown using the density-matrix theory in the frequency domain, taking into account the multiple subband nature near the depletion region of the semiconductor surface. In coupled quantum wells, a three-level

552

model is used and the density-matrix equations can be solved directly in the time domain. Both optical rectification and quantum beat mechanisms have been discussed.

Acknowledgments: This work was done in collaboration with S. Schmitt-Rink, P. Saeta, B. Greene, P. Planken, I. Brener, H. Roskos and M. Nuss. We would like to acknowledge the support of ONR (Grant N00014-90-J-1821).

[1] X. C. Zhang, B. B. Hu, J. T. Darrow, and D. H. Auston, *Appl. Phys. Lett.* **56**, 1011(1990).

[2] X. C. Zhang, J. T. Darrow, B. B. Hu, D. H. Auston, M. T. Schmidt, P. Tham, and E. S. Yang, *Appl. Phys. Lett.* **56**, 2228 (1990); B. B. Hu, X. C. Zhang, and D. H. Auston, *Appl. Phys. Lett.* **57**, 2629 (1990); X. C. Zhang et al., *Appl. Phys. Lett.* **57**, 753 (1990).

[3] B. I. Greene, J. F. Federici, D. R. Dykaar, A. F. J. Levi, and L. Pfeiffer, *Opt. Lett.* **16**, 48 (1991).

[4] S. L. Chuang, S. Schmitt-Rink, B. I. Greene, P. N. Saeta and A. F. J. Levi, *Phys. Rev. Lett.*, **68**, 102 (1992).

[5] H. W. K. Tom, T. F. Heinz, and Y. R. Shen, *Phys. Rev. Lett.* **51**, 1983 (1983).

[6] H. G. Roskos, M. C. Nuss, J. Shah, K. Leo, D. A. B. Miller, S. Schmitt-Rink, and K. Kohler, *Phys. Rev. Lett.* **68**, 2216 (1992).

[7] M. S. C. Luo, S. L. Chuang, P. C. M. Planken, I. Brener, H. G. Roskos, and M. C. Nuss, *IEEE J. Quantum Electron.*, submitted.

[8] P. C. M. Planken, I. Brener, M. C. Nuss, M. S. C. Luo, and S. L. Chuang, *Phys. Rev. B*, Aug. 15, 1993.

[9] M. S. C. Luo, S. L. Chuang, P. C. M. Planken, I. Brener, and M. C. Nuss, *Phys. Rev. B*, Oct. 15, 1993.

[10] X. C. Zhang, Y. Jin, K. Yang, and L. J. Schowalter, *Phys. Rev. Lett.*, **69**, 2303 (1992); *Appl. Phys. Lett.* **61**, 2764(1992).

[11] P. C. M. Planken, M. C. Nuss, W. H. Knox, and D. A. B. Miller, *Appl. Phys. Lett.* **61**, 2009 (1992).

[12] P. C. M. Planken, M. C. Nuss, I. Brener, K. W. Goossen, M. S. C. Luo, and S. L. Chuang, *Phys. Rev. Lett.* **69**, 3800 (1992).

[13] Y.-C. Chang and R. B. James, *Phys. Rev. B* **39**, 12672 (1989).

[14] L. Tsang and S. L. Chuang, *Appl. Phys. Lett.* **60**, 2543 (1992).

[15] B. F. Levine, S. D. Gunapala, J. M. Kuo, S. S. Pei, and S. Hui, Appl. Phys. Lett. 59, 1864 (1991).

Far-infrared Study of an Antenna-coupled Quantum Point Contact

Qing Hu, Rolf A. Wyss, C. C. Eugster, J. A. del Alamo
Department of Electrical Engineering and Computer Science and Research Laboratory of Electronics
Massachusetts Institute of Technology
Cambridge MA 02139

and

Shechao Feng
Department of Physics University of California
Los Angeles CA 90024

ABSTRACT.
  We have developed a theory of photon-assisted quantum transport in quantum point contact devices. According to the theory, ballistic electrons in a quantum point contact can absorb an integer number of far-infrared photons via intersubband transitions. Consequently, a photon-induced drain/source current can be produced by a far-infrared radiation whose frequency is comparable to that of the intersubband spacing at the narrowest constriction of the quantum point contacts. Motivated by this theory, we have studied transport properties of an antenna-coupled quantum point contact under coherent far-infrared (285 GHz) radiation. A pronounced photon-induced drain/source current is observed. The amplitude of the photon-induced current is about 10% of that corresponding to a quantized conductance step, and it oscillates with the gate voltage. Our analysis suggests that the observed photon-induced current is mainly due to heating of the electron gas in the source and drain (a bolometric effect) rather than photon-assisted quantum transport. Further investigation reveals a photon-induced thermopower which oscillates with the gate voltage in a manner that tracks the onset of subband channels. This thermopower is a result of asymmetric heating.

1. Introduction

Ballistic quantum transport through semiconductor quantum point contacts has been an active research area in recent years. In essence, such devices are made of very clean two-dimensional election gas samples with a finely structured split gate on the wafer surface. At low temperatures, electrons can travel from the source to drain without suffering elastic and inelastic scattering. This results in the well-known quantized drain/source conductance behavior, with a conductance quantum of $2e^2/h$.(van Wees et al. 1988,

H. C. Liu et al. (eds.), Quantum Well Intersubband Transition Physics and Devices, 553–564.

Wharam et al. 1988) Extensive work has been done after the discovery of this effect to study various features and extensions of this novel quantum transport device.(Beenakker and van Houten 1991) However, most of the experiments that have been performed on such devices are limited to DC transport measurements.

It is well known in the field of superconducting tunneling that photons can assist the tunneling process, provided that tunneling is elastic.(Tien and Gordon 1963) The theoretical framework for photon-assisted tunneling has been well established and verified in experiments. The effect of photon-assisted tunneling has also been utilized in heterodyne receivers and radiation detectors using superconducting tunnel junctions, which are the best devices in terms of sensitivity and speed.(Richards and Hu 1989) In a broad sense, elastic tunneling is a phase-coherent quantum transport in a classically forbidden region. Therefore, the results of photon-assisted tunneling can be applied to the study of photon-assisted quantum transport in semiconductor devices. If demonstrated successfully, novel devices based on the photon-assisted quantum transport can be developed for applications in the THz frequency range where few devices are currently available.

2. Theory of Photon-assisted Quantum Transport

We consider a standard split-gate quantum point contact device, as shown in Fig. 1. The electrons are confined to two-dimensional movements at the interface of a GaAs/AlGaAs modulation-doped heterojunction. We assume that the 2D electron gas has a high mobility so that no impurity potential near the constriction region needs to be included in our calculation. Two capacitive split gates are patterned on top of the device, which form the constriction for the point contact device. Application of a negative gate voltage $-V_G$ creates subband structures in the direction perpendicular to transport (the y-direction in Fig. 1). We model the effect of this negative gate voltage on the electron gas underneath by an effective channel width of the electron gas, $d(x) \leq D(x)$, which decreases with an increasing negative gate voltage. We also assume that the variation of $d(x)$ is slow enough so that WKB-type adiabatic semiclassical approximations for the electron wave function can be applied. Within this approximation, assume infinite potential sidewalls at $y = \pm d(x)/2$, the electron eigenstates $\Psi_n(x,y)$ can be expressed as products of the longitudinal (in the x direction) eigenfunctions $\phi_n(x)$ and the transverse (in the y direction) standing-wave functions $\chi_n(y)$, where

$$\chi_n(y) = sin[\frac{n\pi y}{d(x)}], \qquad n = even,$$

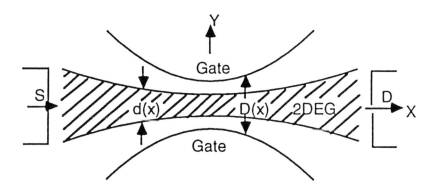

Fig. 1. Geometry of the quantum point contact device under consideration. Under a negative gate voltage, electrons travel from the source (S) to the drain (D) through a channel with an x-dependent width $d(x)$.

$$\chi_n(y) = cos[\frac{n\pi y}{d(x)}], \qquad\qquad n = odd, \qquad\qquad (1)$$

with transverse eigenenergies given by $E_n(x) = \frac{n^2\pi^2\hbar^2}{2md^2(x)}$.

By substituting $\Psi_n(x,y)$ in the time-independent Schrödinger equation, we reduce the original two-dimension equation into one-dimensional equations,

$$[-\frac{\hbar^2}{2m}\frac{d^2}{dx^2} + E_n(x)]\phi_n(x) = E\phi_n(x). \qquad\qquad (2)$$

Eq. (2) implies that the electrons in the nth subband are subject to a potential $E_n(x)$, which increase to their maximum values at the narrowest point of the constriction, i.e. $x = 0$ in Fig. 1. The first few $E_n(x)$'s are sketched in Fig. 2(a). Using WKB approximation, electrons with energies greater than the potential barrier $E_n(0)$ will have a unity transmission coefficient from the source to the drain, while electrons with energies less than $E_n(0)$ will have a zero transmission coefficient. As the electron Fermi energy increases, more subbands will be turned on. Since each subband contributes exactly $2e^2/h$ to the drain/source conductance,(Imry 1986) the total conductance of the device will increase discontinuously as the Fermi energy (or equivalently, the gate voltage) increases. This results in the well-known step structures of the drain/source conductance G_{DS} vs. the gate voltage V_{GS} curves.

The novel idea of photon-assisted quantum transport is to use far-infrared photons to enhance the energies of the 2-D electron gas in the source and drain, so that they will be energetic enough to pass through the subbands and contribute to conduction. (Hu 1993) A schematic illustrating the idea is shown in Fig. 2(a), in which the device is biased in the pinch-off regime where

556

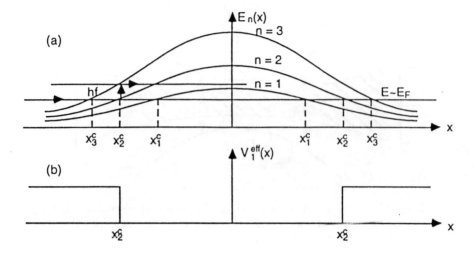

Fig. 2. (a) Effective one-dimensional potential profile for the transport electrons moving from the source to drain in a quantum point contact. The barrier potential arises from the subband energies $E_n(x)$, which is x dependent. Electrons from a particular subband n can propagate adiabatically through the point contact if the electron energy $E \sim E_F$ is greater than the maximum barrier height $E_n(0)$. Under radiation, electrons with energies even less than $E_n(0)$ can propagate over the barriers by absorbing an integer number of photons, as illustrated in the figure. (b) Effective AC potential V_1^{eff} for the electrons in the first subband.

$E_F < E_1(0)$. This photon-assisted transport will result in a photon-induced drain/source current. This idea is similar to photon-assisted tunneling in superconducting tunnel junctions.

In the presence of radiation, the total Hamiltonian will have a time-dependent component. For simplicity, here we will only consider the case in which the radiation field is in the y direction (the transverse case). The results for the longitudinal case (E field in the x direction) is similar to the transverse case, but the mathematics is more complicated.(Feng and Hu 1993) The time-dependent Hamiltonian is

$$H' = -eE_{ac}ycos(\omega t).$$

From selection rules, H' will couple subbands with opposite parities. Thus, intersubband transitions will occur. In general, the solutions of the time-dependent Schrödinger equation should include all the subband wave functions, that is,

$$\Psi(x,y,t) = \sum_n C_n(x,t)\phi_n(x)\chi_n(y),\qquad(3)$$

and

$$[-\frac{\hbar^2}{2m}(\frac{\partial^2}{\partial x^2} + \frac{\partial^2}{\partial y^2}) - eE_{ac}cos\omega t]\Psi(x,y,t) = i\hbar\frac{\partial\Psi(x,y,t)}{\partial t}.\qquad(4)$$

We are most interested in electrons in the first subband where the dc potential barrier is the lowest. Multiplying Eq. (4) by $2\chi_1(y)/d$ and then integrating over y from -d/2 to d/2 leads to:

$$[-\frac{\hbar^2}{2m}\frac{\partial^2}{\partial x^2} + E_1(x)]\phi_1(x)C_1(x) - eE_{ac}(x)cos\omega t \sum_{n\geq 2} D_{1n}C_n(x,t)\phi_n(x)$$
$$= i\hbar\phi_1(x)\frac{\partial C_1(x,t)}{\partial t},\qquad(5)$$

where $D_{1n} = \frac{2}{d(x)}\int_{-d/2}^{d/2} y\chi_1(y)\chi_n(y)dy$. Eq. (5) is the equation of motion for the component of the electron wavefunction in the first subband, which is coupled via "dipole" moment D_{1n} to those of higher subbands with an odd parity. The subband that mixes with the first subband most is the second subband, with a "dipole" moment given by

$$D_{12}(x) = \frac{2}{d(x)}\int_{-d/2}^{d/2} ycos[\pi y/d(x)]sin[2\pi y/d(x))]dy = \frac{16d(x)}{9\pi^2} \approx 0.2d(x).$$

The next subband that mixes with the first subband is the fourth subband. As the dipole moment D_{1n} decays rapidly with n (for example, $|D_{14}| = \frac{2}{25}D_{12}$), we can then make a further simplification by concentrating on the D_{12} term only, and. Eq. (5) then reduces to the form

$$[-\frac{\hbar^2}{2m}\frac{\partial^2}{\partial x^2} + E_1(x) + V_1^{eff}(x,t)]\Psi_1(x,t) = i\hbar\frac{\partial\Psi_1(x,t)}{\partial t},\qquad(6)$$

where we have rewritten $\Psi_1(x,t) = \phi_1(x)C_1(x,t)$, and have introduced an effective time-dependent perturbation potential

$$V_1^{eff} = -eD_{12}(x)E_{ac}(x)cos(\omega t)\frac{\phi_2(x)C_2(x,t)}{\phi_1(x)C_1(x,t)}$$
$$\approx -0.2eV_{ac}\frac{d(x)\phi_2(x)}{D(x)\phi_1(x)}cos(\omega t),\qquad(7)$$

where $V_{ac} = E_{ac}(x)D(x)$. In Eq. (7), we have approximated the two C-functions to be the same, which is certainly valid in regions far away from the classical turning points x_c. Therefore, V_1^{eff} is mainly determined by the ratio of the wave functions of the two subbands $\phi_2(x)$ and $\phi_1(x)$.

Thus, with suitable approximations, we can map the photon-assisted transport problem into an effective one-dimensional Schrödinger equation with an AC and a DC potential. Since $\phi_1(x) \approx \phi_2(x)$ in regions far away from the classical turning point of the second subband x_2^c (x_2^c is shown in Fig. 2(a)), the amplitude of the AC potential V_1^{eff} is approximately constant in the region $|x| > x_2^c$, and V_1^{eff} decreases exponentially for $|x| < x_2^c$. Thus to a good approximation, we can treat V_1^{eff} as a piece-wise potential with $V_1^{eff} = 0.2eV_{ac}d(x_2^c)/D(x_2^c)$ for $|x| \geq x_2^c$, and $V_1^{eff} = 0$ for $|x| < x_2^c$, as shown in Fig. 2(b).

Using this piece-wise ac potential V_1^{eff}, in the region where $|x| < x_2^c$ the solution of the Schrödinger equation is the same as the one without radiation. While in the region $|x| > x_2^c$, since V_1^{eff} is spatially independent, the spatial part of the wave functions remains the same while the temporal part of the wave functions has an additional phase factor (Tien and Gordon 1963)

$$exp[i(eV_1^{eff}/\hbar\omega)sin\omega t)] = \sum_{n=-\infty}^{\infty} J_n(\alpha)e^{-in\omega t}, \qquad (8)$$

where J_n is the nth order Bessel function, and $\alpha = eV_1^{eff}/\hbar\omega$ is a dimensionless parameter which is proportional to the strength of the radiation field.

Eq. (8) implies that the new wave function has components with energies $E+n\hbar\omega$. Physically this means that the photon field creates new eigenstates with energies $E+n\hbar\omega$. Positive $n's$ correspond to states where electrons have absorbed n photons, while negative $n's$ correspond to states where electrons have emitted n photons. The amplitude of these new eigenstates is $J_n(\alpha)$, and the corresponding electron density is $J_n^2(\alpha)$.

Thus, similar to the photon-assisted tunneling current steps in a superconducting tunnel junction, these photon-induced states will produce substeps on top of the staircase G_{DS} vs. V_{GS} curves, as shown in Fig. 3. The width of the ministep is proportional to the frequency, and the height of the ministeps is a function of the radiation power in the form of

$$(2e^2/h)V_{DS} \sum_{n}^{\infty} J_n^2(\alpha). \qquad (9)$$

The lower limit of the summation is determined by the difference between $E_n(0)$ and E_F relative to the photon energy $\hbar\omega$.

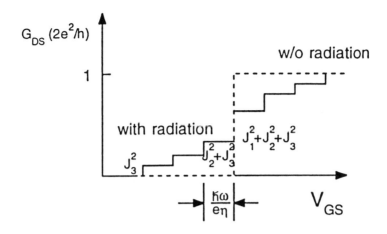

Fig. 3. Predicted G_{DS} vs. V_{GS} curves of a quantum point contact under a coherent far-infrared radiation. The range of the gate voltage is around the onset of the first conductance step. η is a dimensionless factor which is determined by the geometry of the quantum point contact which relates the gate voltage to the energy level of the subbands.

3. Preliminary experimental results

In order to couple far-infrared radiation (whose wavelength is typically on the order of 1 millimeter) to the quantum point contact of submicron size, a planar antenna (Rutledge et al. 1983) is required to concentrate the radiation field into the area where the drain/source conduction takes place. Otherwise, most of the radiation will bypass the quantum point contact. In addition, to excite the intersubband transitions, the E field of the radiation should be polarized in the direction transverse to the drain/source conduction path. This can be achieved by placing the antenna in such a way that its terminals also serve as the split-gate electrodes, as shown in Fig. 4(a). The radiation electric field at the antenna terminals is perpendicular to the antenna electrodes, as shown in Fig. 4(b), which is roughly perpendicular to the drain/source conduction path.

We have successfully fabricated several quantum point contacts which are coupled with planar antennas. (Wyss et al. 1993a) The fabrication was made using a combination of optical and electron-beam lithography on a GaAs/AlGaAs MODFET structure with a two-dimensional electron density $N = 2.8 \times 10^{11} cm^{-2}$, and an electron mobility $\mu = 200,000 cm^2/V \cdot sec$ at 45 K. SEM pictures of one of the devices with a log-periodic antenna (it has a self-complementary pattern for broadband coupling) is shown in Fig. 5. The opening of the split-gate is $0.15\mu m$.

560

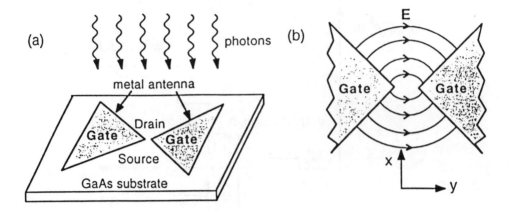

Fig. 4. (a) An antenna-coupled quantum point contact. The planar antenna concentrates the radiation field into the small gap region between the two antenna terminals. (b) Central region in (a), the curved lines are the electric field of the radiation near the antenna/gate terminals.

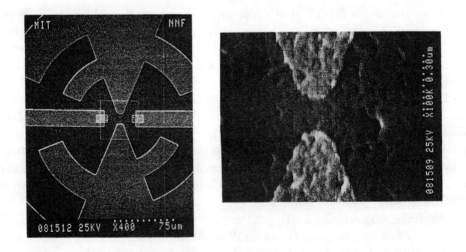

Fig. 5. SEM of a quantum point contact with a log-periodic antenna. The opening of the split-gate is $0.15 \mu m$.

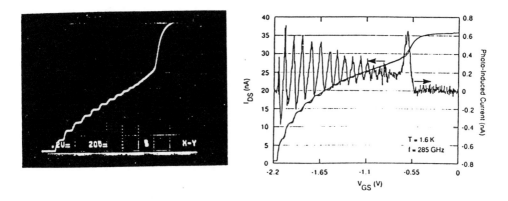

Fig. 6. (a) Drain/source current as a function of the gate voltage. (b) Drain/source current vs. the gate voltage with and without the far-infrared radiation. The difference between the two currents is the photon-induced current, which is the oscillatory curve.

The measurement was performed in a cryostat in which the device is submerged in liquid helium of 1.5 K. For a quick measurement of the drain/source current as a function of voltage, we can apply a dc drain/source voltage of 150 μV and sweep the gate voltage at several ten Hz. The measured $I_{DS} - V_{GS}$ curves is displayed on an oscilloscope, as shown in Fig. 6(a). Fifteen conductance steps are clearly seen which indicates the high quality of the device.

The radiation source is a tunable (75-110 GHz) Gunn oscillator-pumped Schottky frequency tripler with output frequencies at about 300 GHz. The maximum output power is 0.5 mW. The radiation beam of the frequency tripler is launched into free space through a conical horn, and then focused through the window of the cryostat by two TPX lenses. At the maximum radiation power level, a pronounced photon-induced drain/source current is produced throughout the entire gate voltage region where the quantum point contact exhibits the behavior of a 1-D electron system. The amplitude of the photon-induced current is comparable to that corresponding to a quantized conductance step, i.e. $(2e^2/h)V_{DS}$, and it oscillates with the gate voltage, as shown in Fig. 6(b).

In order to compare our measurement with the theoretical predictions shown in Fig. 3, we plot the photon-induced drain/source current and the drain/source conductance measured without radiation as functions of gate

562

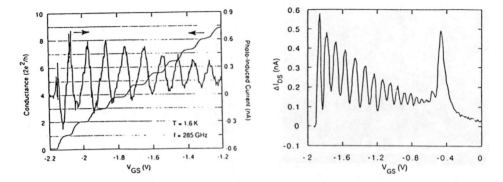

Fig. 7. (a) Drain/source conductance and the photon-induced current as functions of the gate voltage. The photon-induced current is mainly positive below the threshold of the next conductance step and mainly negative above the threshold. (b) Thermally induced drain/source current (which is defined as the difference between the drain/source current at 3.7 K and 1.6 K) as a function of the gate voltage. It exhibits similar behavior as the photon-induced drain/source current, as shown in Fig. 6(b).

voltage in a range where the conductance step structures are sharp, as shown in Fig. 7. The photon-induced current is mainly positive below the threshold of the next conductance step, and negative above the threshold of the step. This behavior qualitatively resembles the prediction from our theory of photon-assisted quantum transport. However, no substep structures such as those shown in Fig. 3 were seen in our measurement.

The most likely cause of the observed far-infrared photon-induced drain/source current is the heating of the 2DEG by the far-infrared radiation, which increases the temperature of the electron gas and the surrounding area. This higher local temperature broadens the Fermi surface and produces an additional drain/source current. To verify this hypothesis, we plot the difference between the drain/source current taken at 3.7 K and 1.6 K (without radiation) as a function of the gate voltage in Fig. 7(b). It appears quite similar to that shown in Fig. 6(b). Thus, this measurement supports our preliminary conclusion that the observed photon-induced drain/source current is mainly due to heating.

The assumption that the observed radiation-induced current is due to heating is further supported by our recent measurement, its results are shown

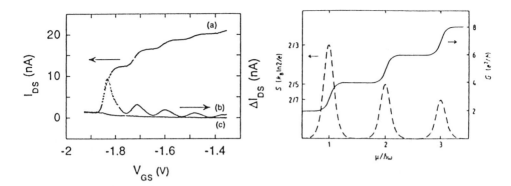

Fig. 8. (a) Drain/source current I_{DS} taken without radiation and with a dc bias of $100\mu V$, (b) I_{DS} taken with radiation and with a zero bias, (c) I_{DS} taken without radiation and with a zero bias, (d) calculated thermopower in a quantum point contact as a function of the Fermi energy (which is proportional to V_{GS}). (Streda 1989)

in Fig. 8. Fig. 8(a) is the dc $I_{DS} - V_{GS}$ curve measured with a dc drain/source bias voltage V_{DS} at $100\mu V$, while Fig. 8(c) is the one at $V_{DS} = 0$. Both were taken without the radiation. Fig. 8(b) is the dc $I_{DS} - V_{GS}$ curve taken with the radiation and at $V_{DS} = 0$. Clearly, this radiation-induced current (at zero bias) oscillates with the gate voltage in such a manner that its peaks track the onset of subband channels. This behavior is what is expected from a thermopower generated in a quantum point contact due to a temperature difference between the source and drain. The calculated thermopower as a function of the Fermi enery (which is proportional to V_{GS}) is shown in Fig. 8(d). (Streda 1989) Further investigation has confirmed this analysis. By shifting the focus of the far-infrared radiation beam from the source to the drain or vice versa, we can reverse the temperature difference across the quantum point contact, and thus reverse the polarity of the radiation-induced current. (Wyss et al. 1993b)

Our preliminary measurement and analysis then raise a difficult issue for experimentally observing the effect of photon-assisted quantum transport. Unlike photon-assisted tunneling in superconducting tunnel junctions, where the superconducting energy gap prevents energy absorption in the leads below the gap frequency, the excitation in the 2-D electron gas in a quantum point contact is, however, gapless. Thus, absorption can occur anywhere in the 2DEG, and the effect of the photon-assisted quantum transport will always be accompanied by a heating effect. In order to observe the predicted effect of photon-assisted quantum transport, we need to enhance the quantum process and minimize the thermal effect. The most crucial step is to use thicker ($\sim 2000\text{Å}$) antenna terminals. According to the previous theoretical discussions, it is crucial to concentrate the radiation field near the classical turning points, which are only several hundred angstroms from the center of the point contacts. Antennas made of high-conductivity films with thicknesses greater than the skin depths are required to achieve this goal.

In summary, we have observed far-infrared photon-induced drain/source current and thermopower in a quantum point contact device which is irradiated with coherent radiation at 285 GHz. Our analysis suggests that the observed photon-induced current is due to heating or a bolometric effect. Our analysis has also identified several improvements need to be made to enhance the photon-assisted quantum transport process. Our theoretical calculation has shown that the current responsivity of an antenna-coupled quantum point contact can approach the quantum efficiency $e/\hbar\omega$ at and above one THz.

Acknowledgements

We would like to thank M. R. Melloch for providing the heterostructure samples, and M. J. Rooks for his help in the electron-beam lithography. This work was supported by the NSF under Grant numbers 9109330-ECS and 9157305-ECS, and by the ARO under Grant number DAAL03-92-G-0251.

References

Beenakker, C. W. and H. van Houten,: 1991, in Solid State Physics, vol. 44, Edited by H. Ehrenreich and D. Turnbull, (Academic, New York), pp. 1-228, and references therein.

Feng, S. and Q. Hu,: 1993, Phys. Rev. **B48**, 5354.

Hu, Q.: 1993, Appl. Phys. Lett. **62**, 837.

Imry, Y.: 1986, in Directions in Condensed Matter Physics, edited by G. Grinstein and G. Mazenko (World Scientific), pp. 101-164.

Richards. P. L. and Qing Hu,: 1989, IEEE Proceedings, 77, 1233.

Rutledge, D. B., D. P. Neikirk, and D. P. Kasilingam,: 1983, in Infrared and Millimeter Waves, Edited by K. J. Button (Academic, New York), pp. 1-90.

Streda, P.: 1989, J. Phys. Condens. Matter 1, 1025.

Tien, P. K. and J.P. Gordon,: 1963, Phys. Rev. **129**, 647.

van Wees, B. J., H. van Houten, C.W.J. Beenakker, J.G. Williamson, L.P. Kouwenhoven, D. van der Marel, and C.T. Foxon,: 1988, Phys. Rev. Lett. **60**, 848.

Wharam, D. A., T.J. Thornton, R. Newbury, M. Pepper, H. Ajmed, J.E.F. Frost, D.G. Hasko, D.C. Peacock, D.A. Ritchie, and G.A.C. Jones,: 1988, J. Phys. C21, L209.

Wyss, R. A., C. C. Eugster, J. A. del Alamo, and Q. Hu,: 1993, to be published in Appl. Phys. Lett., September.

Wyss, R. A., C. C. Eugster, J. A. del Alamo, and Q. Hu,: 1993, in preparation.

Control of Electron Population by Intersubband Optical Excitation in a novel Asymmetric Double Quantum Well Structure

H. Sugawara [(a),(b),(c)],   H. Akiyama [(a),(b),(c)],   Y. Kadoya [(b)],   A. Lorke [(a)],
S. Tsujino[(a),(c)],   T. Matsusue[(c)], and H. Sakaki[(a),(b),(c)]
(a) Research Center for Advanced Science and Technology (RCAST)   University of Tokyo 4-6-1 Komaba Meguro-ku Tokyo 153 Japan
(b) Quantum Wave Project ERATO Research Development Corporation of Japan (JRDC) 4-3-24-302 Komaba Meguro-ku Tokyo 153 Japan
(c) Institute of Industrial Science (IIS) University of Tokyo 7-22-1 Roppongi Minato-ku Tokyo 106 Japan

ABSTRACT. A novel scheme of electron pumping in asymmetric double quantum well (QW) structure is studied, in which an intersubband excitation induces an interwell tunneling via the second subbands, making electrons transfer from one quantum well (QW-A) to the other (QW-B). We have designed and fabricated a novel asymmetric double quantum well structures suitable to this scheme, and demonstrated the electron transfer by measuring photoluminescence (PL) spectra of both QWs while electrons in QW-A are excited by infrared light from CO_2 laser. We have found that PL from QW-B is remarkably enhanced while PL from QW-A is reduced, demonstrating the electron transfer by the intersubband excitation. We discuss requirements in order to achieve more efficient modulation.

1. Introduction

Optical pumping of electrons for intersubband transitions in quantum well (QW) structures can modify electron distributions and make them deviate from the thermal equilibrium. It offers the basis for novel device applications, since optical and transport properties of QWs are to be modulated by the nonthermal distribution of electrons.

It is not easy, however, to populate the excited state with a significant number of optically excited carriers, as the intersubband relaxation process is usually very fast [1–3]. Hence, one must design the structure of QWs so as to reduce the intersubband relaxation rate and enhance the electron lifetime in the excited states. Vagos et al. [4] have achieved nonthermal occupation of the second electron state E_2 by intersubband excitation in an asymmetric coupled quantum wells where the intersubband relaxation was suppressed because of the asymmetry between the wavefunctions of the two lowest states.

In this paper, we report on a novel scheme of electron pumping in a new asymmetric double quantum well structures, in which electrons transfer from one QW to the other via intersubband excitation and relax to the lowest state with a rather long time constant. The excitation is done by a mid infrared light created by CO_2 laser (photon energy $\approx$ 120 meV), and we study photoluminescence (PL) spectra under an illumination of infrared light to prove that the electron distribution is substantially altered. In particular, we investigated how luminescence spectra of two QWs change to prove the importance of electron transfer

H. C. Liu et al. (eds.), Quantum Well Intersubband Transition Physics and Devices, 565–573.

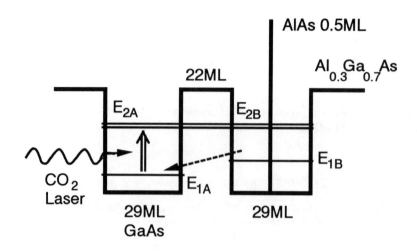

Fig. 1. Schematic of the potential-inserted double quantum well structure. The second subbands E_{2A} and E_{2B} of two quantum wells A and B are coupled whereas the ground subbands E_{1A} and E_{1B} are nondegenerate. An infrared light in resonance with $(E_{2A} - E_{1A})$ can pump electrons from E_{1A} either to E_{2A} or E_{2B}.

process in these changes of luminescence spectra. We also study the strong excitation effect, which leads to a novel phenomenon beyond linear response of intersubband transition.

2. Principal Mechanism of Electron Transfer Induced by Intersubband Transitions

Figure 1 represents the basic idea of potential-inserted double quantum well (PI-DQWs) structures. It consists of two QWs (QW-A, QW-B) of the same thickness, which are separated by a thin barrier. QW-A is a normal QW, whereas QW-B is a potential-inserted QW (PI-QW) [5–7], where a very thin potential layer is inserted in the middle of the QW. The repulsive potential inserted in the PI-QW affects the first level effectively and shifts up its energy. In contrast, the second level is almost unchanged since the wavefunction vanishes at the center of the QW. Thus, the second subbands E_{2A} and E_{2B} of the two wells are coupled, while the first subbands E_{1A} and E_{1B} are nondegenerate ($E_{1A} < E_{1B}$).

Suppose nearly all the electrons are initially in the ground subband E_{1A} of the QW-A. Once some electrons in the ground level E_{1A} are excited to either of the second subbands E_{2A} or E_{2B}, they tunnel through the barrier and spread over QW-A and QW-B before they relax to the ground level, provided that the coupling of the second subbands is strong enough to make the interwell tunneling time τ_{2T} shorter than the intersubband relaxation time. In such a case, some electrons should relax to the ground subband E_{1B} of QW-B, while the rest relax to the ground subband E_{1A} of QW-A. Those electrons that have relaxed to E_{1B} further relax to E_{1A} by off-resonant tunneling. The last process of tunneling escape from QW-B to QW-A can substantially reduced (i) when the subband spacing $E_{1B} - E_{1A}$ is chosen to be less than the optical phonon energy, and/or (ii) when the overlap integral between the wavefunctions of the two ground subbands is reduced by adopting sufficiently thick barrier.

In the steady state the electron population under the intersubband excitation is determined by the balance of all transition processes described above. When the escape rate $1/\tau_T$ is sufficiently small, a substantial portion of electrons remain in E_{1B}. Thus, we should be able to control the electron population by intersubband excitation.

For the efficient modulation of electron population, we must design the structure in such a way that (i) the intersubband excitation rate is high due to the coincidence of the subband spacing with the energy of incoming photons, (ii) the tunneling transfer via the second subbands is fast due to the strong coupling between them, and (iii) the tunnel escape rate from E_{1B} to E_{1A} is low as described earlier. The modulation of electron populations can be theoretically analyzed by solving the coupled rate equation, as discussed in the appendix A.

3. Sample Preparation and Experimental Methods

We grew two samples I and II on semi-insulating (100) GaAs substrates by molecular beam epitaxy. After growing 5000 Å GaAs buffer layer, 5 periods of double quantum well structures were grown, followed by the deposition of 1.2 μm $Al_xGa_{1-x}As$ cladding layer and 200 Å top GaAs layers. Each period of double quantum wells consists of a normal QW (A) and a PI-QW (B), which are sandwiched on both ends by $Al_{0.3}Ga_{0.7}As$ barrier having the thickness of 62 Å. Both QW-A and QW-B have the total well width of 29 monolayer (ML=2.83 Å for GaAs), but in the PI-QW, 0.5 ML AlAs layer is inserted to raise the first subband E_{1B} 25 meV above E_{1A}. The subband spacing ($E_2 - E_{1A}$) is about 120 meV to allow the intersubband optical excitation by CO_2 laser. The splitting of the second subbands is determined by the Al content x and thickness L_{MB} of the middle AlGaAs barrier. For sample I, x and L_{MB} are 0.3 and 22ML, respectively, for which the level splitting is calculated to be ≈ 8 meV. In sample II, the middle barrier layer is thicker ($L_{MB} =24$ ML) and Al mole fraction is higher ($x = 0.37$) to reduce escape rate below that of sample I.

Both samples are undoped but electron-hole pairs are created by the Ar^+ laser light and their concentration was kept relatively low ($\leq 10^{10}$ cm^{-2}) to avoid the excessive carrier heating.

As the luminescence intensity is proportional to the concentration n of electrons or excitons, the measurement of PL intensity is expected to represent the electron concentration of each well as will be discussed later.

The geometry for PL measurements is illustrated in the inset of Fig. 2. The sample was mounted on a cold finger in a cryostat which was set at 10 K. The infrared light from CO_2 laser was focused with a concave mirror on the sample edge which was beveled at 45 degrees. It passes through the sample via the internal reflection with p-polarization for which the optical electric field $\mathcal{E}$ is almost parallel to the dipole moment z_{21} of the intersubband transition. The thick $Al_xGa_{1-x}As$ cladding layer was used so that this component of the electric field is maximized at the double quantum well.

The infrared light used in this work is generated by transversely excited atmospheric (TEA) CO_2 laser, with the pulse width of about 100 ns and the repetition rate of 10 Hz. The visible light from a cw Ar^+ laser ($\lambda = 5145$Å, photon energy = 2.4 eV) was used to generate electron-hole pairs in the sample after passing through an acoust-optic (AO) modulator in which the 500ns light pulses are generated. As the intensity of visible light

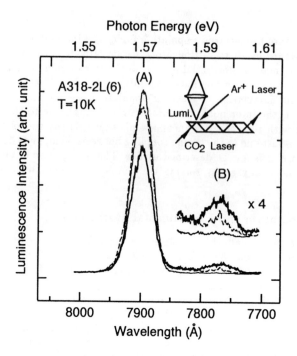

Fig. 2. Photoluminescence spectra of sample-I measured at 10 K under the illumination of infrared light, whose photon energy and the index of CO_2 line are 117 meV and 10P22, respectively. The intensity of infrared light was 1.6 MW/cm^2. Thin curve is a reference spectrum in the absence of infrared light. Two symbols (A) and (B) indicate peak positions of luminenscence associated to the first subbands in QW-A and QW-B, respectively. Thick curve and broken curve are the PL spectra under the infrared excitation with the p-polarization and the s-polarization ($\mathcal{E} \perp z_{21}$). The inset shows the geometry used to irradiate the sample both by visible light and by infrared light at the same position.

was 22 W/cm^2, the electron-hole density created by visible light is estimated to be as low as 6×10^9 cm^{-2} per one period of DQWs, if we assume the radiative lifetime τ_R to be 500 ps.

The photoluminescence from the sample was dispersed in a 25 cm monochromator, and its intensity at a particular timing was detected with a photomultiplier (PMT). Detected PMT signal was fed into a gated Boxcar averager, in which the gate of 30 ns was synchronized to the laser pulses. The averages of every 10 signals were recorded.

4. PL spectra of Double QW Structures and Their Changes

In Figure 2, we show the PL spectra from sample-I measured for infrared light of different intensities. In the absence of infrared light, PL peak A associated with electrons in E_{1A} is much stronger than PL peak B which comes from electrons in E_{1B} state ($I_B/I_A \approx 3$ %). The fact indicates that photogenerated electrons escape efficiently from QW-B to QW-A. When the sample was illuminated by an CO_2 laser light (intensity = 1.6 MW/cm^2) with

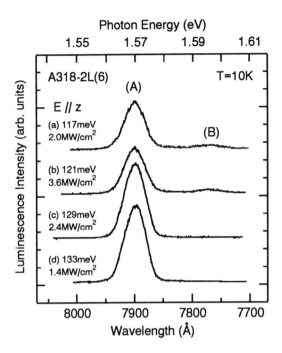

Fig. 3. Photoluminescence spectra of sample-I measured at 10 K under the illumination of infrared light with the intensity ≈ 2 MW/cm² at four different photon energies, (a) 117meV (the index of CO_2 line 10P22), (b) 121meV (10R16), (c) 129meV (9P24), and (d) 133meV (9R10).

the p-polarization, I_B was significantly enhanced, while I_A decreased. Note that no shift of luminescence wavelength was observed, indicating that the lattice heating of the sample is negligible under the infrared light illumination. The enhancement of I_B suggests that the concentration N_{1B} of electrons in E_{1B} state has increased as a result of the electron transfer.

We have also examined how the luminescence intensity depends on the polarization and the photon energy of infrared light. We again refer to Fig. 2 for the polarization dependence. The enhancement of I_B was stronger for the p-polarization than s-polarization for which the intersubband excitation was inhibited. Hence, the observed change of PL intensity seems to be caused mainly by the intersubband excitation.

Figure 3 shows the luminescence spectra measured under the infrared light illumination with four different photon energies. The quenching of PL peak A and the subsequent en- hancement of I_B gets stronger as the photon energy is tuned around 120meV. The observed dependence on the polarization and the photon energy of infrared light agree with what we expect from the electron transfer process. We conclude that electron distribution is indeed modulated by the intersubband excitation.

Next, we investigate sample-II to clarify the effect of escape rate. The photon energy of infrared light was set to be 121meV (10R16), which induced the largest change. Figure

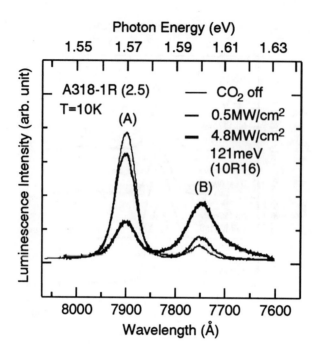

Fig. 4. Photoluminescence spectra from sample-II under the infrared light irradiation with different intensities. The photon energy of the infrared light is 117meV (10R16).

4 shows luminescence spectra measured at different infrared light intensities. Note that PL peak B is observable even in the absence of infrared light irradiation ($I_B/I_A = 0.14$), indicating that photogenerated electrons in QW-B is reasonably high. Once the infrared light is supplied, PL peak B gets substantially enhanced. This increase is much stronger than that in sample-I, and the intensity of PL peak B gets even bigger than peak A ($I_A < I_B$) when the infrared light intensity exceeds 2 MW/cm². These two findings are exactly what has been expected in the sample with lower escape rate and prove the electron transfer through the resonant second subbands to be important.

Note that the total integrated luminescence intensity was kept almost unchanged before the intensity inversion occurred. This fact indicates the invariance of luminescence efficiency, in which only the relative concentrations of electrons changed.

5. Discussion

Through a number of experiments we have found that the electron transfer from one QW to the other has taken place by infrared-light-induced intersubband transition. We now describe a few points that call for further study.

The intensity of infrared light required to modify the luminescence spectra was typically 1 MW/cm². This value is 10 times stronger than our prediction ($\approx$ 100 kW/cm²) which is described in Appendix. The reported onset of nonlinearity in photoinduced intersubband absorption experiments [3, 8, 9] is a few 100 kW/cm^2, which is of the same order of our prediction.

This discrepancy is probably due to our samples being not optimized: Firstly, the intersubband absorption coefficient of the present sample for the photon energy of 121meV is not as high as it should be, since this photon energy is not exactly resonant to the subband energy separation. Secondly, the coupling between the second subbands was not as strong as it should be. This is probably caused by the resonance of two levels being weakened partly by inhomogeneities and partly by the excessively thick barrier. For example, the energy splitting $\Delta E_2 = |E_{2A} - E_{2B}|$ of the second subbands in sample-II is calculated to be as small as 3meV, and the electron transfer gets less efficient.

The second point to note is that PL peak B is substantially enhanced in sample-II even for the s-polarization where the intersubband transition is inhibited. One possible explanation is that our sample might have some roughness or inhomogeneities which generates an electric field vector parallel to z_{21} even for the s-polarization. Another mechanism may be the free carrier absorption of s-polarized light. Although the free carrier absorption is much weaker than the intersubband absorption in usual QW, if the excitation is intense an electric field vector parallel to the QW layer can accelerate electrons to the region of large in-plane wavevector where they can be scattered directly from E_{1A} to E_{1B}, resulting in the population of the E_{1B} state.

The third interesting point is that when the infrared light intensity increases the enhancement of peak B gets saturated, but that PL peak A keeps diminishing. This may be caused by the enhanced contribution of nonradiative processes.

6. Summary

In summary, a novel scheme of electron pumping induced by intersubband optical excitation is demonstrated. Photoluminescence spectra measured under infrared laser irradiation have evidenced that electrons are transferred from one QW to the other through the resonantly coupled second subbands. If sample structure is further optimized, more efficient electron transfer with weaker infrared light should be possible. The free carrier absorption in QW structure and other processes may also play some role in the electron transfer process.

Acknowledgements

The authors acknowledge Prof. Y. Arakawa and Dr. Y. Nagamune for the loan of Ar^+ laser, and Prof. M. Kuwata-Gonokami and Prof. A. Shimizu for stimulating discussions. This work is supported partly by a Grant-in-Aid from the Ministry of Education, Science and Culture, Japan, and partly by JRDC through ERATO program. H. S. and H. A. acknowledge partial financial supports from the Japan Society for the Promotion of Science.

Appendix

A. Rate equations describing the electron transfer process

The efficiency of the electron transfer can be analyzed by employing the rate equations. We assume here a simple three level model which satisfies the following assumptions:

(1) We assume that the linewidth of the incident infrared light is much narrower than that of the intersubband absorption. Hence, the intersubband excitation rate W (number of transitions per unit time) is defined at the single photon energy $\hbar\omega$.

(2) In the steady state, the electron concentrations, N_{2A} and N_{2B}, in the second subbands are determined by a balance of all the transitions. We assume for simplicity that $N_{2A} = N_{2B} = N_2/2$, where N_2 is the total number of electrons in the second subbands.

(3) The intersubband relaxation rates are assumed to be the same ($= \tau_{21}$) in both QWs, since the wavefunctions of two QWs are similar between two QWs and the matrix element of the intersubband relaxation are the same.

(4) The lifetimes, τ_A^* and τ_B^*, of electrons in the ground subbands in two QWs are supposed to be kept constant under the electron pumping.

(5) Photogenerated carriers supplied to each subband are denoted by g_{1A}, g_{1B} and g_2, the number of electrons supplied to each state per unit time.

The populations N_{1A}, N_{1B} and N_2 of electrons in three subbands are then described by rate equations given below.

$$\frac{dN_{1A}}{dt} = -\frac{N_{1A}}{\tau_A^*} - WN_{1A} + \frac{N_{1B}}{\tau_T} + \frac{N_2}{\tau_{21}} + g_{1A}, \tag{1}$$

$$\frac{dN_{1B}}{dt} = -\frac{N_{1B}}{\tau_B^*} \qquad\qquad - \frac{N_{1B}}{\tau_T} + \frac{N_2}{\tau_{21}} + g_{1B}, \tag{2}$$

$$\frac{dN_2}{dt} = \qquad\qquad WN_{1A} \qquad\qquad - \frac{2N_2}{\tau_{21}} + g_2. \tag{3}$$

We assume here that photogenerated electrons are first supplied only to the second subbands ($g_{1A} = g_{1B} = 0, g_2 > 0$). This is valid when the photon energy of the visible light to create electron-hole pairs is much higher than the barrier height of the QWs, since most of electrons first relax to the second subband before they further relax to the ground subband.

For the steady state where $(dN_{1A}/dt) = (dN_{1B}/dt) = (dN_2/dt) = 0$, one finds

$$\frac{N_{1B}}{N_{1A}} = \frac{1/\tau_A^* + W}{1/\tau_B^* + 2/\tau_T} = \frac{\tau_B^*}{\tau_A^*} \cdot \frac{1 + \tau_A^* W}{1 + 2\tau_B^*/\tau_T}. \tag{4}$$

Since the luminescence intensity is approximately proportional to the product of the electron concentration in each QW and the radiative transition rate, $1/\tau_{RA}$ and $1/\tau_{RB}$, the ratio of luminescence intensities is expressed as

$$\frac{I_B}{I_A} = \frac{N_{1B}/\tau_{RB}}{N_{1A}/\tau_{RA}} = \frac{\tau_{RA}}{\tau_{RB}} \cdot \frac{\tau_B^*}{\tau_A^*} \cdot \frac{1 + \tau_A^* W}{1 + 2\tau_B^*/\tau_T} = \frac{\eta_A}{\eta_B} \cdot \frac{1 + \eta_A \tau_{RA} W}{1 + 2\eta_B \tau_{RB}/\tau_T}. \tag{5}$$

Here, we denote the luminescence efficiencies in each QWs by $\eta_A = \tau_A^*/\tau_{RA}$ and $\eta_B = \tau_B^*/\tau_{RB}$. If the radiative recombination dominates the electron lifetime in each ground state ($\eta_A = \eta_B = 1$), the ratio I_B/I_A of luminescence intensities is equal to the ratio N_{1B}/N_{1A} of electron concentrations.

If the luminescence efficiency is the same for two QWs ($\eta_A = \eta_B$), the luminescence intensity is equal ($I_B/I_A = 1$) when the intersubband transition rate W fulfills

$$\tau_{RA}W = 2\tau_{RB}/\tau_T. \tag{6}$$

Next, the infrared light intensity required to invert the luminescence intensities for sample-II will be calculated. Since we need to know the lifetimes, τ_A^* and τ_B^*, of electrons in each subband and escape time τ_T, we assume the luminescence efficiencies of both QWs are unity ($\eta_A = \eta_B = 1$). Then, the lifetimes of electrons are equal to the radiative recombination times ($\tau_A^* = \tau_{RA} = \tau_{PL,A}$ and $\tau_B^* = \tau_{RB}$). From the time resolved luminescence spectroscopy with a dye laser (wavelength $= 7400$ Å) at the excitation power ≈ 4 W/cm^2, we have found $\tau_{PL,A} \approx 600$ ps and $\tau_{PL,B} \approx 300$ ps, respectively,

If we assume that a decay rate of electrons in E_{1B} state is a sum of the radiative transition rate and the tunneling escape rate ($\tau_{PL,B}^{-1} = \tau_B^{*-1} + \tau_T^{-1}$) and use the fact $I_B/I_A = 0.14$ at $W = 0$ (refer to Fig. 4), we obtain $\tau_A^* \approx 600$ ps, $\tau_B^* \approx 460$ ps, and $\tau_T \approx 140$ ps.

Then, the intersubband transition rate W required to achieve the inversion of luminescence intensity was calculated from equation (6) and found to be $W = 1.1 \times 10^{10}$ s^{-1}. It corresponds to the intersubband excitation intensity 100 kW/cm^2, assuming the absorption linewidth $\Gamma = 10$ meV.

References

[1] R. Ferreia and G. Bastard, "Evaluation of some scattering times for electrons in unbiased and biased single- and multiple-quantum-well structures," *Phys. Rev. B*, vol. 40, pp. 1074–86, 1989.

[2] M. C. Tatham, J. F. Ryan, and C. T. Foxon, "Time-Resolved Raman Measurements of Intersubband Relaxation in GaAs Quantum Wells," *Phys. Rev. Lett.*, vol. 63, pp. 1637-40, 1989.

[3] A. Seilmeier, H. J. Hübner, G. Abstreiter, G. Weimann and W. Schlapp, "Intersubband Relaxation in GaAs-Al$_x$Ga$_{1-x}$As Quantum Well Structures Observed Directly by an Infrared Bleaching Technique," *Phys. Rev. Lett.*, vol. 59, pp. 1345-8, 1987.

[4] P. Vagos, P. Boucaud, F. H. Julien, J. M. Lourtioz, and R. Planel, "Photoluminescence up-conversion induced by intersubband absorption in asymmetric coupled quantum wells," *Phys. Rev. Lett.*, vol. 70, pp. 1018–21, 1993.

[5] J. Y. Marzin and J. M. Gérard, "Experimental probing of quantum-well eigenstates," *Phys. Rev. Lett.*, vol. 62, pp. 2172-5, 1989.

[6] W. Trzeciakowski and B. D. McCombe, "Tailoring the intersubband absorption in quantum wells," *Appl. Phys. Lett.*, vol. 55, pp. 891-3, 1989.

[7] H. Sakaki, H. Sugawara, J. Motohisa, and T. Noda, "Intersubband transition and electron transport in potential-inserted quantum well structures and their potentials for infrared photodetector," in *Proceedings of the NATO Advanced Workshop on Intersubband Transitions in Quantum Wells*, pp. 65–72, Plenum (New York), 1992.

[8] M. Olszakier, E. Ehrenfreund, E. Cohen, J. Bajaj, and G. J. Sullivan, "Photoinduced intersubband absorption in undoped multi-quantum-well structures," *Phys. Rev. Lett.*, vol. 62, pp. 2997-3000, 1989.

[9] F. H. Julien, J. M. Lourtioz, N. Herschkorn, D. Delacout, J. P. Pocholle, M. Papuchon, R. Planel, and G. L. Roux, "Optical saturation of intersubband absorption in GaAs-Al$_x$Ga$_{1-x}$As quantum wells," *Appl. Phys. Lett.*, vol. 53, pp. 116-8, 1988; *ibid.* vol. 62, p. 228, 1993.

INDEX

Antenna-coupled 553,560

Background limited 29,39,43,76-77,159-160
Band bending 117-118
Birefringence 321-322,329
Bistability
  optical 392
Bragg reflector 301-304

Capacitance 112
Capture 74,97-98,100,104-106,108,120,433-439
Cavity 44-45
Color
  two 29,31,37,137
  three 123-125
  multi 29,32,37,124,135
Coupling 379
  band-to-band 379,383
  photonic 386
CO_2 laser 207,213-217,219,494,567-570
Crosstalk 45-46

Data coding 533,540
Deep level optical spectroscopy 97-101
Delta doped (δ-doped) 371-375
Depolarization 232,234,240,371-372,374,376
Domain 135,139-147

Effective mass
  position dependent 389
  tensor 221-225
Electron-electron 361
Electron-phonon 443
Emission 511,525-530
Emitter 511
Emission-recombination 97-98,102-103
Escape 16,73,100-101,114-115,498-499
Exchange-correlation 239-240

Fabry-Perot 301-304,512
Focal plane array 20-21,43-53,55,61-62,64-67,78-79,81-82

Free-electron laser (FEL) 261,269,469-470
Fresnel equation 321,323-324

Grating 13-26,29-31,33-45,56-57,363-364
    crossed 13,15-20,22-24,26
    one-dimensional 1,10-11,44-46
    two-dimensional 1-3,43-46

Hartree-Fock 233-234
Harmonic
    generation 266,467,477,483,488
    second 266,268-270,457,462-465,467,471,477,483,488
    third 467,472
Heterodyne 207-210
Hot electron 151-152,154,156-158,160-165,167-168,172,177-178,361

Imaging 55,67,69,78
Impedance spectroscopy 97,101-105
Impurity 403,405-407
Interband 399-402
Intersubband scattering 352-353,421-422,443

Lambertian law 5
Lifetime 74,107,109,114,275,286-288,511,525
Local oscillator (LO) 208-210,216-217
Luttinger-Kohn Hamiltonian 381
L-valley 153-154

Many body 222,233-234,246
Mercury Cadmium Telluride (HgCdTe) 87-95
Miniband 29,31,35,37,141,197-198,201-203,291-293,296,298-299
Mixing 380
Modulation 533,535-536,541-544,567
Multiband 32,35,37,40,55,275-276,281,477-480
Multicolor 37,40

Negative differential resistance 139,141
Noise 46,58,70-71
    dark current 21
    fixed pattern 21,23-24,61,66,70-71
    generation-recombination (g-r) 16,24,46-47,58,161
    Johnson 21,24,46
    photon 21,24

    readout 49-50,70-71
    thermal 49
    1/f 48-49,59-60,71,164
Noise equivalent temperature difference (NEΔT) 10,20-23,25,43,55,61,64,67,70,75-84,157-158,160,184
Non-uniformity 21,69-70
Normal incidence 29-34,38-39,221,225,237,261-262

Phase retardation 321,326-329
Photoinduced 276,283,285,335,339,345,348-349
Photon assisted
    transport 553-556,558,562-564
    tunneling 554,556,558
Photonic gap 387
Photovoltaic 29,31,38-39,187,193
Population inversion 403,512,529
Pseudopotential 412,414,477

Quantum beat 547,549
Quantum interference 313
Quantum point contact 553-564
Quarter-wave stack 301-306

Random scattering reflector 1,4,5,8,11
Random scattering optical coupler 1
Readout 47-48,50
Rectification 548-549
Relaxation 403,525-526
    energy 534,538
    intersubband 353-354,421,425,427,433-434,439,501,565-566
    oscillation 541-544

Saturation 475,501,504-505,507
SiGe 221-234,237-248
Single quantum well 111-121
Stark-ladder 197
Strain 29-30,32,222,225,230,238,262
Sum rule 298
Susceptibility 266,467,472-475,477,479,485,487
    nonlinear 261
    tensor 321-322,459

Terahertz generation 547

Time resolved 275-276,285,423,433
Transistor 151-152,157-165,167-168,177-180
Tunneling 118-121
   interwell 565-566
   resonant 119-121
   sequential resonant 142,527
Two photon 493-500

Waveguide 13-14,17-20,22-24,26,52

X-minimum 38
X-point 32
X-valley 153-154